TEACHER'S EDITION

MCP Mathematics

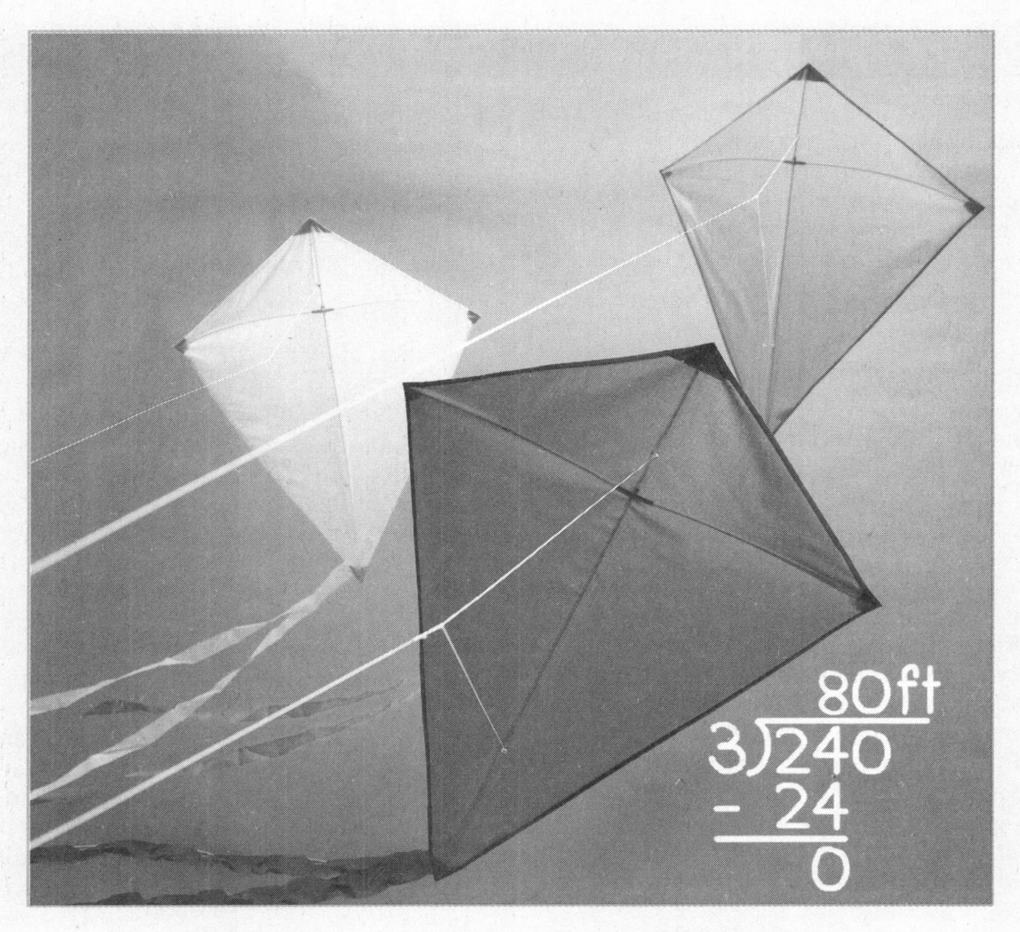

Richard Monnard • Royce Hargrove

TEACH THE COVER: You have 240 feet of string. How much string can be used for each kite?

ISBN 0-7652-6063-8
Printed in the United States of America
3 4 5 6 7 8 9 10 10 09 08 07

1-800-321-3106
www.pearsonlearning.com

MCP Mathematics

About the Program This comprehensive math program will help students in Grades K to 6 develop a solid mathematics background. It is designed to encourage critical-thinking skills, active participation, and mastery of skills within the context of problem-solving situations. The program's developmental sequence introduces and extends skills taught in most mathematics curricula, such as number sense, operations, algebra, geometry, data collection and analysis, logic, and probability.

A Research-Based Approach

The program offers a strong pedagogical approach that is research based. First, students are provided a developed model that introduces the lesson concept. Then, students are given guided practice opportunities to get started. Next, abundant practice is provided for mastery. Finally, students can apply their skills to problem-solving and enrichment activities. An overview of this approach follows.

- **Direct Instruction**
  Each lesson begins with a developed model that demonstrates the algorithm or concept in a problem-solving situation and gets students actively involved in the situation. . . . *Students taught using direct instruction were found to perform better on tests of computation and year-end tests (Crawford and Snider 2000).**

- **Guided Practice**
  A Getting Started section provides samples of the concept or skill being taught and allows you to guide and observe students as they begin to apply the skills learned. *When low-achieving students were taught using such direct instruction methods as . . . guided practice, . . . they showed improved mastery of basic skills, solved computation and word problems correctly in less time, and had higher self-ratings of academic motivation (Kame´enui, Carnine, Darch, and Stein 1986; Din 1998; Ginsburg-Block and Fantuzzo 1998).**

- **Independent Practice**
  The Practice section can be used to develop and master basic skills by allowing students to independently practice the algorithms and to apply learning from the lesson or from previous lessons. *Research indicates that providing students with extended practice appears to serve two purposes: re-teaching of the skill for students who had not yet mastered it and relating of the previously learned skill to new skills, resulting in the formation of interrelationships among concepts that improved retention and yielded higher achievement test results (Hett 1989).**

- **Higher-Order Thinking Skills**
  A collection of real-life word problems in the Problem Solving section provides application opportunities related to the lesson concepts. *Frequent practice with word problems is associated with higher-order skill development (Coy 2001). This finding is especially true when the word problems present familiar real life situations (Coy 2001).**

- **Problem-Solving Strategies**
  Once per chapter, a special lesson introduces students to the techniques of problem solving using Polya's four-step model. The Apply activities in these lessons allow students to use problem-solving strategies in everyday situations. *Students who received instruction in problem-solving processes showed better performance on tests of skills, tasks, and problem solving, as well as on a measure of the transfer of learning (Durnin et al. 1997).**

- **Calculator Usage**
  Calculator lessons in Grades 3 to 6 teach students the functions and the skills needed to use calculators intelligently after they've developed a foundation of competence in pencil-and-paper computation. *. . . Students who used calculators in mathematics instruction had more positive attitudes toward mathematics and a higher math self-concept. . . . A special curriculum developed around calculators has been shown to improve mathematics achievement (Hembree and Dessart 1986).**

- **Frequent and Cumulative Assessment**
  Chapter Tests provide both students and teachers with a checkpoint that tests all the skills taught in a chapter. You will find Alternate Chapter Tests based on the same objectives in the Teacher's Edition. *Assessment has been found to be most effective when it is a frequent and well-integrated aspect of mathematics instruction (Brosnan and Hartog 1993).* Cumulative Assessments maintain skills that have been taught in all previous chapters. A standardized test format is used starting at the middle of the second-grade text. The Teacher's Edition also contains Alternate Cumulative Assessments. . . . *Frequent cumulative tests result in higher levels of achievement than do infrequent tests or tests related only to content since the last test (Dempster 1991).**

- **Remedying Common Errors**
  The comprehensive Teacher's Edition provides abundant additional help for teachers, including a four-step lesson plan that walks the teacher through the lesson, pointing out common errors and providing intervention strategies. *Curricula with an error correction component were found to result in higher scores for computation, math concepts, and problem solving (Stefanich and Rokusek 1992; Crawford and Snider 2000).**

**Research compiled by PRES Associates, Inc. (2004). Research Support for MCP Mathematics (unpublished).*

Use the First Page of a Lesson for Direct Instruction

Each two-page lesson focuses on one main objective. The first page begins with a developed model that students can actively work on as you walk them through it.

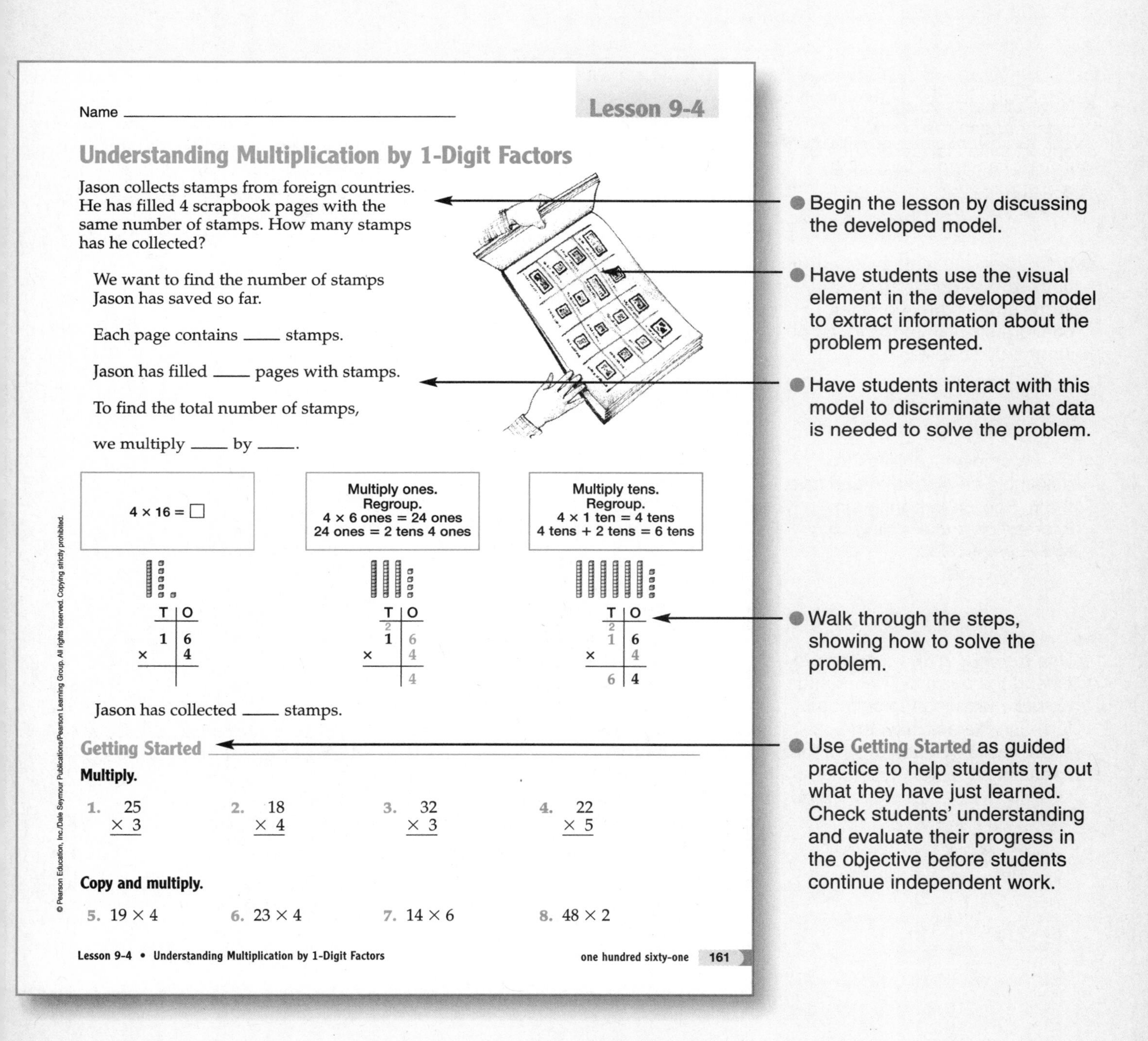

Name ______________________ **Lesson 9-4**

Understanding Multiplication by 1-Digit Factors

Jason collects stamps from foreign countries. He has filled 4 scrapbook pages with the same number of stamps. How many stamps has he collected?

We want to find the number of stamps Jason has saved so far.

Each page contains ____ stamps.

Jason has filled ____ pages with stamps.

To find the total number of stamps,

we multiply ____ by ____.

$4 \times 16 = \square$	Multiply ones. Regroup. 4 × 6 ones = 24 ones 24 ones = 2 tens 4 ones	Multiply tens. Regroup. 4 × 1 ten = 4 tens 4 tens + 2 tens = 6 tens

T	O
1	6
×	4

T	O
²1	6
×	4
	4

T	O
²1	6
×	4
6	4

Jason has collected ____ stamps.

Getting Started

Multiply.

1. 25×3 2. 18×4 3. 32×3 4. 22×5

Copy and multiply.

5. 19 × 4 6. 23 × 4 7. 14 × 6 8. 48 × 2

Lesson 9-4 • Understanding Multiplication by 1-Digit Factors one hundred sixty-one 161

- Begin the lesson by discussing the developed model.
- Have students use the visual element in the developed model to extract information about the problem presented.
- Have students interact with this model to discriminate what data is needed to solve the problem.
- Walk through the steps, showing how to solve the problem.
- Use **Getting Started** as guided practice to help students try out what they have just learned. Check students' understanding and evaluate their progress in the objective before students continue independent work.

Using the Student Edition

Assign Practice from the Second Page of a Lesson

The second page of a lesson provides practice and extension of the lesson's objective. You can begin the process of individual mastery by assigning Practice exercises that students can work on independently.

- Have students Practice independently to provide opportunities for application of basic skills and higher-order thinking.
- Encourage students to work with both vertical and horizontal forms so that they become comfortable with forms found on standardized tests.
- Check students' abilities to assemble an algorithm and give them practice in transferring information by assigning **Copy and . . .** exercises.
- Use Now Try This! activities to extend the basic skill work and to make learning the concepts fun. Use the activities to challenge the minds of the more capable students.

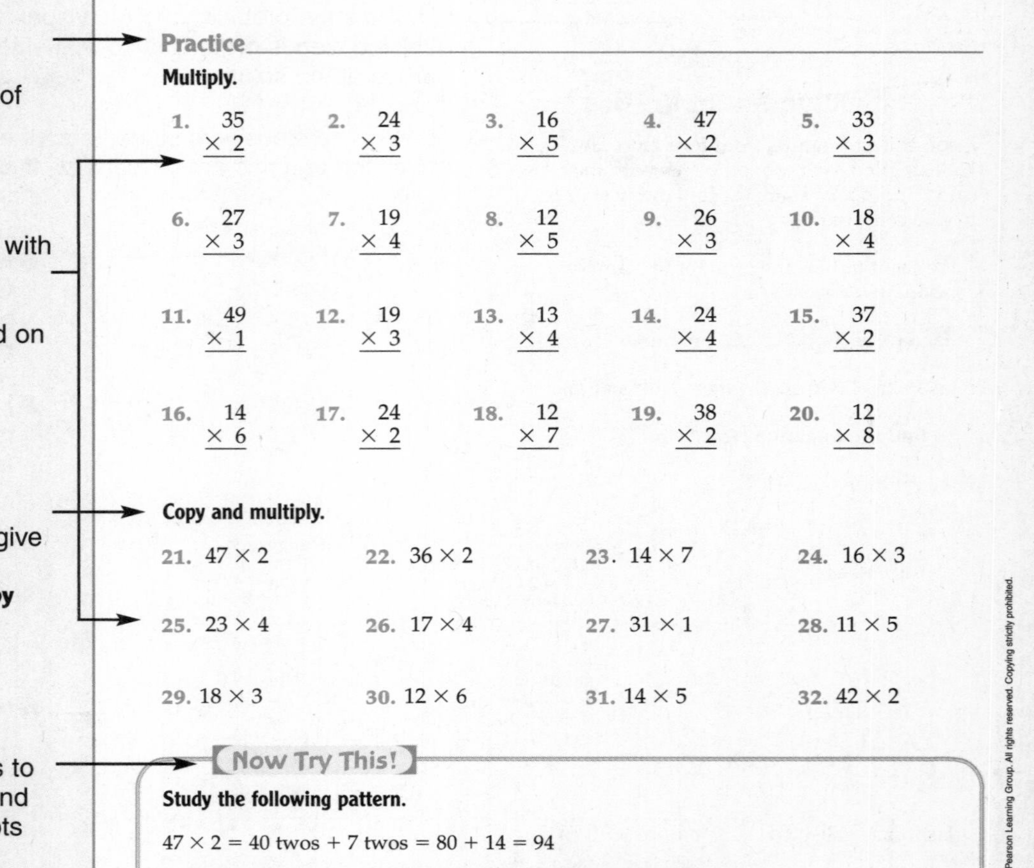

Practice

Multiply.

1. 35 × 2	2. 24 × 3	3. 16 × 5	4. 47 × 2	5. 33 × 3
6. 27 × 3	7. 19 × 4	8. 12 × 5	9. 26 × 3	10. 18 × 4
11. 49 × 1	12. 19 × 3	13. 13 × 4	14. 24 × 4	15. 37 × 2
16. 14 × 6	17. 24 × 2	18. 12 × 7	19. 38 × 2	20. 12 × 8

Copy and multiply.

21. 47 × 2	22. 36 × 2	23. 14 × 7	24. 16 × 3
25. 23 × 4	26. 17 × 4	27. 31 × 1	28. 11 × 5
29. 18 × 3	30. 12 × 6	31. 14 × 5	32. 42 × 2

Now Try This!

Study the following pattern.

47 × 2 = 40 twos + 7 twos = 80 + 14 = 94

Use the pattern to help you complete the rest of the multiplications.

1. 24 × 3 = ____ threes + ____ threes = ____ + ____ = ____
2. 37 × 2 = ____ twos + ____ twos = ____ + ____ = ____
3. 17 × 4 = ____ fours + ____ fours = ____ + ____ = ____
4. 12 × 7 = ____ sevens + ____ sevens = ____ + ____ = ____

162 one hundred sixty-two   Lesson 9-4 • Understanding Multiplication by 1-Digit Factors

Using the Student Edition

Teach Problem Solving and Calculators in Every Chapter

Problem Solving lessons focus on different problem-solving strategies using the chapter concepts and skills. Calculator lessons teach students to use the technology while reinforcing chapter content.

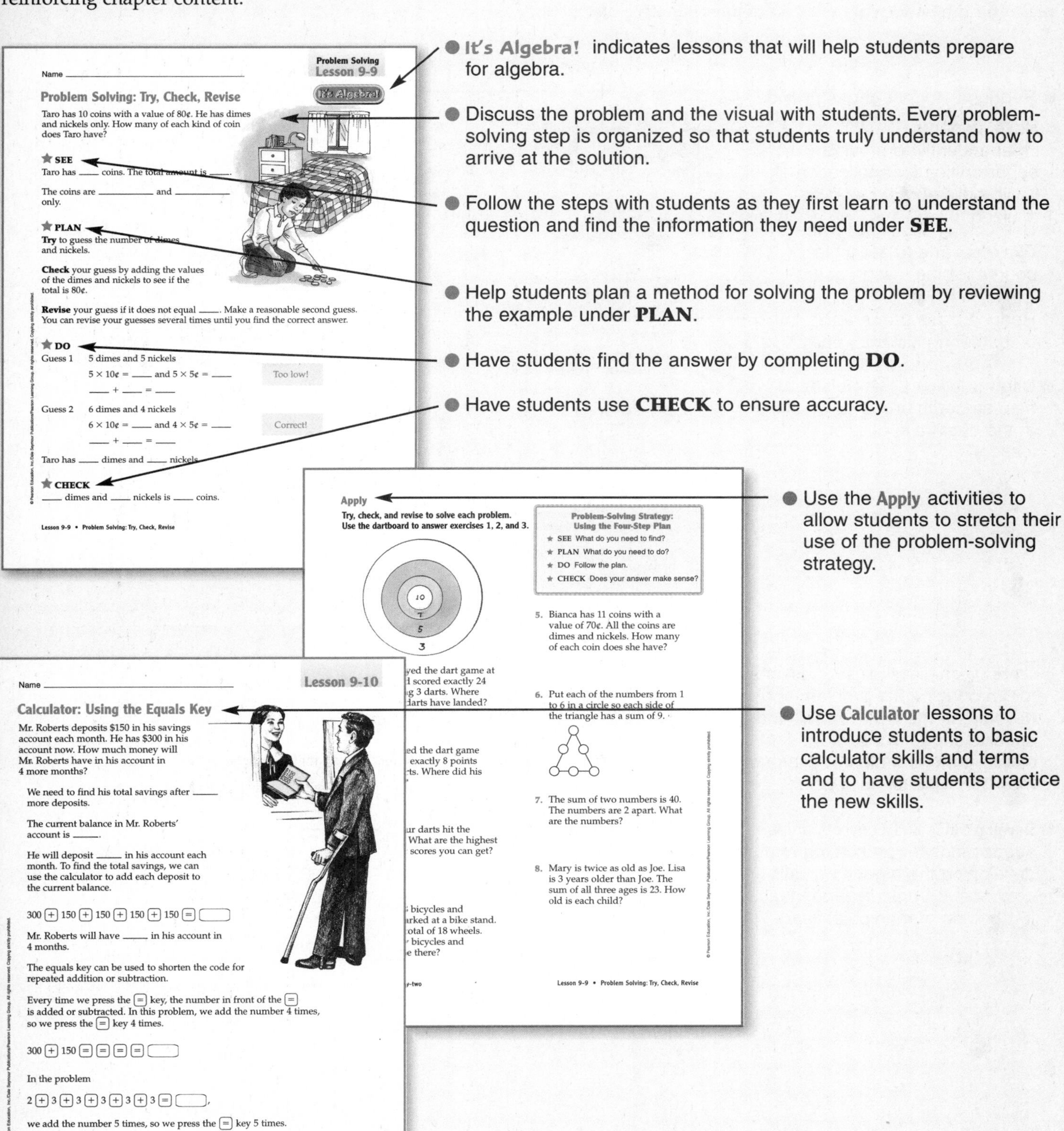

Name ______

Problem Solving
Lesson 9-9

It's Algebra!

Problem Solving: Try, Check, Revise

Taro has 10 coins with a value of 80¢. He has dimes and nickels only. How many of each kind of coin does Taro have?

★ **SEE**

Taro has ___ coins. The total amount is ___.

The coins are ______ and ______ only.

★ **PLAN**

Try to guess the number of dimes and nickels.

Check your guess by adding the values of the dimes and nickels to see if the total is 80¢.

Revise your guess if it does not equal ___. Make a reasonable second guess. You can revise your guesses several times until you find the correct answer.

★ **DO**

Guess 1 5 dimes and 5 nickels

$5 \times 10¢ =$ ___ and $5 \times 5¢ =$ ___ Too low!

___ + ___ = ___

Guess 2 6 dimes and 4 nickels

$6 \times 10¢ =$ ___ and $4 \times 5¢ =$ ___ Correct!

___ + ___ = ___

Taro has ___ dimes and ___ nickels.

★ **CHECK**

___ dimes and ___ nickels is ___ coins.

Lesson 9-9 • Problem Solving: Try, Check, Revise

Apply

Try, check, and revise to solve each problem.
Use the dartboard to answer exercises 1, 2, and 3.

Problem-Solving Strategy: Using the Four-Step Plan

★ SEE What do you need to find?
★ PLAN What do you need to do?
★ DO Follow the plan.
★ CHECK Does your answer make sense?

5. Bianca has 11 coins with a value of 70¢. All the coins are dimes and nickels. How many of each coin does she have?

6. Put each of the numbers from 1 to 6 in a circle so each side of the triangle has a sum of 9.

7. The sum of two numbers is 40. The numbers are 2 apart. What are the numbers?

8. Mary is twice as old as Joe. Lisa is 3 years older than Joe. The sum of all three ages is 23. How old is each child?

Lesson 9-9 • Problem Solving: Try, Check, Revise

Name ______

Lesson 9-10

Calculator: Using the Equals Key

Mr. Roberts deposits $150 in his savings account each month. He has $300 in his account now. How much money will Mr. Roberts have in his account in 4 more months?

We need to find his total savings after ___ more deposits.

The current balance in Mr. Roberts' account is ___.

He will deposit ___ in his account each month. To find the total savings, we can use the calculator to add each deposit to the current balance.

300 [+] 150 [+] 150 [+] 150 [+] 150 [=] [ ]

Mr. Roberts will have ___ in his account in 4 months.

The equals key can be used to shorten the code for repeated addition or subtraction.

Every time we press the [=] key, the number in front of the [=] is added or subtracted. In this problem, we add the number 4 times, so we press the [=] key 4 times.

300 [+] 150 [=] [=] [=] [=] [ ]

In the problem

2 [+] 3 [+] 3 [+] 3 [+] 3 [+] 3 [=] [ ],

we add the number 5 times, so we press the [=] key 5 times.

2 [+] 3 [=] [=] [=] [=] [=] [ ]

Lesson 9-10 • Calculator: Using the Equals Key one hundred seventy-three 173

- **It's Algebra!** indicates lessons that will help students prepare for algebra.
- Discuss the problem and the visual with students. Every problem-solving step is organized so that students truly understand how to arrive at the solution.
- Follow the steps with students as they first learn to understand the question and find the information they need under **SEE**.
- Help students plan a method for solving the problem by reviewing the example under **PLAN**.
- Have students find the answer by completing **DO**.
- Have students use **CHECK** to ensure accuracy.
- Use the **Apply** activities to allow students to stretch their use of the problem-solving strategy.
- Use **Calculator** lessons to introduce students to basic calculator skills and terms and to have students practice the new skills.

Instruct Using a Four-Step Lesson Plan

The Teacher's Edition is designed and organized with you in mind. Consistent four-step lesson plans on two pages will help you make efficient use of your planning time and will help you deliver effective instruction.

- Reduced student pages provide point-of-use information.
- Use Getting Started to review the Objectives and to set a clear course for the lesson goal.
- Reduce class preparation time by gathering Materials early.
- Use the Warm Up exercises to help students brush up on skills at the beginning of each day's lesson.
- Teach gives practical suggestions for introducing the problem and developing the skill. You will find specific suggestions for an effective presentation of the model in **Introduce the Lesson**.
- **Develop Skills and Concepts** gives suggestions for presenting and developing the algorithm, skill, or concept. Some include ideas for the use of manipulatives.

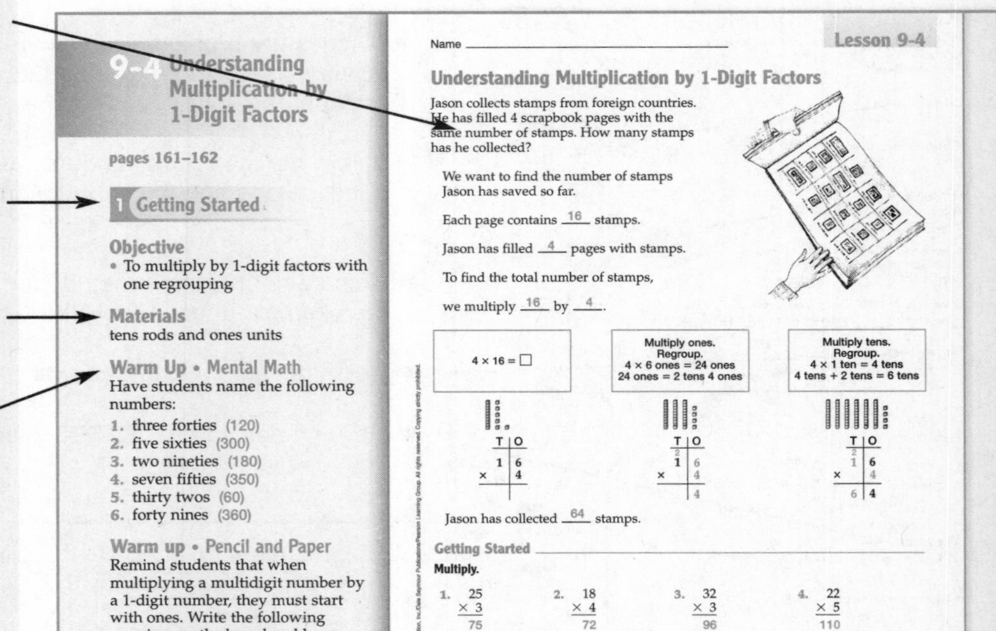

9-4 Understanding Multiplication by 1-Digit Factors

pages 161–162

1 Getting Started

Objective
- To multiply by 1-digit factors with one regrouping

Materials
tens rods and ones units

Warm Up • Mental Math
Have students name the following numbers:
1. three forties (120)
2. five sixties (300)
3. two nineties (180)
4. seven fifties (350)
5. thirty twos (60)
6. forty nines (360)

Warm up • Pencil and Paper
Remind students that when multiplying a multidigit number by a 1-digit number, they must start with ones. Write the following exercises on the board and have students solve them on paper:

1. $121 \times 4 =$ (484)
2. $332 \times 3 =$ (996)
3. $312 \times 2 =$ (624)
4. $634 \times 1 =$ (634)
5. $114 \times 2 =$ (228)

Name ______________ Lesson 9-4

Understanding Multiplication by 1-Digit Factors

Jason collects stamps from foreign countries. He has filled 4 scrapbook pages with the same number of stamps. How many stamps has he collected?

We want to find the number of stamps Jason has saved so far.

Each page contains 16 stamps.

Jason has filled 4 pages with stamps.

To find the total number of stamps,

we multiply 16 by 4.

$4 \times 16 = \square$	Multiply ones. Regroup. 4×6 ones = 24 ones; 24 ones = 2 tens 4 ones	Multiply tens. Regroup. 4×1 ten = 4 tens; 4 tens + 2 tens = 6 tens

Jason has collected 64 stamps.

Getting Started

Multiply.

1. 25×3 = 75
2. 18×4 = 72
3. 32×3 = 96
4. 22×5 = 110

Copy and multiply.

5. 19×4 76
6. 23×4 92
7. 14×6 84
8. 48×2 96

Lesson 9-4 • Understanding Multiplication by 1-Digit Factors — one hundred sixty-one 161

2 Teach

Introduce the Lesson Have students read and identify the problem. Ask students to read and complete the information sentences (16; 4) and the plan sentence. (16, 4)

- Make a quick sketch from Jason's stamp collection book on the board. Point out the *rows* and the *columns* and ask students to repeat. Then draw pages with different numbers of rows and columns and ask: *How many rows (columns) are there?*
- Have students examine the first step of the model. Ask a student to multiply 6 ones by 4. (24) Tell students to write the number of ones—in this case 4—in the ones column, and write the number of tens (2) over the tens column.
- Now, have them look at the next step and ask a student to tell the product of 4×1 ten. (4) Point out that the 2 tens left from the one's multiplication must be added to the 4 tens. (4 tens + 2 tens = 6 tens) Have students complete the solution sentence. (64)

Develop Skills and Concepts Write the following on the board:

24×3

Ask the class where to start multiplying. (in the ones column) Ask a volunteer to multiply ones. (3×4 ones = 12 ones) Ask what number to write in the ones column of the answer (2) and what number will be carried over to the tens column to be added later. (1) Use the base-ten blocks to clarify the regrouping. Display 12 ones units and have a student exchange 10 ones for 1 ten rod. Ask students how many ones are remaining. (2) Let another student multiply tens and add the 1 ten to complete the problem. (72) Repeat with several similar problems.

T161

Practice

Multiply.

1. 35 × 2 = 70	2. 24 × 3 = 72	3. 16 × 5 = 80	4. 47 × 2 = 94	5. 33 × 3 = 99
6. 27 × 3 = 81	7. 19 × 4 = 76	8. 12 × 5 = 60	9. 26 × 3 = 78	10. 18 × 4 = 72
11. 49 × 1 = 49	12. 19 × 3 = 57	13. 13 × 4 = 52	14. 24 × 4 = 96	15. 37 × 2 = 74
16. 14 × 6 = 84	17. 24 × 2 = 48	18. 12 × 7 = 84	19. 38 × 2 = 76	20. 12 × 8 = 96

Copy and multiply.

21. 47 × 2 = 94	22. 36 × 2 = 72	23. 14 × 7 = 98	24. 16 × 3 = 48
25. 23 × 4 = 92	26. 17 × 4 = 68	27. 31 × 1 = 31	28. 11 × 5 = 55
29. 18 × 3 = 54	30. 12 × 6 = 72	31. 14 × 5 = 70	32. 42 × 2 = 84

Now Try This!

Study the following pattern.

47 × 2 = 40 twos + 7 twos = 80 + 14 = 94

Use the pattern to help you complete the rest of the multiplications.

1. 24 × 3 = 20 threes + 4 threes = 60 + 12 = 72
2. 37 × 2 = 30 twos + 7 twos = 60 + 14 = 74
3. 17 × 4 = 10 fours + 7 fours = 40 + 28 = 68
4. 12 × 7 = 10 sevens + 2 sevens = 70 + 14 = 84

3 Practice

Have students solve all the exercises on the page. Remind them to start multiplying in the ones column, and write any tens over the tens column and add them in last.

Now Try This! Discuss with the students how multiplying by decades helps to do problems mentally. Have students complete the exercises.

4 Assess

Ask students where they must begin multiplying a two-digit number. (at the ones)

For Mixed Abilities

Common Errors • Intervention

When students multiply the ones, they may not regroup but write both digits in the answer.

INCORRECT	CORRECT
38 × 2 = 616	38 × 2 = 76 (regroup 1)

Have them work with partners and base-ten blocks to model the problem.

Enrichment • Measurement

Have students think about the 8- and 6-hour clocks they made in the Enrichment on page T160. Ask them to explain the problem with these clocks. (The clock is in the same configuration 3 or 4 times a day.) Ask each student to devise a different way of distinguishing one part of the day from another, as our use of A.M. and P.M. does.

More to Explore • Measurement

Have the students make a simple rain gauge. Ask them to mark off a glass or plastic jar in quarter-inches. Have students place the jar in a safe but open place at home or school. Ask the students to take measurements each time it rains or snows and record the results for a two-week period. Have students make a bar graph to show the amount of rainfall for the two weeks. Have the class compare their results with the official measurements given by the TV weather report or the newspaper.

ESL/ELL STRATEGIES

Make a quick sketch on the board of a page from Jason's stamp collection book. Point out the *rows* and *columns* and ask students to repeat the words. Then, draw pages with different numbers of rows and columns and ask: *How many rows (columns) are there?*

T162

- Use the **Common Errors • Intervention** feature to explore a common error pattern and to provide remediation to struggling students. Collectively, all the Common Errors features in any chapter constitute a complete set of the common errors that students are likely to make when working in that chapter.
- **Enrichment** activities are a direct extension of the skills being taught. Some students may do these activities on their own while you work with those students who need more help.
- **More to Explore** activities are challenging and independent, expanding the mathematical experiences of the students. The More to Explore section encompasses a wide variety of activities and projects and introduces and extends skills taught in the normal curriculum—including data collection and analysis, logic, and probability.
- **ESL/ELL Strategies** are offered twice per chapter. These activities will help you provide insights into English vocabulary and increase comprehension of mathematical concepts. Specific techniques cited ensure that learning is taking place. The techniques also remove potential language barriers for English-language learners at beginning levels of proficiency.
- **Practice** offers you guidelines to assist your students as they practice independently.
- **Assess** provides you with a short question or activity that can be used to quickly evaluate if students have grasped the main objective of the lesson.

Assess Often With Chapter and Cumulative Tests

A variety of assessments help you track your students' mastery of algorithms, basic skills, and problem solving.

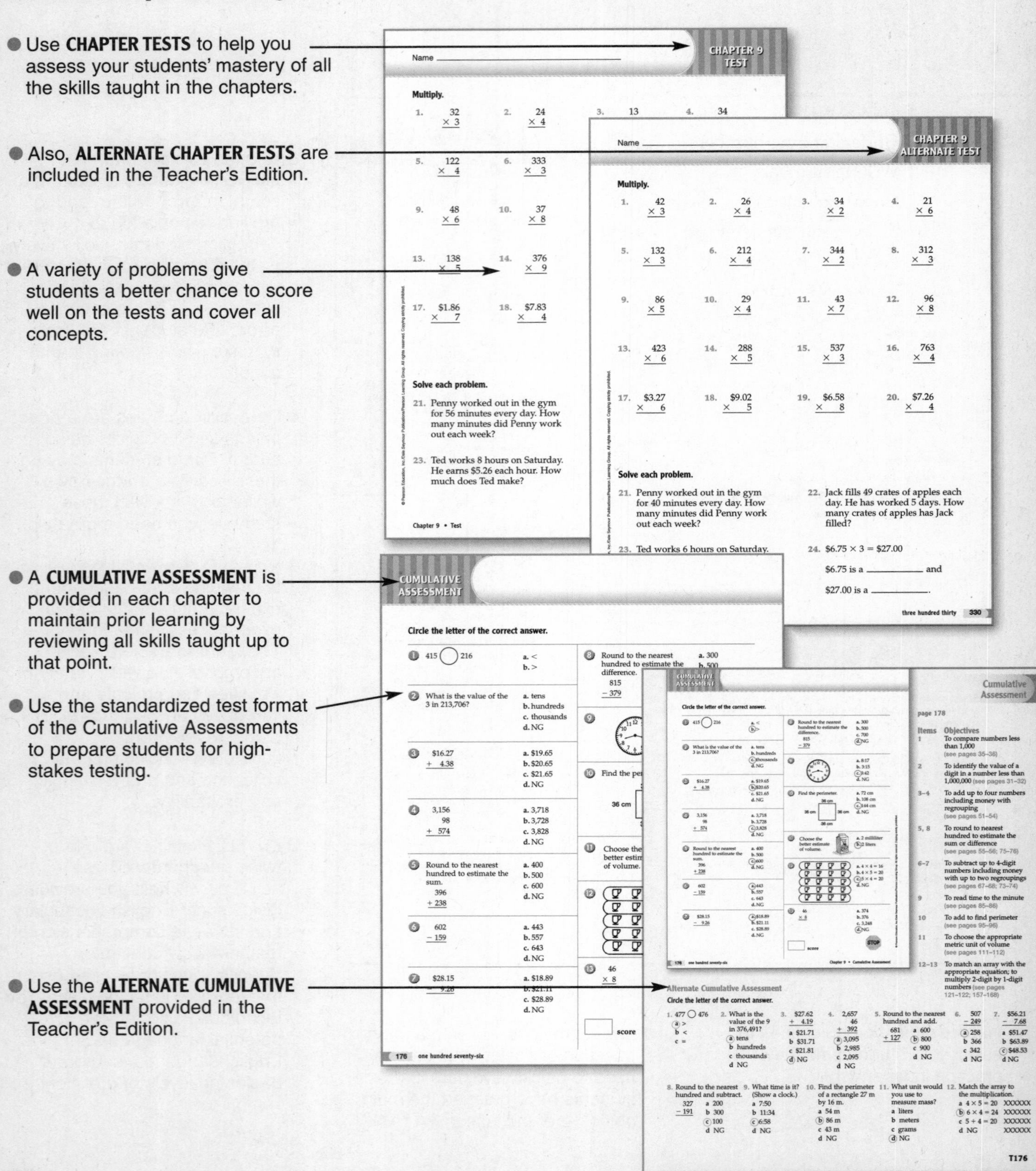

- Use **CHAPTER TESTS** to help you assess your students' mastery of all the skills taught in the chapters.
- Also, **ALTERNATE CHAPTER TESTS** are included in the Teacher's Edition.
- A variety of problems give students a better chance to score well on the tests and cover all concepts.
- A **CUMULATIVE ASSESSMENT** is provided in each chapter to maintain prior learning by reviewing all skills taught up to that point.
- Use the standardized test format of the Cumulative Assessments to prepare students for high-stakes testing.
- Use the **ALTERNATE CUMULATIVE ASSESSMENT** provided in the Teacher's Edition.

Name ______________ CHAPTER 9 TEST

Multiply.

1. 32 × 3 2. 24 × 4 3. 13 4. 34
5. 122 × 4 6. 333 × 3
9. 48 × 6 10. 37 × 8
13. 138 × 5 14. 376 × 9
17. $1.86 × 7 18. $7.83 × 4

Solve each problem.

21. Penny worked out in the gym for 56 minutes every day. How many minutes did Penny work out each week?

23. Ted works 8 hours on Saturday. He earns $5.26 each hour. How much does Ted make?

Chapter 9 • Test

Name ______________ CHAPTER 9 ALTERNATE TEST

Multiply.

1. 42 × 3 2. 26 × 4 3. 34 × 2 4. 21 × 6
5. 132 × 3 6. 212 × 4 7. 344 × 2 8. 312 × 3
9. 86 × 5 10. 29 × 4 11. 43 × 7 12. 96 × 8
13. 423 × 6 14. 288 × 5 15. 537 × 3 16. 763 × 4
17. $3.27 × 6 18. $9.02 × 5 19. $6.58 × 8 20. $7.26 × 4

Solve each problem.

21. Penny worked out in the gym for 40 minutes every day. How many minutes did Penny work out each week?

22. Jack fills 49 crates of apples each day. He has worked 5 days. How many crates of apples has Jack filled?

23. Ted works 6 hours on Saturday.

24. $6.75 × 3 = $27.00
$6.75 is a ________ and
$27.00 is a ________.

three hundred thirty 330

CUMULATIVE ASSESSMENT

Circle the letter of the correct answer.

1. 415 ◯ 216 a. < b. >
2. What is the value of the 3 in 213,706? a. tens b. hundreds c. thousands d. NG
3. $16.27 + 4.38 a. $19.65 b. $20.65 c. $21.65 d. NG
4. 3,156 + 98 + 574 a. 3,718 b. 3,728 c. 3,828 d. NG
5. Round to the nearest hundred to estimate the sum. 396 + 238 a. 400 b. 500 c. 600 d. NG
6. 602 − 159 a. 443 b. 557 c. 643 d. NG
7. $28.15 − 9.26 a. $18.89 b. $21.11 c. $28.89 d. NG
8. Round to the nearest hundred to estimate the difference. 815 − 379 a. 300 b. 500
10. Find the perimeter. 36 cm
11. Choose the better estimate of volume.
13. 46 × 8

score

176 one hundred seventy-six

Cumulative Assessment

page 178

Items	Objectives
1	To compare numbers less than 1,000 (see pages 35–36)
2	To identify the value of a digit in a number less than 1,000,000 (see pages 31–32)
3–4	To add up to four numbers including money with regrouping (see pages 51–54)
5, 8	To round to nearest hundred to estimate the sum or difference (see pages 55–56; 75–76)
6–7	To subtract up to 4-digit numbers including money with up to two regroupings (see pages 67–68; 73–74)
9	To read time to the minute (see pages 85–86)
10	To add to find perimeter (see pages 95–96)
11	To choose the appropriate metric unit of volume (see pages 111–112)
12–13	To match an array with the appropriate equation; to multiply 2-digit by 1-digit numbers (see pages 121–122; 157–168)

Alternate Cumulative Assessment

Circle the letter of the correct answer.

1. 477 ◯ 476 a > b < c =
2. What is the value of the 9 in 376,491? a tens b hundreds c thousands d NG
3. $27.62 + 4.19 a $21.71 b $31.71 c $21.81 d NG
4. 2,657 + 46 + 392 a 3,095 b 2,985 c 2,095 d NG
5. Round to the nearest hundred and add. 681 + 127 a 600 b 800 c 900 d NG
6. 507 − 249 a 258 b 366 c 342 d NG
7. $56.21 − 7.68 a $51.47 b $63.89 c $48.53 d NG
8. Round to the nearest hundred and subtract. 327 − 191 a 200 b 300 c 100 d NG
9. What time is it? (Show a clock.) a 7:50 b 11:34 c 6:58 d NG
10. Find the perimeter of a rectangle 27 m by 16 m. a 54 m b 86 m c 43 m d NG
11. What unit would you use to measure mass? a liters b meters c grams d NG
12. Match the array to the multiplication. a 4 × 5 = 20 b 6 × 4 = 24 c 5 + 4 = 20 d NG

T176

Scope and Sequence

Levels	K	A	B	C	D	E	F
Readiness							
Using attributes of size, shape, and color	●	●	●				
Sorting and classifying	●	●					
Spatial relationships	●						
Numeration							
One-to-one correspondence	●						
Understanding numbers	●	●	●	●	●	●	●
Writing numbers	●	●					
Counting objects	●	●	●				
Sequencing numbers	●	●	●	●	●		
Numbers before and after	●	●	●	●	●		
Ordering numbers	●	●	●	●	●	●	●
Comparing numbers	●	●	●	●	●	●	●
Grouping numbers	●	●	●	●	●		
Ordinal numbers	●	●	●	●			
Number words		●	●	●	●	●	●
Expanded numbers				●	●	●	●
Place value	●	●	●	●	●	●	●
Skip-counting	●	●	●	●	●		
Roman numerals			●	●			●
Rounding numbers			●	●	●	●	●
Squares and square roots							●
Primes and composites				●	●	●	●
Multiples			●	●	●	●	●
Least common multiples						●	●
Greatest common factors						●	●
Exponents						●	●
Addition							
Addition facts	●	●	●	●	●	●	●
Fact families		●	●	●	●	●	●
Missing addends		●	●	●	●	●	●
Adding money	●	●	●	●	●	●	●
Column addition		●	●	●	●	●	●
Two-digit addends		●	●	●	●	●	●
Multidigit addends			●	●	●	●	●
Addition with regrouping		●	●	●	●	●	●
Basic properties of addition				●	●	●	●
Estimating sums			●	●	●	●	●
Addition of fractions				●	●	●	●

Scope and Sequence

Levels	K	A	B	C	D	E	F
Addition *(continued)*							
Addition of mixed numbers				●	●	●	●
Addition of decimals				●	●	●	●
Addition of customary measures						●	●
Addition of integers						●	●
Subtraction							
Subtraction facts	●	●	●	●	●	●	●
Fact families		●	●	●	●	●	●
Missing subtrahends		●	●	●	●		
Subtracting money	●	●	●	●	●	●	●
Two-digit numbers		●	●	●	●	●	●
Multidigit numbers			●	●	●	●	●
Subtraction with regrouping		●	●	●	●	●	●
Zeros in the minuend			●	●	●	●	●
Basic properties of subtraction				●	●	●	●
Estimating differences			●	●	●	●	●
Subtraction of fractions				●	●	●	●
Subtraction of mixed numbers						●	●
Subtraction of decimals				●	●	●	●
Subtraction of customary measures						●	●
Subtraction of integers						●	●
Multiplication							
Multiplication facts			●	●	●	●	●
Fact families			●	●	●	●	●
Missing factors				●	●	●	●
Multiplying money				●	●	●	●
Multiplication by powers of ten				●	●	●	●
Multidigit factors				●	●	●	●
Multiplication with regrouping				●	●	●	●
Basic properties of multiplication			●	●	●	●	●
Estimating products				●	●	●	●
Multiples				●	●	●	●
Least common multiples						●	●
Multiplication of fractions						●	●
Factorization						●	●
Multiplication of mixed numbers						●	●
Multiplication of decimals					●	●	●
Exponents						●	●
Multiplication of integers						●	●

Scope and Sequence

Levels	K	A	B	C	D	E	F
Division							
Division facts			●	●	●	●	●
Fact families			●	●	●	●	●
Divisibility rules				●	●	●	●
Two-digit quotients				●	●	●	●
Remainders				●	●	●	●
Multidigit quotients					●	●	●
Zeros in quotients					●	●	●
Division by multiples of ten					●	●	●
Two-digit divisors					●	●	●
Properties of division					●	●	●
Averages					●	●	●
Greatest common factors						●	●
Division of fractions						●	●
Division of mixed numbers						●	●
Division of decimals						●	●
Division of integers							●
Money							
Counting pennies	●	●	●	●	●		
Counting nickels	●	●	●	●	●		
Counting dimes	●	●	●	●	●		
Counting quarters		●	●	●	●		
Counting half-dollars		●	●	●	●		
Counting dollar bills		●	●	●	●		
Writing dollar and cent signs		●	●	●	●	●	●
Matching money with prices	●	●	●				
Determining amount of change		●	●	●	●		
Determining sufficient amount		●	●				
Determining which coins to use		●	●				
Addition	●	●	●	●	●	●	●
Subtraction	●	●	●	●	●	●	●
Multiplication				●	●	●	●
Division				●	●	●	●
Estimating amounts of money			●	●	●	●	●
Rounding amounts of money			●	●	●	●	●
Buying from a table of prices		●	●	●	●	●	●
Fractions							
Understanding equal parts	●	●	●	●			
One-half	●	●	●	●	●		

Scope and Sequence

Levels	K	A	B	C	D	E	F
Fractions *(continued)*							
One-fourth	●	●	●	●	●		
One-third	●	●	●	●	●		
Identifying fractional parts of figures	●	●	●	●	●	●	●
Identifying fractional parts of sets		●	●	●	●	●	●
Finding unit fractions of numbers				●	●		
Equivalent fractions				●	●	●	●
Comparing and ordering fractions				●	●	●	●
Simplifying fractions					●	●	●
Mixed numbers				●	●	●	●
Addition of fractions				●	●	●	●
Subtraction of fractions				●	●	●	●
Addition of mixed numbers					●	●	●
Subtraction of mixed numbers						●	●
Multiplication of fractions						●	●
Multiplication of mixed numbers						●	●
Division of fractions						●	●
Division of mixed numbers						●	●
Renaming fractions as decimals				●	●	●	●
Renaming fractions as percents						●	●
Decimals							
Place value				●	●	●	●
Reading decimals				●	●	●	●
Writing decimals				●	●	●	●
Converting fractions to decimals					●	●	●
Writing parts of sets as decimals				●	●	●	●
Comparing decimals				●	●	●	●
Ordering decimals				●	●	●	●
Addition of decimals				●	●	●	●
Subtraction of decimals				●	●	●	●
Rounding decimals					●	●	●
Multiplication of decimals					●	●	●
Division of decimals						●	●
Renaming decimals as percents						●	●
Geometry							
Polygons	●	●	●	●	●	●	●
Sides and vertices of polygons		●	●	●	●	●	●
Faces, edges, and vertices		●	●	●	●	●	●
Solid figures	●	●	●	●	●	●	●

Scope and Sequence

Levels	K	A	B	C	D	E	F
Geometry *(continued)*							
Symmetry	●	●	●	●	●	●	●
Lines and line segments				●	●	●	●
Rays and angles				●	●	●	●
Measuring angles						●	●
Transformations			●	●	●	●	●
Congruency				●	●	●	●
Similar figures					●	●	●
Circles					●	●	●
Triangles				●	●	●	●
Quadrilaterals				●	●	●	●
Measurement							
Nonstandard units of measure	●	●					
Customary units of measure		●	●	●	●	●	●
Metric units of measure		●	●	●	●	●	●
Renaming customary measures				●	●	●	●
Renaming metric measures				●	●	●	●
Selecting appropriate units		●	●	●	●		
Estimating measures		●	●	●	●		
Perimeter by counting		●	●				
Perimeter by formula				●	●		●
Area of polygons by counting		●	●	●	●		
Area of polygons by formula					●	●	●
Volume by counting				●	●	●	
Volume by formula					●	●	●
Addition of measures						●	●
Subtraction of measures						●	●
Circumference of circles							●
Area of circles							●
Surface area of space figures							●
Estimating temperatures				●	●	●	●
Reading temperature scales		●	●	●	●	●	●
Time							
Ordering events	●						
Calendars	●	●	●	●	●		
Telling time to the hour	●	●	●	●	●		
Telling time to the half-hour		●	●	●	●		
Telling time to the five minutes		●	●	●	●		
Telling time to the minute			●	●	●		

Scope and Sequence

Levels	K	A	B	C	D	E	F
Time *(continued)*							
Understanding A.M. and P.M.				●	●	●	●
Elapsed time			●	●	●	●	●
Graphing							
Tallies	●	●	●	●	●	●	
Bar graphs		●	●	●	●	●	●
Picture graphs	●	●	●	●	●		
Line graphs				●	●	●	●
Circle graphs						●	
Line plots			●			●	
Stem-and-leaf plots						●	●
Histograms							●
Ordered pairs			●	●	●	●	●
Statistics and Probability							
Understanding probability			●	●	●	●	●
Listing outcomes					●	●	●
Mean, median, and mode				●	●	●	●
Writing probabilities				●	●	●	●
Compound probability							●
Making predictions							●
Tree diagrams					●	●	●
Ratios and Percents							
Understanding ratios					●	●	●
Equal ratios						●	●
Proportions						●	●
Scale drawings						●	●
Ratios as percents						●	●
Percents as fractions						●	●
Fractions as percents						●	●
Finding the percents of numbers						●	●
Integers							
Understanding integers						●	●
Addition of integers						●	●
Subtraction of integers						●	●
Multiplication of integers							●
Division of integers							●
Graphing integers on coördinate planes							●

Scope and Sequence

Levels	K	A	B	C	D	E	F
Problem Solving							
Act it out	●	●	●	●	●	●	●
Choose a strategy	●	●	●	●	●		
Choose the correct operation	●	●	●	●	●		
Collect and use data		●	●	●	●	●	●
Determine missing or extra data		●	●	●	●		
Draw a picture or diagram	●	●	●	●	●	●	●
Identify a subgoal						●	●
Look for a pattern	●	●	●	●	●	●	●
Make a graph	●	●	●	●	●	●	●
Make a list		●	●	●	●	●	●
Make a model					●		
Make a table		●	●	●		●	●
Make a tally graph						●	
Restate the problem					●	●	●
Solve a simpler but related problem				●	●	●	●
Solve multistep problems				●	●	●	●
Try, check, and revise	●	●	●	●	●	●	●
Use a formula						●	●
Use a four-step plan				●	●	●	●
Use an exact answer or an estimate				●	●		
Use logical reasoning		●	●	●	●		●
Work backward				●	●	●	●
Write a number sentence		●	●	●	●		
Calculators							
Calculator codes				●	●	●	●
Equals key				●	●	●	●
Addition/subtraction keys				●	●	●	●
Multiplication/division keys				●	●	●	●
Clear key				●	●		
Calculators: Real-World Applications							
Averages						●	●
Formulas						●	●
Money					●	●	●
Percents						●	●
Repeating decimals						●	●
Statistics						●	●

Scope and Sequence

Levels	K	A	B	C	D	E	F
Algebra							
Patterns	●	●	●	●	●	●	●
Completing number sentences	●	●	●	●	●	●	●
Properties of numbers				●	●	●	●
Numerical expressions				●	●	●	●
Evaluating numerical expressions					●	●	●
Algebraic expressions						●	●
Evaluating algebraic expressions						●	●
Order of operations				●	●	●	●
Integers						●	●
Addition and subtraction of integers						●	●
Multiplication and division of integers							●
Ordered pairs			●	●	●	●	●
Function tables				●	●	●	●
Graphing a rule or an equation						●	●
Variables				●	●	●	●
Equations				●	●	●	●
Solve addition and subtraction equations						●	●
Solve multiplication and division equations						●	●
Model problem situations with equations						●	●
Solve inequalities							●
Formulas						●	●
Properties of equality						●	●

Contents

Chapter 1 Addition and Subtraction

Chapter 2 Place Value

Chapter 3 Addition of Whole Numbers

Chapter 4 Subtraction of Whole Numbers

Chapter 8 Measurement

Chapter 9 Multiplying Whole Numbers

Chapter 10 Division of Whole Numbers

Chapter 11 Geometric Figures

Chapter 12 Using Geometric Figures

Chapter 13 Fractions

1-1 Addition Facts and Subtraction Facts

pages 1–2

1 Getting Started

Objective
- To review basic addition and subtraction facts

Vocabulary
minuend, subtrahend, difference, addend, sum, number sentences

Materials
*carpet squares or masking tape;
*floor number line 0 through 20

Warm Up • Mental Math
Dictate the following problems:

1. 10,000 + 400 (10,400)
2. 60 − 8 (52)
3. 100 + 40 − 60 (80)
4. 7 times 8 plus 3 (59)
5. 10 ÷ 2 (5)
6. 6 + (8 × 2) (22)
7. 3 dozen − 2 dozen (12)

Warm Up • Numeration
Write the number names for 0 through 18 in mixed order on the board. Have students write the number names and their corresponding numbers in order on a sheet of paper.

Name ______________________

Addition and Subtraction

CHAPTER 1

Lesson 1-1

Addition Facts and Subtraction Facts

Rinaldo's goal for this year is to read 12 books. So far, he has read 5 books. How many books must Rinaldo read to reach his goal?

We want to know the number of books Rinaldo must still read to reach his goal.

Rinaldo's goal is to read 12 books.

He has read 5 books so far this year.

To find the number of books he needs to read, we subtract 5 from 12.

12 (minuend) − 5 (subtrahend) = 7 (difference)

12 ← minuend
− 5 ← subtrahend
7 ← difference

Rinaldo needs to read 7 more books this year.

To check the subtraction, add 7 and 5.

7 (addend) + 5 (addend) = 12 (sum)

7 ← addend
+ 5 ← addend
12 ← sum

The subtraction is correct because 7 + 5 = 12.

7 + 5 = 12 and 12 − 5 = 7 are called **number sentences**.

Getting Started

Complete each number sentence. Check your answer.

1. 4 + 2 = 6
2. 7 + 9 = 16
3. 8 + 8 = 16
4. 10 − 8 = 2
5. 11 − 3 = 8
6. 13 − 4 = 9
7. 5 + 6 = 11
8. 9 − 5 = 4

2 Teach

Introduce the Lesson Have a student read the problem aloud. Ask students to tell what they need to find. (how many more books Rinaldo needs to read) Ask students to tell what they already know. (Rinaldo wants to read 12 books and has read 5 so far.) Have students read the sentences aloud as they solve the problem.

- Remind students that they can check their subtraction by adding. Ask a volunteer to read the next section and have the class fill in the information. Finally, read the last sentence and ask students to give additional examples of number sentences.

Develop Skills and Concepts Write 8 − 4 = _____ on the board. Have a student stand on 8 on the number line and then walk 4 spaces back, toward the smaller numbers. Ask another student to tell where the first student stopped. (4)

Write the following on the board:

8 − 4 = 4
minuend subtrahend difference

8 ←minuend
− 4 ←subtrahend
4 ←difference

Explain that 8 − 4 = 4 is a number sentence for subtraction. Tell students that the first number is the **minuend**. It tells where to start on the number line. The second number is the **subtrahend**. It tells how far to walk backward. The stopping place is the **difference**.

- Give several examples using horizontal and vertical notations. Have students act out the problems. Have students use addition to check their subtraction. Ask them to label the addends and the sum.

*** Indicates teacher demonstration materials.**

Practice

Complete each number sentence. Check your answer.

1. $7 + 1 =$ 8 2. $4 + 6 =$ 10 3. $7 + 4 =$ 11 4. $5 + 9 =$ 14

5. $6 + 8 =$ 14 6. $8 + 9 =$ 17 7. $2 + 4 =$ 6 8. $7 + 6 =$ 13

9. $9 - 2 =$ 7 10. $13 - 9 =$ 4 11. $15 - 6 =$ 9 12. $7 - 3 =$ 4

13. $12 - 3 =$ 9 14. $14 - 8 =$ 6 15. $10 - 5 =$ 5 16. $8 - 1 =$ 7

Add or subtract.

17. $8 + 7 = 15$ 18. $6 + 5 = 11$ 19. $4 + 3 = 7$ 20. $9 + 1 = 10$ 21. $5 + 8 = 13$ 22. $2 + 7 = 9$

23. $14 - 7 = 7$ 24. $8 - 3 = 5$ 25. $12 - 9 = 3$ 26. $10 - 7 = 3$ 27. $11 - 6 = 5$ 28. $16 - 7 = 9$

29. $4 + 5 = 9$ 30. $7 + 9 = 16$ 31. $6 + 6 = 12$ 32. $5 + 2 = 7$ 33. $1 + 6 = 7$ 34. $5 + 5 = 10$

35. $13 - 8 = 5$ 36. $5 - 3 = 2$ 37. $7 - 6 = 1$ 38. $12 - 8 = 4$ 39. $9 - 2 = 7$ 40. $17 - 8 = 9$

41. $6 + 9 = 15$ 42. $3 + 7 = 10$ 43. $8 + 9 = 17$ 44. $11 - 9 = 2$ 45. $10 - 4 = 6$ 46. $7 - 2 = 5$

Problem Solving

Solve each problem.

47. Megan bought a wool scarf for \$7 and a pair of mittens for \$6. How much did she spend?
\$13

48. Butch made 9 sandwiches. His brother ate 7 of them for lunch. How many sandwiches does Butch have left to eat?
2

3 Practice

Tell students to complete all the exercises on page 2. Remind students to read carefully to solve the two word problems at the bottom of the page.

4 Assess

Ask students how to check subtraction problems. Then, have them show how to check $12 - 4 = 8$. (add the difference and the subtrahend to make sure it equals the minuend; $8 + 4 = 12$)

For Mixed Abilities

Common Errors • Intervention

Some students may have difficulty mastering their basic addition and subtraction facts. Have students work in pairs to practice the facts by using fact cards that show the fact without the answer on one side and the fact with the answer on the reverse side. When responding, students should say all the numbers in the fact; for example, in response to $5 - 3 = ?$, they should say, "5 minus 3 equals 2." In response to $3 + 5 = ?$, students should say, "3 plus 5 equals 8." The students can each take half the set of fact cards and take turns asking each other to complete the facts.

Enrichment • Spatial Sense

Tell students to draw a map that shows the highway distances between the towns of Paday, Manola, and Ispin if Manola is 7 miles from Ispin and one travels through Manola when driving the 22 miles from Ispin to Paday.

More to Explore • Numeration

Have students bring in the front page of a local newspaper. Tell them to skim the page for any numbers they can find, either in headlines or in stories, and circle them. Have students list the numbers found from least to greatest. Tell students to then write the number words for each number they found. You can also have students exchange newspapers and check each other for numbers they may have missed.

ESL/ELL STRATEGIES

Discuss the words *minuend, subtrahend,* and *difference* with ESL students. Write the words on the board, capitalizing the stressed syllables: MINuend, SUBtrahend, DIFFerence. Have students repeat each word and explain its meaning in simple English.

1-2 Column Addition

pages 3–4

1 Getting Started

Objective
- To add 1-digit numbers in a column

Vocabulary
total

Materials
counters

Warm Up • Mental Math
Ask students to solve the following:

1. $\frac{1}{2}$ of 10 (5)
2. $6 - 4 + 10$ (12)
3. $6 \times (3 + 6)$ (54)
4. $7 \div 1 + 12$ (19)
5. 6 quarters + 8 dimes ($2.30)
6. 45 min before 10:35 (9:50)
7. the place value of the 6 in 168 (tens)
8. $\frac{3}{8} + \frac{5}{8}$ ($\frac{8}{8}$ or 1)

Warm Up • Number Sense
Write the following table on the board:

10	13	15	16
(9 + 1)	(9 + 4)	(9 + 6)	(9 + 7)
(8 + 2)	(8 + 5)	(8 + 7)	(8 + 8)
(7 + 3)	(7 + 6)	(7 + 8)	(7 + 9)
(6 + 4)	(6 + 7)	(6 + 9)	
(5 + 5)	(5 + 8)		
(4 + 6)	(4 + 9)		
(3 + 7)			
(2 + 8)			
(1 + 9)			

Have students write all the basic addition facts under each corresponding sum.

Name ______________________________

Lesson 1-2

It's Algebra!

Column Addition

The first U.S. astronauts orbited Earth in 1962. How many orbits did these Americans complete in that year?

Orbits by U.S. Astronauts in 1962		
Date	**Astronaut**	**Orbits**
February 20	John Glenn	3
May 24	Scott Carpenter	3
October 3	Wally Schirra	6

We want to find the total number of orbits all the astronauts made in 1962.

We know that Glenn orbited __3__ times; Carpenter, __3__ times; and Schirra, __6__ times.

To find this **total** or **sum**, we add __3__, __3__, and __6__.

We can add only two numbers at a time.

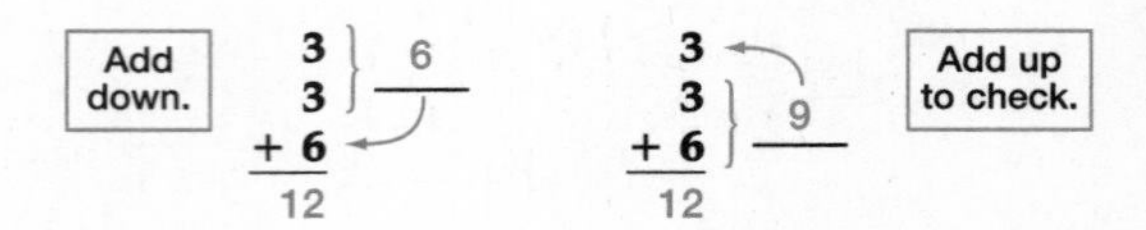

American astronauts completed __12__ orbits in 1962.

Getting Started

Add down. Add up to check.

1.	2.	3.	4.	5.
1	2	3	6	7
3	6	4	3	1
+ 5	+ 3	+ 2	+ 1	+ 7
9	11	9	10	15

6.	7.	8.	9.	10.
6	3	5	7	8
3	2	4	2	1
2	4	5	3	6
+ 4	+ 9	+ 3	5	2
15	18	17	+ 1	+ 2
			18	19

2 Teach

Introduce the Lesson Direct students' attention to the chart as you tell them that on February 20, John Glenn made 3 orbits around Earth. Ask students to tell about Scott Carpenter's adventure in space. (He made 3 orbits on May 24.) Ask students what else the chart tells us. (Wally Schirra made 6 orbits on October 3.)

- Have students read the problem to the left of the chart and tell what it asks them to do. (tell how many times the Americans orbited Earth in 1962) Have students fill in the information and solve the problem in the model. Have them complete the solution sentence.

Develop Skills and Concepts Write **3 + 2 + 5** vertically on the board. Have students place counters in a group of 3, a group of 2, and a group of 5. Tell them to combine the first two groups and tell how many counters there are. (5) Have students combine the two remaining groups and tell how many counters there are in all. (10)

- Tell students to again place their counters in a group of 3, a group of 2, and a group of 5, and combine the last two groups and tell how many counters there are. (7) Have students combine the two remaining groups and tell how many counters there are. (10) Ask students if they have the same sum when they add forward and backward. (yes) Talk through the problem on the board as you add down and then check by adding up the column.
- Repeat for more problems of three addends, being careful that the sum of any two addends is 9 or less and no addends are 0.

Practice

Add down. Add up to check.

1. 6 + 3 + 6 = 15
2. 2 + 3 + 4 = 9
3. 2 + 7 + 2 = 11
4. 8 + 1 + 4 = 13
5. 6 + 3 + 2 = 11
6. 4 + 5 + 3 = 12
7. 1 + 6 + 3 = 10
8. 5 + 2 + 5 = 12
9. 5 + 2 + 1 = 8
10. 7 + 2 + 6 = 15
11. 2 + 5 + 3 = 10
12. 6 + 2 + 3 = 11
13. 4 + 2 + 7 = 13
14. 3 + 1 + 6 = 10
15. 9 + 1 + 5 = 15
16. 8 + 1 + 5 + 1 + 4 = 19
17. 1 + 6 + 3 + 7 = 17
18. 5 + 3 + 4 + 1 = 13
19. 5 + 1 + 6 + 2 = 14
20. 6 + 1 + 3 + 5 + 3 = 18

Now Try This!

It's Algebra!

Complete the boxes by adding each number at the top to each number on the left. Look for patterns.

1.

+	5	3	7
8	13	11	15
18	23	21	25
28	33	31	35
38	43	41	45
48	53	51	55
58	63	61	65

2.

+	2	6	4
9	11	15	13
19	21	25	23
29	31	35	33
39	41	45	43
49	51	55	53
59	61	65	63

3.

+	9	7	5
7	16	14	12
27	36	34	32
47	56	54	52
67	76	74	72
87	96	94	92
107	116	114	112

Now Try This! Students should work down the column and describe the pattern, noting that the numbers in the ones place remain the same in each column. Have them predict other numbers in the series. For example, if 5 + 58 = 63 then 5 + 68 = 73, and so on.

3 Practice

Tell students to complete the exercises on page 4. Remind students to check their work by adding up the column.

4 Assess

Ask students how to check the sum when adding a column of numbers. (If the numbers were added down, check by adding up. If they were added up, check by adding down.)

For Mixed Abilities

Common Errors • Intervention

Some students may have difficulty finding the sum of three or more numbers because they cannot remember the sum of the first two numbers when they go to add the next number. Have students practice by finding the sum of the first two numbers and writing it just above or beside the next addend. Then, have them add this sum to the next addend and write the sum again. They should continue in this manner until all the addends have been used.

Enrichment • Numeration

Tell students to find three different numbers less than 6 whose sum is equal to $\frac{1}{2}$ of 24. (5, 3, 4)

More to Explore • Logic

Give students a copy of this magic circle with 1, 6, and 5 filled in. Have students fill in the empty circles with digits 2 through 9 so that the sum across the circle will always equal 15. Challenge students to devise a magic circle of their own.

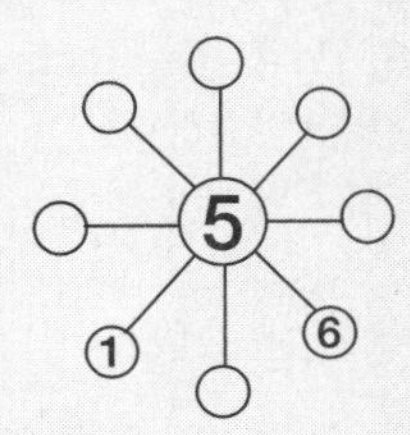

1-3 Basic Properties, Addition

pages 5–6

1 Getting Started

Objective

- To learn the Order, Grouping, and Zero Properties of Addition

Vocabulary

Order or Commutative Property, Grouping or Associative Property, Zero or Identity Property

Warm Up • Mental Math

Ask students what number tells the following:

1. shoes in 7 pairs (14)
2. minutes in $5\frac{1}{2}$ hours (330)
3. $\frac{1}{4}$ of 20 (5)
4. tens in 10,786 (8)
5. inches in 4 feet (48)
6. eggs in $3\frac{1}{4}$ dozen (39)
7. cups in a pint (2)

Warm Up • Numeration

Have students give all the basic addition facts with sums greater than 13 but less than 16. (5 + 9, 9 + 5, 8 + 6, 6 + 8, 7 + 7, 7 + 8, 8 + 7, 9 + 6, 6 + 9) Write the facts on the board as they are given.

Name ______________________

Lesson 1-3

Basic Properties, Addition

It's Algebra!

Finding sums is easy if we remember some important rules.

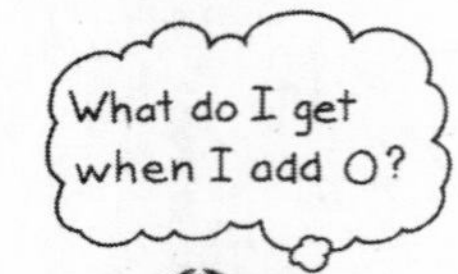

Order or Commutative Property
We can add in any order.

3 + 4 = 7 **4 + 3 =** 7

Grouping or Associative Property
We can change the grouping. Remember to add the numbers in the parentheses first.

(5 + 3) + 6 = ■ **5 + (3 + 6) = ■**
8 **+ 6 =** 14 **5 +** 9 **=** 14

Zero or Identity Property
Adding zero does not affect the answer.

6 + 0 = 6 **0 + 3 =** 3

Getting Started

Complete each number sentence.

1. 5 + 0 = 5
2. (6 + 3) + 2 = 11
3. 0 + 9 = 9
4. 4 + (0 + 6) = 10
5. (2 + 7) + 0 = 9
6. 5 + (3 + 5) = 13

Add down. Add up to check.

7. 6 + 2 + 4 = 12
8. 3 + 9 + 4 = 16
9. 9 + 3 + 0 + 2 = 14
10. 1 + 4 + 5 + 3 + 4 = 17
11. 7 + 5 + 0 + 3 + 2 = 17

2 Teach

Introduce the Lesson Have a student read aloud the thoughts of the children pictured. Read the introductory statement to students. Now, have students read the property names and examples as they complete the number sentences with you.

Develop Skills and Concepts Tell students that they are going to test the three **addition properties**. Have a student write **4 + 2** on the board. Tell students that the **Order or Commutative Property** tells us that we will get the same sum if we add forward or backward. Have a student find the sum both ways and tell if the sums are the same. (The sums are both 6.) Have other students test the Order Property by adding two numbers from 1 to 9.

- Write **4 + 5 + 2 + 3** on the board and tell students that the **Grouping or Associative Property** tells us that we will get the same sum if we add 2 + 3 and then 5 and then 4, as when we add 4 + 5 and then 2 and then 3.

Have a student work the problem aloud and write the sum. (14) Have another student work the problem differently. Now, write the problem vertically and have a student work the problem aloud.

- Write **3 + (4 + 2)** on the board and remind students that they are to always add the numbers in parentheses first.
- Have students similarly test the **Zero or Identity Property**.

3 Practice

Remind students to add numbers in parentheses first when finding the sum in a number sentence. Ask students how they will work the last two rows of exercises on page 6. (find the sum of the first two addends, add the next number to that sum, and add the new sum to the next number) Remind students to check each problem by adding up the column. Have students complete page 6 independently.

Practice

Complete each number sentence.

1. 7 + 0 = 7
2. (4 + 2) + 7 = 13
3. 0 + 8 = 8
4. (8 + 0) + 2 = 10
5. 5 + (8 + 1) = 14
6. (7 + 2) + 3 = 12
7. (0 + 6) + 9 = 15
8. (6 + 0) + 9 = 15
9. 4 + (6 + 3) = 13
10. (2 + 5) + 8 = 15
11. 8 + (5 + 2) = 15
12. 3 + (4 + 5) = 12
13. 6 + (2 + 0) = 8
14. 4 + (3 + 6) = 13
15. (5 + 0) + 5 = 10

Add down. Add up to check.

16. 5, 3, + 2 = 10
17. 2, 7, + 0 = 9
18. 0, 8, + 6 = 14
19. 1, 7, + 2 = 10
20. 6, 0, + 8 = 14
21. 4, 0, + 7 = 11
22. 3, 8, + 0 = 11
23. 8, 1, + 5 = 14
24. 7, 1, + 8 = 16
25. 9, 0, + 4 = 13
26. 0, 5, + 0 = 5
27. 3, 6, + 1 = 10
28. 5, 9, + 1 = 15
29. 2, 2, + 5 = 9
30. 8, 0, + 6 = 14
31. 9, 0, 6, + 4 = 19
32. 1, 4, 3, + 0 = 8
33. 4, 5, 3, + 2 = 14
34. 6, 8, 0, + 5 = 19
35. 7, 2, 5, + 2 = 16
36. 3, 0, 8, 2, + 4 = 17
37. 2, 4, 3, 4, + 2 = 15
38. 5, 4, 2, 1, + 6 = 18
39. 1, 0, 5, 8, + 2 = 16
40. 4, 4, 3, 3, + 3 = 17

4 Assess

Ask students to explain and give an example of the Grouping or Associative Property. (We can change the way numbers are grouped and still get the same sum. We must remember to add the number in parentheses first. Examples of the Grouping or Associative Property will vary.)

For Mixed Abilities

Common Errors • Intervention

Some students may have difficulty believing that three numbers can be added in any order without changing the sum. Have them work in pairs, each with a set of identical addition problems with three or more addends. Ask one student to find the sum by adding from top to bottom. Ask the other student to find the sum by adding from bottom to top. Then, have them compare answers to see that they are the same.

Enrichment • Algebra

Have students write and solve four number sentences that use all three addition properties to find each sum.

More to Explore • Numeration

Ask students to imagine why numbers were first devised. Explain that people survived for thousands of years without numbers. Have students do research to discover when numbers were invented and why. Have students give examples of the use of numbers in early civilizations. (Their answers will vary considerably, depending on the source of information. Look for the ideas that early counting was done with fingers and with written tallies. They may note that numbers were first needed for trade and taxes, activities that accompanied the rise of agriculture and cities, and so on.)

1-4 Basic Properties

pages 7–8

1 Getting Started

Objectives

- To subtract using the Zero Property
- To recognize that addition and subtraction check each other

Vocabulary

fact family

Materials

*floor number line 0 through 20; counters

Warm Up • Mental Math

Have students name the numbers that are 20 more and 20 less than the following:

1. 87 (107, 67)
2. 2 tens 1 one (41, 1)
3. 365 (385, 345)
4. 1,000 (1,020, 980)
5. 705 (725, 685)
6. 194 (214; 174)
7. 8 tens 6 ones (106, 66)

Skill Review • Numeration

Write 7 ◯ 2 = 5 on the board. Ask students what sign would complete this number sentence. (–) Continue for more addition and subtraction sentences in both horizontal and vertical forms. Do not use zero in the problems.

Name ______________________________

Lesson 1-4

It's Algebra!

Basic Properties

Finding **differences** is easy if we remember some important rules.

Zero or Identity Properties

Subtracting zero does not affect the answer.

8 − 0 = 8
3 − 0 = 3

Subtracting a number from itself leaves zero.

7 − 7 = 0
0 − 0 = 0

Opposite or Inverse Operations

Addition and subtraction are related to each other. We can check subtraction by adding. We can check addition by subtracting.

8 − 5 = 3 **8 − 3 =** 5
3 + 5 = 8 **5 + 3 =** 8

These four number sentences are called a **fact family**.

Getting Started

Complete each number sentence.

1. 7 − 7 = 0
2. 5 − 0 = 5
3. 4 − 0 = 4
4. 6 − 6 = 0
5. 8 − 8 = 0
6. 9 − 0 = 9
7. 5 − 5 = 0
8. 3 − 0 = 3

Subtract. Check your answer.

9. 14 − 6 = 8; 8 + 6 = 14
10. 9 − 2 = 7; 7 + 2 = 9
11. 6 − 0 = 6; 6 + 0 = 6
12. 17 − 8 = 9; 9 + 8 = 17
13. 12 − 7 = 5; 5 + 7 = 12
14. 10 − 6 = 4; 4 + 6 = 10

---

2 Teach

Introduce the Lesson Have a student read aloud the thoughts of the children pictured and the statements beside the pictures. Have students complete the number sentences.

Develop Skills and Concepts Have a student stand on 9 on the number line and then move back zero spaces. Write **9 − 0 =** _____ on the board and ask students to tell the difference. (9) Have the student move back 9 spaces and tell the difference. (0) Write **9 − 9 = 0** on the board. Have students practice with more subtraction sentences.

- Write the numbers **5, 3,** and **2** on the board. Have students write addition or subtraction sentences using these three numbers. (5 − 3 = 2, 2 + 3 = 5, 5 − 2 = 3, 3 + 2 = 5) Have some students write each of the sentences in vertical form. Tell students that because these four sentences are the only ones that can be made with 2, 3, and 5, they are called a **fact family**. Write **5 − 3 = 2** on the board and ask students which of the other sentences could be used to check this difference. (2 + 3 = 5 or 3 + 2 = 5) Experiment with subtraction sentences from other fact families.

3 Practice

Tell students to complete all the exercises on page 8. Remind students to check subtraction by adding in reverse order.

Practice

Complete each number sentence.

1. $7 - 0 =$ 7	2. $0 - 0 =$ 0	3. $10 - 3 =$ 7	4. $6 - 6 =$ 0
5. $8 - 3 =$ 5	6. $2 - 1 =$ 1	7. $9 - 0 =$ 9	8. $3 - 0 =$ 3
9. $4 - 4 =$ 0	10. $11 - 3 =$ 8	11. $7 - 4 =$ 3	12. $1 - 1 =$ 0

Subtract. Check your answer.

13. $11 - 6 = 5$; $5 + 6 = 11$	14. $6 - 1 = 5$; $5 + 1 = 6$	15. $14 - 7 = 7$; $7 + 7 = 14$	16. $9 - 5 = 4$; $4 + 5 = 9$	17. $4 - 1 = 3$; $3 + 1 = 4$	18. $12 - 4 = 8$; $8 + 4 = 12$
19. $11 - 7 = 4$; $4 + 7 = 11$	20. $10 - 7 = 3$; $3 + 7 = 10$	21. $8 - 5 = 3$; $3 + 5 = 8$	22. $16 - 8 = 8$; $8 + 8 = 16$	23. $13 - 8 = 5$; $5 + 8 = 13$	24. $10 - 4 = 6$; $6 + 4 = 10$
25. $15 - 9 = 6$; $6 + 9 = 15$	26. $5 - 0 = 5$; $5 + 0 = 5$	27. $13 - 4 = 9$; $9 + 4 = 13$	28. $11 - 8 = 3$; $3 + 8 = 11$	29. $6 - 3 = 3$; $3 + 3 = 6$	30. $14 - 9 = 5$; $5 + 9 = 14$
31. $9 - 2 = 7$; $7 + 2 = 9$	32. $13 - 6 = 7$; $7 + 6 = 13$	33. $12 - 7 = 5$; $5 + 7 = 12$	34. $16 - 9 = 7$; $7 + 9 = 16$	35. $4 - 0 = 4$; $4 + 0 = 4$	36. $8 - 2 = 6$; $6 + 2 = 8$
37. $15 - 8 = 7$; $7 + 8 = 15$	38. $1 - 0 = 1$; $1 + 0 = 1$	39. $11 - 4 = 7$; $7 + 4 = 11$	40. $18 - 9 = 9$; $9 + 9 = 18$	41. $12 - 8 = 4$; $4 + 8 = 12$	42. $9 - 6 = 3$; $3 + 6 = 9$
43. $3 - 3 = 0$; $0 + 3 = 3$	44. $10 - 6 = 4$; $4 + 6 = 10$	45. $8 - 8 = 0$; $0 + 8 = 8$	46. $10 - 9 = 1$; $1 + 9 = 10$	47. $13 - 7 = 6$; $6 + 7 = 13$	48. $17 - 9 = 8$; $8 + 9 = 17$
49. $9 - 4 = 5$; $5 + 4 = 9$	50. $9 - 9 = 0$; $0 + 9 = 9$	51. $10 - 2 = 8$; $8 + 2 = 10$	52. $8 - 0 = 8$; $8 + 0 = 8$	53. $14 - 5 = 9$; $9 + 5 = 14$	54. $12 - 9 = 3$; $3 + 9 = 12$
55. $7 - 5 = 2$; $2 + 5 = 7$	56. $17 - 8 = 9$; $9 + 8 = 17$	57. $9 - 8 = 1$; $1 + 8 = 9$	58. $7 - 7 = 0$; $0 + 7 = 7$	59. $6 - 0 = 6$; $6 + 0 = 6$	60. $13 - 9 = 4$; $4 + 9 = 13$
61. $14 - 8 = 6$; $6 + 8 = 14$	62. $5 - 5 = 0$; $0 + 5 = 5$	63. $2 - 0 = 2$; $2 + 0 = 2$	64. $15 - 6 = 9$; $9 + 6 = 15$	65. $2 - 2 = 0$; $0 + 2 = 2$	66. $12 - 5 = 7$; $7 + 5 = 12$

4 Assess

Ask students to write the fact family with $8 + 8 = 16$. Then, have them explain why this family has only two facts. ($16 - 8 = 8$; one number is added to itself; the other number is the sum.)

For Mixed Abilities

Common Errors • Intervention

Some students may be confused by the two uses of zero in subtraction, for example, as a number being subtracted and as an answer in a subtraction problem. To correct students, group them into pairs and give the pairs sets of exercises similar to those shown below to help them see how zero can be a number to be subtracted and a number that is an answer.

$$\begin{array}{r}5\\-0\\\hline 5\end{array}\quad\begin{array}{r}5\\-1\\\hline 4\end{array}\quad\begin{array}{r}5\\-2\\\hline 3\end{array}\quad\begin{array}{r}5\\-4\\\hline 1\end{array}\quad\begin{array}{r}5\\-5\\\hline 0\end{array}$$

Enrichment • Numeration

Tell students to write all the addition facts that have an addend or sum of 9. Then, write a subtraction sentence to check each fact.

More to Explore • Applications

Have students make math skill clusters. Give each student or pair of students a diagram that shows a center circle as a hub and six more circles on the ends of spokes pointing out from the center.

In the center circle of each cluster, write a math skill such as making change, bookkeeping, telling time, measurement, and so on. Ask students to fill the other six circles around it with pictures of occupations that require that skill. They may cut pictures from magazines or draw them. Have students label each picture. The finished projects may be displayed on a bulletin board and used to discuss what importance math skills might have in their future occupations.

1-5 Finding Missing Addends

pages 9–10

1 Getting Started

Objective
- To find missing addends

Materials
*floor number line 0 through 20

Warm Up • Mental Math
Have students count by 2s, 3s, 5s, 10s, or 25s to tell the total.

1. 7 quarters ($1.75)
2. 15 rows of 10 seats (150)
3. socks in 13 pairs (26)
4. 9 boxes of 25 pencils (225)
5. 8 groups of 3 girls (24)
6. plugs in 16 outlets (32)
7. feet in 9 yards (27)

Warm Up • Numeration
Write _____ + _____ = **10** on the board. Have students complete the sentence with any addition fact for 10. Ask students to write other facts for 10 that can make the sentence true. Write _____ − _____ = **7** and continue the activity. Repeat for other addition and subtraction facts.

Name ______________________

Lesson 1-5

It's Algebra!

Finding Missing Addends

Yan-Wah belongs to the Photography Club and she is saving money to buy a camera. So far, she has saved $9. How much more does Yan-Wah need?

We want to find the amount that Yan-Wah still has to save.

We know that the camera costs $17.

Yan-Wah has saved $9 so far.

We can write this problem as an addition sentence, and let the letter n stand for the missing addend.

$$\$9 + n = \$17$$

To solve this problem, we think of a related subtraction sentence from the same fact family.

$$\$17 - \$9 = n$$

Because **$17 − $9 =** $8, n = $8.

Yan-Wah still needs $8 to buy the camera.

Getting Started

Write the related subtraction sentence.

1. $5 + n = 12$ — $12 - 5 = n$
2. $n + 7 = 16$ — $16 - 7 = n$
3. $6 + n = 14$ — $14 - 6 = n$
4. $n + 7 = 13$ — $13 - 7 = n$

Use a subtraction sentence to help find each missing addend.

5. $6 + n = 10$ — $10 - 6 = 4$ — n = 4
6. $n + 3 = 7$ — $7 - 3 = 4$ — n = 4
7. $n + 4 = 12$ — $12 - 4 = 8$ — n = 8
8. $7 + n = 15$ — $15 - 7 = 8$ — n = 8
9. $8 + n = 11$; $11 - 8 = 3$ — n = 3
10. $n + 9 = 16$; $16 - 9 = 7$ — n = 7
11. $8 + n = 15$; $15 - 8 = 7$ — n = 7
12. $n + 9 = 18$; $18 - 9 = 9$ — n = 9

---

2 Teach

Introduce the Lesson Have a student read the problem aloud. Have students study the picture to tell what they are to find and what they already know. (how much more than $9 is needed to buy a $17 camera)

- Read aloud the information sentences with students as they fill them in. Then, work through the problem with students and have them complete the remaining sentences.

Develop Skills and Concepts Have one student stand on 5 on the number line and another student on 14. Tell students that we want to know how many spaces the first student will have to move to reach the second student. Write **5 + *n* = 14** on the board and tell students that the n stands for the missing number. Have the first student walk forward to 14 as others count the steps. (9) Ask students what number n stands for. (9) Write ***n* = 9** under $5 + n = 14$.

- Now, have the student at 14 walk back 5 spaces. Ask students the number for n. (9) Tell students that the student on 14 found the missing addend by subtracting. Write ***n* = 9** under $14 - 5 = n$. Write **14 − 5 = *n*** on the board and then ***n* = 9** below it.
- Write 5 Xs on the board and draw a circle around them. Tell students that this circle should have 12 Xs in it and that they must find out how many more Xs need to be added. Write **5 + *n* = 12** and ***n* + 5 = 12** on the board. Ask students to give a subtraction problem that will help us find n. ($12 - 5 = n$) Continue for more missing addend problems.

3 Practice

Remind students that a missing addend can be found by identifying a related subtraction sentence. Have students complete page 10 independently.

Practice

Complete the related subtraction sentence.

1. $7 + n = 10$ — $10 - 7 = n$
2. $4 + n = 12$ — $12 - 4 = n$
3. $n + 9 = 15$ — $15 - 9 = n$
4. $n + 6 = 14$ — $14 - 6 = n$
5. $9 + n = 18$ — $18 - 9 = n$
6. $n + 6 = 13$ — $13 - 6 = n$
7. $3 + n = 3$ — $3 - 3 = n$
8. $n + 8 = 10$ — $10 - 8 = n$
9. $6 + n = 14$ — $14 - 6 = n$
10. $8 + n = 14$ — $14 - 8 = n$
11. $9 + n = 16$ — $16 - 9 = n$
12. $8 + n = 12$ — $12 - 8 = n$

Use a subtraction sentence to help find each missing addend.

13. $3 + n = 5$ — $5 - 3 = 2$ — $n = 2$
14. $7 + n = 8$ — $8 - 7 = 1$ — $n = 1$
15. $n + 6 = 12$ — $12 - 6 = 6$ — $n = 6$
16. $4 + n = 10$ — $10 - 4 = 6$ — $n = 6$
17. $7 + n = 14$ — $14 - 7 = 7$ — $n = 7$
18. $n + 9 = 16$ — $16 - 9 = 7$ — $n = 7$
19. $n + 8 = 16$ — $16 - 8 = 8$ — $n = 8$
20. $9 + n = 14$ — $14 - 9 = 5$ — $n = 5$
21. $n + 6 = 13$ — $13 - 6 = 7$ — $n = 7$
22. $n + 8 = 12$ — $12 - 8 = 4$ — $n = 4$
23. $7 + n = 16$ — $16 - 7 = 9$ — $n = 9$
24. $8 + n = 13$ — $13 - 8 = 5$ — $n = 5$
25. $6 + n = 11$ — $11 - 6 = 5$ — $n = 5$
26. $n + 9 = 14$ — $14 - 9 = 5$ — $n = 5$
27. $4 + n = 13$ — $13 - 4 = 9$ — $n = 9$
28. $8 + n = 15$ — $15 - 8 = 7$ — $n = 7$
29. $n + 2 = 5$ — $5 - 2 = 3$ — $n = 3$
30. $8 + n = 9$ — $9 - 8 = 1$ — $n = 1$
31. $3 + n = 7$ — $7 - 3 = 4$ — $n = 4$
32. $n + 5 = 8$ — $8 - 5 = 3$ — $n = 3$
33. $7 + n = 12$ — $12 - 7 = 5$ — $n = 5$
34. $n + 3 = 3$ — $3 - 3 = 0$ — $n = 0$
35. $9 + n = 15$ — $15 - 9 = 6$ — $n = 6$
36. $8 + n = 16$ — $16 - 8 = 8$ — $n = 8$

4 Assess

Ask students to tell how they can solve an addition problem with a missing addend. (think of a related subtraction sentence from the fact family)

For Mixed Abilities

Common Errors • Intervention

If students have difficulty writing a related subtraction sentence to find a missing addend, have them practice writing fact families. Have students work with partners. Give each pair of students an addition fact and have them write the other facts in the fact family. For example,

$7 + 8 = 15$ $15 - 8 = 7$

$8 + 7 = 15$ $15 - 7 = 8$

Enrichment • Algebra

Tell students to write an addition problem and a related subtraction sentence where *n* plus 126 equals 3 more than 200 and then to solve for *n*. ($n + 126 = 3 + 200$; $203 - 126 = n$; $n = 77$)

More to Explore • Measurement

Introduce students to nonstandard units of body measure, especially the **cubit**, the distance from elbow to end of middle finger with arm bent, and the **foot**, the length of a human foot. Have them measure assigned items in the classroom using these nonstandard units and record their results. Have students suggest reasons in favor of using standard units of measure rather than nonstandard units.

1-6 Problem Solving: Use a Four-Step Plan

pages 11–12

1 Getting Started

Objective

- To use a plan to solve problems

Warm Up • Mental Math

Ask students how to get to 186 if they start at the following:

1. 100 (add 86)
2. 198 (subtract 12)
3. 2×90 (add 6)
4. $\frac{1}{2}$ of 300 (add 36)
5. 1,000 (subtract 814)
6. 93 (multiply by 2)
7. $500 \div 2$ (subtract 64)
8. 372 (divide by 2)

2 Teach

Introduce the Lesson Have students think about what they would do if they cut or scraped their skin. Tell students that they would probably wash the cut and apply some ointment and a bandage. Remind students that they would solve a problem of a cut or scrape in this order, as you write the following on the board: **SEE, PLAN, DO,** and **CHECK.** Tell students that this same plan can be used to solve many problems in daily life.

Name ______________________________

Problem Solving: Use a Four-Step Plan

Of the 1,576 students enrolled in Roosevelt School, 757 are girls. Of the 76 students absent today, 27 are girls. How many boys attend Roosevelt? How many boys are present?

★ SEE

We need to find

The number of boys who attend Roosevelt School
The number of boys who are absent
The number of boys who are present

We know

The total number of students is 1,576.
The number of girls is 757.
There are 76 students who are absent.
There are 27 girls who are absent.

★ PLAN

To find out how many boys attend Roosevelt School, we subtract 757 from 1,576. To find out how many boys are absent, we subtract 27 from 76. To find the number of boys who are present, we find the difference between the total number of boys and the number of boys who are absent.

DO

$$1{,}576 - 757 = 819 \text{ boys} \qquad 76 - 27 = 49 \text{ boys absent} \qquad 819 - 49 = 770 \text{ boys present}$$

819 boys attend Roosevelt School. 770 boys are present today.

★ CHECK

$$757 + 819 = 1{,}576 \text{ students enrolled} \qquad 49 + 27 = 76 \text{ students absent} \qquad 770 + 49 = 819 \text{ boys}$$

Develop Skills and Concepts Have a student read the problem aloud. Remind students that the SEE stage is done first as they read and complete the sentences.

- Tell students that we know what the problem is and we know some facts about it, so we can plan what to do. Remind students that just as there were several steps to care for their wound, there are often several steps in working any problem.
- Have students read through the PLAN stage with you and complete the sentences. Tell students to work the problem, write the answers, and then check to see if the solution makes sense. Remind students that just as a sling would be a silly solution to caring for a small cut, sometimes a solution to a problem does not make sense and we need to go back to the SEE and PLAN stages and start again. Tell students to complete the exercises and to check their solution.

3 Apply

Solution Notes

1. Have students find the total of Marcia's savings by adding. This amount is then subtracted from the cost of the camera.
2. First, students need to find the total number of students who either go home to eat or bring their lunches. They then subtract this from the fourth-grade enrollment.
3. First, have students find the number of pages Frank has left to read. To find the average number of pages Frank will need to read each day, we divide the remaining number of pages by the number of days.
4. Several subgoals exist in this problem. Remind students to include the teacher in the number of people getting a treat. Students must also find the number of bars Karen has by multiplying the number of boxes by the number of bars in each. The

Apply

Solve each problem. Remember to use the four-step plan.

1. Marcia wants to buy a camera for $49.99. She has saved $31.27 and will earn $5.00 babysitting on Saturday. How much more will she need to buy the camera?
   $13.72
2. There are 157 fourth graders at Watson School. At lunch, 43 students go home to eat, 27 bring their lunches, and the rest buy their lunches. How many students buy their lunches?
   87 students
3. The book Frank is reading has 135 pages. He has finished reading 81 pages. If Frank is to finish the book in 6 days, how many pages must he read each day?
   9 pages
4. Karen wants to bring a birthday treat for her class. There are 24 students and 1 teacher in the class. She bought 4 boxes of granola bars with 5 granola bars in each. How many more granola bars does she need?
   5 granola bars
5. Read Exercise 3 again. What if Frank is to finish the book in 9 days instead of 6? Now how many pages must he read each day?
   6 pages

Problem-Solving Strategy: Using the Four-Step Plan

- ★ **SEE** What do you need to find?
- ★ **PLAN** What do you need to do?
- ★ **DO** Follow the plan.
- ★ **CHECK** Does your answer make sense?

6. Donuts cost $3.50 a dozen at Aunt Molly's Bakery and 30¢ each at Uncle Don's Diner. At whose place would it cost less to buy 3 dozen donuts?
   Aunt Molly's
7. There are 157 fourth graders at Watson School. Forty-three of them walk to school. Some ride bicycles and others ride school buses. What do you need to know to tell how many fourth-grade students ride school buses?
   the number who ride bicycles
8. You have $10.00 to spend. It costs 50¢ to take the bus downtown. You buy a pencil-and-pen set for $2.98, a notebook for $3.98, and paper for $1.99. Can you buy a ruler for 49¢ and still have enough money left to take the bus home?
   no

For Mixed Abilities

More to Explore • Numeration

Tell students that they are looking for a magic sum. Have students write each of the following numbers on separate cards: 2, 4, 6, 8, 10, 12, 14, 16, 18, 22, and 24. Tell students to arrange the cards in addition pairs so that each pair will have the same magic sum. Have them record their pairs and the magic sum. (All pairs total 26. The pairs should be 2-24, 4-22, 6-20, 8-18, 10-16, and 12-14.)

difference between these two numbers is the number of granola bars Karen still needs.

Higher-Order Thinking Skills

5. **Analysis** Students should recognize that the same basic fact is used in both problems: the original problem (54 pages ÷ 6 days = 9 pages a day) and the "what if" problem (54 pages ÷ 9 days = 6 pages a day).
6. **Evaluation** Students might find the unit cost or the cost of 1 dozen and compare these. Others might find the cost of 3 dozen donuts in each store.
7. **Evaluation** Encourage students to read the problem carefully. Many students will just subtract 43 from 157.
8. **Evaluation** Encourage students to use mental computation to round the sum of $2.98, $3.98, and $1.99 to a whole dollar amount, or about $9. Ten dollars minus $9 leaves about $1, 50¢ of which has already been spent on the bus downtown.

4 Assess

Ask students to name the four steps used to solve problems. (see, plan, do, and check)

Chapter 1 Test

page 13

Items	Objectives
1–11, 23–30	To compute basic addition and subtraction facts (see pages 1–2)
12–16	To add three, four, or five 1-digit addends (see pages 3–4)
17–22	To understand the Grouping and Zero Properties of Addition (see pages 5–6)
31–34	To find missing addends (see pages 9–10)

Name ______________________

CHAPTER 1 TEST

Add.

1. $3 + 2 =$ 5
2. $5 + 5 =$ 10
3. $6 + 4 =$ 10
4. $7 + 5 =$ 12
5. $4 + 4 =$ 8
6. $7 + 3 =$ 10
7. $8 + 7 =$ 15
8. $5 + 9 =$ 14
9. $5 + 6 =$ 11
10. $9 + 9 =$ 18
11. $8 + 3 =$ 11
12. $4 + 6 + 2 =$ 12
13. $3 + 1 + 4 =$ 8
14. $5 + 2 + 6 + 1 =$ 14
15. $8 + 1 + 5 + 3 =$ 17
16. $4 + 5 + 3 + 3 + 6 + 1 =$ 22
17. $(5 + 2) + 3 =$ 10
18. $2 + (8 + 0) =$ 10
19. $(7 + 5) + 2 =$ 14
20. $3 + (6 + 3) =$ 12
21. $(4 + 0) + 9 =$ 13
22. $7 + (5 + 2) =$ 14

Subtract.

23. $9 - 6 =$ 3
24. $14 - 7 =$ 7
25. $11 - 8 =$ 3
26. $15 - 6 =$ 9
27. $8 - 3 =$ 5
28. $16 - 8 =$ 8
29. $12 - 9 =$ 3
30. $7 - 7 =$ 0

Write each missing addend.

31. $5 + n = 8$ $n =$ 3
32. $n + 7 = 13$ $n =$ 6
33. $6 + n = 13$ $n =$ 7
34. $11 + n = 19$ $n =$ 8

Alternate Chapter Test

You may wish to use the Alternate Chapter Test on page 346 of this book for further review and assessment.

Cumulative Assessment

CUMULATIVE ASSESSMENT

Circle the letter of the correct answer.

1. $8 + 8$	(a.) 16 b. 8 c. 15 d. NG	8. $8 + (3 + 5)$	a. 8 (b.) 16 c. 17 d. NG
2. $6 + 3$	a. 3 (b.) 9 c. 12 d. NG	9. $7 - 3$	a. 6 b. 8 c. 10 (d.) NG
3. $3 + 4$	a. 5 b. 6 (c.) 7 d. NG	10. $15 - 7$	a. 6 b. 7 (c.) 8 d. NG
4. $5 + 6$	(a.) 11 b. 12 c. 13 d. NG	11. $14 - 6$	(a.) 8 b. 7 c. 6 d. NG
5. $9 + 6$	a. 12 b. 14 c. 16 (d.) NG	12. $8 - 8$	(a.) 0 b. 4 c. 8 d. NG
6. $4 + 8$	a. 11 (b.) 12 c. 13 d. NG	13. $7 + n = 12$ $n = ■$	a. 7 b. 19 (c.) 5 d. NG
7. $4 + 3 + 5$	a. 11 b. 13 (c.) 12 d. NG	14. $n + 2 = 8$ $n = ■$	(a.) 6 b. 8 c. 10 d. NG

score

STOP

page 14

Items	Objectives
1–6, 9–12	To compute basic addition and subtraction facts (see pages 1–2)
7	To add three 1-digit numbers (see pages 3–4)
8	To understand the Grouping Property of Addition (see pages 5–6)
13–14	To find missing addends (see pages 9–10)

Alternate Cumulative Assessment

Circle the letter of the correct answer.

1. $9 + 9 =$ a 8 b 2 (c) 18 d NG
2. $5 + 3 =$ a 7 (b) 8 c 9 d NG
3. $7 + 4 =$ a 13 (b) 11 c 3 d NG
4. $9 + 3 =$ a 11 b 13 c 15 (d) NG
5. $8 + 6$ a 15 (b) 14 c 13 d NG
6. $5 + 7$ (a) 12 b 13 c 14 d NG
7. $3 + 6 + 2$ a 8 b 9 (c) 11 d NG
8. $(9 + 1) + 4 =$ a 10 b 13 (c) 14 d NG
9. $6 - 4 =$ (a) 2 b 3 c 4 d NG
10. $18 - 9 =$ a 10 (b) 9 c 8 d NG
11. $13 - 6$ a 8 b 9 c 10 (d) NG
12. $7 - 7$ (a) 0 b 1 c 2 d NG
13. $6 + n = 14$, $n = ?$ a 6 (b) 8 c 20 d NG
14. $n + 6 = 12$, $n = ?$ (a) 6 b 12 c 18 d NG

2-1 Whole Numbers

pages 15–16

1 Getting Started

Objective
- To compare whole numbers

Vocabulary
digits

Materials
*number line 0 through 100

Warm Up • Mental Math
Have students solve each of the following exercises:

1. 8 × 7 (56)
2. 6 ÷ 2 (3)
3. 1 dozen + 5 (17)
4. 16 − 11 (5)
5. 1 foot + 11 inches (23 in.)
6. 6 + 7 (13)
7. 6 + 7 + 8 (21)
8. 7¢ less than 8 dimes (73¢)

Warm Up • Place Value
Have a student write a number between 10 and 99 on the board and tell how many tens and ones are in the number. Have another student write the number that is made by reversing those digits and tell how many tens and ones are in the second number. Group students in pairs to continue the activity.

Name ______________________

CHAPTER 2

Place Value

Lesson 2-1

Whole Numbers

Polly is buying dog food for her puppy. Which can will cost Polly the least amount of money?

We want to know which can has the lowest price.

We know the prices for the different dog foods are 49¢, 43¢, and 45¢.

To compare the three prices, we can find them on a number line.

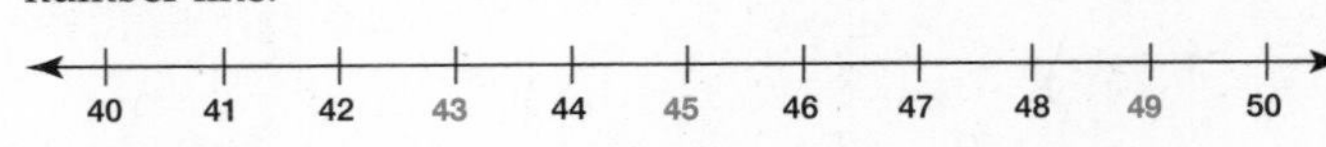

REMEMBER The whole numbers 0, 1, 2, 3, 4, 5, 6, 7, 8, and 9 are called **digits**. On a number line, the number to the right is always greater.

49 is greater than 45. **49 > 45**

REMEMBER A number to the left is always less.

43 is less than 45. **43 < 45**

The dog food that will cost Polly the least costs 43¢.

Getting Started

Write each missing whole number.

1. 26, 27, 28
2. 90 comes after 89.
3. 58 is between 57 and 59.

Compare. Write < or > in the circle.

4. 32 (<) 38
5. 76 (>) 53
6. 27 (<) 72

Write the numbers in order from least to greatest.

7. 32, 46, 15 — 15, 32, 46
8. 13, 43, 29 — 13, 29, 43

---

2 Teach

Introduce the Lesson Have a student read the problem aloud. Tell students to use the picture to tell what they are to find and what facts they already know. (which of the three cans costs the least if the prices are 49¢, 43¢, and 45¢)

- Read with students as they complete the sentences to solve the problem. Have students check the number line to see if 43 comes before 45 and 49. (yes)

Develop Skills and Concepts Display the number line and write **84** and **46** on the board. Ask students to tell which of these whole numbers comes first on the number line (46) and why (4 tens is less than 8 tens). Repeat for more groups of two numbers.

- Remind students that numbers on the number line get larger to the right and smaller to the left. Write **78** and **72** on the board and ask students to compare the tens. (same) Tell students that they must compare the ones when the numbers in the tens column are the same. Ask students to tell which whole number is greater and why. (78 is greater because 8 ones is more than 2 ones.)

- Write **26, 62,** and **21** on the board and ask students to compare the numbers by placing them in order from least to greatest. (21, 26, 62) Review the **greater than** (>) and **less than** (<) signs and remind students that the point of the sign will always be aimed at the smaller number, with the wide end next to the larger number.

Practice

Write each missing whole number.

1. 35, 36, 37, 38
2. 46, 47, 48, 49
3. 63, 62, 61, 60
4. 91, 92, 93, 94
5. 7, 8, 9, 10, 11
6. 17, 18, 19, 20, 21
7. 82, 81, 80, 79, 78
8. 50 comes after 49.
9. 57 is between 56 and 58.
10. 77 is between 76 and 78.

Compare. Write < or > in the circle.

11. 39 (>) 36	12. 73 (>) 17	13. 21 (<) 30
14. 17 (>) 16	15. 81 (<) 89	16. 32 (<) 42
17. 63 (>) 36	18. 22 (<) 33	19. 89 (>) 40
20. 25 (<) 29	21. 57 (>) 51	22. 48 (>) 40
23. 48 (<) 50	24. 96 (<) 99	25. 15 (<) 35

Write the numbers in order from least to greatest.

26. 75, 36, 48 — 36, 48, 75	27. 35, 87, 29 — 29, 35, 87	28. 23, 57, 45 — 23, 45, 57
29. 83, 47, 58 — 47, 58, 83	30. 22, 57, 39 — 22, 39, 57	31. 18, 81, 88 — 18, 81, 88
32. 25, 36, 12, 19 — 12, 19, 25, 36	33. 47, 58, 75, 21 — 21, 47, 58, 75	34. 67, 9, 42, 83 — 9, 42, 67, 83
35. 88, 60, 39, 12 — 12, 39, 60, 88	36. 56, 21, 99, 97 — 21, 56, 97, 99	37. 10, 42, 8, 67 — 8, 10, 42, 67

3 Practice

Remind students to compare 2-digit numbers by comparing the tens and then the ones. Encourage students to use the number line to check their work. Have students complete page 16 independently.

4 Assess

Ask students to write 76, 32, 57, 20, and 15 in order from greatest to least. (76, 57, 32, 20, 15)

For Mixed Abilities

Common Errors • Intervention

Some students may have trouble writing missing numbers in order. Point to 46 on a number line and ask students to name a number that is

1 more (47)

1 less (45)

between 47 and 49 (48)

Enrichment • Application

Tell students that they are to write a beverage menu of five different-priced drinks so that no drink costs less than 29¢ and the purchase of any two drinks is not more than 85¢.

More to Explore • Numeration

Challenge students to think of one thing they do that does not involve numbers. On the board, list all the activities they mention. Encourage students to be critical of each other's suggestions and try to think of a way in which the activity is impossible without numbers. For example, if they suggest that painting or drawing is something they can do without numbers, point out that drawing paper comes in numbered packs and is sold for a given price. See if they can think of a single thing that can be done without numbers.

2-2 Ones, Tens, and Hundreds

pages 17–18

1 Getting Started

Objective
- To use place value to read and write 2- and 3-digit numbers

Vocabulary
ones, tens, hundreds

Materials
*place-value blocks: hundreds, tens, ones

Warm Up • Mental Math
Ask students how to find the following:

1. area (length × width)
2. feet in 12 yards (12 × 3)
3. pints in 20 cups (20 ÷ 2)
4. perimeter (sum of sides)
5. volume (length × width × height)
6. quotient (dividend ÷ divisor)
7. sum (addend + addend)

Warm Up • Numeration
Write **fifty-seven** on the board. Ask a student to write the number. (57) Repeat for forty-six. (46) Now, have a student write a number through 99 on the board and choose a friend to write the number word. Continue the activity until every student has participated.

Name ______________________

Lesson 2-2

Ones, Tens, and Hundreds

Linda bought flower stickers to add to her collection. How many did she buy?

Linda bought 2 pages of 100 stickers each.

She bought 4 strips of 10 stickers each.

She also bought 3 individual stickers.

We say she bought **two hundred forty-three** stickers and write it as 243.

Study the number in this place-value chart.

In the number **243**,

hundreds	tens	ones
2	4	3

the digit 2 is in the **hundreds** place.

the digit 4 is in the **tens** place.

the digit 3 is in the **ones** place.

Linda bought 243 stickers.

Getting Started

Write each number.

1. two hundred seventy-seven 277
2. nine hundred twenty-six 926

Write the place value of each green digit.

3. 276 tens
4. 385 hundreds
5. 620 ones

Write each missing word.

6. 132 one hundred thirty-two
7. 753 seven hundred fifty-three

2 Teach

Introduce the Lesson Have a student read the problem aloud and tell what is to be done. (tell how many stickers Linda bought) Ask students what information is given in the problem and the picture. (Linda bought two 100-sticker sheets plus four 10-sticker sheets plus 3 single stickers.) Discuss with students how 243 fits into the place-value chart in the model. Then, have students complete the solution sentence.

Develop Skills and Concepts Write **35** on the board. Have students lay out 3 tens rods and 5 ones units. Write the following on the board:

tens	ones
3	5

3 tens + 5 ones = 35

Continue the activity for other 2-digit numbers.

- Now, write **167** on the board and repeat the activity using hundreds, tens, and ones blocks. Remind students that 1 hundred equals 10 tens or 100 ones. Ask students how many ones equal 200. (200) Repeat for other equivalents.

3 Practice

Remind students to place a zero in a 3-digit number when there are no tens or ones. Have students complete page 18 independently.

Practice

Write each number.

1. 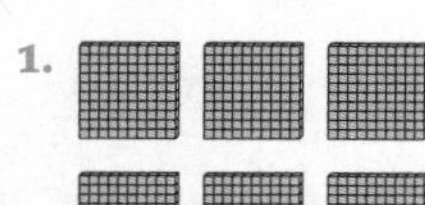 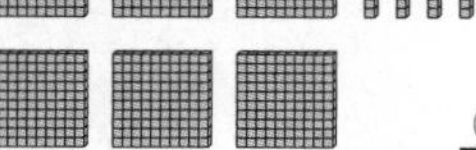  640

2. 806

3. three hundred fifty-two 352
4. six hundred ninety 690
5. two hundred seventy-eight 278
6. five hundred nineteen 519
7. four hundred ninety-nine 499
8. one hundred seven 107
9. eight hundred sixty-five 865
10. seven hundred eleven 711

Write the place value of each green digit.

11. 186 ones
12. 26 ones
13. 159 hundreds
14. 637 tens
15. 575 hundreds
16. 150 tens
17. 320 ones
18. 215 tens
19. 259 hundreds
20. 707 hundreds
21. 60 tens
22. 99 ones
23. 561 tens
24. 800 ones
25. 825 hundreds

Write each missing word.

26. 333 three hundred thirty-three
27. 923 nine hundred twenty-three
28. 600 six hundred
29. 201 two hundred one
30. 575 five hundred seventy-five
31. 815 eight hundred fifteen

4 Assess

Write the number **400** on the board and circle the middle zero. Ask students where on the place-value chart they would write the circled zero. (in the tens column)

5 Mixed Review

1. 3 + 6 + 4 + 1 (14)
2. 15 − 8 (7)
3. (2 + 8) + 7 (17)
4. 9 + 7 (16)
5. 14 − 6 (8)
6. 5 + 4 + 2 + 3 + 4 (18)
7. 8 + (3 + 2) (13)
8. 12 − 7 (5)
9. 7 + 6 + 4 + 9 (26)
10. 17 − 8 (9)

For Mixed Abilities

Common Errors • Intervention

Some students may name the place value of digits incorrectly. Have them place each 3-digit number, such as 294, on a place-value chart and give the place-value names for all 3 digits.

hundreds	tens	ones
2	9	4

2 hundreds
9 tens
4 ones

Enrichment • Numeration

Write three single-digit numbers on the board. Ask students to write all the 3-digit numbers that use those 3 digits. Then, have students arrange the numbers in order from least to greatest.

More to Explore • Applications

Provide students with a list of five major cities in the world. Using an atlas, have them find the distance from their hometown to each of these cities. Have them list the cities in order from the farthest away, to the closest. Then, divide students into groups and assign each group a city. Tell groups to find out as much numerical information as they can about their city and report it to the class. Students should see that they could report on everything from monetary units to annual rainfall and population.

2-3 Money

pages 19–20

1 Getting Started

Objective
- To read and write amounts less than $10

Vocabulary
pennies, dimes, dollars

Materials
*figure with six equal sides; play or real money

Warm Up • Mental Math
Dictate the following numbers for students to round to the nearest ten:

1. 87 (90)
2. 218 (220)
3. 61 (60)
4. 612 (610)
5. 439 (440)
6. 1,317 (1,320)
7. 999 (1,000)

Warm Up • Geometry
Trace a six-sided figure on the board. Label each side 5 inches. Have students count by 5s to find the perimeter. (30 inches) Change the length of each side to 10 centimeters and have students count by 10s to find the perimeter. (60 centimeters) Repeat for 25 miles to count by 25s, 100 yards to count by 100s, and so on.

Name ____________________

Lesson 2-3

Money

Artiss is the treasurer for the Drama Club. He is sorting the coins he has collected for membership dues. He puts the pennies into penny tubes. How much money, in pennies, has Artiss collected?

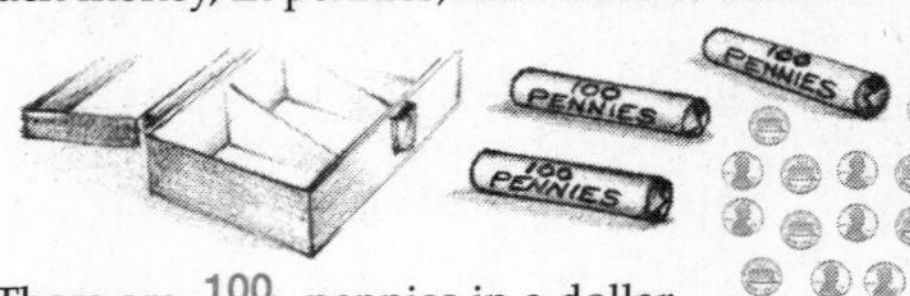

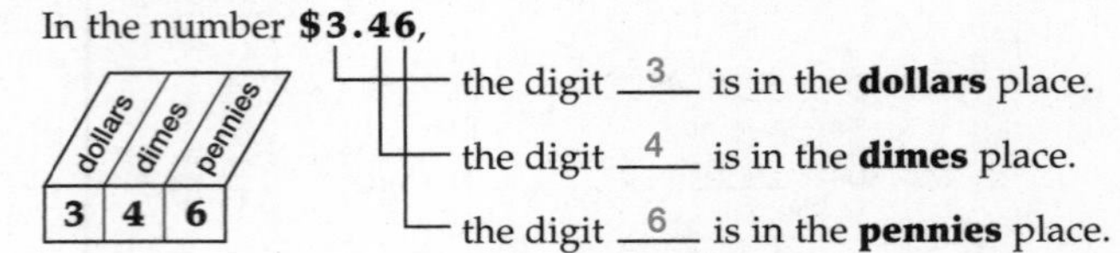

There are 100 pennies in a dollar.

Artiss has 300 pennies in the penny tubes.

He has 46 more pennies that are not in a tube.

We say Artiss has collected **three dollars and forty-six cents**, and write it as $3.46.

To understand the total amount of money that Artiss has, study this place-value chart.

In the number **$3.46**,

dollars	dimes	pennies
3	4	6

the digit 3 is in the **dollars** place.

the digit 4 is in the **dimes** place.

the digit 6 is in the **pennies** place.

REMEMBER Include the dollar sign and decimal point when writing amounts of money.

Artiss has collected $3.46 in pennies.

Getting Started

Write each amount.

1.  $3.43
2. 8 dollars and 30 pennies $8.30
3. 85 cents $0.85
4. 4 dollars $4.00

2 Teach

Introduce the Lesson Have a student read the problem and tell what question is to be answered. (the total amount of money Artiss collected in pennies) Ask students what information is given in the problem and the picture. (Artiss has 3 groups of 100 pennies and 46 more pennies.)

- Discuss with students how $3.46 fits into the place-value chart in the model. Then, have students complete the solution sentence.

Develop Skills and Concepts Write 327¢ on the board. Ask a student to read the amount. (three hundred twenty-seven cents) Ask students to name the place value of each number. (3 hundreds, 2 tens, 7 ones)

Draw the following charts on the board:

hundreds	tens	ones
3	2	7

dollars	dimes	pennies
(3)	(2)	(7)

Ask students how many pennies equal 7 ones. (7) Ask how many dimes equal 2 tens. (2) Repeat for 3 hundreds and write each in the money chart. Write **327¢ = $3.27** on the board and have students read with you the dollar amount. (three dollars and twenty-seven cents) Repeat for other 3-digit amounts.

- Write **nine cents, $0.09, seventeen cents**, and **$0.17** on the board and have students read each amount. Repeat for more amounts under $1.00.

Practice

Write each amount.

1.   $4.58

2. 3 dollars and 15 cents $3.15
3. 2 dollars and 4 dimes $2.40
4. 5 dimes and 6 cents $0.56
5. 7 dollars and 27 cents $7.27
6. 4 dollars, 2 dimes, and 8 pennies $4.28
7. 6 dollars and 2 dimes $6.20
8. 9 dollars and 99 cents $9.99
9. 1 dollar and 5 cents $1.05
10. 5 dollars and 89 cents $5.89
11. 2 dollars, 7 dimes, and 0 pennies $2.70

Now Try This!

Where is the noisy drummer going?

To find out, match the letters in the chart with the money values below. The first one is done for you.

A = forty cents	K = eighteen cents
B = twenty-seven cents	N = four cents
E = five cents	O = ninety-three cents
H = zero	T = fifty cents

T	O	T	H	E	B	A	N	K
50¢	93/100	1 half-dollar	$0.00	5¢	2 dimes 7 pennies	$0.40	$0.04	18¢

T	O	T	H	E	B	A	N	K
50 pennies	$0.93	50/100	00/100	$0.05	2 dimes 1 nickel 2 pennies	40¢	4 pennies	$0.18

T	O	T	H	E	B	A	N	K
$0.50	93¢	5 dimes	0¢	05/100	27¢	40/100	04/100	1 dime 1 nickel 3 pennies

B	A	N	K	B	A	N	K
$0.27	4 dimes	4¢	18/100	27/100	8 nickels	$0.04	18¢

Now Try This! Money is written in various ways throughout this exercise. The fraction form will probably be least familiar to students. Explain how this is used in writing amounts on checks. Have students read the solution orally in rhythm to realize its full significance.

3 Practice

Remind students that we say the word *and* for a decimal point when reading an amount of money. Have students complete page 20 independently.

4 Assess

Write **643¢** on the board and ask students how many dollars, dimes, and pennies it represents. (6 dollars, 4 dimes, and 3 pennies)

For Mixed Abilities

Common Errors • Intervention

Some students may place the dollar sign and decimal point incorrectly when writing amounts of money. Have them make and use a chart for dollars, dimes, and pennies. Stress that the decimal point separates the dollars from the dimes, that is, the dollars from the cents.

$5.49

dollars	dimes	pennies
5	4	9

Enrichment • Money

Ask students to count out $3.62 in play money using the least number of coins possible. Then, have them cut priced items from catalogs or newspapers to show two or more items they could purchase with $3.62. Repeat for three different shopping trips.

More to Explore • Geometry

There are many craft kits in the market that use the concept of geometric designs. String art and geoboards are two examples. Students can make their own geoboards by hammering nails into a square or rectangular board or by pressing golf tees into ceiling tile squares. Multicolored rubber bands can then be used to create a variety of geometric pictures. As students create a design or picture, have them record it on a 3-by-5-inch index card and file it. Encourage students to exchange cards to reproduce each other's designs.

2-4 Counting Money

pages 21–22

1 Getting Started

Objective
- To identify money, count money, and write amounts greater than $10

Materials
play or real money

Warm Up • Mental Math
Have students complete each comparison.

A nickel is to 5 as a

1. dime is to ____ (10)
2. dollar is to ____ (100)
3. quarter is to ____ (25)
4. penny is to ____ (1)
5. $5 bill is to ____ (500)
6. $10 bill is to ____ (1,000)
7. half-dollar is to ____ (50)
8. $2 bill is to ____ (200)

Warm Up • Numeration
Write **10, 20,** and **30** on the board. Ask students to read the numbers aloud and decide how they are counting. (by 10s) Write **35, 40,** and **45** and ask students how they are counting now. (by 5s) Continue the activity for counting by 20s. Repeat for more practice with counting by 1s, 5s, 10s, 20s, and 100s in any order.

Name ____________________

Counting Money

Help Onida count the money she collected on her paper route.

We count	We say
three $20 bills	twenty forty sixty
$10 bill	seventy
two $5 bills	seventy-five eighty
three $1 bills	eighty-one eighty-two eighty-three
coins	eighty-three fifty eighty-three seventy-five eighty-three eighty-five eighty-three ninety eighty-three ninety-one eighty-three ninety-two eighty-three ninety-three

REMEMBER Include the dollar sign and decimal point.

Onida has collected $83.93.

Getting Started

Write each amount.

1.  $50.95

2.  $85.71

---

2 Teach

Introduce the Lesson Have students tell about the picture. (A girl is counting money.) Have a student read the problem aloud.

- Ask students to tell what they are to find. (total amount of money Onida has collected) Ask students to identify the bills and coins pictured. Have students read with you to count the money. Tell students to complete the solution sentence, giving the amount of money Onida has collected.

Develop Skills and Concepts Have students examine a $20 bill and tell its value. (20 dollars) Continue to have students identify the following bills and coins and tell the value of each: $10 bill, $5 bill, $1 bill, half-dollar, quarter, dime, nickel, and penny.

- Have students lay out nickels as they count by 5s through 100. Repeat with dimes for counting by 10s.

Now, have students count aloud by 20s and then by 25s through 200. Ask students to lay out 2 quarters, 3 dimes, and 3 nickels and count with you, *25, 50, 60, 70, 80, 85, 90, 95¢*. Have students name other amounts, lay out the money, and then write the amounts on the board.

3 Practice

Remind students to use a dollar sign and decimal point in each money amount. Have students complete page 22.

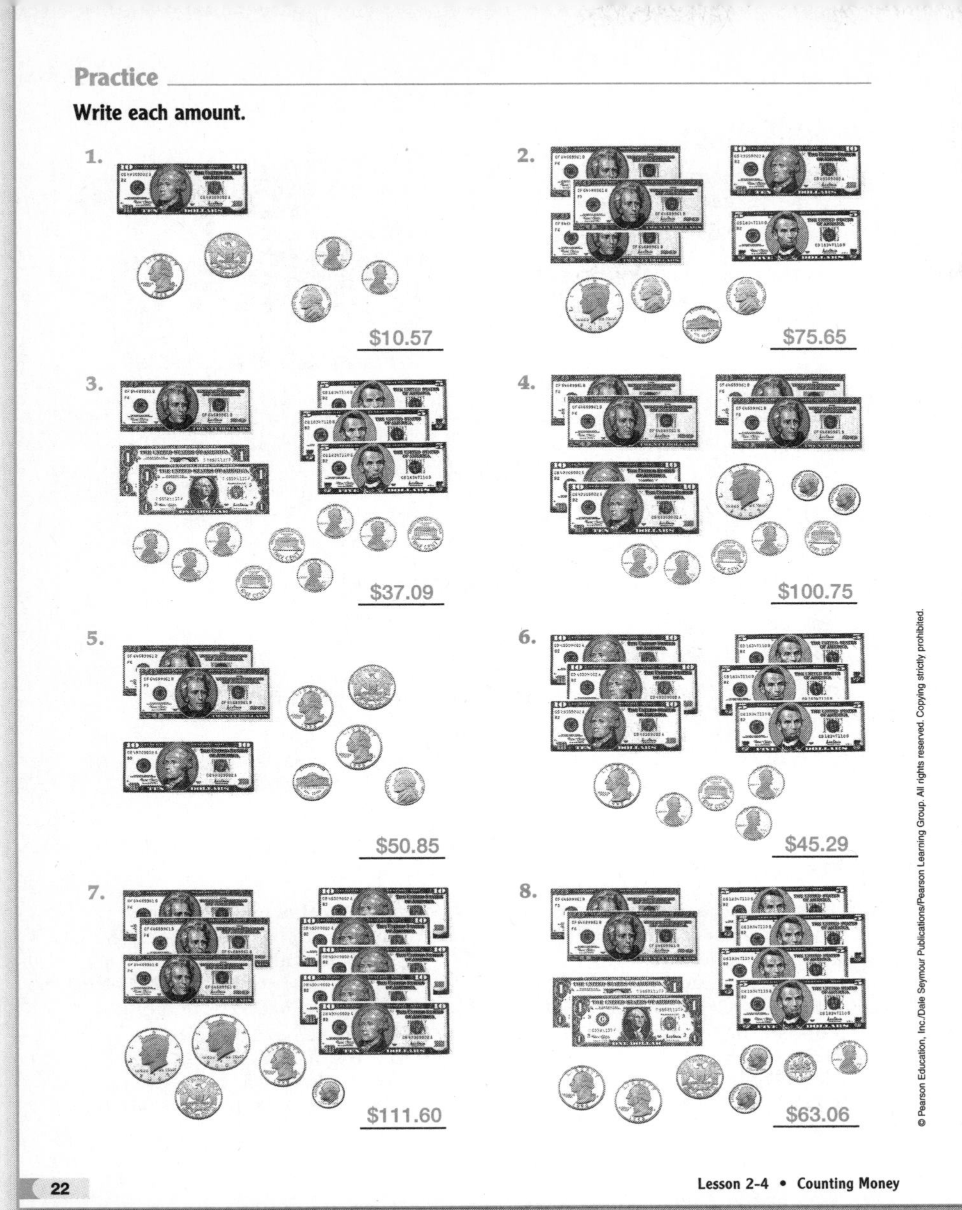

Practice

Write each amount.

1. $10.57
2. $75.65
3. $37.09
4. $100.75
5. $50.85
6. $45.29
7. $111.60
8. $63.06

4 Assess

Ask students to skip-count by 5s and to tell how many nickels are in $1.00. (20 nickels)

For Mixed Abilities

Common Errors • Intervention

Some students will have difficulty counting to find the total amount for a group of bills and coins. Have students work in cooperative groups, giving each group two $20 bills, one $5 bill, three $1 bills, two quarters, and one penny.

Have the group write the addition problem for the bills and coins vertically and find the sum as shown on the right. Then, have them count together down the column to find the total: "$20, $40, $45, $46, $47, $48, $48.25, $48.50, $48.51."

$20.00
20.00
5.00
1.00
1.00
1.00
0.25
0.25
0.01
$48.51

Enrichment • Money

Have students use the fewest bills and coins possible, and draw the change they would receive from a $100 bill after purchasing two items costing $26.73 and $46.94.

More to Explore • Numeration

Ask students if anyone remembers what a palindrome is. (A palindrome is a number that reads the same forward and backward, for example, 727.) Tell students that there is a mathematical method to finding a palindrome. Tell them to write any 3-digit number. Tell them to reverse the digits of the number and add it to the original number. Have them continue the process, reversing each digit and adding it to the previous number until a palindrome is reached. Have them try several examples.

ESL/ELL STRATEGIES

Before asking students to complete the activities, review the meaning of the words *penny, nickel, dime, quarter, half-dollar*, and *bill* as in *dollar bill*. Give each student several different coins and have pairs take turns telling each other how many of each coin they are holding.

2-5 Making Change

pages 23–24

1 Getting Started

Objective
- To make change for amounts of money up to $100

Materials
play money

Warm Up • Mental Math
Have students find the sum of the following amounts of money:

1. 1 dime plus 1 nickel (15¢)
2. 1 dime plus 1 penny (11¢)
3. 1 quarter plus 1 penny (26¢)
4. 1 quarter plus 1 nickel (30¢)
5. 1 quarter plus 1 dime (35¢)

Warm Up • Addition
Have students find the sum.

1. $0.25 + $0.25 ($0.50)
2. $0.50 + $0.10 ($0.60)
3. $0.50 + $0.25 ($0.75)
4. $0.60 + $0.35 ($0.95)
5. $0.45 + $0.45 ($0.90)

Name ______________________

Making Change

Kyle bought a computer game. He handed the cashier $60.00. How much change did he receive?

We want to know how much change Kyle got. The game cost $48.39.

Kyle paid for the game with $60.00.

Start by counting on from $48.39.

$48.40 $48.50 $48.75 $49.00 $50.00 $60.00

Count the bills and coins from greatest to least value.

$10.00 $11.00 $11.25 $11.50 $11.60 $11.61

Kyle's change was $11.61.

Getting Started

Find the amount of change.

1. Linda bought a computer program that cost $23.45. She gave the cashier $30. How much change did Linda receive? What coins and bills could she have received?
   $6.55; one possible answer: 1 five-dollar bill, 1 one-dollar bill, 2 quarters, 1 nickel

2. Mrs. Martinez went to the grocery store and spent $72.38. She paid with a $100 bill. How much change did Mrs. Martinez receive? What coins and bills could she have received?
   $27.62; one possible answer: 1 twenty-dollar bill, 1 five-dollar bill, 2 one-dollar bills, 2 quarters, 1 dime, 2 pennies

2 Teach

Introduce the Lesson Have students read the problem aloud and tell what they are to find. (the amount of change Kyle received from the cashier) Ask students to complete the information sentences.

- Have students focus on the first set of coins and bills in the text. Ask a volunteer to identify how the money is ordered. (from least to greatest value) Have students count on from $48.40 to $60.
- Ask students to look at the next set of coins and bills. Ask a volunteer to identify how the money is ordered. (from greatest to least value) Have students count the change and complete the solution sentence.

Develop Skills and Concepts Have students examine a $100 bill and tell its value. (100 dollars) Ask volunteers to give different combinations of bills that add up to $100. (five $20 bills, ten $10 bills, and so on) Record the answers on the board. Repeat with $50 and $20. Then, have students name the different coins they can use to make $1.

3 Practice

Have students complete all the exercises on page 24. Remind them to use a dollar sign and a decimal point in each money amount.

Practice

Find the amount of change.

1. Sally spent $17.25 for a ticket to the amusement park. How much change did she receive from a $20 bill? What bills and coins could she have received?
   $2.75; one possible answer: 2 one-dollar bills, 3 quarters

2. Brian went to the dentist and paid $85.85 for a teeth cleaning. If he paid with $90, how much change did he receive? What bills and coins could he have received?
   $4.15; one possible answer: 4 one-dollar bills, 1 dime, 1 nickel

3. Tania bought her mother a birthday present for $61.92. She gave the cashier four $20 bills. How much change did she receive? What bills and coins could she have received?
   $18.08; one possible answer: 1 ten-dollar bill, 1 five-dollar bill, 3 one-dollar bills, 1 nickel, 3 pennies

4. Warren bought a book for $22.15. He paid with a $20 bill and a $10 bill. How much change did he receive? If the cashier did not have any quarters, what bills and coins could Warren have received?
   $7.85; one possible answer: 1 five-dollar bill, 2 one-dollar bills, 1 half-dollar, 3 dimes, 1 nickel

5. April bought a DVD player for $49.65. She paid with three $20 bills. How much change did she receive? What bills and coins could she have received?
   $10.35; one possible answer: 1 ten-dollar bill, 1 quarter, 1 dime

6. Tyrone bought a large bag of dog food for $33.60. He paid with two $20 bills. How much change did he receive? If he received back in change the least number of bills and coins possible, what bills and coins did he receive?
   $6.40; 1 five-dollar bill, 1 one-dollar bill, 1 quarter, 1 dime, 1 nickel

4 Assess

Ask students how to tell in which order they should count change. (from least to greatest value)

5 Mixed Review

1. $4 + 7 = 7 + n$ (4)
2. $6 + n = 6$ (0)
3. $5 + (5 + 4) = (n + 5) + 4$ (5)
4. $7 + n = 13$ (6)
5. $5 + n = 12$ (7)
6. $4 + 4 + n = 12$ (4)
7. $3 + 5 = n + 3$ (5)
8. $9 - n = 2$ (7)
9. $12 - n = 6$ (6)
10. $n - 5 = 6$ (11)

For Mixed Abilities

Common Errors • Intervention

Some students will have difficulty knowing where to start to make change. Have these students work in pairs, with one student as the customer and the other student as the cashier. Give students play money and have them practice counting on from the cost of the item. Have pairs count the change needed to get to the next dollar and then count the dollars until they reach the cost of the item purchased.

Enrichment • Algebra

Write the following exercises on the board and have students fill in the missing numbers:

1. $0.32 + ($0.51) = $0.83
2. $0.53 + ($0.43) = $0.96
3. $0.72 − ($0.46) = $0.26
4. $0.93 − ($0.28) = $0.65
5. $0.59 − ($0.31) = $0.28
6. $0.27 + ($0.48) = $0.75

More to Explore • Measurement

Have students order a set of bus schedules from your local transit system. Discuss and demonstrate how to follow the schedules. Invite a representative of the transit system to visit and talk to students about how schedules are made. Ask students to bring in a variety of other schedules such as sports season schedules, television guides, school calendars, and so on. Make a display of them on a bulletin board. Finally, have students make a schedule of their school day for the whole week.

2-6 Thousands, Ten Thousands, and Hundred Thousands

pages 25–26

1 Getting Started

Objective

- To read and write numbers in the thousands, ten thousands, and hundred thousands

Materials

*number cards 0 through 9

Warm Up • Mental Math

Have students name the following number:

1. the reverse of 1,632 (2,361)
2. 2 times 6 + 9 (21)
3. 14 less than 16 + 14 (16)
4. 200 more than $\frac{1}{4}$ of 8 (202)
5. less than 415 but more than 413 (414)
6. tenth if counting by 100s (1,000)
7. 0.1 of 100 (10)

Warm Up • Numeration

Dictate numbers having 1, 2, 3, or 4 digits for students to write on the board. Have students name the value of each digit. Now, have students write the number representing 2 thousands 4 hundreds 6 tens 8 ones and have a student write the number on the board. Repeat for similar problems.

Name ______________________

Lesson 2-6

Thousands, Ten Thousands, and Hundred Thousands

A human heart normally beats 72 times every minute. How many times does a normal heart beat in 1 day?

Normal Heart Rates	
Beats	**Time**
72	minute
4,320	hour
103,680	day

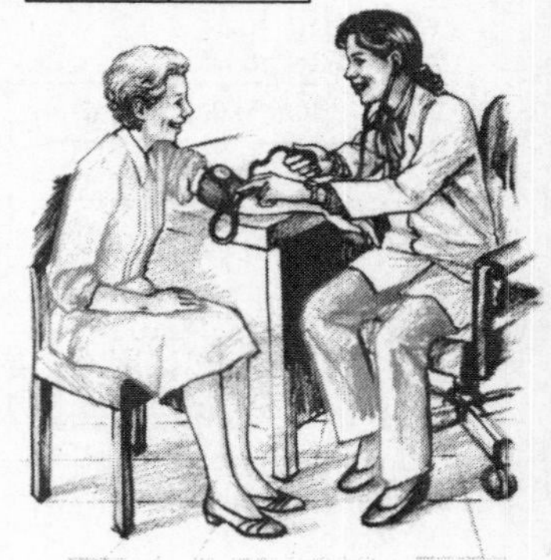

We know that the normal heart beats 103,680 times in 1 day. To understand this number, study this place-value chart.

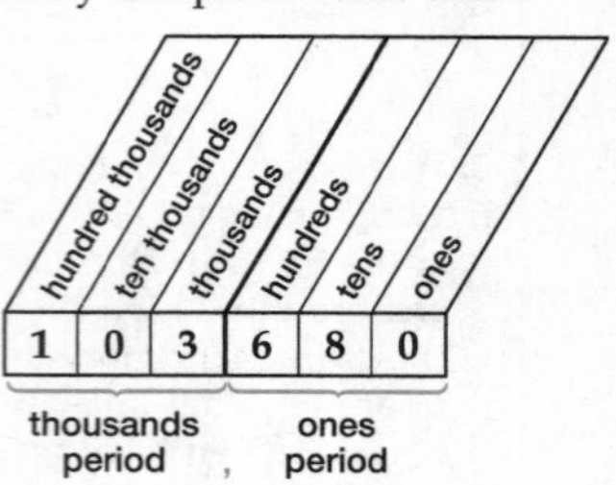

A comma separates the periods.

We say a normal heart beats **one hundred three thousand, six hundred eighty** times a day.

We can write the number as 103,680 in standard form.

We can also write it in expanded form, **100,000 + 3,000 + 600 + 80**.

Getting Started

Write the place value of each green digit.

1. 341,219 ten thousands
2. 156,035 hundred thousands

Write the number.

3. two hundred forty-six thousand, five hundred fifteen 246,515

Write the missing word.

4. 470,050 four hundred seventy thousand, fifty

2 Teach

Introduce the Lesson Have a student read the problem and tell what question is to be answered. (how many times a normal heart beats per day) Tell students to study the table, reread the problem, and determine what fact is needed to complete the answer. (The heart beats 103,680 times per day.)

- Next, help students study the place-value chart until they understand how 103,680 fits in. Point out that ones, tens, and hundreds together compose the ones period. Thousands, ten thousands, and hundred thousands compose the thousands period. Have students complete the last sentence with the correct number.

Develop Skills and Concepts Write **8,928** on the board and have students read it with you as you write **eight thousand, nine hundred twenty-eight**. Ask students where the comma is written in both the number and its number name. (after the thousands digit and the word thousand) Go through the place value of each digit, introducing the thousands place.

- Write **42,081** and ask students the place value of the 4. (ten thousands) Write **436,962** and ask students the place value of the 3. (ten thousands) Tell students that the 4 is now in the hundred thousands place. Write **four hundred thirty-six thousand, nine hundred sixty-two** on the board and read it together. Be sure students do not use the word *and* to separate the periods. Repeat the procedure for more examples of 6-digit numbers.

3 Practice

Remind students to write a comma after the thousands digit and after the word *thousand* when writing a number name. Have students complete page 26 independently.

Practice

Write the place value of each green digit.

1. 63,215 thousands
2. 139,073 hundreds
3. 21,875 ten thousands
4. 370,150 hundred thousands
5. 929,175 ten thousands
6. 836,207 tens
7. 180,000 thousands
8. 57,329 ones

Write each number.

9. four hundred twenty-seven thousand, three hundred twelve 427,312
10. two hundred nine thousand, one hundred fifty-six 209,156
11. eighty-seven thousand, nine hundred nine 87,909
12. ten thousand, two hundred eighty-three 10,283

Write each missing word.

13. 64,419 sixty-four thousand, four hundred nineteen
14. 406,060 four hundred six thousand, sixty
15. 515,291 five hundred fifteen thousand, two hundred ninety-one

Problem Solving

Write the number given in each problem.

16. The highest mountain in the world is Mt. Everest. It is twenty-nine thousand, twenty-eight feet high.
29,028 feet

17. In 1962, Major Robert M. White flew an X-15 plane to the altitude of three hundred fourteen thousand, seven hundred fifty feet.
314,750 feet

18. Earth is about two hundred thirty-eight thousand, eight hundred sixty miles from the Moon.
238,860 miles

19. Earth orbits the Sun at about sixty-six thousand, six hundred miles per hour.
66,600 miles per hour

4 Assess

Ask students the place value of the 2, 9, and 6 in the number 296,341. (hundred thousands, ten thousands, thousands)

For Mixed Abilities

Common Errors • Intervention

Students may name the place value of a digit incorrectly because they have difficulty reading large numbers. Write the numbers shown below on the board and have students read them, noticing each time how many digits the number contains. Change the 1s to 7s and have them read the numbers again.

10
100
1,000
10,000
100,000

Enrichment • Application

Have students work with a friend to determine their pulse rates for 30 seconds. They should then find their pulse or heartbeat rate per minute, hour, and day.

More to Explore • Numeration

Test students' visual sense of numbers. Prepare small sets of objects and put each set in a paper bag, for example, seven buttons, three table tennis balls, ten paper clips, and so on. Have students work in pairs. One student is to empty the bag into a tray and immediately cover the tray with a sheet of black paper. The second student is to guess how many objects were seen before the tray was covered. Have the pairs keep track of each guess and count the actual number of objects in each bag at the end of the activity.

Ask students to explain why a visual memory of numbers would have been important before written numbers were devised.

ESL/ELL STRATEGIES

Ask students to repeat some long numbers. Point out that we do not usually use the word *and*. Say, *It's not two hundred forty-six thousand, and five hundred and fifteen*. Also, we don't say the zero. Say, *It's not one hundred fifty-six thousand, zero hundred thirty-five.*

2-7 Millions

pages 27–28

1 Getting Started

Objective
- To read and write numbers in the millions

Materials
cards with numbers 0 through 9; 2 comma cards

Warm Up • Mental Math
Have students name a number that is

1. evenly divisible by 2 (2, 4, and so on)
2. 1,000 more than 100 (1,100)
3. odd and less than 25 (1, 3, 5, . . . , 23)
4. 27 pairs (54)
5. 1,000 less than 100,000 (99,000)
6. 47 more than 20,100 (20,147)
7. 64 when multiplied by itself (8)

Warm Up • Numeration
Write **100,000** on the board. Ask students to tell the next number if counting by 10,000s. (110,000) Have a student write **110,000** under 100,000. Continue counting and writing the numbers through 200,000. Now, have students count and write the numbers by 20,000s through 300,000. Continue to count and write by 50,000s through 500,000 and then by 100,000s through 900,000.

Name ______________________

Millions

Chicago is the home of the world's busiest airport. More than 20 million passengers take off from O'Hare Airport each year. How many travelers have departed from O'Hare according to the meter?

Passenger Departures							
2	5,	6	3	6,	4	8	3

We know the meter shows 25,636,483 passengers. To understand this number, study the following place-value chart.

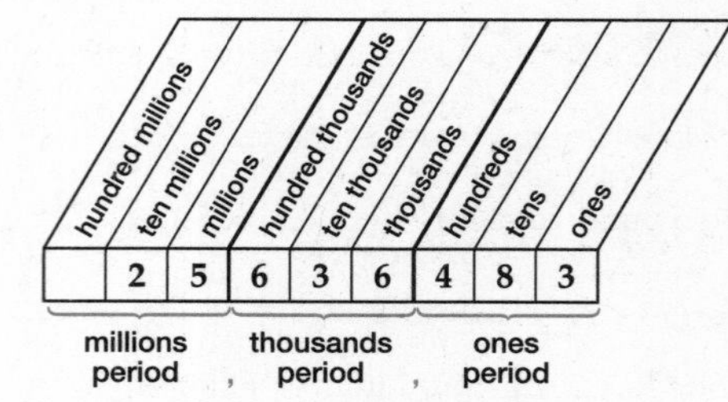

Commas separate the periods.

We say that **twenty-five million, six hundred thirty-six thousand, four hundred eighty-three** passengers departed from O'Hare Airport according to the meter. We write this number as 25,636,483.

Getting Started

Write the number.

1. two hundred twelve million, four hundred sixty-five thousand, twenty-nine 212,465,029

Write the place value of each green digit.

2. 23,465,183 millions
3. 706,341,212 ten thousands
4. 5,962,159 hundred thousands

2 Teach

Introduce the Lesson Have a student read the problem and tell what is to be found. (the exact number of travelers who have flown out of O'Hare Airport) Ask students if the exact number is given in the problem or meter. (meter) Tell students to write the number in the information sentence.

- Help students study the place-value chart until they understand how 25,636,483 fits in. Read the number word with students and then have them complete the last sentence with the correct number. Have students complete the sentences as they read aloud with you.

Develop Skills and Concepts Have students count by 10,000s and write the numbers from 900,000 to 990,000. Tell students that 1 million is 1,000 thousands and is written 1,000,000. Write **990,000** and **1,000,000** under it.

Ask students if they can guess how long it would take to spend $1,000,000 if they spent $10 each day. (100,000 days, or almost 274 years)

- Draw a place-value chart for 9-digit numbers. Write **273,646,901** in the chart. Help students tell the value of each digit. (2 hundred millions, 7 ten millions, 3 millions, 6 hundred thousands, 4 ten thousands, 6 thousands, 9 hundreds, 0 tens, 1 one) Write **two hundred seventy-three million, six hundred forty-six thousand, nine hundred one** on the board. Repeat the activity for more 8- and 9-digit numbers. Tell students that a comma separates the millions period from the thousands period, and the thousands period from the ones period.

3 Practice

Remind students to use commas to separate the periods in numbers of 1,000 or more. Have students complete page 27 independently.

Practice

Write each number.

1. three hundred seventy-five million, two hundred five thousand, sixty-seven 375,205,067

2. four hundred ten million, five hundred sixteen thousand, four hundred twenty 410,516,420

3. thirty-eight million, sixty-three thousand, eight hundred forty-nine 38,063,849

Write the place value of each green digit.

4. 6,241,573 millions
5. 14,903,124 hundred thousands
6. 136,248,796 ten millions
7. 828,297,159 hundred millions
8. 16,274,302 hundreds
9. 460,760,858 ten thousands
10. 3,926,575 ten thousands
11. 47,239,105 hundred thousands
12. 906,400,000 millions

Now Try This!

Rearrange each group of letters to write a number.

1. enevs seven; lloimni million; hteer three; nudrhde hundred; tduhonas thousand
2. tyiff-wto fifty-two; sdntuoha thousand; owt two; rddheun hundred; iffneet fifteen
3. thgie eight; dhernud hundred; neo one; nmliloi million; inne nine; shudtoan thousand

Now Try This! Have students rearrange the letters into numerical words. Extend the activity by having them write the numbers as numerals and by writing some scrambled number words of their own.

4 Assess

Ask students to name the place value of each digit in 835,976,034. (hundred millions, ten millions, millions, hundred thousands, ten thousands, thousands, hundreds, tens, ones)

For Mixed Abilities

Common Errors • Intervention

Students may omit zeros as placeholders or write too many zeros when they are writing numbers in the millions. Have them write the numbers in a place-value chart using period names—millions, thousands, ones—to guide them. Every place on the chart to the right of the leftmost digit must also have a digit.

Enrichment • Numeration

Have students write four numbers, each having 9 digits. They should then round each number to the nearest million and thousand.

More to Explore • Measurement

Conduct a classroom "Meter Olympics" spread over several class periods. Students choose a variety of athletic events whose results can be measured in meters. Some examples might be the standing broad jump, softball throw, and running long jump. In the schoolyard or gymnasium, have students take turns trying each event. Use a meterstick to record the results. Keep a chart of the number of meters each person jumped or threw. Continue this activity for several days and keep a daily record. Students may wish to compare their first efforts with later jumps or throws and note their individual improvement.

2-8 Comparing Numbers

pages 29–30

1 Getting Started

Objective

- To compare numbers with up to 7 digits

Materials

place-value blocks: hundreds, tens, ones; large place-value charts

Warm Up • Mental Math

Dictate the following problems:

1. $\frac{1}{8} + \frac{4}{8}$ ($\frac{5}{8}$)
2. 40 × 9 (360)
3. 5,000 − 100 (4,900)
4. 63 ÷ 7 (9)
5. 2.1 + 6.1 (8.2)
6. $\frac{1}{6}$ of 24 (4)
7. $\frac{1}{5}$ of (20 + 5) (5)
8. 18 ÷ 9 (2)

Warm Up • Numeration

Write **79** ◯ **26** on the board and have a student write the greater than or less than sign in the circle. (>) Continue writing more pairs of 2-digit numbers on the board. Have students write the correct sign to compare the numbers.

Name ____________________

It's Algebra!

Comparing Numbers

Which country, Eritrea or Moldova, has a greater population?

Estimated Populations of Countries		
Country	**Population**	**Year**
Costa Rica	3,940,000	2001
Croatia	4,780,000	1991
Eritrea	4,136,000	2000
Moldova	4,431,000	2000

We want to know which of the two countries has a greater population.

We know the population of Eritrea is about 4,136,000.

We know the population of Moldova is about 4,431,000.

To find the greater population, we compare both populations.

	Start with the digits on the left. Compare.	Look at the next place. Compare the digits.
Eritrea	4,136,000	4,136,000
Moldova	4,431,000	4,431,000
	4 = 4	4 > 1

We say 4,431,000 is greater than 4,136,000.

We write 4,431,000 > 4,136,000.

The population of Moldova is greater than the population of Eritrea.

Getting Started

Compare. Write < or > in the circle.

1. 583 (>) 581
2. 7,919 (<) 8,215
3. 15,275 (<) 15,725
4. 47,362 (>) 6,395
5. 295,363 (>) 288,481
6. 3,528,219 (<) 3,547,129

Write the numbers in order from least to greatest.

7. 3,715; 4,210; 1,650
   1,650, 3,715, 4,210
8. 675,283; 597,126; 604,322
   597,126, 604,322, 675,283

Write the names of the countries in the above chart in order from the lowest population to the greatest population.

9. Costa Rica, Eritrea, Moldova, Croatia

---

2 Teach

Introduce the Lesson Have a student read the problem and tell what is to be found. (which country, Eritrea or Moldova, has the greater population) Ask students to tell what information is given in the table. (Costa Rica has 3,940,000 people; Croatia has 4,780,000 people; Eritrea has 4,136,000 people; and Moldova has 4,431,000 people.) Ask students if all the information is needed to solve the problem. (no) Ask what information they will use. (Moldova has 4,431,000 people; Eritrea has 4,136,000 people.)

- Work through the problem in the model, comparing the populations of the countries. Then, have students complete the solution sentence.

Develop Skills and Concepts Write **267** and **276** on the board. Give groups of four students place-value blocks to model each number on large place-value charts. Ask them how many hundreds each number has. (2) Ask how many tens each number has. (267 has 6 tens and 276 has 7 tens.) Ask which number has the greatest number of tens. (276) Write **276** (>) **267** on the board.

- Repeat this process for 1,763 and 1,784, helping students understand that both numbers have the same number of thousands and hundreds. Continue with pairs of greater numbers up to 10,000.
- Refer students to the chart of populations on page 29 and have students compare the populations for more practice.

3 Practice

Have students complete the exercises on page 30. Remind them to always start comparing with the highest place value of the numbers.

Practice

Compare. Write < or > in the circle.

1. 468 (>) 463
2. 297 (<) 300
3. 897 (>) 879
4. 3,246 (<) 3,252
5. 6,485 (>) 6,481
6. 8,296 (>) 8,290
7. 26,175 (<) 27,675
8. 47,283 (>) 8,835
9. 35,175 (>) 35,157
10. 285,248 (>) 258,428
11. 68,385 (<) 175,238
12. 417,361 (<) 472,358
13. 1,952,359 (>) 1,875,324
14. 3,851,183 (>) 3,815,792
15. 485,272 (<) 3,602,105

Write the numbers in order from least to greatest.

16. 468, 686, 560
    468, 560, 686
17. 212, 235, 210
    210, 212, 235
18. 376, 372, 378
    372, 376, 378
19. 3,210; 3,240; 3,260
    3,210, 3,240, 3,260
20. 8,512; 7,416; 7,800
    7,416, 7,800, 8,512
21. 5,286; 5,179; 5,280
    5,179, 5,280, 5,286
22. 472,385; 427,835; 473,285
    427,835, 472,385, 473,285
23. 5,917,252; 5,971,522; 5,975,122
    5,917,252, 5,971,522, 5,975,122

Problem Solving

Solve each problem.

24. The Ohio River is 975 miles long. The Red River is 1,270 miles long. Which river is longer?
    Red River
25. The Lincoln Tunnel is 8,216 feet long. The Holland Tunnel is 8,557 feet long. Which tunnel is longer?
    Holland Tunnel
26. In the 2000 presidential election, Al Gore received 2,912,353 votes in Florida. George W. Bush received 2,912,790 votes in Florida. Which candidate received more votes in Florida?
    George W. Bush
27. North Dakota has 68,976 square miles of land. Missouri has 68,886 square miles of land, and Oklahoma has 68,667 square miles of land. Which of these states has the most land?
    North Dakota

4 Assess

Ask students to tell which number is greater, 200,002 or 202,002. Have them explain their answer. (202,002 is greater. Students should explain how they compared each place.)

For Mixed Abilities

Common Errors • Intervention

Some students may have difficulty when comparing numbers in a side-by-side position. Have them write the numbers in a place-value chart, one under the other, and compare the places from left to right.

Enrichment • Application

Have students use the World Almanac to find the average depth in feet of the South China Sea, Mediterranean Sea, Bering Sea, and Sea of Japan. Then, have them make a table showing the depths in order from the deepest to the most shallow.

More to Explore • Numeration

Give students an incomplete sequence of numbers such as 5, 11, 17, 23, 29, ____, ____, ____. Have them fill in the missing three numbers by finding the pattern of the sequence. Explain that these are called *arithmetic progressions*. The number added or subtracted each time is called a *common difference*. Provide another progression such as 2, 6, 5, 15, 14, ____, ____, ____ and have them supply the next three numbers. Have students devise their own sequences of numbers, exchange them with a partner to complete the missing numbers, and explain the common differences in the sequences.

2-9 Rounding Numbers

pages 31–32

1 Getting Started

Objective

- To round numbers up to the nearest million

Materials

number cards 20 through 30; newspaper grocery ads

Warm Up • Mental Math

Have students name the year that came first.

1. 1776 or 1976 (1776)
2. 1986 or 1968 (1968)
3. 1072 or 865 (865)
4. 1615 or 1516 (1516)
5. 1326 or 1218 (1218)
6. 1812 or 1912 (1812)
7. 1971 or 1969 (1969)
8. 1421 or 1399 (1399)

Warm Up • Numeration

Write **26 + 48 + 137** vertically on the board. Have a student find the sum. (211) Have another student write the problem and the sum as money amounts. ($0.26 + $0.48 + $1.37 = $2.11) Repeat for more problems.

Name ____________________

Rounding Numbers

About how many people live in Kentucky and Indiana?

State	Population	Year
Kentucky	4,041,769	2000
Indiana	6,080,485	2000

We want to know about how many people live in Kentucky and in Indiana.

We know the population of Kentucky is 4,041,769 and the population of Indiana is 6,080,485.

We can round these populations to millions, hundred thousands, ten thousands, and thousands.

To round a number to a particular place value, locate the digit to be rounded. If the digit to the right is 0, 1, 2, 3, or 4, the digit we are rounding stays the same. If the digit to the right is 5, 6, 7, 8, or 9, the digit we are rounding is raised one.

	Population	Rounded to Millions	Rounded to Hundred Thousands	Rounded to Ten Thousands	Rounded to Thousands
Kentucky	4,041,769	4,000,000	4,000,000	4,040,000	4,042,000
Indiana	6,080,485	6,000,000	6,100,000	6,080,000	6,080,000

The number of people living in Kentucky rounded to the nearest
- million is 4,000,000
- hundred thousand is 4,000,000
- ten thousand is 4,040,000
- thousand is 4,042,000

The number of people living in Indiana rounded to the nearest
- million is 6,000,000
- hundred thousand is 6,100,000
- ten thousand is 6,080,000
- thousand is 6,080,000

Getting Started

Round each number to the nearest million.

1. 8,275,259 8,000,000
2. 6,524,920 7,000,000
3. $3,752,496 $4,000,000

Round each number to the nearest hundred thousand

4. 8,275,259 8,300,000
5. $6,524,920 $6,500,000
6. 3,752,496 3,800,000

2 Teach

Introduce the Lesson Have a student read the problem and tell what is to be done. (estimate how many people live in Kentucky and Indiana) Ask students to use the chart to tell what facts are given and then fill in the information sentences. (The population of Kentucky is 4,041,769; the population of Indiana is 6,080,485.)

- Work through the rounding process in the model with students. Then, have them complete the solution sentences.

Develop Skills and Concepts Tell students that we often need to know a sum quickly but do not need to know the exact amount. We can then **round** the numbers to get an **estimate**. Tell students that an example might be if you had $2 and wanted to quickly know if you could afford to buy three items costing 79¢, 59¢, and 83¢. Help students name other times when rounding numbers could be helpful. (referring to the speed traveled, pages in a book, and so on)

- Draw a number line from 1,000 to 2,000 on the board. Write **1,600** on the board. Ask students if 1,600 is closer to 1,000 or to 2,000. (2,000) Show students that 1,600 is rounded to 2,000 rather than to 1,000 because it is closer to 2,000. Tell students that we always round a number that is exactly in the middle up, so 1,500 is rounded to 2,000. Repeat the procedure for rounding 162,000 to the nearest hundred thousand.
- Now, draw a number line from 100,000 to 200,000 and show students how 162,000 is rounded to the nearest hundred thousand. Give more examples of rounding 5- and 6-digit numbers.

3 Practice

Have students complete the exercises on page 32.

Practice

Round each number to the nearest million.

1. 2,657,239 3,000,000
2. $8,902,432 $9,000,000
3. 4,285,230 4,000,000
4. 9,425,295 9,000,000
5. 1,289,392 1,000,000
6. 3,333,333 3,000,000

Round each number to the nearest hundred thousand.

7. 425,389 400,000
8. $819,259 $800,000
9. 558,219 600,000
10. 962,495 1,000,000
11. 3,283,425 3,300,000
12. 7,529,382 7,500,000

Round each number to the nearest ten thousand.

13. $38,253 $40,000
14. 74,528 70,000
15. 673,952 670,000
16. 528,258 530,000
17. 397,295 400,000
18. 4,875,258 4,880,000

Round each number to the nearest thousand.

19. 8,375 8,000
20. $27,839 $28,000
21. 39,438 39,000
22. 552,450 552,000
23. 985,836 986,000
24. 7,058,294 7,058,000

Problem Solving

Solve each problem.

25. The attendance at the River Dogs' last game was 27,648. What was the attendance rounded to the nearest thousand?
28,000

26. The population of Jacksonville, Florida, is 762,461. What is Jacksonville's population rounded to the nearest ten thousand people?
760,000

27. Jesse said that Jacksonville's population is about 800,000. Which place did Jesse round to?
hundred thousands

28. At 9,166,601 square kilometers, the United States has the fourth greatest area of any country on Earth. What is the area of the United States rounded to the nearest million?
9,000,000

4 Assess

Ask students to name the greatest whole number that when rounded to the nearest million is 1,000,000.
(999,999)

For Mixed Abilities

Common Errors • Intervention

Some students may round to the wrong place when they are working with large numbers. Have students first draw an arrow above the place to which they are rounding. Next, have them look at the digit to the right to determine whether the digit under the arrow should stay the same or be increased by 1. Then, have them replace all the digits to the right of the arrow with zeros.

Enrichment • Application

Tell students to bring in some grocery store ads from a newspaper. Have students select five items they would like to purchase. Have them round each price to the nearest ten or nearest dollar and estimate the amount of money they would pay.

More to Explore • Logic

Tell students to draw a box that is divided into nine squares. Using the numbers 1 through 9, have students put one number in each square, arranging them in such a way that when adding the numbers in any row—vertical, horizontal, or diagonal—the answer will be the same.

(6)	(5)	(4)
(3)	(5)	(7)
(6)	(5)	(4)

Tell them they have created a magic square.

2-10 Problem Solving: Make an Organized List

pages 33–34

1 Getting Started

Objective

- To solve problems by making an organized list

Warm Up • Mental Math

Dictate the following exercises and ask students to round each number as directed:

1. 685 to the nearest hundred (700)
2. 2,358 to the nearest hundred (2,400)
3. 5,389 to the nearest thousand (5,000)
4. 27,645 to the nearest thousand (28,000)
5. 395,436 to the nearest ten thousand (400,000)
6. 517,383 to the nearest hundred thousand (500,000)

Warm Up • Numeration

Write the following exercises on the board. Tell students to compare each pair of numbers using < or >.

1. 325 ◯ 352 (<)
2. 4,163 ◯ 3,975 (>)
3. 35,385 ◯ 5,937 (>)
4. 53,284 ◯ 53,293 (<)
5. 183,285 ◯ 183,825 (<)
6. 754,385 ◯ 743,385 (>)

Name ______________________

Problem Solving: Make an Organized List

Karim is going to the sports store to buy clothes. He has $56.25. He wants to buy at least two items. How many combinations of items can Karim buy?

Price List	
Item	**Cost**
Sweat shirt	$27.50
T-shirt	$8.99
Sneakers	$48.95
Baseball cap	$12.55
Team jacket	$42.50

★SEE

We know Karim has $56.25 to take to the store.

We are looking for the different ways Karim can buy at least two items.

★PLAN

We can make an **organized list**. We can start with the T-shirt and the baseball cap because they cost the least.

★DO

Fill in the list for all of the combinations that cost a total of $56.25 or less.

Choices	Item 1	Item 2	Item 3	Total Amount
First	$8.99 T-shirt	$12.55 baseball cap		$21.54
Second	$8.99 T-shirt	$12.55 baseball cap	$27.50 sweat shirt	$49.04
Third	$8.99 T-shirt	$27.50 sweat shirt		$36.49
Fourth	$8.99 T-shirt	$42.50 team jacket		$51.49
Fifth	$12.55 baseball cap	$42.50 team jacket		$55.05

Karim can buy five different combinations of two or more items.

★CHECK

We can check by comparing each amount to $56.25.

First choice $21.54 < $56.25
Second choice $49.04 (<) $56.25
Third choice $36.49 (<) $56.25
Fourth choice $51.49 (<) $56.25
Fifth choice $55.05 (<) $56.25

2 Teach

Introduce the Lesson Have students read the problem aloud and tell what they are to find. (how many combinations of items Karim can buy) Remind students of the four-step strategy used to solve problems.

Develop Skills and Concepts Ask a student to read the SEE section. Explain that in this section, they will restate the problem and fill in the information needed to solve the problem. Ask students to fill in the answers and for volunteers to give the missing information.

- Ask another volunteer to read the PLAN sentences. Tell students that there are several ways to solve this problem, but an organized list will help them work through the problem and find the combinations of items that cost a total of $56.25 or less. Explain that by starting with the items that cost the least, they can keep on increasing the value and number of the items to get a good idea of which items can be bought for the given amount.
- Work through the DO step, helping students fill in the information when needed. Then, have students complete the solution sentence. Finally, work through the CHECK step with students. Make sure students understand that the total amount of each combination of items is being compared to the amount of money Karim has.

3 Apply

Solution Notes

1. Ask students to explain how they organized the numbers they made.
2. Ask students to show the addition sentences they made to find the combinations. (10¢ + 10¢, 10¢ + 5¢ + 5¢, 5¢ + 5¢ + 5¢ + 5¢)

Apply

Make an organized list to help you solve each problem.

1. How many different 3-digit numbers can you make using the digits 2, 3, and 4?
   6 numbers; 234, 243, 324, 342, 423, 432
2. How many different ways can you count 20¢ without using pennies?
   3 ways; 2 dimes, 1 dime and 2 nickels, 4 nickels
3. How many different ways can you count $20 without using coins? Also, do not use $2 bills.
   10 ways
4. Tim saw people and dogs while sitting on a park bench. He counted 22 legs. How many people and dogs were there?
   Possible answers: 1 person and 5 dogs, 3 people and 4 dogs, 5 people and 3 dogs, 7 people and 2 dogs, 9 people and 1 dog
5. How many combinations of two letters can Paul make from his name?
   12 combinations; PA, PU, PL, AU, AL, AP, UP, UA, UL, LP, LA, LU
6. How many different amounts can you make using only coins and not more than one of each coin? Do not use half-dollars or dollar coins.
   16 different amounts

Problem-Solving Strategy: Using the Four-Step Plan

★ **SEE** What do you need to find?
★ **PLAN** What do you need to do?
★ **DO** Follow the plan.
★ **CHECK** Does your answer make sense?

7. Ice cream comes in vanilla and chocolate, and the sizes come in large, medium, and small. How many different ways can you order one flavor of ice cream in one size?
   6 ways; large vanilla, medium vanilla, small vanilla, large chocolate, medium chocolate, small chocolate
8. Sixteen contestants started an air hockey tournament. Contestants play one game, and if they win, they play again against a different opponent. If they lose, they do not play again. How many games will the contestants have to win to become the air hockey champion?
   4 games

For Mixed Abilities

More to Explore • Application

Have students record the names of three students in the class as well as their own name. Tell students that they are the coach of a swimming team and they have to decide in which order the swimmers will swim. Students should make an organized list to find all of the combinations that they can make to order the four swimmers. (24) Encourage students to present their findings to the class for display and discussion.

3. Ask students which bills they can use to make $20. ($1, $5, $10, $20)
4. Make sure students understand that there is more than one possible answer. Challenge students to find all of the possible answers.
5. If students have an easier time making combinations with letters, tell students that they can substitute the letters of PAUL with the numbers 1, 2, 3, and 4. Use this problem as a springboard to ask students whether it is a good idea to make an organized list when they are required to find combinations.

Higher-Order Thinking Skills

6. **Comprehension:** Some students may want to use combinations with more than one of a coin. Remind students to read, and even reread, these problems very carefully.
7. **Synthesis:** Students may say that there are 5 combinations because there are 2 flavors and 3 sizes. Challenge students to figure out why they cannot add the two components to calculate the number of combinations. (There are 3 different ways to serve each flavor and there are 2 flavors.)
8. **Analysis:** Students must recognize that after the first round there will be 8 teams, after the second round there will be 4 teams, after the third round there will be 2 teams, and after the fourth round, there will only be 1 team left.

4 Assess

Ask students how to make an organized list to find how many different outfits they can make from 2 shirts, 1 red and 1 blue, and 2 pairs of pants, 1 blue and 1 green. (Combine the first shirt [red] with the first pair of pants [blue] and the second shirt [blue] with the second pair of pants [green] and then swap them.)

Chapter 2 Test

page 35

Items	Objectives
1–8	To identify the place value of digits in numbers up to 9-digits (see pages 17–18, 25–28)
9–13	To read and write numbers less than 1 billion (see pages 17–18, 25–28)
14–16	To compare and order numbers less than 10,000 (see pages 29–30)
17–19	To round numbers to the nearest ten (see pages 31–32)
20–25	To round numbers to the nearest hundred or dollar (see pages 31–32)

Name ____________________

CHAPTER 2 TEST

Write the place value of each green digit.

1. 6,508 hundreds
2. 13,279 thousands
3. 8,111 tens
4. 149,000,432 hundred millions
5. 3,406 ones
6. 126,031,210 ten millions
7. 2,479,036 millions
8. 75,493,000 hundred thousands

Write each number.

9. three hundred sixty-five 365
10. four thousand, two hundred twenty-four 4,224
11. three million, three hundred thousand, sixty-two 3,300,062

Write each missing word.

12. 56,483 fifty-six thousand, four hundred eighty-three
13. 216,050,000 two hundred sixteen million, fifty thousand

Compare. Write < or > in the circle.

14. 6,247 (>) 4,396
15. 7,743 (<) 7,748
16. 5,029 (<) 5,039

Round each number to the nearest ten.

17. 1,463 1,460
18. 3,555 3,560
19. 4,096 4,100

Round each number to the nearest hundred or dollar.

20. 398 400
21. 456 500
22. $7.48 $7
23. $8.21 $8
24. 1,612 1,600
25. 550 600

Alternate Chapter Test

You may wish to use the Alternate Chapter Test on page 348 of this book for further review and assessment.

Circle the letter of the correct answer.

1. 53 ◯ 35
   a. <
   (b.) >

2. 4 + 8
   a. 4
   b. 10
   (c.) 12
   d. NG

3. $8 + 7$
   a. 14
   (b.) 15
   c. 16
   d. NG

4. $3 + 2 + 6$
   a. 5
   b. 8
   c. 13
   (d.) NG

5. 7 + (6 + 3)
   a. 2
   b. 9
   (c.) 16
   d. NG

6. 14 − 8
   (a.) 6
   b. 7
   c. 8
   d. NG

7. $12 - 9$
   (a.) 3
   b. 4
   c. 5
   d. NG

8. $9 - 0$
   a. 0
   (b.) 9
   c. 10
   d. NG

9. $n + 6 = 13$
   $n = ■$
   a. 5
   b. 6
   (c.) 7
   d. NG

10. What is the place value of the 5 in 4,256?
    a. ones
    (b.) tens
    c. hundreds
    d. NG

11. 3,279 ◯ 3,289
    (a.) <
    b. >

12. Round 896 to the nearest hundred.
    a. 800
    (b.) 900
    c. NG

13. Round 6,750 to the nearest thousand.
    a. 6,000
    b. 6,800
    (c.) 7,000
    d. NG

14. What is the place value of the 9 in 29,067?
    a. tens
    b. hundreds
    (c.) thousands
    d. NG

score

STOP

Cumulative Assessment

page 36

Items	Objectives
1	To compare and order numbers less than 100 (see pages 15–16)
2–3	To compute basic addition facts (see pages 1–2)
4	To add three 1-digit numbers (see pages 3–4)
5	To understand the Grouping Property of Addition (see pages 5–6)
6–8	To compute basic subtraction facts (see pages 1–2)
9	To find missing addends (see pages 9–10)
10, 14	To identify the place value of digits in numbers less than 1,000,000 (see pages 25–28)
11	To compare and order numbers less than 10,000 (see pages 29–30)
12–13	To round numbers to the nearest hundred or thousand (see pages 31–32)

Alternate Cumulative Assessment

Circle the letter of the correct answer.

1. 67 ◯ 76
   a >
   (b) <
   c =

2. 3 + 9 =
   a 9
   b 11
   c 13
   (d) NG

3. $9 + 6$
   a 13
   b 14
   (c) 15
   d NG

4. $5 + 4 + 3$
   (a) 12
   b 9
   c 7
   d NG

5. 3 + (7 + 2) =
   a 9
   b 6
   (c) 12
   d NG

6. 18 − 7 =
   (a) 11
   b 10
   c 9
   d NG

7. $14 - 8$
   a 5
   b 7
   c 9
   (d) NG

8. $7 - 7$
   a 14
   b 1
   (c) 0
   d NG

9. $n + 7 = 16$
   $n = ?$
   a 8
   (b) 9
   c 23
   d NG

10. What is the place value of the 3 in 3,621?
    (a) thousands
    b hundreds
    c tens
    d NG

11. 6,927 ◯ 6,279
    (a) >
    b <
    c =

12. Round 723 to the nearest hundred.
    a 600
    (b) 700
    c 800
    d NG

13. Round 4,869 to the nearest thousand.
    a 4,900
    b 4,000
    (c) 5,000
    d NG

14. What is the value of the 2 in 37,256?
    a tens
    (b) hundreds
    c thousands
    d NG

CHAPTER 3

3-1 Adding 2-Digit Numbers

pages 37–38

1 Getting Started

Objective

- To add 2-digit numbers with regrouping

Materials

*3 clear plastic shoeboxes, one labeled "ones," another "tens," and the last "hundreds;" place-value blocks: hundreds, tens, and ones

Warm Up • Mental Math

Remind students that $4 + 4 = 8$ is called a double. Ask students to name the sum of that double plus 1. ($4 + 5 = 9$) Continue until all doubles, doubles plus 1, and their sums are given.

Warm Up • Algebra

Have students find the missing addend.

1. $23 + n = 30$ (7)
2. $n + 15 = 24$ (9)
3. $35 + n = 41$ (6)
4. $n + 8 = 16$ (8)
5. $41 + n = 45$ (4)
6. $n + 37 = 46$ (9)

Name ______

Addition of Whole Numbers

CHAPTER 3

Lesson 3-1

Adding 2-Digit Numbers

Carla eats fruit for breakfast each morning. This morning, she ate a peach and half a grapefruit. How many calories does Carla get from the fruit?

We want to know the total number of calories that Carla gets from the fruit.

We know a peach provides 35 calories and half a grapefruit provides 39 calories.

To find the total, we add 35 and 39.

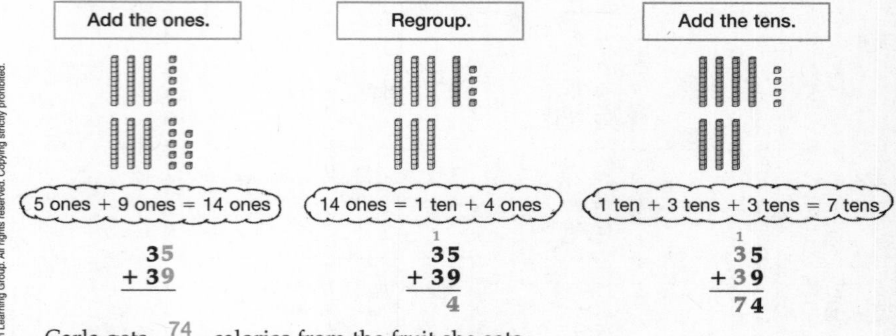

Add the ones.	Regroup.	Add the tens.
35 + 39	35 + 39 = 4 (regroup 1)	35 + 39 = 74

Carla gets 74 calories from the fruit she eats.

Getting Started

Add.

1. 43 + 18 = 61
2. 79 + 35 = 114
3. 26 + 53 = 79
4. 86 + 58 = 144

Copy and add.

5. 17 + 59 = 76
6. 76 + 8 = 84
7. 61 + 48 = 109
8. 87 + 67 = 154

2 Teach

Introduce the Lesson Have a student tell about the picture. (a chart of the calories in fruits) Have a student read the problem aloud and tell what is to be solved. (how many calories Carla gets from a peach and half a grapefruit) Ask students what information they need to use from the picture. (A peach has 35 calories and half a grapefruit has 39 calories.)

- Have students read and complete the sentences to solve the problem. Guide students through the three-step algorithm, explaining each regrouping depicted.

Develop Skills and Concepts Write **29 + 16** on the board. Ask students to tell how many tens and ones there are in 29. (2, 9) Put 2 tens rods in the shoebox labeled "tens" and 9 ones cubes in the shoebox labeled "ones." Ask, *How many ones will there be when we add 6 ones to the ones shoebox?* (15) Dump the ones cubes out and regroup 10 ones cubes for 1 tens rod. Ask, *How many ones are left?* (5) Put the 5 ones cubes in the ones shoebox and 1 tens rod in the tens shoebox. Ask, *How many tens rods are there in all?* (4) Ask students to tell the sum of 29 and 16. (45) Repeat for more sums of two 2-digit numbers and use the hundreds shoebox similarly when sums are more than 99.

3 Practice

Have students complete page 38 independently.

Practice

Add.

1. 38 + 26 = 64	2. 49 + 8 = 57	3. 36 + 81 = 117	4. 9 + 86 = 95	5. 78 + 57 = 135
6. 76 + 43 = 119	7. 45 + 55 = 100	8. 26 + 9 = 35	9. 59 + 63 = 122	10. 71 + 65 = 136
11. 6 + 58 = 64	12. 67 + 96 = 163	13. 80 + 17 = 97	14. 46 + 74 = 120	15. 98 + 96 = 194

Copy and add.

16. 32 + 19 51
17. 75 + 48 123
18. 82 + 57 139
19. 25 + 96 121
20. 43 + 58 101
21. 22 + 67 89
22. 9 + 86 95
23. 82 + 53 135
24. 74 + 8 82
25. 82 + 48 130
26. 93 + 19 112
27. 57 + 70 127

Problem Solving

Solve each problem.

28. Robert poured a cup of skim milk in his bowl of oatmeal. The oatmeal contained 99 calories and the skim milk had 85. How many calories did Robert eat?
184 calories

29. Juanita ate a scrambled egg and a slice of toast. The egg contained 96 calories and the toast had 65. How many calories did Juanita eat?
161 calories

4 Assess

Ask students which column to add first when adding 2-digit numbers. (the ones column)

For Mixed Abilities

Common Errors • Intervention

Students may add incorrectly because they add each column separately, failing to regroup.

Incorrect	Correct
47 + 68 = 1015	(1) 47 + 68 = 115

Correct students by having them use place-value materials to model the problem.

Enrichment • Application

Tell students to find the number of cans of fruit punch you would need to buy to serve the 27 students in your class and 28 students across the hall if each student will have 3 ounces and 1 can of fruit punch contains 55 ounces. (3 cans)

More to Explore • Logic

Provide students with the following observations collected after a class field trip to the park. Ask them to use these facts to figure out what color the tanager was and which bird the class liked best.

The blue jay was blue and the goldfinch was yellow.

The hawk was not the red one.

The oriole and the brown bird were in the same tree as the blue jay and the red bird.

They all liked the orange bird best.

(red; oriole)

3-2 Adding 3-Digit Numbers

pages 39–40

1 Getting Started

Objective

- To add 3-digit numbers with regrouping

Materials

*4 clear plastic shoeboxes (one labeled "ones," another "tens," the next "hundreds," and the last "thousands")

Warm Up • Mental Math

Have students name the pattern.

1. 7, 10, 13, 16, 19 (+ 3)
2. 92, 88, 84, 80, 76 (− 4)
3. 4, 3, 5, 4, 6, 5, 7 (− 1, + 2)
4. 5, 10, 20, 40, 80 (× 2)
5. 36, 27, 18, 9 (− 9)
6. $\frac{1}{8}, \frac{2}{8}, \frac{3}{8}, \frac{4}{8}$ $(+ \frac{1}{8})$

Warm Up • Addition

Have students identify facts whose sums would require regrouping. (any fact with a sum of 10 or more) Have students name addition facts whose sums would not require regrouping. (any fact with a sum of 9 or less) Now, ask students to tell whether they would regroup if the fact 8 + 7 were in the tens column. (yes) Ask what regrouping would be made. (10 tens for 1 hundred) Continue for other facts.

Name ________________

Lesson 3-2

Adding 3-Digit Numbers

Jerry is the left tackle for Saratoga High School. Bill is the left guard. Jerry and Bill work well when they block together. What is their total weight?

We want to know the total weight of the two players.

Jerry weighs 235 pounds and Bill weighs 177 pounds.

To find their total weight, we add 235 and 177.

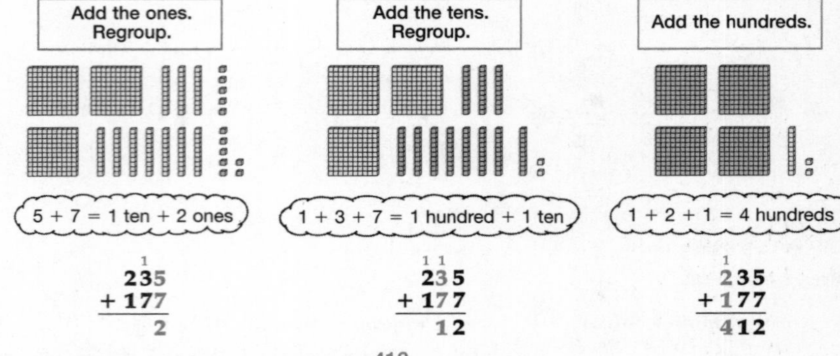

235 + 177 = 2	235 + 177 = 12	235 + 177 = 412

The total weight for both boys is 412 pounds.

Getting Started

Add.

1. 138 + 147 = 285
2. 384 + 135 = 519
3. 473 + 798 = 1,271
4. 157 + 60 = 217

Copy and add.

5. 683 + 294 = 977
6. 64 + 743 = 807
7. 811 + 496 = 1,307
8. 426 + 788 = 1,214

2 Teach

Introduce the Lesson Have students tell what the picture shows. (Jerry is number 17 and weighs 235 pounds; Bill is number 66 and weighs 177 pounds.) Have a student read the problem aloud and tell what we are to find. (the total weight of the boys)

- Discuss with students that in football there is a guard and a tackle on each side of the center on the offensive line. Have students complete the sentences and guide students through the three-step model problems as they find the boys' total weight.

Develop Skills and Concepts Write **164 + 749** vertically on the board. Talk through the problem as you add and make the regroupings. Write **606 + 587** vertically on the board and talk through the problem as you work it. Remind students that a regrouping is not needed as you add and record the tens. Ask students if regrouping is necessary in the hundreds column. (yes) Record the regrouping for 1 thousand and complete the problem.

- Remind students that we use a comma in numbers with 4 or more digits. Have a student place a comma in the answer and read the sum. (1,193; one thousand, one hundred ninety-three) Repeat for more problems of two 3-digit addends.

3 Practice

Remind students that they must regroup for 1 thousand if the sum of the hundreds column is 10 or more. Tell students that they may use addition or subtraction for the four word problems on page 40. Have students complete the page independently.

Practice

Add.

1. 256 + 129 = 385	2. 627 + 438 = 1,065	3. 581 + 276 = 857	4. 875 + 89 = 964	5. 67 + 483 = 550
6. 373 + 876 = 1,249	7. 541 + 212 = 753	8. 347 + 96 = 443	9. 709 + 584 = 1,293	10. 295 + 99 = 394
11. 560 + 348 = 908	12. 883 + 910 = 1,793	13. 396 + 264 = 660	14. 79 + 389 = 468	15. 543 + 377 = 920

Copy and add.

16. 386 + 409 795	17. 718 + 534 1,252	18. 475 + 221 696	19. 256 + 908 1,164
20. 415 + 344 759	21. 690 + 283 973	22. 596 + 19 615	23. 753 + 188 941
24. 93 + 815 908	25. 126 + 568 694	26. 836 + 684 1,520	27. 314 + 375 689

Problem Solving

Solve each problem.

28. The Amelia Earhart School library contains 472 fiction books. The librarian is ordering 255 more. How many fiction books will the library have?
727 fiction books

29. The boys at Clara Barton School collected 267 cans of food for the Thanksgiving food drive. The girls brought in 278 cans. Which group collected the greater number of cans?
the girls

30. Madison scored 225 points in her first game. She played three games and scored a total of 483 points. Her lowest score was an 85. What was her other score?
173

31. George Washington was born in 1732. He became the first president of the United States in 1789. He died 10 years later. How old was he when he died?
67 years old

4 Assess

Ask students how they know when to regroup ones, tens, and hundreds. (when there are 10 or more ones, tens, or hundreds)

For Mixed Abilities

Common Errors • Intervention

Students may regroup incorrectly because they reverse the tens and ones or the hundreds and tens.

Incorrect	Correct
$\begin{array}{r} ^{4} \\ 345 \\ +\ 819 \\ \hline 1191 \end{array}$	$\begin{array}{r} ^{1} \\ 345 \\ +\ 819 \\ \hline 1164 \end{array}$

Incorrect	Correct
$\begin{array}{r} ^{2} \\ 671 \\ +\ 353 \\ \hline 1114 \end{array}$	$\begin{array}{r} ^{1} \\ 671 \\ +\ 353 \\ \hline 1024 \end{array}$

Correct students by having them use place-value materials to model the problem and show the regrouping.

Enrichment • Addition

Tell students to write a problem with two addends and a sum of 1,876, where regrouping is required in the ones and hundreds columns but not in the tens column.

More to Explore • Applications

A weatherman needs to be able to figure temperature *range*. Ask students to keep track of the high and low temperatures for one full week. They should use the local newspaper for their reference. The teacher may record the temperatures on the board or a student may make a chart on cardboard. List the high and low temperatures in either Fahrenheit or Celsius degrees for all 7 days. Explain and demonstrate how to figure the range by subtracting the lowest number from the highest. Students should use the chart to find out the following:

What is the range of temperatures for each day?

What is the range of low temperatures for the week?

What is the range of high temperatures for the week?

Remind students that answers are expressed in degrees.

3-3 Adding 4-Digit Numbers

pages 41–42

1 Getting Started

Objective

- To add 4-digit numbers with regrouping

Warm Up • Mental Math

Tell students to name the following number:

1. $\frac{1}{4}$ inches in 1 foot (48)
2. area of a square if one side is 4 (16)
3. half of a 98-page book (49)
4. milliliters in 1 liter (1,000)
5. dimes in $4 (40)
6. ounces in $2\frac{1}{2}$ pounds (40)
7. pencils in $4\frac{1}{4}$ dozens (51)

Warm Up • Addition

Write **60 + 90** on the board. Tell students that the sum of these numbers requires us to regroup 1 ten for 1 hundred. Show the regrouping as you work the problem. Ask a student to write and solve a problem of two 3-digit addends whose sum would require the regrouping of 1 hundred and 1 thousand. (600 + 800 = 1,400) Continue to have students write and solve similar problems that do or do not require regrouping.

Name ______________________

Adding 4-Digit Numbers

The Missouri-Mississippi River system is the third longest river in the world. The Missouri starts at Three Forks, Montana, and joins the lower Mississippi River just north of St. Louis. It empties into the Gulf of Mexico near New Orleans. How long is the Missouri-Mississippi River system?

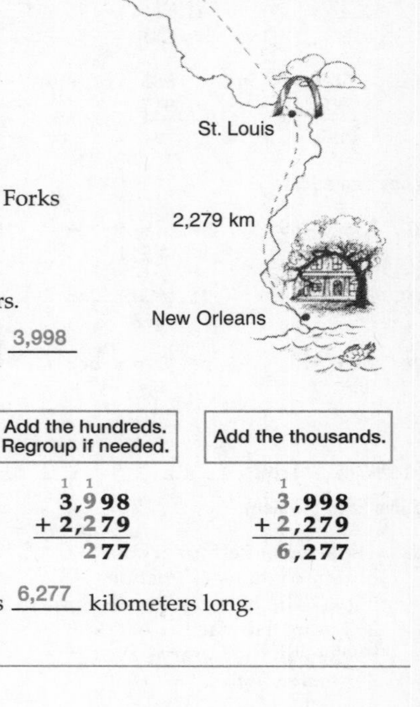

We want to know the total length of the Missouri-Mississippi River system.

We know the Missouri River from Three Forks to St. Louis measures <u>3,998</u> kilometers.

The Mississippi River from St. Louis to New Orleans measures <u>2,279</u> kilometers.

To find the system's total length, we add <u>3,998</u> and <u>2,279</u>.

Add the ones. Regroup if needed.	Add the tens. Regroup if needed.	Add the hundreds. Regroup if needed.	Add the thousands.
1 3,998 + 2,279 7	1 1 3,998 + 2,279 77	1 1 3,998 + 2,279 277	1 3,998 + 2,279 6,277

The Missouri-Mississippi River system is <u>6,277</u> kilometers long.

Getting Started

Add.

1. 7,836 + 1,193 = 9,029
2. 4,827 + 9,162 = 13,989
3. 6,256 + 3,498 = 9,754

Copy and add.

4. 4,275 + 907 — 5,182
5. 8,246 + 3,766 — 12,012
6. 659 + 5,941 — 6,600

2 Teach

Introduce the Lesson Discuss the map with students. Ask students what river is shown. (Missouri-Mississippi) Discuss the importance of this river to the early settlers of the United States. Use a map of the central states to show how a river and its tributaries contribute to the development of cities. A tributary is a stream that flows into a larger stream or other body of water.

- Have a student read the problem aloud and tell what is to be found. (the length of the Missouri-Mississippi River) Ask students what data are given on the map. (The Missouri River is 3,998 kilometers from Three Forks to St. Louis and the Mississippi is 2,279 kilometers from St. Louis to New Orleans.) Ask students how they can find the total length of the river system. (add 3,998 kilometers and 2,279 kilometers) Talk through the addition process and have students complete the solution sentence.

Develop Skills and Concepts Write **9,063 + 957** vertically on the board. Remind students that we begin with the ones and work to the left. Have a student record the work on the board as other students talk through each step. Remind students to place a comma between the thousands and hundreds in the sum. Repeat for other problems having addends of 5 or fewer digits.

- Write **3,725 + 4,328** on the board. Have students find the sum. (8,053) Then, ask them to check the reasonableness of their answer by rounding each addend and finding the sum. Students can determine that because 8,000 is close to the actual answer their answer makes sense.

3 Practice

Have students complete page 42 independently.

Practice

Add.

1. 4,263 + 3,514 = 7,777	2. 3,482 + 6,279 = 9,761	3. 6,257 + 5,767 = 12,024	4. 8,546 + 1,829 = 10,375
5. 4,395 + 7,826 = 12,221	6. 2,054 + 9,975 = 12,029	7. 8,370 + 7,851 = 16,221	8. 6,975 + 6,648 = 13,623

Copy and add.

9. 5,483 + 807 6,290	10. 4,563 + 8,789 13,352	11. 7,625 + 9,498 17,123
12. 3,748 + 6,943 10,691	13. 673 + 4,287 4,960	14. 8,496 + 7,827 16,323
15. 5,356 + 7,409 12,765	16. 4,358 + 97 4,455	17. 2,195 + 3,284 5,479

Problem Solving

Solve each problem.

18. The enrollment of Colgate University in 1 year included 1,423 men and 1,127 women. What was the total number of students?
2,550 students

19. It is 1,411 miles from Pittsburgh to Denver, and 1,174 miles from Denver to Los Angeles. How far is it from Pittsburgh to Los Angeles through Denver?
2,585 miles

Now Try This!

Fill in the blanks on this magic square so that the sum of each row down, across, and diagonally equals 65. You may use the numbers 1 through 25 only once.

17	24	1	8	15
23	5	7	14	16
4	6	13	20	22
10	12	19	21	3
11	18	25	2	9

Now Try This! Tell students to first determine which numbers from 1 to 25 are unused.

Ask students to find the sum of all the numbers in the completed square. Because they know the sum of each row, a shortcut would be to multiply, 5 × 65 = 325.

4 Assess

Ask students how they know when to regroup thousands.
(when there are 10 or more hundreds)

For Mixed Abilities

Common Errors • Intervention

Some students may be regrouping incorrectly because they are adding from left to right.

Incorrect	Correct
111	1 1
7,652	7,652
+ 9,463	+ 9,463
6,126	17,115

Have students use counters and a place-value chart to show the two addends. Then, have them join the counters in each place, from right to left, regrouping as they go along.

Enrichment • Numeration

Tell students that 464, 383, and 4,664 are palindromes. Tell them to write and solve an addition problem of three different palindromic addends.

More to Explore • Measurement

Tell students that at one time people used parts of their body to take measurements and that our system of measurement comes from these ancient people. Ask students to research and report on the origins of these words: *cubit, span, hand, inch, yard, fathom,* and *foot*. Ask them to explain whether they think these types of measurements were accurate.

ESL/ELL STRATEGIES

As you encounter them, make sure that students understand all the words that describe various aspects of rounding numbers. These include *rounding, exact, estimating, nearest,* and *approximately*. Model examples of rounding on the board, using these words to describe the process.

3-4 Estimating Sums by Rounding

pages 43–44

1 Getting Started

Objective

- To estimate sums of 2-, 3-, and 4-digit numbers by rounding

Vocabulary

overestimate, underestimate

Warm Up • Mental Math

Tell students to find the cost of one if items are sold

1. 10 for $1 (10¢)
2. 3 for 99¢ (33¢)
3. 5 for 75¢ (15¢)
4. 4 for $10 ($2.50)
5. 6 for $1.50 (25¢)
6. 2 for $1.98 (99¢)
7. 20 for $100 ($5)
8. 4 for $6 ($1.50)

Warm Up • Numeration

Dictate numbers of 9 or fewer digits. Have students write each number on the board and then round it to the nearest ten, hundred, and thousand.

Name ______________________________

Estimating Sums by Rounding

Sometimes we do not need an exact answer to a problem. We only need to estimate the answer. For example, approximately how many gallons of gasoline are stored in these two tanks?

We want an estimate of the total number of gallons in both tanks.

Tank 1 contains 1,934 gallons of gasoline.

Tank 2 contains 3,587 gallons.

To estimate the sum, we can round each addend to the nearest thousand and add the rounded numbers.

1,934 rounds to 2,000.

3,587 rounds to 4,000.

$$\begin{array}{r} 2{,}000 \\ +\,4{,}000 \\ \hline 6{,}000 \end{array}$$

There are about 6,000 gallons of gasoline stored in the two tanks.

Getting Started

Estimate the sum after rounding each addend to the nearest ten.

1.	56 + 87	60 + 90 = 150
2.	34 + 415	30 + 420 = 450
3.	163 + 589	160 + 590 = 750
4.	3,019 + 2,936	3,020 + 2,940 = 5,960

Estimate the sum after rounding each addend to the nearest hundred.

5.	368 + 526	400 + 500 = 900
6.	1,946 + 758	1,900 + 800 = 2,700
7.	968 + 412	1,000 + 400 = 1,400
8.	3,750 + 1,895	3,800 + 1,900 = 5,700

Estimate the sum after rounding each addend to the nearest thousand.

9.	5,260 + 1,680	5,000 + 2,000 = 7,000
10.	8,576 + 2,750	9,000 + 3,000 = 12,000
11.	6,749 + 7,284	7,000 + 7,000 = 14,000
12.	8,500 + 3,850	9,000 + 4,000 = 13,000

2 Teach

Introduce the Lesson Have a student describe the picture. (There are 1,934 gallons of gas in one tank and 3,587 gallons in the other.) Have a student read the problem aloud and tell what is to be found. (an estimate of the total gallons in both tanks) Ask students why an estimate is going to be used. (It is not necessary to have an exact answer.)

- Remind students about the rules for rounding. When rounding to the nearest thousand, if the hundreds digit is 0, 1, 2, 3, or 4, the thousands digit stays the same, and the hundreds, tens, and ones digits are replaced by zero. If the hundreds digit is 5, 6, 7, 8, or 9, the thousands digit is raised one, and the hundreds, tens, and ones digits are replaced by zero. Ask students which place they should round to. (thousands) Have students read aloud with you as they complete the sentences to solve the problem.

Develop Skills and Concepts Tell students that we use rounding to estimate an answer in certain situations. Give students an example of 6,858 people attending a rock concert. Tell students that we can estimate this attendance to be 7,000. If we wanted to know the total number of fans at that concert and another concert, we would round the number of fans at the second concert to the nearest thousand also and then add the two estimates.

- Write **4,950** and **6,201** on the board. Ask students to round the two numbers to the nearest thousand and add the numbers to find an estimate of the number of fans. (5,000 + 6,000 = 11,000) Now, have students round the numbers to the nearest hundred and add them. (5,000 + 6,200 = 11,200) Repeat for rounding to the nearest ten. (4,950 + 6,200 = 11,150)

3 Practice

Tell students that they are to work each problem in the first three rows by rounding each addend and then finding the sum of the rounded numbers. Tell students to check their work in the Copy and Add exercises by estimating. Have students complete page 44 independently.

Practice

Estimate the sum after rounding each addend to the nearest ten.

1. 23 + 58 → 20 + 60 = 80
2. 275 + 88 → 280 + 90 = 370
3. 546 + 682 → 550 + 680 = 1,230
4. 7,367 + 4,431 → 7,370 + 4,430 = 11,800

Estimate the sum after rounding each addend to the nearest hundred.

5. 485 + 836 → 500 + 800 = 1,300
6. 1,941 + 287 → 1,900 + 300 = 2,200
7. 1,755 + 4,628 → 1,800 + 4,600 = 6,400
8. 2,808 + 1,576 → 2,800 + 1,600 = 4,400

Estimate the sum after rounding each addend to the nearest thousand.

9. 3,965 + 2,646 → 4,000 + 3,000 = 7,000
10. 8,397 + 6,845 → 8,000 + 7,000 = 15,000
11. 5,545 + 7,851 → 6,000 + 8,000 = 14,000

Copy and add.

12. 73 + 86 — 159
13. 93 + 18 — 111
14. 27 + 76 — 103
15. 41 + 88 — 129
16. 375 + 916 — 1,291
17. 876 + 652 — 1,528
18. 248 + 796 — 1,044
19. 850 + 525 — 1,375
20. 5,964 + 8,572 — 14,536
21. 8,863 + 1,650 — 10,513
22. 9,452 + 3,925 — 13,377
23. 229 + 4,076 — 4,305

Problem Solving

Solve each problem.

24. One storage tank can hold 4,246 gallons of gasoline. Another tank can hold 3,575 gallons. About how many gallons of gasoline can both tanks hold?
about 8,000 gallons

25. It is 1,717 miles from Albany, New York, to Dallas, Texas. Dallas is another 2,151 miles from Seattle, Washington. About how far is it from Albany to Seattle if you drive through Dallas?
about 4,000 miles

4 Assess

Ask students to explain how to round numbers to the nearest thousand. (When rounding to the nearest thousand, if the hundreds digit is 0, 1, 2, 3, or 4, the thousands digit stays the same, and the hundreds, tens, and ones digits are replaced by zero. If the hundreds digit is 5, 6, 7, 8, or 9, the thousands digit is raised one, and the hundreds, tens, and ones digits are replaced by zero.)

5 Mixed Review

1. 492 + 127 (619)
2. 13 − 8 (5)
3. 437 + 2,725 (3,162)
4. 7 + 4 + 8 + 7 + 2 (28)
5. 11 − 5 (6)
6. 6,208 + 1,395 (7,603)
7. 89 + 2,176 (2,265)
8. 18 − 9 (9)
9. 4,382 + 7,438 (11,820)
10. (8 + 6) + 3 (17)

For Mixed Abilities

Common Errors • Intervention

Sometimes students will find the exact sum first, which they then round to give their estimate. Have students discuss instances where estimates are used as the only answer required, encouraging them to see that once numbers are rounded, it is much easier to use mental computation to find the answer—the estimate—than to add the original addends.

Enrichment • Underestimate/Overestimate

Write **2,895 + 4,505** vertically on the board. Have a volunteer demonstrate how to estimate the answer using rounding. (8,000) Ask a student to find the actual answer. (7,400) Have students compare the estimate to the actual answer. Explain that the estimate is greater than the actual sum because both numbers were replaced with greater numbers. Tell students that this is called an *overestimate*.

Write **6,439 + 2,143** vertically on the board. Ask one volunteer to estimate the sum and another to find the actual sum. (8,000; 8,582) Explain to students that we call this estimate an *underestimate* because it is less than the actual sum. Ask students why it is an underestimate. (because both numbers were replaced with lesser numbers)

More to Explore • Geometry

Show students a pair of gloves. Explain that many things have a handedness just like the gloves. Have one student identify the right-handed glove and another the left-handed one. Tell students to find several objects that have a handedness. Explain that this property is called bilateral symmetry. Examples might include shoes, boots, mittens, and so on. Many toys and dolls are bilaterally symmetrical. The human body has a right-hand and left-hand side. Many houses are symmetrical about a line down the middle. Have students explain why bilateral symmetry is sometimes called mirror symmetry.

3-5 Front-End Estimation

pages 45–46

1 Getting Started

Objective

- To estimate sums using front-end estimation

Vocabulary

front-end digits

Warm Up • Mental Math

Have students find the sum.

1. 300 + 400 (700)
2. 400 + 700 (1,100)
3. 600 + 500 (1,100)
4. 8,000 + 7,000 (15,000)
5. 3,000 + 6,000 (9,000)
6. 40,000 + 50,000 (90,000)

Warm Up • Numeration

Ask students to round

1. 285 to the nearest ten (290)
2. 642 to the nearest ten (640)
3. 3,853 to the nearest hundred (3,900)
4. 4,214 to the nearest hundred (4,200)
5. 8,538 to the nearest thousand (9,000)
6. 5,385 to the nearest thousand (5,000)

Name ____________________

Front-End Estimation

Derek is flying from New York to Rome, Italy, and then from Rome to Tokyo, Japan. Estimate the number of miles Derek will fly to get from New York to Tokyo.

Derek's Flights	
Flight	**Distance in Miles**
New York to Rome	4,298
Rome to Tokyo	6,120

We want to **estimate** the number of miles Derek will fly from New York to Tokyo.

To estimate the distance using **front-end digits**, add the front digits and write zeros for the other digits.

$$\begin{array}{r} 4{,}298 \\ +\ 6{,}120 \\ \hline 10{,}000 \end{array} \qquad \begin{array}{r} 4{,}000 \\ +\ 6{,}000 \\ \hline 10{,}000 \end{array}$$

Derek will fly about 10,000 miles.

Getting Started

Estimate each sum using front-end estimation.

1. 825 + 648 → 800 + 600 = 1,400
2. 2,385 + 5,642 → 2,000 + 5,000 = 7,000
3. 3,097 + 9,253 → 3,000 + 9,000 = 12,000
4. 681 + 538 → 600 + 500 = 1,100
5. 5,325 + 3,789 → 5,000 + 3,000 = 8,000
6. 7,784 + 4,138 → 7,000 + 4,000 = 11,000
7. 735 + 783 → 700 + 700 = 1,400
8. 1,487 + 6,844 → 1,000 + 6,000 = 7,000
9. 3,072 + 4,193 → 3,000 + 4,000 = 7,000
10. 4,990,125 + 9,570,312 → 4,000,000 + 9,000,000 = 13,000,000
11. 7,899,621 + 3,697,368 → 7,000,000 + 3,000,000 = 10,000,000
12. 9,050,793 + 2,986,849 → 9,000,000 + 2,000,000 = 11,000,000

---

2 Teach

Introduce the Lesson Have students study the problem and the chart and tell what is being asked. (to estimate how many miles Derek will travel from New York to Tokyo) Ask what information is needed. (the number of miles for each part of the trip)

- Read the problem to the class and have a volunteer state the required operation. (addition) Explain that the problem requires using a different type of estimation called **front-end** estimation. Read the instructions and do the problem with the class. Tell students to fill in the solution sentence. (10,000)

Develop Skills and Concepts Explain that estimating sums is useful for getting a general idea of the sum. Write the following examples on the board:

$$\begin{array}{r} 3{,}958 \\ +\ 1{,}531 \\ \hline (4{,}000) \end{array} \qquad \begin{array}{r} 3{,}958 \\ +\ 1{,}531 \\ \hline (5{,}489) \end{array} \qquad \begin{array}{r} 2{,}145 \\ +\ 8{,}045 \\ \hline (10{,}000) \end{array} \qquad \begin{array}{r} 2{,}145 \\ +\ 8{,}045 \\ \hline (10{,}190) \end{array}$$

- Have one student find the estimated sum using front-end digits and have another find the actual sum. Explain that when using front-end estimation the estimate will always be less than the actual answer.

3 Practice

Have students complete all the exercises on page 46. Remind them that they are only to use front-end estimation.

Practice

Estimate each sum using front-end estimation.

	Problem	Estimate		Problem	Estimate		Problem	Estimate
1.	465 + 712	400 + 700 = 1,100	2.	925 + 637	900 + 600 = 1,500	3.	811 + 532	800 + 500 = 1,300
4.	3,289 + 6,038	3,000 + 6,000 = 9,000	5.	7,125 + 7,822	7,000 + 7,000 = 14,000	6.	9,284 + 2,354	9,000 + 2,000 = 11,000
7.	6,628 + 1,277	6,000 + 1,000 = 7,000	8.	3,500 + 8,015	3,000 + 8,000 = 11,000	9.	5,721 + 6,127	5,000 + 6,000 = 11,000
10.	4,329 + 3,825	4,000 + 3,000 = 7,000	11.	2,111 + 6,352	2,000 + 6,000 = 8,000	12.	9,130 + 8,415	9,000 + 8,000 = 17,000
13.	5,005,000 + 8,341,000	5,000,000 + 8,000,000 = 13,000,000	14.	7,945,000 + 9,253,000	7,000,000 + 9,000,000 = 16,000,000	15.	1,753,000 + 4,236,000	1,000,000 + 4,000,000 = 5,000,000

Problem Solving

Solve each problem. Use front-end estimation.

16. There are 8,785 people living in Dallas County, Arkansas. Monroe County, Arkansas, has a population of 9,689. About how many people altogether live in the two counties?
about 17,000 people

17. A truck driver drove from Chicago, Illinois, to Los Angeles, California, and then to Washington, D.C. It is 2,054 miles from Chicago to Los Angeles. It is 2,631 miles from Los Angeles to Washington, D.C. About how far did the truck driver drive?
about 4,000 miles

4 Assess

Ask students to explain how to estimate sums using front-end estimation. (add the front digits and write zeros for the other digits)

5 Mixed Review

1. 35 + 42 (77)
2. 29 + 63 (92)
3. 77 + 54 (131)
4. 275 + 432 (707)
5. 629 + 198 (827)
6. 458 + 344 (802)
7. 563 + 752 (1,315)
8. 3,645 + 4,782 (8,427)
9. 2,905 + 3,591 (6,496)
10. 6,392 + 857 (7,249)

For Mixed Abilities

Common Errors • Intervention

Some students may add just the digit in the greatest place. Remind students that in front-end estimation, they add the front digits and write zeros for the other digits. For those students having difficulty, have them make a place-value chart when using front-end estimation.

More to Explore • Measurement

Have students use a large measuring instrument such as a tape measure or yardstick to measure large items in the classroom such as the chalkboard, the door, the floor, etc. Have students measure using feet and inches and then explain how they could use front-end estimation to approximate the lengths of each of their measures.

Enrichment • Number Sense

Explain to students that because the sum using front-end estimation will always be less than the actual sum, they can use front-end estimation with adjustment. Students can adjust the sums by adding the digits in the second greatest place and then use rounding to adjust. For example, the sum of 375 + 428 using front-end estimation is 700. However, by adding the tens, the front-end estimation with adjustment is 800, which is much closer to the actual sum of 803. Have students use front-end estimation with adjustment to resolve the problems on page 45.

ESL/ELL STRATEGIES

Ask a volunteer to explain the meaning of *digits* and *front end*. *Front end* means "at the beginning" and a digit is a single numeral that is part of a number. A front-end digit is the first numeral in a number.

3-6 Adding 5-Digit Numbers

pages 47-48

1 Getting Started

Objective

- To add 5-digit numbers with regrouping

Warm Up • Mental Math

Tell students to complete each comparison. 10 is to 100 as

1. 1 is to ____ (10)
2. 100 is to ____ (1,000)
3. ____ is to 200 (20)
4. 4,000 is to ____ (40,000)
5. ____ is to 10 (1)
6. 60,000 is to ____ (600,000)
7. ____ is to 1,000,000 (100,000)
8. 50 is to ____ (500)

Warm Up • Numeration

Write several 4- to 9-digit numbers on the board. Have students tell the place value of each digit and then read each number. Have students arrange the numbers in order from least to greatest.

Name ____________________

Adding 5-Digit Numbers

The Dallas Cowboys and the Washington Redskins play each other twice each year. What was the total attendance for this year's games?

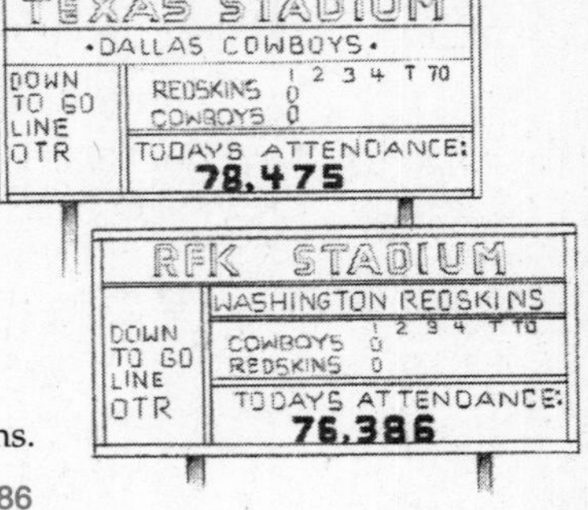

We want to know the total number of fans who attended both games.

We know that 78,475 fans attended the game in the Texas Stadium.

The game in RFK Stadium drew 76,386 fans.

To find the sum, we add 78,475 and 76,386.

> To add two 5-digit numbers, begin by adding the ones. Then, add each column to the left. Regroup if needed.

78,475	78,475	78,475	78,475	78,475
+ 76,386	+ 76,386	+ 76,386	+ 76,386	+ 76,386
1	61	861	4,861	154,861

The total attendance for both games was 154,861.

Getting Started

Add.

1. 17,836 + 1,295 = 19,131
2. 4,827 + 19,254 = 24,081
3. 16,875 + 58,596 = 75,471
4. 39,454 + 12,926 = 52,380
5. 49,025 + 38,484 = 87,509
6. 22,456 + 67,381 = 89,837

Copy and add.

7. 16,214 + 8,295 = 24,509
8. 2,897 + 56,850 = 59,747
9. 37,488 + 55,746 = 93,234

2 Teach

Introduce the Lesson Have students tell about the picture. (78,475 fans were at Texas Stadium and 76,386 fans were at RFK Stadium.) Discuss events that often draw large crowds. Tell students that soccer is the most watched sport in the world and often draws more than 100,000 fans a game in other countries.

- Have a student read the problem and tell what is to be found. (total attendance at both games of the Dallas Cowboys vs. the Washington Redskins) Ask students if they are given the necessary data. (yes)
- Have students complete the sentences. Guide students through each step of the addition of the 5-digit numbers to solve the problem.

Develop Skills and Concepts Write **2,061 + 7,853** vertically on the board. Have a student talk through the problem as another student records the work. Remind students to work the ones column first and to work to the left.

- Now, write **42,061 + 47,853** on the board and tell students that we have added 40,000 to each of the numbers. Again, have a student talk through the problem as another student records the work. Provide more problems of 5-digit addends for students to talk through and solve.

3 Practice

Have students work the problems on page 48 independently.

Practice

Add.

1. 16,354 + 2,865 = 19,219	2. 4,967 + 15,096 = 20,063	3. 17,858 + 8,392 = 26,250	4. 11,456 + 56,729 = 68,185
5. 96,254 + 39,687 = 135,941	6. 9,560 + 83,785 = 93,345	7. 26,586 + 76,253 = 102,839	8. 18,609 + 25,996 = 44,605
9. 35,694 + 47,828 = 83,522	10. 83,467 + 25,704 = 109,171	11. 75,496 + 86,894 = 162,390	12. 48,182 + 69,778 = 117,960

Copy and add.

13. 56,965 + 38,758
95,723

14. 41,853 + 26,348
68,201

15. 77,486 + 67,398
144,884

16. 47,836 + 9,548
57,384

17. 64,495 + 40,867
105,362

18. 87,187 + 56,838
144,025

19. 14,390 + 78,956
93,346

20. 60,086 + 88,956
149,042

21. 47,539 + 26,211
73,750

Now Try This!

It's Algebra!

Write each missing number.

1. [4]6,50[7] + 1[2],9[2]4 = 59,[4]31

2. [2]3,[8]74 + 25,42[9] = 4[9],3[0]3

3. 38,[0]5[3] + [2]6,1[9]7 = 6[4],250

4. [1]7,0[4]3 + 5[4],218 = 71,[2]6[1]

Now Try This! Encourage students to check their work by subtracting the bottom addend from the sum.

4 Assess

Ask students to explain how to add 56,908 + 49,395. (Students should follow the step-by-step plan in their texts.)

For Mixed Abilities

Common Errors • Intervention

Students may regroup incorrectly by dropping the regrouped value.

Incorrect	Correct
	1 11
24,867	24,867
+ 31,985	+ 31,985
55,742	56,852

Have students put their problems on a place-value chart and add, showing their regrouped numbers each time.

More to Explore • Applications

Take a trip to a local hardware store. Have students make a list of items in the store that are sold by length, weight, or count. Arrange for the store manager to show students how to use the scales and other measuring tools that employees would have to use. Have students notice how many of these tools use the metric system. Ask whether some items are counted by weighing them. Have students list ways in which knowledge of measurement systems is used by someone who works in a hardware store.

Enrichment • Geometry

Tell students to draw a triangle, label its sides with 5-digit numbers, and then find the perimeter.

3-7 Column Addition

pages 49–50

1 Getting Started

Objective

- To add multidigit numbers in a column

Materials

*deck of cards

Warm Up • Mental Math

Ask students to give the answer for each of the following:

1. Change 4 cups to pints. (2)
2. Add 200 to 6 × 4. (224)
3. 9 days after October 24 (November 2)
4. 35 years before 1984 (1949)
5. 4° below freezing F (28°)
6. 2 hours after 2:40 (4:40)
7. 3 decades (30 years)
8. shape of a ball (sphere)

Warm Up • Addition

Have a student write **2 + 6 + 8 + 1** vertically on the board and find the sum. (17) Have another student write **7¢ + 6¢ + 2¢** on the board and find the sum. (15¢) Repeat for other column addition problems of 1-digit numbers with and without cent signs.

Name ____________________

Column Addition

The Metro All-Stars celebrated their tournament win by going out for pizza. They ordered one medium deep-dish pizza, one large supreme pizza, and one medium vegetable pizza. How much was their bill?

Pizza			
	Deep-Dish	Supreme	Vegetable
Large	$12.95	$11.59	$10.29
Medium	$9.95	$8.59	$7.29
Small	$6.95	$5.59	$4.29

We want to know the total bill.

We know that a medium deep-dish pizza costs $9.95.

A large supreme pizza costs $11.59.

A medium vegetable pizza costs $7.29.

To find the sum of all the prices, we add $9.95, $11.59, and $7.29.

REMEMBER When adding numbers in a column, add two digits at a time and then add that sum to the next digit.

REMEMBER When adding amounts of money, line up the decimal points. Include the dollar sign and decimal point in the sum.

```
   2                         1 2        1 1        1
 $ 9.95 | 5 + 9 = 14      $ 9.95     $ 9.95     $ 9.95
  11.59 |                  11.59      11.59      11.59
+  7.29   14 + 9 = 23    +  7.29   +  7.29   +  7.29
      3                      .83       8.83     $28.83
```

The total pizza bill was $28.83.

Getting Started

Add.

1. 43 + 96 + 83 = 222
2. $37.46 + 158.49 + 65.53 = $261.48

Copy and add.

3. 756 + 4,096 + 8,749 13,601
4. $52.46 + $96.75 + $43.38 $192.59

2 Teach

Introduce the Lesson Have a student read the problem and tell what is to be solved. (total pizza bill) Ask students which pizzas were ordered. (one medium deep-dish, one large supreme, and one medium vegetable) Ask students how we find the cost of each pizza. (across the Medium row and down the Deep-Dish column to $9.95; across the Large row and down the Supreme column to $11.59; across the Medium row and down the Vegetable column to $7.29)

- Give added practice using the table by asking the cost of other sizes and types of pizza. Have students read and complete the sentences. Guide students through the column addition steps in the model to find the total pizza bill.

Develop Skills and Concepts Write **64,356 + 2,491 + 984** vertically on the board. Remind students that when adding 6, 1, and 4 we find the sum of two numbers and then add the third number to that sum. Talk through the problem as you call attention to the regroupings. (67,831)

- Now, write **$643.56 + 24.91 + 9.84** vertically on the board and show students how to line up the decimal points and then add. ($678.31) Remind students that we can check the sum by adding up the column. Have students check the problem with you. Repeat for more problems of 3 or 4 addends of 3 to 5 digits each.

3 Practice

Tell students that they are to use the menu on page 49 to complete Exercises 23 and 24. Remind students to check their work by adding up the columns. Have students complete page 50 independently.

Practice

Add.

1. 86 + 94 + 27 = 207
2. 215 + 48 + 135 = 398
3. $2.47 + 6.29 + 5.15 = $13.91
4. $8.37 + 0.96 + 4.15 = $13.48
5. 168 + 457 + 816 = 1,441
6. $4.86 + 2.95 + 6.18 = $13.99
7. 738 + 915 + 2,316 = 3,969
8. $25.10 + 6.15 + 8.85 = $40.10
9. 1,643 + 851 + 1,175 = 3,669
10. $126.21 + 27.39 + 436.51 = $590.11
11. $ 16.43 + 537.82 + 4.57 + 89.75 = $648.57
12. 37,265 + 15,586 + 39,509 + 4,228 = 96,588

Copy and add.

13. 48 + 87
135
14. 64 + 168 + 452
684
15. 869 + 47 + 1,750
2,666
16. 984 + 5,465 + 3,750
10,199
17. $41.56 + $1.97 + $562.27
$605.80
18. $39.75 + $18.56 + $27.38
$85.69
19. $50.37 + $392.47 + $88.50
$531.34
20. $65.46 + $39.75 + $132.59 + $3.40
$241.20
21. 13,750 + 27,967 + 37,875
79,592
22. 27,296 + 8,275 + 11,750 + 2,068
49,389

Problem Solving

Use the pizza order board on page 49 to find the total cost of each order.

23. 1 large vegetable
1 medium vegetable
1 small vegetable
$21.87

24. 1 large deep-dish
1 large supreme
1 medium supreme
1 small vegetable
$37.42

4 Assess

Ask students to explain how to add four numbers in a column. (add two digits at a time, then add that sum to the next digit)

5 Mixed Review

1. 15 − 9 (6)
2. $127.31 + $16.29 + $325.40 ($469.00)
3. 3,542 + 6,497 (10,039)
4. (9 + 7) + 8 (24)
5. 9 − 5 (4)
6. 65,278 + 2,468 + 12,176 (79,922)
7. 17 + 21 + 35 + 47 (120)
8. 11 − 7 (4)
9. 137 + 65 + 48 (250)
10. 2,095 + 4,076 (6,171)

For Mixed Abilities

Common Errors • Intervention

Some students may get incorrect sums in column addition because they align the digits incorrectly under one another. Have these students write the problems on grid paper, using the vertical lines to keep the digits placed in their proper columns.

More to Explore • Applications

Ask students to draw a picture that depicts their favorite book and share it with the class. Conduct a class discussion on how the books might be classified into groups or sets. Have students make suggestions for classification headings and list them on the board. Then, have students write the titles of their favorite books under the appropriate heading. Tell students to use this information to make a pictograph showing the reading interests of the class. Display the graphs. Discuss any differences that occurred and why each graph should look similar.

Enrichment • Application

Tell students that they own a pizza parlor. Have them make a table with prices of three sizes of their favorite three pizzas. Tell them to write a bill showing the total cost if a customer ordered one of each pizza they sell.

3-8 Problem Solving: Use an Exact Answer or an Estimate

pages 51–52

1 Getting Started

Objective
- To solve problems that require an exact answer or an estimate

Vocabulary
exact answer, estimate

Warm Up • Mental Math
Dictate the following exercises. Tell students to estimate the sum using front-end estimation.

1. 489 + 625 (1,000)
2. 729 + 636 (1,300)
3. 849 + 306 (1,100)
4. 4,382 + 8,872 (12,000)
5. 3,138 + 4,320 (7,000)
6. 6,711 + 8,428 (14,000)

Warm Up • Pencil and Paper
Write the following exercises on the board, asking students to round each number to the nearest hundred or dollar and compute the sum.

1. 458 + 625 (1,100)
2. 847 + 428 (1,200)
3. 329 + 384 (700)
4. $9.85 + $5.29 ($15)
5. $2.35 + $5.83 ($8)
6. $4.53 + $1.89 ($7)

Name ________________________

Problem Solving
Lesson 3-8

Problem Solving: Use an Exact Answer or an Estimate

The students in Mr. Roberts's class are selling boxes of fruit to raise money for a class trip. The students want to find out about how many boxes of fruit they sold so far?

Boxes of Fruit Sold by Mr. Roberts's Class	
Boys	375
Girls	398

★SEE

The boys in Mr. Roberts's class sold 375 boxes of fruit.

The girls in Mr. Roberts's class sold 398 boxes of fruit.

★PLAN

The boys and girls in Mr. Roberts's class want to know about how many boxes of fruit they sold so far.

We do not need an **exact answer**. We need to make an **estimate** of the total number of boxes of fruit sold.

★DO

To estimate the total number of boxes of fruit sold by Mr. Roberts's class, we can round each number to the nearest hundred and add.

375 rounds to → 400
398 rounds to → + 400
800

The boys and girls in Mr. Roberts's class sold about 800 boxes of fruit.

★CHECK

Find the actual number of boxes of fruit sold.

375
+ 398
773

2 Teach

Introduce the Lesson Point out the four-step problem-solving strategy and have students describe how to use each step.

Develop Skills and Concepts Ask students to read the problem aloud and tell what they are to find. (about how many boxes of fruit Mr. Roberts's class has sold)

- Read each sentence under the SEE step, asking for volunteers to give the missing information. Ask another volunteer to read the PLAN sentences.
- To help students decide whether to use an estimate or an exact number, tell them that often a problem will use the word *about* when only an estimate is required. Write the words *Estimate* and *Exact Answer* on the board. List a few situations in which each might be used. (Estimate: whether you have saved enough money to purchase something; Exact: actually paying for something, giving change, a score, and so on)
- Work through the DO step, telling students that for this lesson they will be using rounding to estimate. Finally, work through the CHECK step with students. Make sure students understand how and when an estimate may be used instead of an exact answer to solve a problem.

3 Apply

Solution Notes

1. Ask students to explain why it is appropriate to give an estimate for this answer. (We only need to know about how many caps are needed.)
2. Ask students why an exact answer might be needed to solve this problem. (The principal wants everyone to have a T-shirt without having to buy extra.)

Apply

Use the table below to solve each problem. Give an estimate. Then, find the actual amount.

Students at Washington Elementary	
Grade	**Students**
First	258
Second	341
Third	272
Fourth	318
Fifth	384
Sixth	329

1. Each second- and third-grade student at Washington Elementary School will be receiving a baseball cap for field day. About how many caps are needed?
about 600; there are actually 613 baseball caps needed

2. The principal of Washington Elementary wants to buy a T-shirt for each fourth- and fifth-grade student. About how many T-shirts does she need to buy?
about 700; she needs to buy 702 T-shirts

3. Tyra thinks that there are approximately 600 third- and fourth-grade students altogether at Washington. Is she correct?
Yes, there are actually 590 third- and fourth-graders altogether.

4. About how many students are there in the first and third grades?
about 600 students; there are actually 530 students

Problem-Solving Strategy: Using the Four-Step Plan

★ **SEE** What do you need to find?
★ **PLAN** What do you need to do?
★ **DO** Follow the plan.
★ **CHECK** Does your answer make sense?

5. There are 234 fourth graders at Jefferson. The fourth-grade students at Jefferson and Washington are going to go on a class trip together. If every fourth-grade student from both schools goes on the trip, about how many students will be going?
about 500; actually 552 students will be going

6. Kaitlyn thinks that there are about 1,000 students in the fourth, fifth, and sixth grades at Washington Elementary. Is she correct?
Yes, there are actually 1,031 students.

7. Jefferson Elementary has 525 students in the first and second grades. About how many students does Washington Elementary have? Which school has more students in the first and second grades?
about 600 students; Washington Elementary; there are actually 599 students

For Mixed Abilities

More to Explore• Application

Have students look through the *World Almanac* to find situations in which exact numbers and estimates are used to describe those situations. They might also record situations they find in newspapers and magazines. Encourage students to present their findings to the class for display and discussion.

3. Make sure students know which numbers they need to use. Remind students that they must find the actual amount to check their answer.
4. Have students read the problem carefully. Ask them what the word *about* means? (close, near, approximate)

Higher-Order Thinking Skills

5. **Comprehension:** Some students may not see that an estimate is needed. Help these students understand that the word *about* indicates estimation in the question.
6. **Analysis:** Make sure students select the correct numbers from the table. Remind students that they must find the actual amounts to check their answer.
7. **Synthesis:** The exact answer for the number of students was found to be 599. Challenge students to figure out how they could have come close to the exact answer by estimating before adding the actual numbers.

4 Assess

Ask students what words are used when an estimate is required. (about, almost, approximately)

Chapter 3 Test

page 53

Items	Objectives
1–4	To add two 2-digit numbers (see pages 37–38)
5–8	To add two 3-digit numbers (see pages 39–40)
9–16	To add two 4-digit numbers (see pages 41–42)
17–24	To estimate sums up to 4 digits (see pages 43–44)
25–28	To add more than two numbers (see pages 49–50)

Name ______________________

CHAPTER 3 TEST

Add.

1. 63 + 31 = 94	2. 52 + 26 = 78	3. 84 + 29 = 113	4. 75 + 86 = 161
5. 225 + 134 = 359	6. 604 + 138 = 742	7. 541 + 183 = 724	8. 958 + 476 = 1,434
9. 2,426 + 1,273 = 3,699	10. 4,625 + 2,419 = 7,044	11. 7,219 + 5,825 = 13,044	12. 4,926 + 8,286 = 13,212
13. 4,952 + 6,098 = 11,050	14. 7,846 + 9,752 = 17,598	15. 9,476 + 3,697 = 13,173	16. 6,543 + 9,758 = 16,301

Estimate the sum after rounding to the nearest hundred or dollar.

17. 459 + 675 → 500 + 700 = 1,200	18. 352 + 687 → 400 + 700 = 1,100	19. 5,290 + 7,500 → 5,300 + 7,500 = 12,800	20. $18.96 + 37.25 → $19 + 37 = $56

Estimate the sum after rounding to the nearest thousand.

21. 5,296 + 3,475 → 5,000 + 3,000 = 8,000	22. 1,256 + 7,185 → 1,000 + 7,000 = 8,000	23. 6,295 + 8,921 → 6,000 + 9,000 = 15,000	24. 8,377 + 2,934 → 8,000 + 3,000 = 11,000

Add.

25. 892 + 437 + 186 = 1,515	26. 7,275 + 4,687 + 2,651 = 14,613	27. 11,751 + 2,627 + 37,583 = 51,961	28. $29.47 + 7.55 + 58.43 = $95.45

Alternate Chapter Test

You may wish to use the Alternate Chapter Test on page 350 of this book for further review and assessment.

Cumulative Assessment

Circle the letter of the correct answer.

1. 5 + 9 — a. 13 (b.) 14 c. 15 d. NG
2. 4 + 8 — (a.) 12 b. 13 c. 14 d. NG
3. 4 + 3 + 8 — a. 7 b. 11 (c.) 15 d. NG
4. 16 − 9 — a. 8 (b.) 7 c. 6 d. NG
5. 11 − 8 — a. 4 b. 5 c. 6 (d.) NG
6. $7 + n = 9$, $n =$ ■ — (a.) 2 b. 5 c. 16 d. NG
7. What is the place value of the 3 in 2,436? — (a.) tens b. hundreds c. thousands d. NG
8. 2,358 ◯ 2,538 — (a.) < b. >
9. Round 750 to the nearest hundred. — a. 600 b. 700 (c.) NG
10. Round 5,285 to the nearest thousand. — (a.) 5,000 b. 6,000 c. NG
11. What is the place value of the 5 in 357,286? — (a.) ten thousands b. thousands c. hundreds d. NG
12. 54 + 87 — a. 131 (b.) 141 c. 1,311 d. NG
13. 257 + 189 — a. 336 b. 346 (c.) 446 d. NG
14. 3,279 + 5,726 — (a.) 9,005 b. 9,015 c. 9,105 d. NG

STOP

score

Cumulative Assessment

page 54

Items	Objectives
1–2, 4–5	To compute basic addition and subtraction facts (see pages 1–2)
3	To add three 1-digit numbers (see pages 3–4)
6	To find missing addends (see pages 9–10)
7, 11	To identify the place value of digits in numbers less than 1,000,000 (see pages 25–26)
8	To compare and order numbers less than 10,000 (see pages 29–30)
9–10	To round numbers to the nearest hundred or thousand (see pages 31–32)
12	To add two 2-digit numbers (see pages 37–38)
13	To add two 3-digit numbers (see pages 39–40)
14	To add two 4-digit numbers (see pages 41–42)

Alternate Cumulative Assessment

Circle the letter of the correct answer.

1. 6 + 8 = — a 16 b 15 (c) 14 d NG
2. 7 + 9 — (a) 16 b 17 c 18 d NG
3. 5 + 6 + 3 — a 9 b 11 (c) 14 d NG
4. 18 − 9 = — a 10 (b) 9 c 8 d NG
5. 14 − 8 — a 9 b 8 c 7 (d) NG
6. $9 + n = 15$, $n = ?$ — a 24 (b) 6 c 9 d NG
7. What is the place value of the 7 in 6,723? — a tens (b) hundreds c thousands d NG
8. 4,697 ◯ 4,679 — (a) > b < c =
9. Round 345 to the nearest hundred. — a 200 (b) 300 c 400 d NG
10. Round 6,586 to the nearest thousand. — a 6,000 b 6,500 (c) 7,000 d NG
11. What is the place value of the 2 in 267,954? — a ten thousands (b) hundred thousands c millions d NG
12. 67 + 35 — a 912 (b) 102 c 92 d NG
13. 366 + 257 — a 513 b 613 (c) 623 d NG
14. 4,368 + 4,635 — a 8,993 (b) 9,003 c 9,113 d NG

4-1 Subtracting 2-Digit Numbers

pages 55-56

1 Getting Started

Objective
- To subtract 2-digit numbers with regrouping

Materials
*3 clear plastic shoeboxes (labeled ones, tens, and hundreds); addition and subtraction fact cards; place-value blocks (ones, tens, and hundreds)

Warm Up • Mental Math
Tell students to complete the following statements with >, <, or =:

1. 82 ◯ 9 × 9 (>)
2. 6 + 14 ◯ $\frac{1}{2}$ of 40 (=)
3. 12 pints ◯ $6\frac{1}{2}$ quarts (<)
4. 400 − 6 ◯ 90 × 2 (>)
5. 5 feet ◯ 2 yards (<)
6. 18 ÷ 3 ◯ 36 ÷ 6 (=)
7. days in November ◯ days in December (<)

Name ______________________________

Subtraction of Whole Numbers

CHAPTER 4

Lesson 4-1

Subtracting 2-Digit Numbers

Abby, who is on the swim team, has saved $50 to buy a waterproof watch. How much will she have left after she buys the watch?

We want to know how much Abby will have left after her purchase.

We know that Abby has saved $50.

The watch costs $37.

To find the amount left, we subtract $37 from $50.

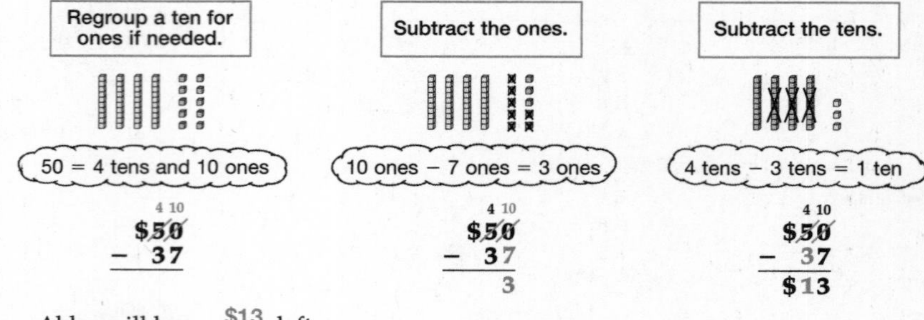

4 10 $50 − 37	4 10 $50 − 37 = 3	4 10 $50 − 37 = $13

Abby will have $13 left.

Getting Started

Subtract.

1. 79 − 16 = 63
2. 25 − 9 = 16
3. 97 − 28 = 69
4. 52 − 14 = 38

Copy and subtract.

5. 80 − 14 = 66
6. 57 − 20 = 37
7. 63 − 61 = 2
8. 84 − 27 = 57

Warm Up • Number Sense
Have students work in pairs. As one student shows an addition fact, the second student gives the sum of the addition fact and a related subtraction fact and its difference. Have students change roles after 20 facts. It is valuable practice to give the 100 addition and subtraction facts tests often.

2 Teach

Introduce the Lesson Have students tell about the picture. (A girl looks at a $37 watch.) Have a student read the problem aloud and tell what is being asked (how much money Abby will have left if she buys the watch) Ask students what information is given. (Abby has saved $50 and the watch costs $37.) Have students read and complete the sentences to solve the problem.

Develop Skills and Concepts Write **35 − 19** on the board. Put 3 tens rods in the tens box and 5 ones cubes in the ones box. Ask students how to take away 9 ones. (regroup 1 ten for 10 ones) Make the regrouping and ask how many ones there are now. (15) Have students continue the subtraction process with the manipulatives to find the difference of 1 ten and 6 ones, or 16. Ask students to check the work by adding. Repeat for more problems of 2-digit minuends and subtrahends.

3 Practice

Have students complete page 56 independently.

Practice

Subtract.

1. 43 − 13 = 30
2. 72 − 34 = 38
3. 58 − 36 = 22
4. 84 − 29 = 55
5. 29 − 5 = 24
6. 74 − 25 = 49
7. 60 − 28 = 32
8. 37 − 18 = 19
9. 82 − 26 = 56
10. 94 − 67 = 27
11. 39 − 17 = 22
12. 76 − 9 = 67
13. 41 − 38 = 3
14. 57 − 14 = 43
15. 63 − 45 = 18

Copy and subtract.

16. 19 − 12 7
17. 43 − 16 27
18. 70 − 15 55
19. 47 − 17 30
20. 37 − 20 17
21. 61 − 43 18
22. 58 − 17 41
23. 30 − 19 11
24. 72 − 36 36
25. 15 − 12 3
26. 82 − 37 45
27. 96 − 58 38

Problem Solving

Solve each problem.

28. Marge spent $24 on a new blouse. Alice spent $38 on a new outfit. How much more did Alice spend for clothes? $14

29. Chen earned $53 in 1 month. He spent $38 on birthday gifts for three friends. How much did Chen have left? $15

30. Randi bought a blouse for $22.95 and a skirt for $31.72. She had $68.40 in her checking account before she went shopping. How much money does she have left? $13.73

31. The Spanish Club had $1,273 in its account before it held a bake sale. Now, there is $1,859 in the account. How much money did the Spanish Club earn at the bake sale? $586

4 Assess

Ask students how to check subtraction problems. (by adding)

For Mixed Abilities

Common Errors • Intervention

Students may not regroup but subtract the lesser digit from the greater digit in a subtraction problem.

Incorrect	Correct
83 − 45 = 42	$\overset{7\,13}{\cancel{8}\cancel{3}}$ − 45 = 38

Correct students by having them work in pairs and use place-value materials to model the problems.

Enrichment • Money

Have students solve this problem: If you pay 6¢ tax for every dollar you spend, how much more tax would you pay if you bought a skateboard that cost $47 than if you bought one that cost $39? (48¢)

More to Explore • Logic

Have students draw a grid containing four rows of four squares each. Next, have them make four red markers and four black markers.

Tell students that they are to place markers on the squares so that no two of the same color are in a line either horizontally, vertically, or diagonally. Have the first person to find the correct placement draw his or her grid on the board.

	R	B	
B			R
R			B
	B	R	

ESL/ELL STRATEGIES

Review the meaning of *regrouping* by having volunteers explain the process in their own words. For example, "When you regroup, you break a number up into different pieces before you subtract another number from it." Discuss the examples on the Student Edition page.

4-2 Subtracting 3-Digit Numbers

pages 57–58

1 Getting Started

Objective

- To subtract 3-digit numbers with regrouping

Warm Up • Mental Math

Find the total cost of the following stamps:

1. twelve 5¢ (60¢)
2. three 22¢ (66¢)
3. twenty 10¢ ($2)
4. fifteen 3¢ (45¢)
5. eleven 10¢ ($1.10)
6. eight 20¢ ($1.60)
7. one hundred 10¢ ($10.00)
8. thirty 5¢ ($1.50)

Warm Up • Place Value

Draw a 5-digit place-value chart on the board. Have one student write a 5-digit number in the chart for another student to read and tell the value of each digit. Repeat for other numbers.

Name ______________________

Subtracting 3-Digit Numbers

The Lopez family drove from Phoenix to the Grand Canyon for a vacation. Their California relatives, the Ruiz family, drove from San Francisco to Yosemite National Park. How much farther did the Lopez family drive than the Ruiz family?

Driving Distances to Parks	
Denver—Pikes Peak	134 kilometers
Los Angeles—Death Valley	416 kilometers
Phoenix—Grand Canyon	351 kilometers
Portland—Crater Lake	394 kilometers
San Francisco—Yosemite	293 kilometers

We want to compare the distances to find how much farther the Lopez family drove.

We know that Phoenix is 351 kilometers from the Grand Canyon.

San Francisco is 293 kilometers from Yosemite.

To compare these distances, we subtract 293 from 351.

Subtract the ones. Regroup if needed.	Subtract the tens. Regroup if needed.	Subtract the hundreds.	Check by adding.
4 11 35̸1̸ − 293 8	14 2 4̸ 11 3̸5̸1̸ − 293 58	14 2 4̸ 11 3̸51 − 293 58	58 + 293 351

The Lopez family drove 58 kilometers farther than the Ruiz family.

Getting Started

Subtract.

1. 836 − 275 = 561
2. $4.57 − 1.39 = $3.18
3. 983 − 259 = 724
4. 821 − 637 = 184
5. 654 − 406 = 248
6. $7.93 − 1.87 = $6.06
7. 519 − 483 = 36
8. $9.42 − 0.96 = $8.46

Copy and subtract.

9. 341 − 170 = 171
10. 523 − 487 = 36
11. $8.51 − $3.75 = $4.76
12. $7.12 − $6.88 = $0.24

2 Teach

Introduce the Lesson Have a student read the problem and tell what is to be solved. (how much farther the Lopez family drove than the Ruiz family) Ask students what information is needed from the problem and the table. (The Lopez family drove 351 kilometers from Phoenix to the Grand Canyon and the Ruiz family drove 293 kilometers from San Francisco to Yosemite.) Ask students if they need the rest of the information in the table to solve the problem. (no)

- Ask if any students have visited any of the places named. Have students complete the sentences to solve the problem. Guide them through the three-step subtraction method in the model, and have them complete the solution sentence.

Develop Skills and Concepts Write **946 − 182** vertically on the board. Ask if regrouping is needed in the ones column. (no) Ask if regrouping is needed in the tens column. (yes) Ask why. (8 tens is greater than 4 tens.) Have a student work the problem and talk through each step. Check the problem by adding. Repeat for more problems of 3-digit numbers.

3 Practice

Remind students to use dollar signs and decimal points in the answers to money problems. Have students complete page 58 independently.

Practice

Subtract.

1. 645 − 437 = 208
2. 592 − 358 = 234
3. 728 − 283 = 445
4. $4.16 − 1.53 = $2.63
5. 721 − 258 = 463
6. 926 − 305 = 621
7. 823 − 647 = 176
8. 615 − 339 = 276
9. 752 − 283 = 469
10. $4.35 − 0.97 = $3.38
11. 632 − 108 = 524
12. 723 − 659 = 64

Copy and subtract.

13. 275 − 88 (187)
14. 732 − 486 (246)
15. $6.26 − $1.54 ($4.72)
16. 924 − 808 (116)
17. 437 − 158 (279)
18. 821 − 480 (341)
19. 515 − 326 (189)
20. 740 − 196 (544)
21. 527 − 83 (444)
22. $4.73 − $1.29 ($3.44)
23. 752 − 481 (271)
24. 916 − 548 (368)
25. 626 − 206 (420)
26. 816 − 539 (277)
27. 420 − 396 (24)
28. $9.38 − $8.42 ($0.96)

Problem Solving

Use the chart on page 57 to solve each problem.

29. The Peterson family is driving from San Francisco to Yosemite. They stop for lunch after driving 157 kilometers. How much farther do the Petersons have to drive?
136 kilometers

30. How far is it from Portland to Crater Lake and back again?
788 kilometers

4 Assess

Ask students how they can tell when to regroup in subtraction. (when the number we are subtracting from is not large enough to subtract from)

5 Mixed Review

1. 4 + (9 + 2) (15)
2. 725 − 432 (293)
3. 157 + 2,651 + 65 (2,873)
4. 9 + 5 + 4 + 8 + 8 (34)
5. $8.43 − $5.67 ($2.76)
6. 748 − 309 (439)
7. 6,427 + 3,722 (10,149)
8. 80 − 23 (57)
9. $27.95 + $14.35 ($42.30)
10. 967 − 432 (535)

For Mixed Abilities

Common Errors • Intervention

Students may regroup but forget to decrease the digit to the left after regrouping.

Incorrect	Correct
12	212
43~~2~~	4~~3~~~~2~~
− 315	− 315
127	117

Have students first determine if regrouping is necessary and then have them rewrite the portion of the problem that has been regrouped; for example, 3 tens 2 ones = 2 tens 12 ones.

Enrichment • Application

Have students use a map to locate each park listed in the chart on page 57. Tell them to find the population of the city near each park to the nearest thousand.

More to Explore • Application

Review how to make flowcharts. Ask students to design a chart for tasks that they do often, such as going through the lunch line, playing a record, multiplying a 2-digit number, and so on. Tell students to choose a specific task from a career they are familiar with and design a flowchart for that task. For example, they could chart a short-order cook making a cheeseburger. Students may need to observe or interview a person in that career before they can make the chart.

4-3 Subtracting, Minuends With Zeros

pages 59–60

1 Getting Started

Objective
- To subtract 3-digit numbers with zeros in the minuend

Materials
place-value blocks

Warm Up • Mental Math
Have students find the total cost.

1. 5 @ 21¢ ($1.05)
2. 2 @ $1.14 ($2.28)
3. 15 @ $4 ($60)
4. 110 @ 2¢ ($2.20)
5. 40 @ $3 ($120)
6. 6 @ $1.50 ($9)
7. 18 @ $1 per dozen ($1.50)
8. 3 @ $2.50 ($7.50)

Warm Up • Number Sense
Write **$4.26** on the board. Have students tell how many hundreds, tens, and ones are in the number. (4, 2, 6) Have students tell how many dollars, dimes, and pennies are in the number. (4, 2, 6) Repeat for more amounts through $9.99.

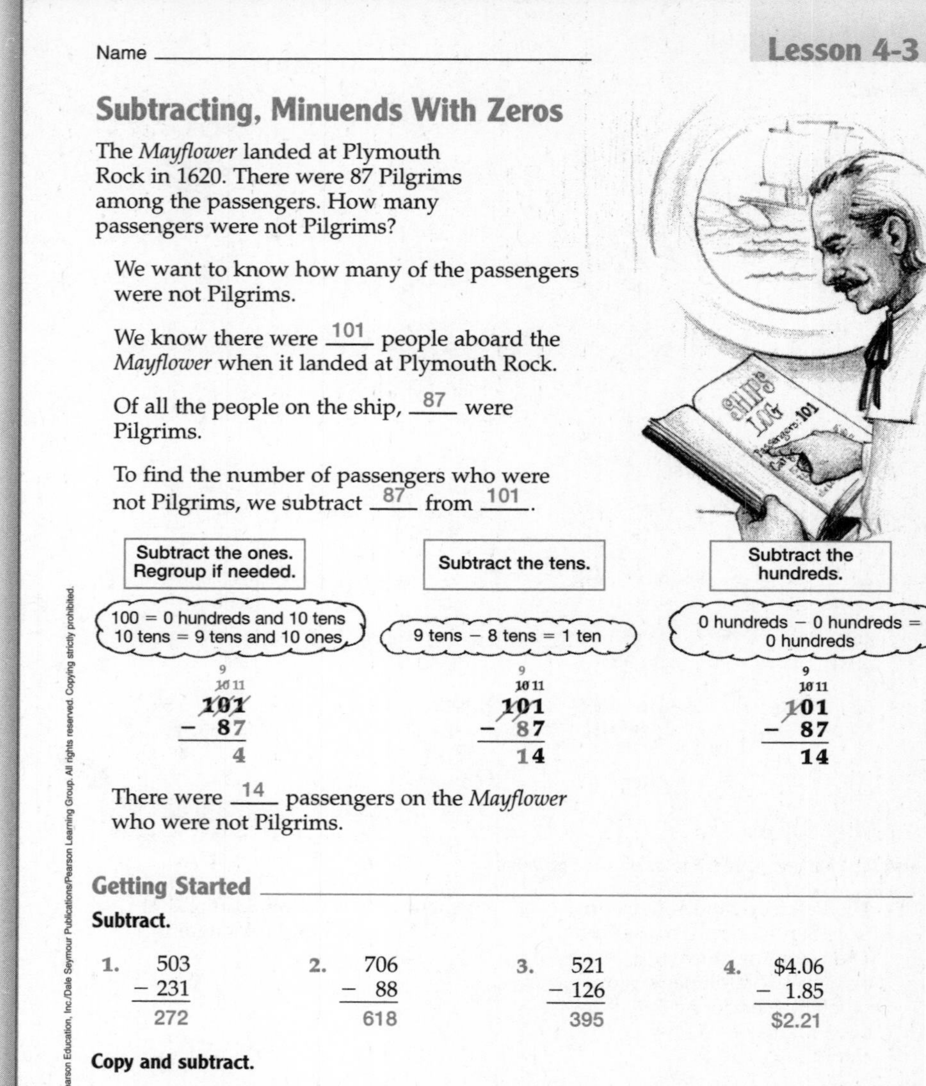

Name ______________________

Lesson 4-3

Subtracting, Minuends With Zeros

The *Mayflower* landed at Plymouth Rock in 1620. There were 87 Pilgrims among the passengers. How many passengers were not Pilgrims?

We want to know how many of the passengers were not Pilgrims.

We know there were 101 people aboard the *Mayflower* when it landed at Plymouth Rock.

Of all the people on the ship, 87 were Pilgrims.

To find the number of passengers who were not Pilgrims, we subtract 87 from 101.

Subtract the ones. Regroup if needed.	Subtract the tens.	Subtract the hundreds.
100 = 0 hundreds and 10 tens; 10 tens = 9 tens and 10 ones	9 tens − 8 tens = 1 ten	0 hundreds − 0 hundreds = 0 hundreds
101 − 87 = 4	101 − 87 = 14	101 − 87 = 14

There were 14 passengers on the *Mayflower* who were not Pilgrims.

Getting Started

Subtract.

1. 503 − 231 = 272
2. 706 − 88 = 618
3. 521 − 126 = 395
4. $4.06 − 1.85 = $2.21

Copy and subtract.

5. $3.05 − $2.95 = $0.10
6. 603 − 245 = 358
7. 905 − 728 = 177
8. $7.04 − $5.28 = $1.76

2 Teach

Introduce the Lesson Have students describe the picture. (The captain of the *Mayflower* sees that there are 101 passengers on board.) Discuss the landing of the *Mayflower* with students. Have a student read the problem and tell what is to be found. (the number of passengers who were not Pilgrims) Ask students what information is needed to solve the problem. (total number of passengers and number of Pilgrims)

- Have students complete the sentences, and work through each step of the subtraction problem in the model with them. Have students complete the solution sentence.

Develop Skills and Concepts Write **403 − 69** vertically on the board. Tell students that because 9 ones cannot be taken from 3 ones, we need to regroup 1 ten for 10 ones. Ask students if there are any tens. (no) Tell students that we then regroup 1 of the 4 hundreds for 10 tens. Make the regrouping, showing 3 hundreds left. Now, ask students if we can regroup 1 ten for 10 ones. (yes) Make the regrouping, showing 9 tens left. Continue the subtraction to find the difference. (334) Remind students to check their work by adding the subtrahend and the difference and that this sum must equal the minuend 403. Repeat for more problems requiring regrouping in the tens or hundreds column or both.

3 Practice

Remind students that some of the exercises on this page require more than one regrouping. Have students complete page 60 independently.

Practice

Subtract.

1. 604 − 136 = 468	2. 502 − 185 = 317	3. 988 − 358 = 630	4. $6.05 − 4.83 = $1.22
5. 408 − 391 = 17	6. $6.00 − 4.75 = $1.25	7. 201 − 128 = 73	8. 708 − 220 = 488
9. $3.05 − 2.95 = $0.10	10. 613 − 245 = 368	11. 905 − 728 = 177	12. $7.04 − 5.28 = $1.76

Copy and subtract.

13. 307 − 184 123	14. 509 − 436 73	15. 802 − 656 146	16. $9.05 − $6.47 $2.58
17. 408 − 227 181	18. 506 − 309 197	19. 328 − 253 75	20. $4.07 − $2.28 $1.79
21. 107 − 98 9	22. $8.00 − $5.53 $2.47	23. 604 − 586 18	24. 705 − 489 216

Problem Solving

Solve each problem.

25. There are 206 bones in the human body. Of these, 29 bones are in the head. How many bones are there in the rest of the body?
177 bones

26. A bunch of daisies costs $3.79. If you give the clerk a $5 bill, how much change will you receive?
$1.21

Now Try This!

Build two more magic word squares like the first one. Use the same words on the horizontal as you do on the vertical. Answers will vary.

T	A	P
A	R	E
P	E	T

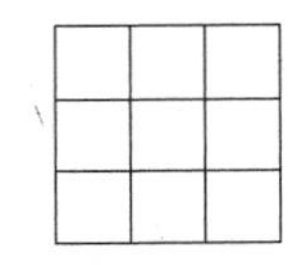

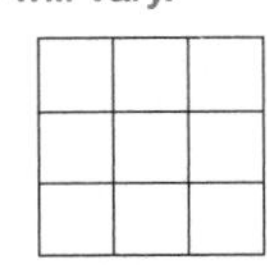

Now Try This! In a numerical magic square, the horizontal, vertical, and diagonal lines produce the same numerical answer. These word magic squares work on a similar principle for the horizontal and vertical lines. Have students study the first word magic square. Encourage them to work out a strategy for creating the other two. (for example, using two words that begin and end in consonants and one word that begins and ends in vowels)

4 Assess

Ask students to find the difference of 700 − 516. (184)

For Mixed Abilities

Common Errors • Intervention

Some students may forget to regroup twice when they are subtracting with a zero in the minuend.

Incorrect	Correct
4 ~~0~~ ~~3~~ (regrouped: 10 13) − 1 6 8 3 4 5	~~4~~ ~~0~~ ~~3~~ (regrouped: 3, ~~10~~ 9, 13) − 1 6 8 2 3 5

Have students practice with exercises such as 100 − 16 in which they use place-value models to rename 100 as 10 tens and then rename one of the tens giving them 9 tens 10 ones.

Enrichment • Application

Ask students to write and solve problems to show their change from a $20 bill if they buy a book that costs $4.69, a shirt that costs $14.97, or a bag of candy that costs $2.14. Then, ask how much change they would receive if they bought two of each item.

More to Explore • Statistics

Ask students to research and list the areas of the United States that have the most and the least amount of rainfall in 1 year. Suggest that they use an atlas to do their research. Have them also find the amount of rainfall for their own city or state. Ask students to make a bar graph illustrating their findings.

4-4 Subtracting Multidigit Numbers

pages 61–62

1 Getting Started

Objective

- To subtract multidigit numbers

Materials

place-value materials

Warm Up • Mental Math

Ask students to name the number.

1. the smallest 3-digit number (100)
2. the largest 4-digit number (9,999)
3. the smallest 5-digit number (10,000)
4. the largest 3-digit number (999)
5. the smallest 6-digit number (100,000)
6. the smallest 9-digit number (100,000,000)
7. the largest 8-digit number (99,999,999)
8. the largest 6-digit number (999,999)

Warm Up • Place Value

Write several 4-, 5-, and 6-digit numbers on the board. Have students rename the numbers by telling the place-value word for each digit.

Name ______________________________

Subtracting Multidigit Numbers

How much larger is Craters of the Moon National Monument than Organ Pipe Cactus National Monument?

National Monuments		
Name	**Opened**	**Size**
Craters of the Moon	1924	714,727 acres
Dinosaur	1915	210,844 acres
Organ Pipe Cactus	1937	330,689 acres
Sunset Crater	1930	3,040 acres
Walnut Canyon	1915	3,541 acres

We want to know how much larger Craters of the Moon is than Organ Pipe Cactus.

We know that Organ Pipe Cactus National Monument is 330,689 acres.

Craters of the Moon National Monument measures 714,727 acres.

To find the difference in size between the two parks, we subtract 330,689 from 714,727.

To subtract multidigit numbers, begin by subtracting the ones. Regroup if needed. Continue subtracting each column to the left.

REMEMBER Regroup one place value at a time. Sometimes you will need to regroup a regrouping.

```
        11
 6 11  6 1 17
 714,727
-330,689
 384,038
```

Craters of the Moon National Monument is 384,038 acres larger than Organ Pipe Cactus National Monument.

Getting Started

Subtract.

1. 8,475 − 2,243 = 6,232
2. $123.27 − 86.43 = $36.84
3. 176,043 − 29,228 = 146,815

Copy and subtract.

4. 48,758 − 21,475 = 27,283
5. 192,175 − 39,857 = 152,318
6. $436.25 − $348.99 = $87.26

2 Teach

Introduce the Lesson Have a student read the problem and tell what is to be solved. (find out how much larger Craters of the Moon National Monument is than Organ Pipe Cactus National Monument) Ask students what information is needed from the problem and the table. (sizes of the two parks) Have students describe their visits to any of the parks listed. Read with students as they complete the sentences and solve the algorithm in the model. Point out the regrouping rule.

Develop Skills and Concepts Write **111 − 62** vertically on the board and have a student solve the problem, showing the regrouping. (49) Next, write **1,110 − 620** vertically on the board and tell students that we have added a zero to each place and now the minuend has a thousands place and the subtrahend has a hundreds place. Show students the regrouping as you work the problem. (490) Help students see that one regrouping is made at a time and that they may need to regroup from a regrouping. Check the problem by adding. Repeat for more problems with minuends through 6 digits until students are comfortable working the regroupings.

3 Practice

Remind students to place commas between periods in answers of four or more digits, and dollar signs and decimal points where needed. Have students complete page 62 independently.

Practice

Subtract.

1. 6,456 − 2,329 = 4,127
2. 7,502 − 3,296 = 4,206
3. $57.43 − 18.59 = $38.84
4. 18,275 − 3,596 = 14,679
5. 4,253 − 1,327 = 2,926
6. 6,851 − 5,926 = 925
7. $91.15 − 63.75 = $27.40
8. 8,273 − 4,829 = 3,444
9. 127,243 − 16,158 = 111,085
10. 40,871 − 29,137 = 11,734
11. 281,275 − 29,469 = 251,806
12. $173.84 − 86.96 = $86.88

Copy and subtract.

13. 7,243 − 1,165
6,078
14. $24.75 − $13.72
$11.03
15. 8,527 − 695
7,832
16. 36,741 − 5,196
31,545
17. 20,275 − 6,889
13,386
18. $543.78 − $275.81
$267.97
19. 296,258 − 26,579
269,679
20. $877.73 − $639.86
$237.87
21. 47,868 − 21,975
25,893
22. 79,246 − 37,865
41,381
23. 430,173 − 14,688
415,485
24. 93,182 − 81,996
11,186

Problem Solving

Use the chart on page 61 to solve Problems 25 and 26.

25. How many acres larger is Craters of the Moon than Walnut Canyon National Monument?
711,186 acres
26. Is Craters of the Moon National Monument larger than the other four listed national monuments combined?
yes
27. Queen Victoria ruled Great Britain for 64 years. She was born in 1819 and died in 1901. How old was she when she died?
82 years old
28. The United States admitted 398,613 immigrants in 1976 and 596,600 in 1981. How many more immigrants entered the United States in 1981 than in 1976? 197,987 immigrants

4 Assess

Ask students to find the difference of 866,719 − 798, 927. Then, have them check their subtraction. (67,792)

5 Mixed Review

1. $40.25 − $25.37 ($14.88)
2. 31,276 − 4,148 (27,128)
3. $27.95 + $17.28 + $19.94 ($65.17)
4. 3 + (9 + 2) (14)
5. 16,472 + 28,799 (45,271)
6. 392,165 − 18,176 (373,989)
7. 28 + 47 + 93 (168)
8. 307 + 1,756 (2,063)
9. 905 − 107 (798)
10. 123,470 + 17,630 (141,100)

For Mixed Abilities

Common Errors • Intervention

Some students may forget to add the digit in the minuend after they regroup and before they subtract.

Incorrect	Correct
1 10 7 10	1 13 7 14
2,384 (crossed out)	2,384 (crossed out)
−1,569	−1,569
511	815

Have students work in pairs with place-value materials to do exercises such as 84 − 69 to recognize that when they regroup, 8 tens 4 ones becomes 7 tens 14 ones.

Enrichment • Numeration

Tell students to use the information in the table on page 61 to arrange the national parks in order by age and by size. Then, have them use the *World Almanac* or maps to locate each of the parks.

More to Explore • Creative Drill

Make a jumbo-size dollar bill from a large sheet of paper. Discuss with students the flow of money from one business or individual to another. Let them share ideas about where they become involved in this flow, such as in the lunch line, in a grocery store, or at a basketball game. Play a game called Passing the Buck. One student takes the dollar, selects an occupation, and says, for example, "I am a student, and I gave this dollar to the dentist for cleaning my teeth." That student passes the dollar to another student who might say, "I am the dentist who gave the dollar to a gas station attendant when I bought gas." The game and dialogue continues until the "buck" has been passed around the room. List the occupations on the board as they are suggested in the game. Ask students what they learned about the importance of money skills in their everyday lives.

4-5 Subtracting, More Minuends With Zeros

pages 63–64

1 Getting Started

Objective

- To subtract 4- or 5-digit numbers when minuends have zeros

Materials

*place-value blocks: thousands, hundreds, tens, and ones; play money

Warm Up • Mental Math

Tell students to answer *true* or *false*.

1. 776 has 77 tens. (T)
2. 10 hundreds $<$ 1,000 (F)
3. 46 is an odd number. (F)
4. 926 can be rounded to 920. (F)
5. 72 hours = 3 days (T)
6. $42 \div 6 > 7 \times 1$ (F)
7. $\frac{1}{3}$ of 18 = $\frac{1}{2}$ of 12 (T)
8. perimeter = $L \times W$ (F)

Warm Up • Numeration

Write four numbers of 3- to 5-digits each on the board. Have students arrange the numbers in order, from the least to the greatest, and then read the numbers as they would appear if written from the greatest to the least. Repeat for more sets of four or five numbers.

Name ______________________

Subtracting, More Minuends With Zeros

The Susan B. Anthony School held its annual fall read-a-thon.

Class Reading Records

Fourth Grade	3,795 pages
Fifth Grade	5,003 pages
Sixth Grade	4,056 pages

How many more pages did the fifth grade read than the second-place class?

We want to know how many more pages the fifth grade read than the second-place class.

We know the fifth grade read 5,003 pages.

The second-place class is the sixth grade.

It read 4,056 pages.

To find the difference between the number of pages, we subtract 4,056 from 5,003.

REMEMBER Regroup one place value at a time.

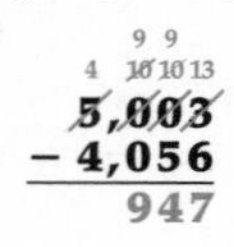

```
   9 9
 4 10 10 13
  5,003
- 4,056
    947
```

The fifth grade read 947 more pages than the sixth grade.

Getting Started

Subtract.

1. 3,005 − 1,348 = 1,657
2. \$40.09 − 9.75 = \$30.34
3. 3,300 − 1,856 = 1,444
4. \$50.00 − 27.26 = \$22.74
5. 8,512 − 7,968 = 544
6. \$90.17 − 20.87 = \$69.30

Copy and subtract.

7. 26,007 − 18,759 = 7,248
8. 70,026 − 23,576 = 46,450
9. \$900.05 − \$267.83 = \$632.22

2 Teach

Introduce the Lesson Have a student read the problem. Ask students what two problems are to be solved. (which class read the second-highest number of pages and how many more pages the first-place fifth graders read) Ask students how we can find out which class came in second place. (arrange numbers from the table in order from greatest to least) Ask a student to write the numbers from greatest to least on the board. (5,003; 4,056; 3,795) Have students complete the sentences and work through the model problem with them.

Develop Skills and Concepts Write **4,000 − 2,875** vertically on the board. Ask students if a regrouping is needed to subtract the ones column. (yes) Tell students that because there are no tens and no hundreds, we must regroup 1 thousand for 10 hundreds. Show the 3 thousands and 10 hundreds left. Now, tell students that we can regroup 1 hundred for 10 tens. Show the regrouping with 9 hundreds and 10 tens left. Tell students that we can now regroup 1 ten for 10 ones. Show the regrouping so that 9 tens and 10 ones are left. Tell students that we can now subtract each column beginning with the ones column and working to the left.

- Show students the subtraction with a solution of 1,125. Remind students to add the subtrahend and the difference to check the work. Repeat for more problems with zeros in the minuend.

3 Practice

Remind students to begin with the ones column, work to the left, and regroup from one place value at a time. Have students complete page 64 independently.

Practice

Subtract.

1. 3,004 − 2,356 = 648	2. 8,002 − 5,096 = 2,906	3. 3,891 − 1,750 = 2,141	4. $20.08 − 15.99 = $4.09
5. 4,020 − 1,865 = 2,155	6. $87.00 − 28.59 = $58.41	7. 3,007 − 2,090 = 917	8. $50.06 − 37.08 = $12.98
9. 19,006 − 8,275 = 10,731	10. 20,006 − 14,758 = 5,248	11. $400.26 − 236.58 = $163.68	12. $793.42 − 253.87 = $539.55

Copy and subtract.

13. 4,001 − 2,756
1,245

14. $70.05 − $26.59
$43.46

15. 8,060 − 7,948
112

16. 7,007 − 2,468
4,539

17. 21,316 − 12,479
8,837

18. 14,000 − 8,396
5,604

19. $100.21 − $93.50
$6.71

20. 60,004 − 51,476
8,528

21. 52,006 − 9,037
42,969

22. $800.00 − $275.67
$524.33

23. 34,612 − 29,965
4,647

24. 50,010 − 36,754
13,256

Problem Solving

Use the chart on page 63 to solve Problems 25 and 26.

25. How many pages did the three classes read all together?
12,854 pages

26. How many more pages did the sixth grade read than the fourth grade?
261 pages

27. Roger has 3,208 baseball cards in his collection. Karen has 2,572 baseball cards in her collection. How many more cards does Roger have?
636 cards

28. Nikki's family had $6,000 to spend on vacation. They returned from vacation with $2,989. How much money did they spend on vacation?
$3,011

4 Assess

Ask students to subtract $4,137.54 from $5,000.00 and explain the regrouping. ($862.46)

For Mixed Abilities

Common Errors • Intervention

Some students may bring down the numbers that are being subtracted when there are zeros in the minuend.

Incorrect	Correct
3,006	2 10 10 (tens regrouped to 9) — 3,0 0 6
− 1,425	− 1, 4 2 5
2,421	1, 5 8 1

Have students work in pairs and use play money to model a problem such as $300 - $142, where they see that they must regroup 3 hundreds for 2 hundreds, 9 tens, and 10 ones before they can subtract.

Enrichment • Statistics

Tell students to find out the year in which each member of their family was born and to make a chart to show how old each will be in the year 2020.

More to Explore • Biography

An American inventor, Samuel Morse, struggled for many years before his inventions, the electric telegraph and Morse code, were recognized. Morse was born in Massachusetts in 1791 and studied to be an artist. On a trip home from Europe, Morse heard his shipmates discussing the idea of sending electricity over wire. Intrigued, Morse spent the rest of the voyage formulating his ideas about how this could be accomplished. Morse taught at a university in New York City and used his earnings to develop the telegraph. After 5 years, Morse demonstrated his invention but found very little support. After years of requests for support, Congress finally granted Morse $30,000 to test his invention. He dramatically strung a telegraph wire from Washington, D.C., to Baltimore, Maryland, and relayed a message using Morse code. Morse's persistence finally won him wealth and fame. A statue honoring him was unveiled in New York City 1 year before his death in 1872.

4-6 Estimating Differences

pages 65–66

1 Getting Started

Objective

- To estimate differences by rounding

Warm Up • Mental Math

Have students find the following:

1. the volume of a 4 × 4 × 4 cube (64)
2. the number of minutes in $2\frac{1}{5}$ hours (132)
3. the corners in 3 pentagons (15)
4. the date that is 26 days after May 8 (June 3)
5. 4 tens more than 1,084 (1,124)
6. the number of dimes to equal $6.70 (67)
7. the number of years in $\frac{1}{2}$ century (50)

Warm Up • Estimation

Remind students that we use estimation often in our everyday lives. Ask students to tell the present time to the nearest hour and then to the nearest half-hour and quarter-hour. Ask students to tell the number of pages in this book to the nearest ten and nearest hundred. Ask the number of students in the class to the nearest ten. Ask why we would not estimate the number of students in the class to the nearest hundred. (It would be zero and that is not meaningful.)

Name ____________________

Estimating Differences

Dominic earned $46.75 delivering papers. He wants to buy his own stereo. About how much more money does Dominic need to earn?

We want to estimate how much money Dominic still needs to earn.

We know the stereo costs $89.29.

So far, Dominic has earned $46.75.

To estimate what he still needs, we round both amounts of money to the nearest dollar and find the difference between the two.

$89.29 —rounds to→ **$89**
$46.75 ————→ **− 47**
$42

Dominic still needs about $42 to buy the stereo.

Getting Started

Round to the nearest hundred. Estimate each difference.

1. 845 − 236 → 800 − 200 = 600
2. 906 − 483 → 900 − 500 = 400
3. 726 − 413 → 700 − 400 = 300
4. 586 − 275 → 600 − 300 = 300

Round to the nearest thousand. Estimate each difference.

5. 4,796 − 1,926 → 5,000 − 2,000 = 3,000
6. 6,500 − 2,375 → 7,000 − 2,000 = 5,000
7. 4,975 − 1,610 → 5,000 − 2,000 = 3,000
8. 8,279 − 3,758 → 8,000 − 4,000 = 4,000

Round to the nearest dollar. Estimate each difference.

9. $29.35 − 12.50 → $29 − 13 = $16
10. $76.21 − 48.76 → $76 − 49 = $27
11. $94.39 − 81.56 → $94 − 82 = $12
12. $62.83 − 46.15 → $63 − 46 = $17

2 Teach

Introduce the Lesson Have a student describe the picture. (A boy is looking at a camera and a stereo in a store window.) Have a student read the problem and tell what is to be solved. (about how much more money Dominic needs to buy the stereo) Ask students what information is needed from the problem and the picture. (Dominic has $46.75 and the stereo costs $89.29.)

- Ask students if we need the other information that is given. (no) Ask if we need to know the exact amount of money Dominic needs. (no) Have students read and complete the sentences. Guide them through the estimation in the model, and have them complete the solution sentence.

Develop Skills and Concepts Write **2,762 − 1,128** vertically on the board. Tell students that these numbers could represent the number of days two people have lived. Tell students that we would like to know about how many more days the older person has lived, so we round each number and then subtract. Tell students that we can round each number to the nearest ten, hundred, or thousand. Have students round each number to the nearest ten. (2,760; 1,130)

- Have a student subtract the numbers. (2,760 − 1,130 = 1,630) Have a student round each number to the nearest hundred and subtract. (2,800 − 1,100 = 1,700) Have students check their work by finding the exact difference. Repeat for more problems.

3 Practice

Remind students to round each number and then subtract. Have students complete page 66 independently.

Practice

Round to the nearest hundred. Estimate each difference.

	Problem	Estimate
1.	579 − 346	600 − 300 = 300
2.	481 − 276	500 − 300 = 200
3.	825 − 279	800 − 300 = 500
4.	921 − 735	900 − 700 = 200
5.	791 − 347	800 − 300 = 500
6.	426 − 139	400 − 100 = 300
7.	776 − 438	800 − 400 = 400
8.	861 − 475	900 − 500 = 400

Round to the nearest thousand. Estimate each difference.

	Problem	Estimate
9.	7,351 − 2,686	7,000 − 3,000 = 4,000
10.	9,251 − 4,460	9,000 − 4,000 = 5,000
11.	6,865 − 4,956	7,000 − 5,000 = 2,000
12.	8,753 − 1,829	9,000 − 2,000 = 7,000
13.	4,629 − 1,510	5,000 − 2,000 = 3,000
14.	8,475 − 6,832	8,000 − 7,000 = 1,000
15.	4,500 − 3,723	5,000 − 4,000 = 1,000
16.	8,216 − 5,006	8,000 − 5,000 = 3,000

Round to the nearest dollar. Estimate each difference.

	Problem	Estimate
17.	$43.27 − 21.95	$43 − 22 = $21
18.	$71.38 − 18.46	$71 − 18 = $53
19.	$92.50 − 46.89	$93 − 47 = $46
20.	$52.89 − 46.75	$53 − 47 = $6
21.	$38.27 − 14.58	$38 − 15 = $23
22.	$86.89 − 71.76	$87 − 72 = $15
23.	$97.17 − 42.45	$97 − 42 = $55
24.	$66.12 − 51.87	$66 − 52 = $14

Problem Solving

Use the picture on page 65 to solve each problem.

25. About how much money would you have left if you bought the camera with a $100 bill?
$3

26. About how much do the stereo and camera cost together?
$186

4 Assess

Ask students to explain how to round a number to the nearest thousand. (Look at the hundreds place. If the number is 5 or more, add one to the thousands place and replace hundreds, tens, and ones digits with zeros.)

5 Mixed Review

1. 225 − 38 (187)
2. $90.00 − $45.76 ($44.24)
3. 276,158 + 338,942 (615,100)
4. 343 + 658 + 72 (1,073)
5. (9 + 6) + 8 (23)
6. 4,080 − 2,324 (1,756)
7. $600.03 − $275.48 ($324.55)
8. 4,375 + 6,087 (10,462)
9. 70,018 − 27,399 (42,619)
10. 7 + 2 + 8 + 2 + 5 (24)

For Mixed Abilities

Common Errors • Intervention

Students may estimate incorrectly because they round incorrectly. Use a number line to estimate the answer to $13.75 − $12.12 by rounding. Show the whole-dollar amounts of $12, $13, and $14 on the number line and have students discuss whether the point for $13.75 is closer to $13 or to $14 and whether the point for $12.12 is closer to $12 or $13. Then, have them use the whole-dollar amounts they chose, $14 − $12, to estimate the answer.

Enrichment • Numeration

Tell students to find out how many days they have lived by multiplying 365 by their age and adding the days from their last birthday to today. Have them estimate how many more days one of their parents has lived than they have.

More to Explore • Logic

Have students solve the following problem by finding the actual cost of each item and the profit made.

Linda took some apples to school to sell at lunchtime. She told Susan that she would sell her 1 apple for 15¢ or 2 apples for 25¢. Either way, Linda's profit would be the same. The next day, she took cookies to sell. She told Lucy that she would sell her 2 cookies for 20¢ or 3 cookies for 28¢. Linda would still make the same profit. The next day, Linda took peanut butter and jelly sandwiches. She told Tommy she would sell him 2 sandwiches for 38¢ or 4 sandwiches for 62¢. Again, Linda would still make the same profit.

(actual cost for 1 apple, 10¢; profit, 5¢; actual cost for 1 cookie, 8¢; profit, 4¢; actual cost for 1 sandwich, 12¢; profit, 14¢)

4-7 Problem Solving: Too Much or Too Little Information

pages 67–68

1 Getting Started

Objective

- To determine if there is too much or too little information to solve problems

Warm Up • Mental Math

Dictate the following:

1. $726 - (10 \div 2)$ (721)
2. $5{,}000 \times 3$ (15,000)
3. $(84 \times 2) + 100$ (268)
4. $18 - 6 + (4 \times 3)$ (24)
5. $(45 \div 9) + 2\frac{1}{2}$ ($7\frac{1}{2}$)
6. $333 - 300 - 30 - 3$ (0)
7. $\frac{2}{10}$ of 100 (20)
8. $71 + 29 \div 4$ (25)

Warm Up • Pencil and Paper

Write the following exercises on the board. Tell students to find the difference.

1. $75 - 39$ (36)
2. $91 - 18$ (73)
3. $849 - 306$ (543)
4. $711 - 483$ (228)
5. $382 - 120$ (262)
6. $928 - 342$ (586)

Name ______________________

Problem Solving: Too Much or Too Little Information

Doug is working on a project for his Social Studies class. He needs to find out how many more people live in Boone County than Cass County.

Population of Counties in Illinois	
County	**Population**
Boone County	41,786
Lee County	36,062
Clay County	14,560
Cass County	13,665

★ **SEE**

We know that 41,786 people live in Boone County, 36,062 people live in Lee County, 14,560 people live in Clay County, and 13,665 people live in Cass County.

★ **PLAN**

Find the facts we need to solve the problem. Cross out the facts we do not need. Subtract the number of people living in Cass County from the number of people living in Boone County.

★ **DO**

The population of Boone County is 41,786.

The population of Cass County is 13,665.

Cross out Clay County's population of 14,560 and Lee County's population of 36,062.

$$\begin{array}{r} 41{,}786 \\ -\ 13{,}665 \\ \hline 28{,}121 \end{array}$$

★ **CHECK**

Add to check the difference.

28,121 + 13,665 = 41,786

2 Teach

Introduce the Lesson Review with students the four-step problem-solving strategy.

Develop Skills and Concepts Have students read the problem aloud and tell what they are to find. (how many more people live in Boone County than in Cass County) Read the sentence under the SEE step, asking volunteers to give the missing information. Be sure all students fill in the answers. Ask another volunteer to read the PLAN step.

- Help students determine what information is needed to solve the problem (the number of people in Boone County, 41,786, and the number of people in Cass County, 13,665) and what information is extra (the number of people in Clay County, 14,560, and the number of people in Lee County, 36,062).
- Finally, work through the DO and CHECK steps with students. Make sure students use only the information that is needed to solve the problem.

3 Apply

Solution Notes

1. Ask students to explain how they would answer the question if they had enough information. (subtract the year that George I became king from 1830)
2. Ask students why they did not need to know about the National Football League teams. (The question asks only about baseball teams.)
3. Ask students to explain how they would answer the question if they had enough information. (add 32 to the electoral votes that Lincoln received in 1860)
4. After students have determined the extra information, ask students to determine other questions that could have been asked from the data given in the problem.

Apply

Solve the problems that have enough information. Tell what is missing in the problems that have too little information.

1. In 1830, Great Britain's King George IV died. The previous three kings were also named George. How many years in a row was a George the king of Great Britain?
   need to know when George I became king
2. In 2003, there were 30 teams in Major League Baseball and 32 teams in the National Football League. In 1960, there were 16 teams in Major League Baseball. How many more baseball teams were there in 2003 than in 1960?
   14 more teams
3. In the 1860 presidential election, Abraham Lincoln defeated three other candidates. In the 1864 election, Lincoln received 32 more electoral votes than he did in 1860. How many electoral votes did Lincoln receive in 1864?
   need to know how many electoral votes Lincoln received in 1860

Problem-Solving Strategy: Using the Four-Step Plan

* ★ **SEE** What do you need to find?
* ★ **PLAN** What do you need to do?
* ★ **DO** Follow the plan.
* ★ **CHECK** Does your answer make sense?

4. During basketball season, Kristen scored 325 points. The team scored 548 points, and Mary was the team's second-leading scorer with 81 points. How many more points did Kristen score than Mary?
   244 points
5. Alaska became the forty-ninth state in the United States in 1959. It had been 47 years since Arizona became the forty-eighth state. In what year did Arizona become a state?
   1912
6. Granite Peak is the highest point in Montana. It has an elevation of 12,799 feet. The lowest point in Montana is the Kootenai River. What is the difference in elevations between Granite Peak and the Kootenai River?
   need to know the elevation of Kootenai River

For Mixed Abilities

More to Explore • Biography

Have students use a reference book or the Internet to find the missing information to Exercises 1, 3, and 6 on page 68. Then, have students use the new information to solve the problems. Have students pick one of the questions and have them write a paragraph detailing information from the question.

Higher-Order Thinking Skills

5. **Evaluation:** Have students write the sentence they used to solve this problem. (1,959 - 47 = 1,912) Ask students what is different about this question than the others in this lesson. (There is exactly enough information to answer the question.)
6. **Analysis:** Students must recognize that they need to know the elevation of Kootenai River to answer the question. Ask students to determine which operation they would use if they knew the elevation. (subtraction)

4 Assess

Ask students how they can determine if there is too much, too little, or enough information to solve a problem. (restate the problem)

4-8 Calculator Codes

pages 69–70

1 Getting Started

Objective

- To use a calculator to add and subtract numbers

Materials

*large drawing of a simple calculator

Warm Up • Mental Math

Have students name the numbers that round to

1. 30 (25, . . . , 34)
2. 100 (95, . . . , 104; 50, . . . , 149)
3. 460 (455, . . . , 464)
4. 700 (695, . . . , 704; 650, . . . , 749)
5. 990 (985, . . . , 994)
6. 4,040 (4,035, . . . , 4,044)
7. 810 (805, . . . , 814)

Warm Up • Estimation

Write **47,059 + 786** on the board. Ask students how to estimate the sum to the nearest thousand. (round each number to the nearest thousand and add) Ask a student to write and solve the problem. (47,000 + 1,000 = 48,000) Repeat for more problems with addends of 5 digits or less. Have students find estimated sums to the nearest ten, hundred, or thousand.

Name ______________________

It's Algebra!

Calculator Codes

A calculator has **number keys**, **operation keys**, and **special keys**. The calculator also has an **on/off key**. When you turn your calculator on, a zero should show on the screen. When you press a number or operation key, an **entry** is made in the calculator. The entry shows on the screen. When you press the [C] key, all entries are cleared from the calculator. When you press the [CE] key, the last entry is cancelled, but the previous entries are remembered.

Number Keys	Operation Keys	Special Keys
[7] [8] [9]	[÷] [×]	[C]
[4] [5] [6]	[−] [+]	[CE]
[1] [2] [3]	[=] [√]	[.]
[0]		

Getting Started

A calculator code gives the order for pressing the keys on your calculator.

Follow each code and write the result on the screen.

1. 5162 [+] 463 [+] 2000 [=] [7625]
2. 18036 [+] 6259 [+] 2475 [+] 72405 [=] [99175]
3. 69421 [−] 39704 [−] 2184 [=] [27533]

REMEMBER When you enter an amount of money into the calculator, the zeros to the far right of the decimal point do not have to be entered. The calculator will not print them, or a dollar sign, in the answer.

$9.20 will be entered as [9] [.] [2] [0] and appear as [9.2].

Enter this code and write the answer on the screen:

65 [.] 45 [+] 72 [.] 05 [+] 6 [.] 4 [=] [143.9]

---

2 Teach

Introduce the Lesson Refer to the large drawing of a calculator as you read the paragraph to students. Have students find each number key on the drawing. Remind students that all numbers are made by using combinations of these ten numbers.

- Have students find each operation key on the drawing. Ask students how many operations there are. (6) Have students find each special key on the drawing as you repeat its use. Tell students that the decimal point is used when working with amounts of money. Tell students that the % key will not be used at this time.

Develop Skills and Concepts Have students find each **number key, operation key**, and **special key** on their calculator keyboard. Read about the calculator code to students. Have students press the **on/off key**, work the three problems, and write the answers. Remind students to press **C** to clear the screen before working each new problem.

- Write more calculator codes on the board for extra practice. Have students estimate each sum by rounding the addends to the nearest hundred or thousand. Now, continue reading with students to learn about entering money amounts into a calculator. Have students enter the code that is shown at the bottom of the page and write the answer. Write more codes with money amounts on the board for continued practice.

3 Practice

Have students work Exercises 1 to 13 on page 70 independently. Remind students to estimate each sum or difference to check their work. Then, tell students to read and tell what is to be solved in the word problem. (total feet climbed if one were to climb all three mountains)

Practice

Complete each code. Write the answer on the screen.

1. 5 [.] 16 [+] 9 [.] 84 [=] [15]
2. 7691 [+] 12465 [=] [20156]
3. 36615 [−] 9475 [=] [27140]
4. 6458 [+] 605 [+] 11927 [=] [18990]

Use a calculator to find each sum or difference.

5. 247 + 629 + 515 = 1,391
6. 11,643 − 9,754 = 1,889
7. 99,346 − 89,739 = 9,607
8. 42,621 + 27,436 + 16,948 + 11,753 = 98,758
9. 16,941 + 40,432 + 895 + 6,748 = 65,016

Copy and add or subtract.

10. $3.57 + $2.96 + $8.38
$14.91
11. $7.95 + $24.63 + $87.51
$120.09
12. 8,207 + 28,416 + 3,796
40,419
13. 14,256 − 3,901
10,355

Problem Solving

Use a calculator to solve the problem.

14. The three tallest mountains in the United States are Mt. McKinley, St. Elias, and Mt. Foraker. These mountains are all in Alaska. If you climbed all three mountains, how many feet of mountain would you have scaled? 55,728 feet

Tallest Mountains in the U.S.

Mountain	Height
Mt. McKinley	20,320 ft
St. Elias	18,008 ft
Mt. Foraker	17,400 ft

Now Try This!

What does Sally do at the seashore?

To figure out this mystery, complete each code on the calculator. Then, turn the calculator upside down and print each word that shows on the screen.

1. 113 [+] 96 [+] 136 [=] S H E
2. 44772 [+] 12963 [=] S E L L S
3. 175596 [+] 39501 [+] 362248 [=] S H E L L S

Have students tell what is to be done. (add the three measurements in the chart) Ask what operation keys will be used. (+ and =) Tell students that the comma is not entered as there is no key marked comma on their calculators. Have a student write the calculator code on the board. (20320 + 18008 + 17400 =) Have students find the sum. (55,728 feet)

Now Try This! Encourage students to make up their own code words by determining which numbers spell the desired word and create a corresponding calculator problem. Remind them that decimal placement may affect the desired answer.

4 Assess

Ask students to name the special keys on a calculator and describe how to use them. ([C], [CE], [.]; [C] means that all entries are cleared from the calculator. [CE] means that the last entry is cancelled. The [.] is used as the decimal point.)

For Mixed Abilities

Common Errors • Intervention

Some students may get the incorrect answer using a calculator because they are not careful when they enter the numbers and operations. Have them practice by first entering problems with one operation, always estimating first to be sure that the answer displayed is reasonable.

Enrichment • Calculator

Tell students to use their calculators to find the total number of days from January 1 to August 26.

More to Explore • Application

Tell students that they must become "super shoppers" to find the best sale price of various items. Have students find advertisements that give the regular price and the amount of discount for some sale items. If they cannot find suitable ads, have them make some of their own for things they commonly purchase. Have them use a calculator to find the sale price of these items. Also, have students find ads that offer a rebate. Discuss what a rebate is and how it can make a sale price look better than it really is.

ESL/ELL STRATEGIES

When you introduce calculators, point out and name all the operation keys and the special keys. Describe procedures and answer any questions students may have. Provide sample language for students to repeat. For example, "You use the 'clear' key when you make a mistake."

Chapter 4 Test

page 71

Items	Objectives
1–4	To subtract two 2-digit numbers, check subtraction with addition (see page 55–56)
5–8	To subtract two 3-digit numbers, check subtraction with addition (see pages 57–58)
9–12	To subtract when the middle digit in a minuend is zero, check subtraction with addition (see pages 59–60)
13–20	To subtract two 4- or 5-digit numbers, check subtraction with addition (see pages 61–62)
21–28	To estimate differences of 3- and 4-digit numbers (see pages 65–66)

Name ____________________

CHAPTER 4 TEST

Subtract and check.

1. 68 − 34 = 34	**2.** 96 − 21 = 75	**3.** 61 − 37 = 24	**4.** 85 − 19 = 66
5. 375 − 123 = 252	**6.** 485 − 149 = 336	**7.** 718 − 253 = 465	**8.** 826 − 198 = 628
9. 503 − 322 = 181	**10.** 601 − 427 = 174	**11.** \$9.03 − 5.27 = \$3.76	**12.** \$4.00 − 3.72 = \$0.28
13. 7,294 − 3,485 = 3,809	**14.** \$85.37 − 29.18 = \$56.19	**15.** 16,375 − 9,887 = 6,488	**16.** \$483.29 − 136.84 = \$346.45
17. 4,006 − 2,988 = 1,018	**18.** \$30.00 − 11.76 = \$18.24	**19.** 70,004 − 27,268 = 42,736	**20.** \$300.25 − 129.37 = \$170.88

Round to the nearest thousand. Estimate each difference.

21. 8,275 − 3,927 → 8,000 − 4,000 = 4,000	**22.** 8,069 − 4,275 → 8,000 − 4,000 = 4,000	**23.** 7,523 − 2,130 → 8,000 − 2,000 = 6,000	**24.** 6,135 − 1,852 → 6,000 − 2,000 = 4,000

Round to the nearest hundred. Estimate each difference.

25. 796 − 249 → 800 − 200 = 600	**26.** 816 − 257 → 800 − 300 = 500	**27.** 923 − 375 → 900 − 400 = 500	**28.** 695 − 96 → 700 − 100 = 600

Alternate Chapter Test

You may wish to use the Alternate Chapter Test on page 352 of this book for further review and assessment.

Circle the letter of the correct answer.

1. 6 + 7 — a. 1 b. 11 c. 14 (d.) NG
2. 5 + 3 + 6 — a. 8 b. 9 (c.) 14 d. NG
3. 12 − 4 — a. 4 (b.) 8 c. 16 d. NG
4. What is the place value of the 0 in 3,076? — a. ones b. tens (c.) hundreds d. NG
5. 5,265 ◯ 4,265 — a. < (b.) >
6. Round 873 to the nearest hundred. — a. 700 b. 800 (c.) NG
7. Round 6,750 to the nearest thousand. — a. 6,000 (b.) 7,000 c. NG
8. What is the place value of the 8 in 823,075? — (a.) hundred thousands b. ten thousands c. thousands d. NG
9. 62 + 49 — a. 27 (b.) 111 c. 1,011 d. NG
10. 456 + 324 — a. 770 (b.) 780 c. 880 d. NG
11. 4,327 + 1,495 — a. 5,722 b. 5,812 c. 6,822 (d.) NG
12. 41,615 + 29,256 — (a.) 70,871 b. 70,881 c. 71,871 d. NG
13. 626 − 359 — (a.) 267 b. 333 c. 367 d. NG

score

STOP

Cumulative Assessment

page 72

Items	Objectives
1, 3	To compute basic addition and subtraction facts (see pages 1–2)
2	To add three 1-digit numbers (see pages 5–6)
4	To identify the place value of digits in numbers less than 10,000 (see pages 27–28)
5, 8	To compare and order numbers less than 1,000,000 (see pages 25–26)
6–7	To round numbers to the nearest hundred or thousand (see pages 31–32)
9	To add two 2-digit numbers (see pages 37–38)
10	To add two 3-digit numbers (see pages 39–40)
11	To add two 4-digit numbers (see pages 41–42)
12	To add two 5-digit numbers (see pages 51–52)
13	To subtract two 3-digit numbers (see pages 57–58)

Alternate Cumulative Assessment

Circle the letter of the correct answer.

1. 8 + 7 = — a 13 b 5 c 17 (d) NG
2. 6 + 4 + 7 — a 10 b 11 (c) 17 d NG
3. 14 − 6 = — a 7 b 9 c 11 (d) NG
4. What is the place value of the 9 in 6,497? — a ones (b) tens c hundreds d NG
5. 9,732 ◯ 6,541 — (a) > b < c =
6. Round 643 to the nearest hundred. — a 500 (b) 600 c 700 d NG
7. Round 8,965 to the nearest thousand. — a 7,000 b 8,000 (c) 9,000 d NG
8. What is the place value of the 4 in 649,823? — (a) ten thousands b thousands c tens d NG
9. 53 + 39 = — a 82 b 812 (c) 92 d NG
10. 624 + 217 — a 831 (b) 841 c 941 d NG
11. 7,616 + 2,295 — a 9,801 b 10,911 (c) 9,911 d NG
12. 62,427 + 18,244 — a 81,661 (b) 80,671 c 81,771 d NG
13. 414 − 248 — a 234 b 136 c 266 (d) NG

5-1 Understanding Multiplication

pages 73–74

1 Getting Started

Objective
- To understand multiplication

Vocabulary
factor, product

Materials
counters

Warm Up • Mental Math
Tell students to estimate the following to the nearest ten or dime:
1. \$451 − \$149 (\$300)
2. 18 years + 61 years (80 years)
3. $\frac{1}{2}$ of 164 (80)
4. 111 + 80 (190)
5. \$16.16 + \$0.45 (\$16.70)
6. 58 ÷ 2 (30)
7. 10:42 A.M. (10:40)
8. 72 − 18 (50)

Warm Up • Numeration
Write **2 + 2 + 2** on the board. Ask students the sum. (6) Write **3 + 3 + 3 + 3** and ask students the sum. (12) Continue to present problems of the same number added to itself several times and ask the sums. Alternate the problems so that students add horizontally and vertically.

Name ______________________

Multiplication Facts and Concepts

CHAPTER 5

Lesson 5-1

Understanding Multiplication

Charlie packs groceries after school at Food Mart. How many soup cans can he fit into a box?

We want to know how many soup cans fit into a box.

We know there are 4 rows with 3 cans in each row.

We can add the number of cans in each row.

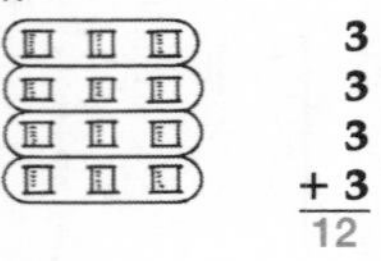

3
3
3
+ 3
12

We can add the number of cans in each column.

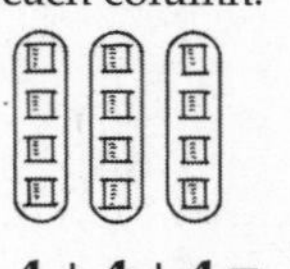

4 + 4 + 4 = 12

We can multiply the number of rows by the number of cans in each.

4 threes = 12

4 × 3 = 12

factor × factor = product

We can multiply the number of columns by the number of cans in each.

3 fours = 12

3 × 4 = 12

factor × factor = product

Charlie can fit 12 cans into one box.

Getting Started

Use both addition and multiplication to show how many items are in each picture.

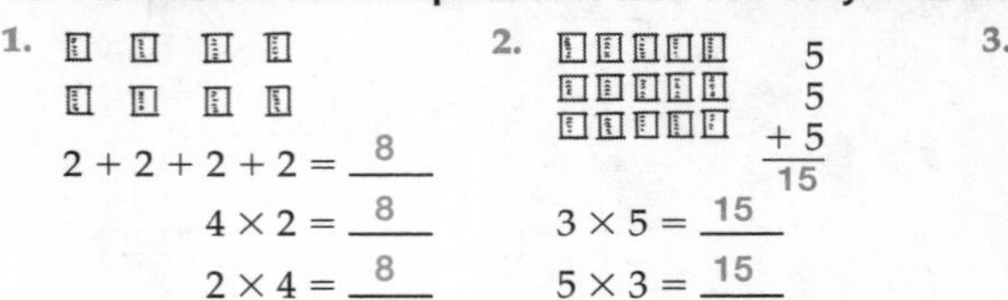

1. 2 + 2 + 2 + 2 = 8
   4 × 2 = 8
   2 × 4 = 8

2. 5 + 5 + 5 = 15
   3 × 5 = 15
   5 × 3 = 15

3. 4 + 4 + 4 + 4 = 16
   4 × 4 = 16

2 Teach

Introduce the Lesson Have a student read the problem aloud and tell what is to be solved. (the number of soup cans Charlie can pack into a box) Tell students that they can count all the cans to solve this simple problem, but sometimes it takes too long to count every item.

- Work through the two models with students to make sure they understand the relationship between addition and multiplication, and that multiplication is the quicker method of finding the total. Have students complete the solution sentence.

Develop Skills and Concepts Draw 4 rows of 5 Xs on the board. Have students count the Xs one by one to find the total. (20) Tell students that we could also add 5 + 5 + 5 + 5 or 4 + 4 + 4 + 4 + 4 to get a total of 20. Tell students that this adding method also takes a lot of time. Ask students how many rows of 5 Xs there are. (4)

- Write **4 fives = 20** and **4 × 5 = 20** on the board. Tell students that the 4 and the 5 are called **factors** and the 20 is called the **product**. Tell students that we can also write 5 fours = 20 and 5 × 4 = 20 because we could say that there are 5 rows of 4 Xs in each row.
- Write **2 × 6 = 12** on the board and have students tell the factors (2 and 6) and the product (12). Have a student draw 2 rows of 6 Xs to verify the product of 12. Have a student write **6 × 2 = 12** and verify the answer by showing 6 rows of 2 Xs in each row. Have students name the factors and products for more multiplication facts.

3 Practice

Have students complete page 74 independently.

Practice

Use both addition and multiplication to show how many items are in each picture.

1. $3 + 3 + 3 + 3 + 3 + 3 =$ 18
$6 \times 3 =$ 18
$3 \times 6 =$ 18

2. 5 + 5 = 10
$2 \times 5 =$ 10
$5 \times 2 =$ 10

3. $1 + 1 + 1 + 1 + 1 + 1 + 1 =$ 7
$7 \times 1 =$ 7
$1 \times 7 =$ 7

4. 6 + 6 + 6 + 6 = 24
$4 \times 6 =$ 24
$6 \times 4 =$ 24

5. 9 + 9 = 18
$2 \times 9 =$ 18
$9 \times 2 =$ 18

6. $4 + 4 + 4 + 4 + 4 + 4 + 4 =$ 28
$7 \times 4 =$ 28
$4 \times 7 =$ 28

7. $7 + 7 + 7 =$ 21
$3 \times 7 =$ 21
$7 \times 3 =$ 21

8. 5 + 5 + 5 + 5 + 5 = 25
$5 \times 5 =$ 25

4 Assess

Write **5 × 2 = ?** on the board. After students find the product, have them label each factor and the product. (10; 5 and 2 should be labeled "factor" and 10 should be labeled "product.")

For Mixed Abilities

Common Errors • Intervention

Some students may not make the connection between multiplication and repeated addition. Have them work with counters. For example, to find four 3s, or 4 × 3, have them lay out 4 sets of 3 counters and skip-count to find the total: 3, 6, 9, 12.

Enrichment • Spatial Sense

Have students draw a seating arrangement for 72 people in a long, narrow room. Half of the people should be on each side of an aisle.

More to Explore • Logic

Dictate the following problem for students to solve: A man found that he could read 80 pages of a book in 80 minutes when he wore his red shirt. However, when he wore his blue shirt, it took him an hour and 20 minutes to read the same number of pages. How could this be true? (The shirt, of course, has nothing to do with it. Eighty minutes and an hour and 20 minutes are the same length of time.)

5-2 Multiplying, the Factors 2 and 3

pages 75–76

1 Getting Started

Objective
- To multiply by the factor 2 or 3

Materials
counters; $\frac{1}{4}$-in. grid paper

Warm Up • Mental Math
Tell students to halve the number and subtract 9.

1. 120 (51)
2. 200 (91)
3. 1,050 (516)
4. 52 (17)
5. 300 (141)
6. 98 (40)
7. 60 (21)
8. 76 (29)

Warm Up • Numeration
Write the following on the board:

3 × 5 = 15

5 × 3 = 15

Now, write **3 + 3 + 3 + 3 = 12** on the board and have students write the two multiplication facts for four 3s. (4 × 3 = 12, 3 × 4 = 12) Repeat for other addition sentences. Vary the activity by writing more multiplication facts and have students write the corresponding addition problems.

Name ______________________

Multiplying, the Factors 2 and 3

Rose helped her uncle pick cherries. She saved some of the cherries for her lunch. How many cherries did Rose save for lunch?

We want to know how many cherries Rose saved for lunch.

We know there are __4__ bunches of __3__ cherries each on the plate.

To find out how many cherries she saved, we multiply __4__ by __3__.

4 × 3 = __12__

$$\begin{array}{r} 3 \\ \times\ 4 \\ \hline 12 \end{array}$$

We say **four times three equals twelve.**

Rose saved __12__ cherries for lunch.

Getting Started

Complete the table.

1.

Multiplication Facts for 2									
0	1	2	3	4	5	6	7	8	9
0	2	4	6	8	10	12	14	16	18

Multiply.

2. 5 × 3 = __15__
3. 4 × 2 = __8__
4. $\begin{array}{r} 2 \\ \times 9 \\ \hline 18 \end{array}$
5. $\begin{array}{r} 3 \\ \times 7 \\ \hline 21 \end{array}$
6. 3 × 9 = __27__
7. 2 × 2 = __4__
8. $\begin{array}{r} 2 \\ \times 6 \\ \hline 12 \end{array}$
9. $\begin{array}{r} 3 \\ \times 2 \\ \hline 6 \end{array}$

2 Teach

Introduce the Lesson Have a student read the problem aloud and tell what is to be solved. (how many cherries Rose saved for lunch) Ask students to count each cherry to tell the total. (12) Next, have them count the number of cherries in each group and the number of groups and fill in the information sentences. Work through the model with students, showing them how multiplication can be used to quickly solve the problem. Then, have students complete the solution sentence.

Develop Skills and Concepts Have students lay out 1 group of 2 counters and tell how many counters there are in all. (2) Write **1 + 1 = 2, one 2 = 2, 1 times 2 equals 2,** and **1 × 2 = 2** across the board. Have students read each with you. Have students tell the total number of counters. (2)

- Tell students to lay out another group of 2 counters. Have students write **2 + 2 = 4, two 2s = 4, 2 times 2 equals 4**, and **2 × 2 = 4** under the facts for 1 group of 2. Have students read each with you and then count by 2s to tell the total number of 2s. (2, 4)
- Continue to develop the facts for 2 through 9 × 2. As students complete the table, ask them what pattern they see in the products. (Each product is 2 greater than the previous product.) Repeat the procedure for the facts for 3. Remind students that in 2 × 2 = 4, each 2 is a factor and 4 is the product. When all facts for 2 and 3 are developed, have students read some of the facts and tell the factors and the product of each.

3 Practice

Have students complete the table and then work the exercises on page 76. Tell students that they must decide which operation to use in each of the word problems.

Practice

Complete the table.

1.

Multiplication Facts for 3									
0	1	2	3	4	5	6	7	8	9
0	3	6	9	12	15	18	21	24	27

Multiply.

2. $2 \times 3 =$ 6
3. $3 \times 4 =$ 12
4. $3 \times 5 =$ 15
5. $2 \times 6 =$ 12
6. $2 \times 2 =$ 4
7. $3 \times 9 =$ 27
8. $2 \times 8 =$ 16
9. $2 \times 7 =$ 14
10. $3 \times 8 =$ 24
11. $3 \times 2 =$ 6
12. $2 \times 5 =$ 10
13. $3 \times 3 =$ 9
14. $2 \times 9 =$ 18
15. $3 \times 7 =$ 21
16. $3 \times 6 =$ 18
17. $2 \times 4 =$ 8

18. 2×8 = 16
19. 3×9 = 27
20. 2×4 = 8
21. 2×5 = 10
22. 3×6 = 18
23. 3×7 = 21
24. 3×8 = 24
25. 2×9 = 18
26. 2×5 = 10
27. 3×5 = 15
28. 3×4 = 12
29. 2×7 = 14

Problem Solving

Solve each problem.

30. Fred has outgrown 4 pairs of tennis shoes in one year. How many shoes has he outgrown?
8 shoes

31. There are 9 vases, each containing 3 daisies. How many daisies are there altogether?
27 daisies

32. There are 5 study tables in the library. Each table has 2 chairs. How many chairs are there?
10 chairs

33. Bill has 9 model cars to build. Don has 3 cars. How many more cars does Bill have?
6 cars

34. Betty has 2 records. Each record has 8 songs. How many songs can Betty listen to if she plays both records?
16 songs

35. Ilonda ate 3 apples on Tuesday, 5 apples on Wednesday, and 2 apples on Friday. How many apples did she eat altogether?
10 apples

4 Assess

Ask students to name the products in the following exercises: 2×1 (2), 2×6 (12), 3×3 (9), 3×6 (18), and 2×3 (6).

For Mixed Abilities

Common Errors • Intervention

Some students may have trouble learning their facts for 3. Have them use grid paper. In the first row, have them color 3 squares and write $1 \times 3 = 3$. In the next row, have them color 3 squares with one color and the adjacent 3 squares with another color and write $2 \times 3 = 6$. In the third row, have them show 3 sets of 3, alternating colors, and write $3 \times 3 = 9$. Have them continue in this manner to show all the facts of 3 through $9 \times 3 = 27$.

Enrichment • Multiplication

Have students make multiplication fact cards for 2 and 3 through 15×2, 2×15, 15×3, and 3×15.

More to Explore • Numeration

Have students review the tables they made for the facts for 2 and 3. Then, have students develop rules for when a product will be an even number and when a product will be an odd number.

5-3 Multiplying, the Factors 4 and 5

pages 77–78

1 Getting Started

Objective
- To multiply by the factor 4 or 5

Materials
*multiplication fact cards for 2 and 3

Warm Up • Mental Math
Ask if the following statements are true or false:

1. 1912 was 40 years before 1952. (T)
2. 20 nickels = $1.50 (F)
3. $\frac{1}{4}$ of $1.00 = 25¢ (T)
4. 5 cups > 1 pint (T)
5. $n = 2$ if $n + 3 + 8 = 15$ (F)
6. 1 year = 52 weeks (T)
7. 230 = 22 tens and 10 ones (T)
8. $500 < \frac{1}{2}$ of 1,000 (F)

Warm Up • Numeration
Show the 3 × 4 fact card and have students tell a related addition problem to find the total. (4 + 4 + 4 = 12) Repeat for other facts.

Name ______________________________

Multiplying, the Factors 4 and 5

Mr. Kerry is making pickles. How many canning jars did he buy?

We want to know how many jars Mr. Kerry bought.

We know Mr. Kerry bought __5__ boxes of __4__ canning jars each.

To find out how many jars are in the boxes, we think of 5 sets of 4 each.

We can count by __4s__.

4 8 12 16 20

We can also multiply __5__ by __4__.

1 × 4 = __4__ **2 × 4 =** __8__ **3 × 4 =** __12__ **4 × 4 =** __16__ **5 × 4 =** __20__

Mr. Kerry bought __20__ canning jars.

Getting Started

Complete the table.

1.

Multiplication Facts for 4									
0	1	2	3	4	5	6	7	8	9
0	4	8	12	16	20	24	28	32	36

Multiply.

2. 6 × 4 = __24__
3. 9 × 5 = __45__
4. 5 × 2 = 10
5. 4 × 7 = 28

2 Teach

Introduce the Lesson Have a student read the problem aloud and tell what is to be solved. (how many canning jars Mr. Kerry bought) Ask students what information is given. (the number of boxes and the number of jars in each) Have students fill in the information sentences. Work through the model with students, having them supply answers for each fact. Then, have them complete the solution sentence.

Develop Skills and Concepts Draw a number line across the board. Mark intervals to accommodate the numbers from 0 to 32. Have a student write the numbers 1 through 4 along the line and tell how many 4s there are. (1) Write **1 × 4 = 4** on the board. Have another student write the numbers from 5 to 8 along the line and tell how many 4s there are in all. (2) Have the student write 2 × 4 = 8 on the board. Have students continue to develop multiples of 4 along the line and write the facts.

Have students count by 4s as you draw arcs from 0 to 4 to 8 to 12 and so on. Now, point to the products of 4 at random as students tell the two factors that equal each product. Repeat this process with the factor 5.

3 Practice

Have students complete the table and then work the exercises on page 78. Tell students that they must decide which operation they will need to use to solve each word problem.

Practice

Complete the table.

1.

Multiplication Facts for 5									
0	1	2	3	4	5	6	7	8	9
0	5	10	15	20	25	30	35	40	45

Multiply.

2. $2 \times 4 =$ 8
3. $4 \times 5 =$ 20
4. $5 \times 5 =$ 25
5. $7 \times 4 =$ 28
6. $5 \times 4 =$ 20
7. $6 \times 5 =$ 30
8. $8 \times 5 =$ 40
9. $3 \times 4 =$ 12
10. $4 \times 4 =$ 16
11. $2 \times 5 =$ 10
12. $3 \times 5 =$ 15
13. $6 \times 4 =$ 24
14. $9 \times 4 =$ 36
15. $7 \times 5 =$ 35
16. $9 \times 5 =$ 45
17. $8 \times 4 =$ 32
18. 5×5 = 25
19. 5×8 = 40
20. 2×2 = 4
21. 4×7 = 28
22. 3×4 = 12
23. 2×9 = 18
24. 4×9 = 36
25. 5×4 = 20
26. 3×6 = 18
27. 5×7 = 35
28. 5×2 = 10
29. 4×6 = 24
30. 5×9 = 45
31. 3×3 = 9
32. 4×8 = 32
33. 3×7 = 21
34. 2×3 = 6
35. 4×5 = 20

Problem Solving

Solve each problem.

36. Each package holds 5 sticks of gum. How many sticks of gum are in 8 packages?
40 sticks

37. Michelle has 8 seedlings. She bought 5 more from the florist. How many seedlings does Michelle have?
13 seedlings

38. Brian had $5. He spent $4 at the movies. How much does Brian have left?
$1

39. Sam and Mike each bought 4 cans of apple juice. How many cans of juice did they buy together? 8 cans

4 Assess

Ask students to name the products in the following exercises: 4×3 (12), 4×9 (36), 5×4 (20), 5×7 (35), and 5×3 (15).

5 Mixed Review

1. 7,040 − 3,236 (3,804)
2. $169.32 + $38.78 ($208.10)
3. 3×8 (24)
4. 378 + 265 + 674 (1,317)
5. 6,483 − 4,252 (2,231)
6. $570.00 − $376.28 ($193.72)
7. 5×8 (40)
8. 7 + 3 + 5 + 8 + 3 + 2 (28)
9. $17.56 + $21.74 + $31.58 ($70.88)
10. 9×3 (27)

For Mixed Abilities

Common Errors • Intervention

Some students may have difficulty remembering the facts for 4 and 5. Have them work with a partner and flashcards to quiz each other, first with the facts in order and then with the facts in random order.

Enrichment • Geometry

Have students draw squares or rectangles on grid paper to show the multiplication facts for 5. They should then find the area and perimeter of each figure.

More to Explore • Measurement

Write **millimeter, centimeter,** and **kilometer** on the board. Ask students to copy the words and underline the root word. Ask students to find out what the origin of each prefix is, including both its language and meaning.

5-4 Basic Properties, Multiplication

pages 79–80

1 Getting Started

Objective

- To understand the Commutative, Zero, and Identity Properties of Multiplication

Warm Up • Mental Math

Ask students to tell whether they would use inches, feet, or yards to measure the following:

1. a room (feet)
2. a notebook (inches)
3. fabric (yards)
4. a television screen (inches)
5. scissors (inches)
6. a football field (yards)
7. a screwdriver (inches)
8. a parking space (feet)

Warm Up • Algebra

Review with students that any number plus zero equals that number and any number minus zero equals that number. Write sample addition and subtraction problems with zeros as addends or subtrahends on the board. Have students supply the answers. Remind students that zero always means "nothing."

Name ______________________

Lesson 5-4

It's Algebra!

Basic Properties, Multiplication

There are some ideas that are important when we multiply.

Order or Commutative Property

We can multiply factors in any order.

6 × 4 = 24

4 × 6 = 24

Zero Property

The product of a number and 0 is always 0.

4 × 0 = 0

Identity Property

The product of a number and 1 is that number.

4 × 1 = 4

Getting Started

Write *yes* or *no*.

1. Does a number times zero equal that number? no
2. Is a number times 5 the same as 5 times the number? yes
3. Does the Commutative Property tell us that numbers can be multiplied in any order? yes

Multiply.

4. $1 \times 7 =$ 7 5. $7 \times 1 =$ 7 6. $0 \times 8 =$ 0 7. $8 \times 0 =$ 0

2 Teach

Introduce the Lesson Read the first two sentences on the page to students. Then, read each property one by one and discuss the examples with the class.

Develop Skills and Concepts Have a student draw 3 circles of 5 Xs each on the board. Have students tell the total. (15) Have another student draw 5 circles of 3 Xs each. Ask the total. (15) Ask students if 3 × 5 has the same product as 5 × 3. (yes) Tell students that two numbers can be multiplied in any order and the product is the same.

Now, have 2 students stand with nothing in their hands. Ask students how many books each student is holding. (none) Ask how many books the 2 students are holding in all. (none) Write **2 × 0 = 0** on the board. Have 1 more student join the 2 students and ask how many students are holding no books. (3) Ask the product of 3 × 0. (0) Tell students that 3 times nothing, or zero, is always zero. Now, give each student 1 book to hold. Ask how many books each student has. (1) Ask how many students there are in all. (3) Ask the product of 3 × 1. (3) Add 2 more students with 1 book each. Ask what 5 times 1 equals. (5) Tell students that any number times 1 equals that number.

3 Practice

Tell students to answer each of the questions and then solve the exercises on page 80.

Practice

Write *yes* or *no*.

1. If you multiply a number by zero, will the product always be zero? yes
2. Is 6 times a number the same as the number times 6? yes
3. If you multiply 1 by 0, is the answer 1? no
4. If you multiply a number by 1, is the product always the same as the other factor? yes
5. If you multiply 1 by 1, is the product 2? no

Multiply.

6. $3 \times 3 = 9$
7. $8 \times 2 = 16$
8. $3 \times 1 = 3$
9. $9 \times 0 = 0$
10. $6 \times 5 = 30$
11. $5 \times 6 = 30$
12. $0 \times 2 = 0$
13. $4 \times 4 = 16$
14. $8 \times 3 = 24$
15. $3 \times 8 = 24$
16. $1 \times 7 = 7$
17. $7 \times 1 = 7$
18. $5 \times 3 = 15$
19. $3 \times 4 = 12$
20. $9 \times 5 = 45$
21. $1 \times 7 = 7$
22. $0 \times 8 = 0$
23. $2 \times 5 = 10$
24. $3 \times 7 = 21$
25. $1 \times 9 = 9$
26. $8 \times 5 = 40$
27. $0 \times 4 = 0$
28. $4 \times 6 = 24$
29. $1 \times 2 = 2$
30. $0 \times 0 = 0$
31. $4 \times 5 = 20$
32. $7 \times 4 = 28$
33. $2 \times 4 = 8$
34. $1 \times 5 = 5$
35. $8 \times 5 = 40$
36. $3 \times 2 = 6$
37. $5 \times 9 = 45$
38. $2 \times 9 = 18$
39. $3 \times 6 = 18$
40. $5 \times 6 = 30$
41. $9 \times 4 = 36$
42. $6 \times 0 = 0$
43. $1 \times 1 = 1$
44. $9 \times 9 = 81$
45. $0 \times 6 = 0$
46. $8 \times 9 = 72$
47. $9 \times 8 = 72$

4 Assess

Ask students which property says that you can multiply factors in any order. (Commutative or Order Property)

For Mixed Abilities

Common Errors • Intervention

Some students may confuse multiplication by 1 with multiplication by 0 and think that 0 times any number is that number. Have students draw 5 stick figures on paper. Then, tell them to draw zero tennis balls for each figure and tell how many balls in all they drew and have them write the number sentence $0 \times 5 = 0$ to represent this.

Enrichment • Application

Have students draw a picture to illustrate the following looks-can-be-deceiving situation: The coach had 9 tennis-ball cans, each of which can hold 3 tennis balls. The cans looked new, so the coach assumed that he had 27 new balls. However, he found that he had only 13 tennis balls because 3 cans were empty and only 2 cans were full.

More to Explore • Measurement

Using newspapers and magazines, working in groups or individually, have students find as many examples of units of weight and measure as they can. Emphasize that they must find examples of both metric and customary measures. Each example should be cut out and pasted on a chart. Have students compare their findings and identify the examples that are the least and greatest for each unit of measure.

ESL/ELL STRATEGIES

Explain the term *properties* and describe each multiplication property in simple terms. Then, ask students to give original examples of each one. For example, for the Commutative Property, a student might say, "Four times three is twelve and three times four is twelve."

5-5 Understanding the Rule of Order

pages 81–82

1 Getting Started

Objective
- To understand the rule of order

Materials
*counters; *cards numbered 0 through 9

Warm Up • Mental Math
Have students solve the following:

1. 6 tens plus 308 (368)
2. value of n if $10 + 7 - n = 5$ (12)
3. 10 tens minus 47 (53)
4. nine 3s + 3 (30)
5. 5 minus 1.5 (3.5)
6. $\frac{1}{5}$ of 50 (10)
7. $48 \div 6 + 12$ (20)
8. $1 + 9 - (6 \div 2)$ (7)

Warm Up • Multiplication
Have students tell the product of 4×6 as you write the fact on the board. (24) Ask students to tell the product of 6×4. (24) Continue reviewing the facts for 2 through 5.

Name ______________________________

Lesson 5-5

Understanding the Rule of Order

Working an operation in a mathematical sentence to find the value of n is called solving for n. Solve for n in the sentence on the board.

$9 - 4 \times 2 = n$

If we subtract and then multiply, the value of n is 10.

If we multiply and then subtract, the value of n is 1.

Both answers cannot be correct. We must follow the **rule of order** to know how to solve this sentence correctly.

First, work all multiplications left to right. Then, work all additions and subtractions left to right.

$\mathbf{9 - 4 \times 2 = n}$
$\mathbf{9 -}$ 8 $\mathbf{= n}$
1 $\mathbf{= n}$

In the rule of order, operations within parentheses should be worked before multiplications.

$\mathbf{(9 - 4) \times 2 = n}$
5 $\mathbf{\times 2 = n}$
10 $\mathbf{= n}$

The correct answer for the sentence on the board is 1.

The rule of order has been followed in these three mathematical sentences. Solve for n.

$3 \times 4 - 3 = n$	$(3 \times 4) - 3 = n$	$3 \times (4 - 3) = n$
12 $- 3 = n$	12 $- 3 = n$	$3 \times$ 1 $= n$
9 $= n$	9 $= n$	3 $= n$

Getting Started

Solve for n. Follow the rule of order.

1. $5 + (3 \times 4) = n$ — 17 $= n$
2. $3 + 4 \times 5 = n$ — 23 $= n$
3. $(6 - 0) \times 4 = n$ — 24 $= n$

2 Teach

Introduce the Lesson Write the following on the board: **Slow Cattle Crossing**. Discuss the confusion one could have when seeing this sign. Help students see that it could mean that slow cattle are crossing or that one should go slowly because cattle are crossing. Ask students to suggest ways to punctuate the words to help the reader understand its meaning.

- Have students read the problem in the picture of their texts. Tell students that math problems can also be confusing unless we know whether to subtract first or to multiply first. Read the introduction to students. Work through the rule of order with students. Have them complete the solution sentence. Then, read the last section, working through each problem with students.

Develop Skills and Concepts Tell students that parentheses and the rule of order are used in math to help us understand how to work a problem. Write $\mathbf{4 + 2 \times 6 =}$ ____ on the board. Tell students that the **rule of order** tells us to multiply from left to right first, so we have $4 + 12 = 16$. Write $\mathbf{4 + (2 \times 6) =}$ ____ on the board. Ask students if the parentheses changed this problem. (No, the rule of order tells us to multiply first anyway.) Now, write $\mathbf{(4 + 2) \times 6 = 36}$ on the board and tell students that the parentheses have changed this problem because the rule of order also tells us to work the problem within parentheses first. Present more problems with and without parentheses.

3 Practice

Remind students to look for parentheses and work that part of the problem first. Tell students that if there are no parentheses, they must multiply left to right and then do additions or subtractions from left to right. Have students complete the exercises on page 82 independently.

Practice

Solve for n. Follow the rule of order.

1. $(2 \times 5) + 3 = n$ — 13 $= n$
2. $8 + (3 \times 4) = n$ — 20 $= n$
3. $(7 \times 3) - 9 = n$ — 12 $= n$
4. $4 \times 3 + 7 = n$ — 19 $= n$
5. $(7 \times 4) + 15 = n$ — 43 $= n$
6. $(6 \times 5) - 18 = n$ — 12 $= n$
7. $(24 - 16) \times 4 = n$ — 32 $= n$
8. $(2 \times 3) \times 4 = n$ — 24 $= n$
9. $(5 - 3) \times 7 = n$ — 14 $= n$
10. $8 - (6 \times 1) = n$ — 2 $= n$
11. $5 + 3 \times 6 = n$ — 23 $= n$
12. $9 \times (8 - 8) = n$ — 0 $= n$
13. $(7 - 6) \times 4 = n$ — 4 $= n$
14. $(7 \times 5) + 26 = n$ — 61 $= n$
15. $56 - (4 \times 8) = n$ — 24 $= n$
16. $46 + (9 \times 0) = n$ — 46 $= n$
17. $(3 \times 9) + 46 = n$ — 73 $= n$
18. $(5 \times 2) + 30 = n$ — 40 $= n$
19. $7 \times 2 - 8 = n$ — 6 $= n$
20. $72 + (8 \times 4) = n$ — 104 $= n$
21. $(95 - 95) \times 4 = n$ — 0 $= n$

Now Try This!

How many numbers can you make using four 4s?

$(4 \times 4) \times (4 \times 4) = 256$
$(4 \times 4) + (4 \times 4) = 32$
$4 \times (4 + 4) \times 4 = 128$
$(4 + 4) \times (4 + 4) = 64$
$4 + (4 \times 4) + 4 = 24$
$(4 \times 4) - (4 \times 4) = 0$
$(4 \times 4) - (4 + 4) = 8$
$(4 + 4 + 4) \times 4 = 48$
Give credit for any other reasonable answer.

Now Try This! Through the use of addition, subtraction, multiplication, and parentheses, students are to make different numbers using four 4s. Tell students to place parentheses around the operation to be solved first in each problem.

4 Assess

Ask students what the rule of order tells us about operations within parentheses. (They should be worked before multiplications.)

For Mixed Abilities

Common Errors • Intervention

Some students may do the operations in order from left to right instead of using the rule of order. Have them work in pairs to solve three number sentences such as the following:

$3 + (8 \times 5) =$ (43)

$(3 + 8) \times 5 =$ (55)

$3 + 8 \times 5 =$ (43)

After they find the answers, have them discuss how this activity illustrates that a rule of order is necessary so that the last sentence has just one answer.

Enrichment • Numeration

Have students use multiplication and addition to write and solve a problem that tells how many days are in a year.
$(30 \times 12 + 5 = 365;$
$7 \times 52 + 1 = 365)$

More to Explore • Numeration

Have a student bring in a box of toothpicks. Tell the student who brought in the toothpicks to solve the following toothpick puzzles by carefully following directions.

1. Take away two toothpicks to make this solution true:
   VII − I = I
   (Remove the V, then II − I = I)
2. You are given nine toothpicks. Can you make ten out of them? No, you can't break them! (The toothpicks spell the word *TEN*.)

5-6 Multiplying, the Factors 6 and 7

pages 83–84

1 Getting Started

Objective
- To multiply by the factor 6 or 7

Materials
*counters; *sheets of paper

Warm Up • Mental Math
Ask students how many there are.

1. 4s in 32 − 8 (6)
2. 5s in 40 + 5 (9)
3. 3s in 16 − 13 (1)
4. 4s in 125 − 25 (25)
5. 6s in 24 ÷ 4 (1)
6. halves in 2 wholes (4)
7. quarts in 4 gallons (16)
8. legs on 3 octopi (24)

Warm Up • Numeration
Write **(3 × 2) + (4 × 2) =** _____ on the board. Have students work the problem to tell what 7 × 2 equals. (14) Continue to present more fact-plus-fact problems to have students find the total of three 4s, four 6s, five 7s, and three 6s.

Name ______________________

Multiplying, the Factors 6 and 7

It is exactly 8 weeks from New Year's Day to Opal's birthday. How many days does Opal have to wait to celebrate her birthday?

We want to find the number of days Opal has to wait for her birthday.

We know that Opal's birthday is __8__ weeks from New Year's Day.

There are __7__ days in 1 week.

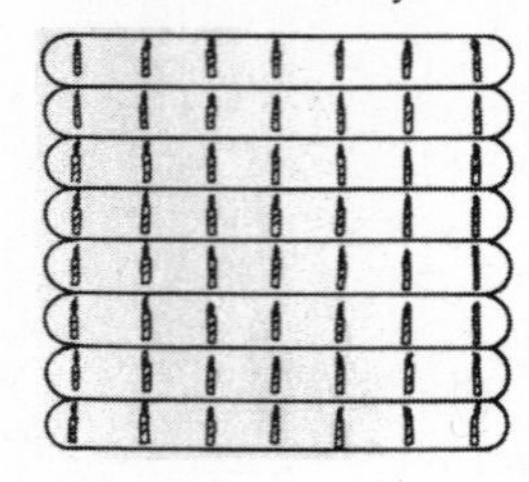

7 × 8 = __56__

8 × 7 = __56__

Opal has to wait __56__ days to celebrate her birthday.

Getting Started

Complete the table.

1.

Multiplication Facts for 6									
0	1	2	3	4	5	6	7	8	9
0	6	12	18	24	30	36	42	48	54

Multiply.

2. 3 × 6 = __18__
3. 5 × 6 = __30__
4. 4 × 6 = __24__
5. 2 × 6 = __12__
6. 6 × 9 = 54
7. 7 × 6 = 42
8. 6 × 6 = 36
9. 6 × 8 = 48

2 Teach

Introduce the Lesson Have students describe the picture. (A girl is pointing to January 1 on a calendar.) Have a student read the problem aloud and tell what the significance of January 1 is. (Opal's birthday is 8 weeks from January 1.) Ask students what holiday is on January 1. (New Year's Day) Ask what problem is to be solved. (how many days until Opal's birthday) Ask students what information is known. (January 1 is 8 weeks from Opal's birthday.) Work through the model with students and then have them complete the solution sentence.

Develop Skills and Concepts Write the following on the board:

1 × 6 = 6	2 × 6 = 12	3 × 6 = 18	4 × 6 = 24
(6 × 1 = 6)	(6 × 2 = 12)	(6 × 3 = 18)	(6 × 4 = 24)

Have students write the related multiplication fact for each. Ask students what (5 × 6) + (1 × 6) would equal. (36) Continue to develop the 6s in this way through 6 × 9 = 54. Repeat the activity for the facts for 7 through 7 × 9 = 63.

3 Practice

Have students complete the table and work the multiplication exercises on page 84. Tell students they must decide which operation to use to solve each word problem.

Practice

Complete the table.

1.

Multiplication Facts for 7									
0	1	2	3	4	5	6	7	8	9
0	7	14	21	28	35	42	49	56	63

Multiply.

2. $5 \times 6 =$ 30
3. $1 \times 7 =$ 7
4. $2 \times 6 =$ 12
5. $7 \times 7 =$ 49
6. $3 \times 7 =$ 21
7. $9 \times 7 =$ 63
8. $0 \times 6 =$ 0
9. $4 \times 6 =$ 24
10. $6 \times 7 =$ 42
11. $7 \times 6 =$ 42
12. $4 \times 7 =$ 28
13. $1 \times 6 =$ 6
14. $3 \times 6 =$ 18
15. $9 \times 6 =$ 54
16. $8 \times 7 =$ 56
17. $0 \times 7 =$ 0
18. 6×6 = 36
19. 7×5 = 35
20. 7×2 = 14
21. 6×8 = 48
22. 6×1 = 6
23. 7×4 = 28
24. 7×5 = 35
25. 6×0 = 0
26. 6×4 = 24
27. 7×0 = 0
28. 6×5 = 30
29. 7×8 = 56
30. 7×1 = 7
31. 7×7 = 49
32. 7×9 = 63
33. 6×2 = 12
34. 7×3 = 21
35. 6×7 = 42

Problem Solving

Solve each problem.

36. I have 9 key rings. Each key ring holds 6 keys. How many keys do I have?
54 keys

37. There are 15 apples in a bag. Seven apples are rotten. How many apples are not rotten?
8 apples

38. One ticket for the ringtoss game costs 5¢. How much do 7 tickets cost?
35¢

39. Jessica waters 6 of her plants each day, in rotation. By the end of the week, she has watered all her plants. How many plants does Jessica have?
42 plants

4 Assess

Ask students to explain how knowing the product of 6×4 helps them know the product of 4×6. (The factors are the same, so the product is also the same.)

5 Mixed Review

1. 587 − 379 (208)
2. 6,020 − 4,317 (1,703)
3. 6×7 (42)
4. $60 - (7 \times 6)$ (18)
5. 26,176 + 6,275 (32,451)
6. 5×3 (15)
7. \$495.20 − \$168.76 (\$326.44)
8. 295 + 15,438 (15,733)
9. 7×8 (56)
10. 396 + 478 + 209 (1,083)

For Mixed Abilities

Common Errors • Intervention

Some students may have difficulty with facts for 6 and 7 because they think they must learn 20 new facts. Use the Order Property to show that they already know many of these facts, such as $2 \times 7 = 7 \times 2 = 14$. Thus, they only need to learn 7 new facts. They can draw arrays or use flashcards to help learn these new facts.

Enrichment • Geometry

Have students draw and label the side measurements of rectangles or squares whose areas are 54 square yards, 60 square miles, 49 square centimeters, and 56 square feet.

More to Explore • Applications

Have students estimate the cost of a pizza to go. Get a copy of pizza prices from a local pizzeria. Discuss how prices are based on the size of the pizza and the number of toppings ordered. Have each student practice figuring the price of a pizza by first letting them write their own favorite pizza order on an order slip. Put all of the orders in a box and let everyone take turns being the store manager. Students draw an order from the box and calculate the price of the order from the price list. Then, have students suppose they were splitting the cost of their pizza among four friends. How would they decide what each of them should pay?

5-7 Multiplying, the Factors 8 and 9

pages 85–86

1 Getting Started

Objective
- To multiply by the factor 8 or 9

Warm Up • Mental Math
Have students name the century for the following years:

1. 1216 (13th)
2. 1977 (20th)
3. 80 years ago (20th)
4. 20 years before 1702 (17th)
5. 2 centuries ago (19th)
6. 50 years after 1776 (19th)
7. 100 years into the future (22st)

Warm Up • Algebra
Write **7 + 2 × 7** on the board. Have students put parentheses in this problem to show which operation is to be done first. [7 + (2 × 7)] Have a student work the problem. (21) Continue with more problems having multiplication and addition or subtraction.

Name ____________________

Multiplying, the Factors 8 and 9

The Crosby County baseball team plays 8 games each summer. If all the games are complete, how many innings does the team play?

We want to know the number of innings the Crosby County baseball team plays over the summer.

A regular baseball game lasts for 9 innings.

We know that Crosby County plays 8 games.

To find the total number of innings, we multiply 9 by 8.

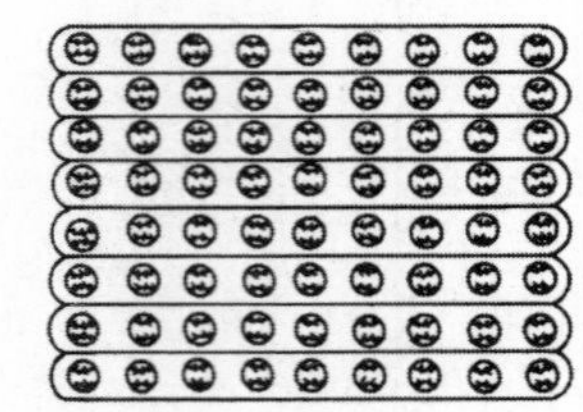

9 × 8 = 72

8 × 9 = 72

The Crosby County baseball team plays 72 innings.

Getting Started

Complete the table.

1.

Multiplication Facts for 8									
0	1	2	3	4	5	6	7	8	9
0	8	16	24	32	40	48	56	64	72

Multiply.

2. 3 × 8 = 24
3. 7 × 8 = 56
4. 0 × 8 = 0
5. 5 × 8 = 40
6. 8 × 9 = 72
7. 8 × 8 = 64
8. 8 × 7 = 56
9. 6 × 8 = 48

2 Teach

Introduce the Lesson Have a student read the problem aloud and tell what is to be solved. (how many innings the Crosby County baseball team plays each summer) Ask students what fact is given in the problem. (The team plays 8 complete games.) Ask students what fact is needed. (how many innings are in a game) Ask students if this information is given. (yes) Ask students if we need the scores of each team to answer this problem. (no) Work through the model with students and then have them complete the solution sentence.

Develop Skills and Concepts Write the following on the board:

1 × 8 = 8	2 × 8 = 16	3 × 8 = 24	4 × 8 = 32
(8 × 1 = 8)	(8 × 2 = 16)	(8 × 3 = 24)	(8 × 4 = 32)

Have students write the related multiplication fact for each. Remind students that (4 × 8) + (1 × 8) = 5 × 8, or 40. Help students develop the remaining facts for 8 through 8 × 9 = 72. Write each fact on the board and have students write the related multiplication fact for each. Repeat the activity to develop the facts for 9 through 9 × 9 = 81.

3 Practice

Have students complete the table for the facts for 9 and work the exercises on page 86.

Practice

Complete the table.

1.

Multiplication Facts for 9									
0	1	2	3	4	5	6	7	8	9
0	9	18	27	36	45	54	63	72	81

Multiply.

2. $0 \times 9 =$ 0
3. $8 \times 9 =$ 72
4. $1 \times 8 =$ 8
5. $3 \times 8 =$ 24
6. $7 \times 8 =$ 56
7. $4 \times 9 =$ 36
8. $9 \times 8 =$ 72
9. $6 \times 9 =$ 54
10. $4 \times 8 =$ 32
11. $0 \times 8 =$ 0
12. $7 \times 9 =$ 63
13. $3 \times 9 =$ 27

14. 9×5 = 45
15. 8×6 = 48
16. 9×9 = 81
17. 8×2 = 16
18. 8×9 = 72
19. 8×8 = 64
20. 8×0 = 0
21. 8×5 = 40
22. 9×6 = 54
23. 9×1 = 9
24. 8×2 = 16
25. 9×7 = 63
26. 8×4 = 32
27. 8×7 = 56
28. 9×4 = 36
29. 9×0 = 0
30. 8×1 = 8
31. 8×3 = 24

Problem Solving

Solve each problem.

32. Walter practiced his drums for 9 hours each week. How many hours did he practice in 6 weeks?
54 hours

33. Mary bought 8 vases. Each vase cost $8. How much did the vases cost Mary?
$64

34. It costs $8 for a ticket to see Bill's favorite musical group. He has saved $5. How much more does he need for a ticket?
$3

35. Paper plates for a picnic are packed in packages of 8 each. How many plates are there in 7 packages?
56 plates

4 Assess

Ask students to make a drawing that shows 9×2 and then have them explain how to use that drawing to find 9×4. (draw the same thing again)

For Mixed Abilities

Common Errors • Intervention

Some students may have difficulty with facts for 8 and 9 because they think they must learn 20 new facts. Use the Order Property to show that they already know many of these facts, such as $2 \times 8 = 8 \times 2 = 16$. Thus, they only need to learn 3 new facts. They can draw arrays or use flashcards to help learn these new facts. It also might help to remember facts for 9 if they recognize that the sum of the digits in the product for any fact is 9. For example, the sum of the digits in the product 72 for the multiplication fact 8×9 is 9.

Enrichment • Numeration

Have students work in pairs. Each student should toss two dice a total of 20 throws. Have students list a multiplication fact for each throw, using each die as a factor. They should then exchange lists of facts and work each problem.

More to Explore • Numeration

This activity could be used over several days. Have students solve the following math riddles by answering which two numbers have

1. a product of 48, a sum of 14, and a difference of 2 (6, 8)
2. a product of 60, a sum of 19, and a difference of 11 (15, 4)
3. a product of 75, a sum of 28, and a difference of 22 (25, 3)
4. a product of 100, a sum of 25, and a difference of 15 (20, 5)
5. a product of 36, a sum of 20, and a difference of 16 (2, 18)
6. a product of 0, a sum of 7, and a difference of 7 (0, 7)
7. a product of 90, a sum of 33, and a difference of 27 (3, 30)
8. a product of 144, a sum of 24, and a difference of 0 (12, 12)

Have students make up five math riddles of their own. Tell them to exchange the riddles with a partner and then solve them.

5-8 Multiplying, the Factors 10, 11, and 12

pages 87–88

1 Getting Started

Objective

- To multiply by the factor 10, 11, or 12

Warm Up • Mental Math

Have students find the product.

1. 3×4 (12)
2. 4×7 (28)
3. 6×5 (30)
4. 8×9 (72)
5. 3×6 (18)
6. 4×5 (20)

Warm Up • Pencil and Paper

Have students copy and determine the correct symbol: <, >, or =.

1. $2 \times 7 \bigcirc 3 \times 5$ (<)
2. $8 \times 3 \bigcirc 6 \times 4$ (=)
3. $9 \times 4 \bigcirc 7 \times 5$ (>)
4. $6 \times 7 \bigcirc 9 \times 5$ (<)
5. $7 \times 3 \bigcirc 9 \times 2$ (>)
6. $5 \times 5 \bigcirc 3 \times 9$ (<)

Name ______________________

Multiplying, the Factors 10, 11, and 12

There are 11 players on the field at a time for a soccer team. If there are 12 teams on a field at a time, how many people are playing soccer?

We want to know how many people are playing soccer.

We know that there are 12 teams on the field.

There are 11 people on a soccer team.

To find the total number of people, we multiply 12 by 11.

$12 \times 11 = $ 132

$11 \times 12 = $ 132

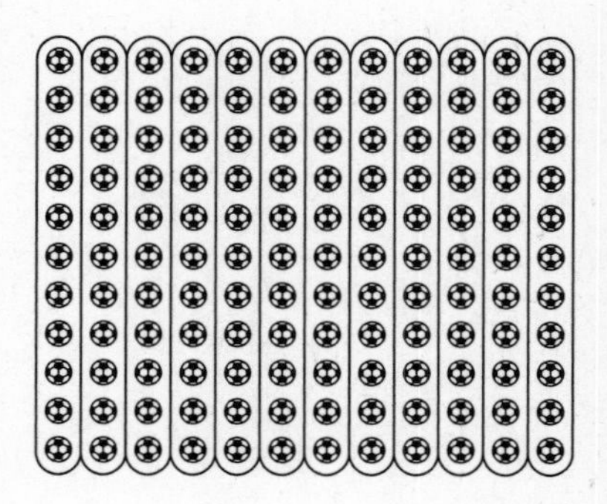

Getting Started

Complete the table.

1.

Multiplication Facts for 10												
0	1	2	3	4	5	6	7	8	9	10	11	12
0	10	20	30	40	50	60	70	80	90	100	110	120

Multiply.

2. $5 \times 10 = $ 50
3. $7 \times 10 = $ 70
4. $2 \times 10 = $ 20
5. $3 \times 10 = $ 30
6. $10 \times 6 = $ 60
7. $10 \times 4 = $ 40
8. $10 \times 11 = $ 110
9. $10 \times 9 = $ 90

2 Teach

Introduce the Lesson Have students study the problem and tell what is being asked. (how many people are playing soccer if there are 12 teams on a field at a time) Ask students what facts are given in the problem. (There are 11 players on a soccer team and there are 12 teams on the field.) Have students fill in the information sentences.

- Read the plan to the class and have a volunteer state the required operation. (multiplication) Work through the model with students and have them complete the solution sentence.

Develop Skills and Concepts Explain to students that multiplying by 10 is like counting dimes. Write the following example on the board: **9 × 10**. Have students skip-count by 10s as though they were counting the number of cents that they would have when counting the value of dimes. Then, have them write the multiplication sentences that they have just skip-counted. Next, have students skip-count by 11s. Ask students if they see a pattern in the numbers. (Until reaching 110, the tens and ones digits are the same.) Have students write the multiplication sentences that they have skip-counted by 11s. Then, repeat this process for 12. It may help students to think of 12 as 6×2 because they already know the multiples of 6.

3 Practice

Have students complete the tables of facts for 10, 11, and 12, and the exercises on page 88.

Practice

Complete the table.

1.

Multiplication Facts for 11												
0	1	2	3	4	5	6	7	8	9	10	11	12
0	11	22	33	44	55	66	77	88	99	110	121	132

Multiply.

2. $6 \times 11 =$ 66
3. $4 \times 11 =$ 44
4. $0 \times 11 =$ 0
5. $8 \times 11 =$ 88
6. $11 \times 2 = 22$
7. $11 \times 10 = 110$
8. $11 \times 9 = 99$
9. $11 \times 7 = 77$

Complete the table.

10.

Multiplication Facts for 12												
0	1	2	3	4	5	6	7	8	9	10	11	12
0	12	24	36	48	60	72	84	96	108	120	132	144

Multiply.

11. $5 \times 12 =$ 60
12. $8 \times 12 =$ 96
13. $1 \times 12 =$ 12
14. $10 \times 12 =$ 120
15. $12 \times 4 = 48$
16. $12 \times 12 = 144$
17. $12 \times 9 = 108$
18. $12 \times 3 = 36$
19. $12 \times 6 = 72$
20. $12 \times 0 = 0$

Problem Solving

Solve each problem.

21. Mrs. Harding bought 4 dozen eggs for a brunch that she is hosting. There are 12 eggs in a dozen. How many eggs did Mrs. Harding buy?
48 eggs

22. José bought 8 compact discs. Each compact disc cost $10. How much did the compact discs cost José?
$80

4 Assess

Ask students to explain how they can multiply by 10 without having to compute. (insert a zero after the factor being multiplied by 10)

5 Mixed Review

1. 72 − 38 (34)
2. 64 + 57 (121)
3. 228 − 39 (189)
4. 652 + 879 (1,531)
5. 6 × 8 (48)
6. 9 × 5 (45)
7. 7 × 7 (49)
8. 3 × 5 (15)
9. 4 × 8 (32)
10. 8 × 8 (64)

For Mixed Abilities

Common Errors • Intervention

Some students may have difficulty multiplying by 11 and 12 because they are 2-digit numbers. Remind students that the product of a 1-digit number multiplied by 11 will always be the same as writing the other factor twice. For 12, students can break 12 into two parts and multiply a factor times 10 and add it to the factor times 2. Students can also draw arrays or use flashcards to help them learn the new facts.

Enrichment • Number Sense

Tell students to expand their tables of the facts for 10, 11, and 12 until the other factor is 20. Ask students what patterns they see.

More to Explore • Numeration

Assign each letter of the alphabet a number. Give the students problems such as these to solve using the numbers you have chosen. Tell them that the answers will match a letter in the alphabet and spell out the name of a state.

6 + 3 = 9	30 − 15 = 15	53 − 30 = 23	3 − 2 = 1
I	O	W	A

When they have solved the first problem, tell students to devise their own state code problems, using addition, subtraction, multiplication, or division. Have them exchange the problems with a partner and solve each other's code.

5-9 Missing Factors

pages 89–90

1 Getting Started

Objective
- To find missing factors

Materials
counters

Warm Up • Mental Math
Have students name the number.
1. nine 7s minus two 10s (43)
2. three 100s plus seven 8s (356)
3. 4 boxes of 7 plus six 1s (34)
4. $\frac{1}{2}$ of eight 8s (32)
5. $\frac{1}{4}$ of 2 hours (30 minutes)
6. 6 rows of 12 each (72)
7. 8 nickels plus 1 dime (50¢)
8. $2 \times 6 + 8$ (20)

Warm Up • Numeration
Write $(7 \times 4) + (2 \times 4) =$ _____ on the board. Have students work the problem to tell what 9×4 equals. (36) Present more fact-plus-fact problems for students to find the total of nine 6s, eight 7s, seven 6s, and nine 8s.

Name ______________________

Lesson 5-9

It's Algebra!

Missing Factors

Mike and his sister caught 63 fish. Mike will clean the fish and put 7 of them in each freezer bag. How many freezer bags will Mike need?

We want to know the number of bags Mike will need to freeze all the fish.

We know he and his sister caught __63__ fish.

He is putting __7__ fish in each freezer bag.

To find the total number of bags needed, we can write a multiplication sentence.

We let n represent the number of bags.

$$7 \times n = 63$$

We think, 7 times what number equals 63?

$$7 \times \underline{9} = 63$$

$$n = \underline{9}$$

Mike needs __9__ freezer bags.

Getting Started

Solve for n.

1. $n \times 3 = 15$ $n =$ __5__
2. $4 \times n = 28$ $n =$ __7__
3. $9 \times n = 0$ $n =$ __0__
4. $n \times 7 = 42$ $n =$ __6__
5. $8 \times n = 56$ $n =$ __7__
6. $n \times 2 = 16$ $n =$ __8__

2 Teach

Introduce the Lesson Have a student read the problem aloud and tell what is to be solved. (how many bags are needed) Ask students what facts are needed to solve the problem. (the total number of fish and the number of fish to go into each bag) Ask if the information is given. (yes) Work through the model with students and then have them complete the solution sentence.

Develop Skills and Concepts Write $8 \times 6 =$ _____ on the board. Have a student write the product to complete the fact. (48) Cover the 6 with a sheet of paper or your hand and ask students what is missing in the fact. (6) Write $8 \times n = 48$ on the board and ask students what n stands for. (the missing number, or 6) Tell students that we can ask ourselves what number times 8 equals 48 or 8 times what equals 48. Write $n = 6$ under the problem. Now, cover the 8 in $8 \times 6 = 48$ and ask students what number is missing in the fact. (8) Write $n \times 6 = 48$ and $n = 8$ on the board. Write $7 \times n = 63$ on the board and ask students what times 7 equals 63. (9) Have a student write $n = 9$ on the board. Repeat for more problems to solve for n.

3 Practice

Have students complete the exercises on page 90 by solving for n.

Practice

Solve for n.

1. $7 \times n = 56$ $n =$ 8
2. $5 \times n = 30$ $n =$ 6
3. $n \times 4 = 8$ $n =$ 2
4. $5 \times n = 45$ $n =$ 9
5. $n \times 8 = 64$ $n =$ 8
6. $n \times 7 = 35$ $n =$ 5
7. $8 \times n = 40$ $n =$ 5
8. $4 \times n = 36$ $n =$ 9
9. $7 \times n = 42$ $n =$ 6
10. $n \times 4 = 36$ $n =$ 9
11. $4 \times n = 12$ $n =$ 3
12. $4 \times n = 0$ $n =$ 0
13. $n \times 8 = 48$ $n =$ 6
14. $8 \times n = 16$ $n =$ 2
15. $n \times 1 = 8$ $n =$ 8
16. $9 \times n = 45$ $n =$ 5
17. $9 \times n = 27$ $n =$ 3
18. $n \times 4 = 32$ $n =$ 8
19. $8 \times n = 56$ $n =$ 7
20. $9 \times n = 54$ $n =$ 6
21. $n \times 5 = 0$ $n =$ 0

Now Try This!

Make the following sentences true by filling in the circle with either an addition symbol or a subtraction symbol.

Example: 8 (+) 3 (−) 1 = 10

1. 9 (−) 6 (+) 2 = 5
2. 15 (−) 7 (−) 2 = 6
3. 8 (+) 9 (−) 4 = 13
4. 6 (+) 0 = 0 (+) 6
5. 12 (−) 7 = 3 (+) 2
6. 14 (−) 9 = 11 (−) 7 (+) 1

Now Try This! In these exercises, it is important that operations be done from left to right. No parentheses should be inserted.

4 Assess

Ask students to tell what the letter n stands for in the following: $12 \times n = 144$. (12)

5 Mixed Review

1. $8 + 5 \times 9$ (53)
2. 20,175 − 8,254 (11,921)
3. ____ × 4 = 32 (8)
4. \$65.93 − \$0.78 (\$65.15)
5. 9 + 9 + 9 + 9 (36)
6. 9×7 (63)
7. 362 + 1,758 + 27 (2,147)
8. ____ × 6 = 54 (9)
9. 3,688 − 1,496 (2,192)
10. $(98 - 97) \times 5$ (5)

For Mixed Abilities

Common Errors • Intervention

If students have difficulty finding missing factors, have them use counters to model problems. For example, to solve $3 \times n = 21$, they separate 21 counters into 3 equal groups and count to find that there are 7 counters in each group. Hence, $3 \times 7 = 21$.

Enrichment • Application

Tell students to total the veterinarian bill that covers a visit for their three pets for the following: all their distemper shots were \$54, combined rabies shots cost \$27, and 2 flea baths totaled \$32. What was the cost of the distemper and rabies shots for each pet? (\$81; \$27 each)

More to Explore • Numeration

Have students create simple division word problems involving missing factors for division sentences. For example:

There are 6 boxes of golf balls. If there are 24 balls in all, how many balls are there in each box? $(24 \div 6 = 4)$

Tell students to illustrate their word problems by drawing sets of objects to represent the division sentence. Have students exchange word problems to solve.

ESL/ELL STRATEGIES

Ask students to explain the meaning of the term *solve for n* in their own words. A student might say, "It means find out what the number n stands for."

5-10 Problem Solving: Look for a Pattern

pages 91–92

1 Getting Started

Objective

- To find a pattern to solve a problem

Warm Up • Mental Math

Ask students how many flat surfaces are on the following:

1. 2 cubes (12)
2. 3 rectangular prisms (18)
3. 4 unopened soda cans (8)
4. 6 coffee cups (6)
5. a kitchen drawer (10)
6. a triangular prism (4)
7. a box of cereal (6)
8. two 8-pane windows (32)

It's Algebra!

Name ______________________

Problem Solving: Look for a Pattern

This arrangement of numbers is called Pascal's triangle. What are the numbers in the next three rows?

1
1 1
1 2 1
1 3 3 1
1 4 6 4 1

★SEE

We need to find the pattern in Pascal's triangle.

We know

The end number in each row is <u>1</u>

The other numbers in the row are the sum of the two closest numbers in the row above

Each row has <u>1</u> more number in it than the row above it

★PLAN

We predict the numbers in the next three rows by extending the pattern.

★DO

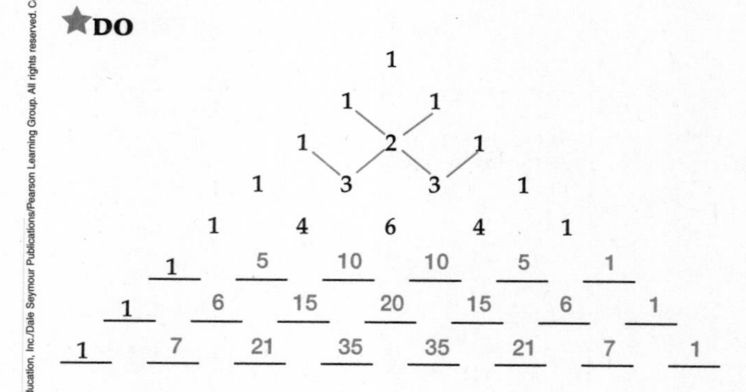

★CHECK

We check to see if the numbers continue the pattern.

2 Teach

Introduce the Lesson Remind students of the four-step problem-solving strategy.

Develop Skills and Concepts Have a student read the problem aloud. Tell students that Pascal's triangle is used in higher mathematics. Read the steps with students as they complete the problem. Guide students through the DO step. Tell students that the method they used for determining each new row of Pascal's triangle is the most common but is not the only method. Allow interested students to discover their own patterns for continuing the triangle. Note: If some students are further intrigued, information should be available in any second-year algebra text under the topics binomial expansion, probability, combinations, or permutations.

3 Apply

Solution Notes

1. If necessary, refer students to p. 91 to see the pattern again. (7 in seventh row and so on)
2. Help students see that the pattern is a 90-degree rotation plus an added dot. Students may find it helpful to draw the figure and act this problem out.
3. Have students multiply $6 \times 6 \times 6$ and $6 \times 6 \times 6 \times 6$ and continue until they see the pattern of 6 being in the ones place in each product and therefore 6 ones are multiplied by 6 again and again.
4. Students should see that the pattern is each addend being 1 more than the previous one.

Apply

Look for a pattern.

1. How many numbers are needed for the fiftieth row of Pascal's triangle?
   50

2. Complete the last picture.

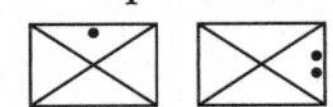 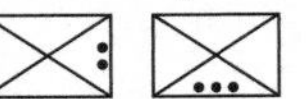 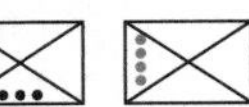

   See Solution Notes.

3. If two 6s are multiplied together, the product is 36. If twelve 6s are multiplied together, what is the number in the ones place?
   6

4. What number is missing?
   1, 2, 4, 7, 11, 16, 22, 29, 37

5. The first three rows of Pascal's triangle use six numbers.

   1
   1 1
   1 2 1

   The first four rows use ten numbers. How many numbers are needed for the first ten rows? How can you prove that your answer is correct?
   See Solution Notes.

Problem-Solving Strategy: Using the Four-Step Plan

★ **SEE** What do you need to find?
★ **PLAN** What do you need to do?
★ **DO** Follow the plan.
★ **CHECK** Does your answer make sense?

6. Think of each row in Pascal's triangle as one number. You can get the number in Row 2 by multiplying 11×1. You can get Row 3 by multiplying 11×11. Explain how to use 11 as a factor to get more rows.
   See Solution Notes.

7. Make five rows of another triangle by multiplying each digit in five rows of Pascal's triangle by 3. How is the pattern in your new triangle like that in Pascal's triangle? How is it different?
   See Solution Notes.

8. The sequence of numbers shown below is known as the Fibonacci sequence.

   1, 1, 2, 3, 5, 8, 13, . . .

   Tell what the next number in the sequence is and explain how you found it.
   21 See Solution Notes.

For Mixed Abilities

More to Explore • Measurement

Have students measure the length and width of their math textbooks in inches and find the area. Have them repeat the activity for each of their other textbooks. Have students list their textbooks in order, from smallest in area to largest. Have students circle the length of their largest textbook. Have one student calculate the number of inches in a mile and write this on the board. Now, have each student calculate how many of their largest textbook, laid end to end, it would take to form a "book mile."

Higher-Order Thinking Skills

5. **Evaluation:** A possible answer is to show the pattern:
   +2 +3 +4 +5 +6 +7 +8 +9 +10
   1, 3, 6, 10, 15, 21, 28, 36, 45, 55
   Another possible solution is to actually show the first ten rows of Pascal's triangle and count.
6. **Synthesis:** Multiply the number in any row by 11 to get the number in the next row.
7. **Analysis:** It is the same in that the end numbers in each row are the same, the middle numbers are sums of the two numbers above, and each row has one more number than the row above. It is different in that the numbers in the rows are different.
8. **Analysis:** Add the two preceding numbers to get the next number.

4 Assess

Refer students to the completed Pascal's triangle on page 91. Have them add two more rows to the triangle.

Chapter 5 Test

page 93

Items	Objectives
1–35	To recall multiplication facts through 9 (see pages 73–86)
36–41	To find the value of expressions with parentheses (see pages 81–82)
42–47	To find missing factors (see pages 89–90)

Name ______________________

CHAPTER 5 TEST

Multiply.

1. $3 \times 2 = 6$	2. $8 \times 5 = 40$	3. $7 \times 9 = 63$	4. $4 \times 3 = 12$	5. $2 \times 8 = 16$
6. $9 \times 6 = 54$	7. $5 \times 8 = 40$	8. $5 \times 5 = 25$	9. $6 \times 7 = 42$	10. $5 \times 9 = 45$
11. $8 \times 7 = 56$	12. $6 \times 0 = 0$	13. $6 \times 5 = 30$	14. $8 \times 9 = 72$	15. $7 \times 8 = 56$
16. $3 \times 9 = 27$	17. $9 \times 7 = 63$	18. $6 \times 6 = 36$	19. $8 \times 6 = 48$	20. $4 \times 5 = 20$
21. $6 \times 9 = 54$	22. $1 \times 5 = 5$	23. $7 \times 7 = 49$	24. $8 \times 8 = 64$	25. $5 \times 7 = 35$
26. $0 \times 4 = 0$	27. $9 \times 9 = 81$	28. $7 \times 5 = 35$	29. $9 \times 8 = 72$	30. $4 \times 6 = 24$
31. $7 \times 6 = 42$	32. $3 \times 2 = 6$	33. $9 \times 5 = 45$	34. $5 \times 6 = 30$	35. $6 \times 8 = 48$

Solve for *n*.

36. $(5 \times 3) + 6 = n$ $n = 21$

37. $(8 - 4) \times 5 = n$ $n = 20$

38. $9 \times (3 + 5) = n$ $n = 72$

39. $45 + (9 \times 5) = n$ $n = 90$

40. $36 - (9 \times 4) = n$ $n = 0$

41. $(8 \times 7) - 38 = n$ $n = 18$

42. $7 \times n = 49$ $n = 7$

43. $n \times 3 = 15$ $n = 5$

44. $n \times 8 = 64$ $n = 8$

45. $n \times 4 = 36$ $n = 9$

46. $6 \times n = 42$ $n = 7$

47. $6 \times n = 0$ $n = 0$

Alternate Chapter Test

You may wish to use the Alternate Chapter Test on page 354 of this book for further review and assessment.

CUMULATIVE ASSESSMENT

Circle the letter of the correct answer.

1. What is the place value of the 9 in 9,058?
   a. tens
   b. hundreds
   (c.) thousands
   d. NG

2. 3,651 ◯ 3,615
   (a.) >
   b. <

3. Round 850 to the nearest hundred.
   a. 800
   (b.) 900
   c. NG

4. Round 6,786 to the nearest thousand.
   a. 8,000
   (b.) 7,000
   c. NG

5. What is the place value of the 3 in 632,461?
   a. tens
   b. hundreds
   c. thousands
   (d.) NG

6. 57 + 86
   a. 133
   (b.) 143
   c. 1,313
   d. NG

7. 359 + 283
   a. 532
   b. 542
   c. 552
   (d.) NG

8. 29,468 + 36,875
   (a.) 66,343
   b. 66,433
   c. 67,343
   d. NG

9. $16.48 + 37.19
   a. $43.67
   b. $53.57
   (c.) $53.67
   d. NG

10. 83 − 27
   (a.) 56
   b. 64
   c. 66
   d. NG

11. 926 − 458
   a. 432
   (b.) 468
   c. 532
   d. NG

12. 40,276 − 29,867
   (a.) 10,409
   b. 20,409
   c. 29,611
   d. NG

13. 9 × 6
   (a.) 54
   b. 56
   c. 63
   d. NG

STOP

score

Cumulative Assessment

page 94

Items	Objectives
1, 5	To identify the place value of digits in numbers less than 1,000,000 (see pages 25–26)
2	To compare and order numbers less than 10,000 (see pages 29–30)
3–4	To round numbers to the nearest hundred or thousand (see pages 31–32)
6–7	To add two 2- or 3-digit numbers (see pages 37–40)
8	To add two 5-digit numbers (see pages 47–48)
9	To add money amounts (see pages 49–50)
10–11	To subtract two 2- or 3-digit numbers (see pages 55–58)
12	To subtract two 5-digit numbers (see pages 61–62)
13	To multiply using 9 as a factor (see pages 85–86)

Alternate Cumulative Assessment

Circle the letter of the correct answer.

1. What is the place value of the 6 in 4,692?
   a tens
   (b) hundreds
   c thousands
   d NG

2. 4,321 ◯ 4,231
   (a) >
   b <
   c =

3. Round 731 to the nearest hundred.
   a 600
   (b) 700
   c 800
   d NG

4. Round 8,463 to the nearest thousand.
   (a) 8,000
   b 8,400
   c 9,000
   d NG

5. What is the place value of the 4 in 421,653?
   a hundreds
   b thousands
   c ten thousands
   (d) NG

6. 38 + 65
   a 913
   b 93
   (c) 103
   d NG

7. 647 + 274 =
   a 821
   (b) 921
   c 811
   d NG

8. 46,365 + 13,947
   a 32,418
   b 59,202
   c 60,202
   (d) NG

9. $25.67 + 46.24
   (a) $71.91
   b $61.91
   c $61.81
   d NG

10. 94 − 37 =
   a 67
   b 63
   (c) 57
   d NG

11. 632 − 467 =
   (a) 165
   b 235
   c 275
   d NG

12. 50,385 − 18,568
   a 42,827
   b 41,827
   c 48,223
   (d) NG

13. 9 × 7
   a 48
   b 54
   (c) 63
   d NG

CHAPTER 6

6-1 Multiples

pages 95–96

1 Getting Started

Objective

- To name multiples of whole numbers

Vocabulary

multiple, skip counting

Materials

*multiplication fact cards

Warm Up • Mental Math

Have students tell the number that comes after.

1. 47 − 19 (29)
2. 1,719,000 (1,719,001)
3. $28 \times \frac{1}{2}$ (15)
4. 356 − 4 tens (317)
5. 7 × 20 (141)
6. 26 ÷ 2 (14)
7. $\frac{1}{4}$ of 200 (51)
8. 20 tens + 268 (469)

Warm Up • Numeration

Have students count by 2s and 5s to 100 and then write those sequences on the board.

Name ____________________

CHAPTER 6

Multiplication of Whole Numbers

Lesson 6-1

Multiples

The 25 students in Miss Lane's class are counting off to form squares for square dancing. Every fifth person will stand in the center of a square. What numbers will the students in the centers have?

We want to know the numbers of the students who will stand in the centers of the squares.

We know there are __25__ students in Miss Lane's class.

We know that every __fifth__ student will stand in the center of a square.

We can count from 1 to 25, marking off every fifth number.

1, 2, 3, 4, (5)　6, 7, 8, 9, (10)　11, 12, 13, 14, (15)
16, 17, 18, 19, (20)　21, 22, 23, 24, (25)

We say __5__, __10__, __15__, __20__, and __25__ are multiples of 5. A **multiple** of a number is a product that has that number for at least one of its factors.

The least multiple of any number is 0 because any number times 0 equals 0. Naming multiples of a number is called **skip counting**. We skip-count by 5s by saying 0, 5, 10, 15, 20, 25, and so on.

Students with the numbers __5__, __10__, __15__, __20__, and __25__ will stand in the centers of the squares.

Getting Started

Write the first nine multiples of each number.

1. 6 __0__, __6__, __12__, __18__, __24__, __30__, __36__, __42__, __48__
2. 8 __0__, __8__, __16__, __24__, __32__, __40__, __48__, __56__, __64__

2 Teach

Introduce the Lesson Have a student read the problem and tell about the picture. Tell students that they need to first decide what information is known. (There are 25 students and every fifth person will stand in the middle of a square of 4.) Have students complete the sentences and skip-counting to solve the problem. Students may want to check their solution by acting it out.

Develop Skills and Concepts Place the multiplication fact cards for 4 × 0 through 4 × 9 in order across the chalk tray. Have students write the product of each fact above it. Tell students that these products are called **multiples** of 4. Have students say the multiples of 4 in order. Have students write the numbers **1, 2,** and **3** between the first two fact cards to show that these numbers were skipped. Continue to have students write the skipped numbers between the multiples. Tell students that **skip counting** is saying the multiples of a number in order. Place the fact cards for 6 × 0 through 6 × 9 across the chalk tray and repeat the activity. Group students in pairs to continue the activity for more skip counting.

3 Practice

Remind students that the least multiple of any number is zero. Have students complete page 96 independently.

Practice

Write the first nine multiples of each number.

1. 2 0, 2, 4, 6, 8, 10, 12, 14, 16
2. 5 0, 5, 10, 15, 20, 25, 30, 35, 40
3. 7 0, 7, 14, 21, 28, 35, 42, 49, 56
4. 4 0, 4, 8, 12, 16, 20, 24, 28, 32
5. 9 0, 9, 18, 27, 36, 45, 54, 63, 72

Skip-count by 7.

6. 56, 63, 70, 77, 84, 91, 98, 105, 112

Skip-count by 4.

7. 32, 36, 40, 44, 48, 52, 56, 60, 64

Skip-count by 9.

8. 72, 81, 90, 99, 108, 117, 126, 135, 144

Skip-count by 2.

9. 16, 18, 20, 22, 24, 26, 28, 30, 32

Skip-count by 6.

10. 48, 54, 60, 66, 72, 78, 84, 90, 96

Skip-count by 5.

11. 40, 45, 50, 55, 60, 65, 70, 75, 80

Skip-count by 3.

12. 24, 27, 30, 33, 36, 39, 42, 45, 48

4 Assess

Ask students to explain what a multiple of a number is. (A multiple of a number is a product that has that number for at least one of its factors.)

For Mixed Abilities

Common Errors • Intervention

Some students may have difficulty skip-counting. Draw a number line across the board and show the numbers 0 to 81. Have students work together in a group. Students may take turns being a rabbit who stands at the board and places a finger at 0. As the other students say ten multiples of a number in order, the rabbit hops with his or her finger to the multiples on the number line. Each time a new rabbit goes to the board, a new set of multiples should be used.

Enrichment • Numeration

Tell students to skip-count the books on several classroom or library shelves. Have them record their totals and compare with a classmate's work.

More to Explore • Logic

Dictate the following problem to students to build skills in multiplication, division, and factoring. It can be solved by large or small groups.

Mrs. Carson, the sixth-grade science teacher, kept a cage full of gerbils in her homeroom. One morning, she and her students found that someone had left the cage door open and all the gerbils had escaped. The students volunteered to find them and asked how many were missing. Mrs. Carson didn't remember how many gerbils there had been. She did remember that when she counted them by 2s, 3s, or 4s, she always had one left over. When she counted them by 5s, however, her count came out even. Her students quickly calculated how many gerbils were missing and returned them all safely to the cage. What is the smallest possible number of gerbils in the cage? (25)

6-2 Multiplication Properties

pages 97–98

1 Getting Started

Objective

- To understand the Grouping, Multiplication-Addition, and Multiplying by 10 Properties

Vocabulary

Associative or Grouping Property, Multiplication-Addition Property

Materials

small paper pieces

Warm Up • Mental Math

Have students tell each best buy.

1. 3/89¢ or 4/$1 (4/$1)
2. 6/$9 or 3/$3 (3/$3)
3. 2/$1 or 45¢ each (45¢)
4. 8/$2 or 10/$3 (8/$2)
5. 4/20¢ or 6/60¢ (4/20¢)
6. 3/99¢ or 8/$2 (8/$2)
7. 10¢ each or 12/$1 (12/$1)
8. 2/$9 or $5 each (2/$9)

Warm Up • Numeration

Write the numbers 1 through 9 across the board. Have students skip-count by 10s as you write the numbers 10 through 90 under 1 through 9. Have students write the facts for 10 on the board in a column. ($1 \times 10 = 10$ through $9 \times 10 = 90$) Leave this work on the board for later use.

Name ____________________

Lesson 6-2

It's Algebra!

Multiplication Properties

Besides the properties we have already studied, there are other properties that help us find shortcuts in multiplication.

Associative or Grouping Property
Factors can be grouped in any way.

$(3 \times 2) \times 5 = 3 \times (2 \times 5)$

$6 \times 5 = 3 \times$ 10

30 = 30

Distributive Property
Multiplication can be distributed over addition.

$5 \times (4 + 6) = (5 \times 4) + (5 \times 6)$

$5 \times$ 10 = 20 + 30

50 = 50

Multiplying by 10
If one factor is 10, the product is the other number followed by a zero.

$2 \times 10 =$ 20

$10 \times 7 =$ 70

Getting Started

Solve for *n*. Use the properties to help you.

1. $4 \times (7 + 3) = n$ $n =$ 40	2. $8 \times 10 = n$ $n =$ 80	3. $6 \times (4 \times 10) = n$ $n =$ 240
4. $9 \times 10 = n$ $n =$ 90	5. $(2 \times 3) \times 10 = n$ $n =$ 60	6. $(7 \times 2) + (7 \times 8) = n$ $n =$ 70

2 Teach

Introduce the Lesson Have a student read aloud the thoughts of the two students pictured. Have a student read the introductory paragraph. Guide students through each property definition as they fill in the examples.

Develop Skills and Concepts Refer students to the work on the board from the above Warm Up • Numeration Activity. Help students see that when a number is multiplied by 10, the product is that number followed by a zero. Write **(2 × 4) × 6** on the board and remind students that the operation in the parentheses is done first. Have a student write the product on the board. (48) Now, write **2 × (4 × 6)** on the board and have a student write the product. (48) Ask if the product changed when the factors were grouped differently. (no) Have a student write the problem to show another way to regroup the factors. [(6 × 2) × 4 or (6 × 4) × 2] Tell students that they can use the Multiplying by 10 and the Multiplication-Addition Properties to find a shortcut in multiplication.

- Write the following on the board: **(9 × 2) + (9 × 8) = 9 × (2 + 8) = 9 × 10 = 90**. Have students read the problem with you. Write **(8 × 3) + (8 × 7)** on the board and have students follow the previous examples to work the problem on the board.

3 Practice

Remind students to use the three properties to solve the exercises in the first section and to check the operation sign in the second section. Have students complete page 98 independently.

Practice

Solve for n. Use the properties to help you.

1. $3 \times (0 \times 5) = n$ — $n = 0$
2. $5 \times (4 + 3) = n$ — $n = 35$
3. $(6 \times 3) + (6 \times 7) = n$ — $n = 60$
4. $7 \times 10 = n$ — $n = 70$
5. $2 \times (5 \times 8) = n$ — $n = 80$
6. $9 \times (0 \times 3) = n$ — $n = 0$
7. $4 \times 10 = n$ — $n = 40$
8. $8 \times (3 \times 1) = n$ — $n = 24$
9. $(8 \times 5) \times 0 = n$ — $n = 0$
10. $(1 \times 1) \times 1 = n$ — $n = 1$
11. $10 \times 9 = n$ — $n = 90$
12. $(3 \times 4) + (3 \times 6) = n$ — $n = 30$
13. $7 \times (9 + 1) = n$ — $n = 70$
14. $0 \times (8 \times 10) = n$ — $n = 0$
15. $(4 \times 0) + (4 \times 0) = n$ — $n = 0$

Add, subtract, or multiply.

16. $6 + 9 = 15$
17. $8 \times 7 = 56$
18. $15 - 9 = 6$
19. $3 + 7 = 10$
20. $5 \times 8 = 40$
21. $6 - 6 = 0$
22. $4 \times 7 = 28$
23. $9 \times 6 = 54$
24. $17 - 8 = 9$
25. $7 + 6 = 13$
26. $0 \times 8 = 0$
27. $5 + 8 = 13$
28. $13 - 6 = 7$
29. $9 - 0 = 9$
30. $7 \times 1 = 7$
31. $6 \times 6 = 36$
32. $4 \times 3 = 12$
33. $8 + 1 = 9$
34. $5 + 5 = 10$
35. $5 \times 5 = 25$
36. $5 - 5 = 0$
37. $14 - 9 = 5$
38. $8 \times 6 = 48$
39. $8 - 3 = 5$
40. $8 \times 8 = 64$
41. $8 - 8 = 0$
42. $8 + 8 = 16$
43. $16 - 7 = 9$
44. $3 + 9 = 12$
45. $9 - 6 = 3$

4 Assess

Ask students to explain the Associative or Grouping Property. (Factors can be grouped in any way.)

5 Mixed Review

1. 15 + 18 + 21 + 36 (90)
2. \$57.43 − \$28.72 (\$28.71)
3. (6 × 5) + (7 × 8) (86)
4. 6,076 − 3,257 (2,819)
5. 9 × 7 (63)
6. 23,053 + 62,898 (85,951)
7. 8 × (3 + 5) (64)
8. 700 − 358 (342)
9. 2 + 6 × 3 (20)
10. 7,258 + 761 (8,019)

For Mixed Abilities

Common Errors • Intervention

Some students may have difficulty with the Multiplication-Addition Property, also called the Distributive Property. Have them use counters to model a problem such as (2 × 3) + (2 × 4) by showing 2 sets of 3 and 2 sets of 4. Then, have them combine the counters and see how many sets of 7 they can make. Write the following on the board :

(2 × 3) + (2 × 4) = 2 × (3 + 4)

Enrichment • Measurement

Have students find the combined areas of 10 rectangles if each is 7 centimeters long and 9 centimeters wide.
(630 centimeters)

More to Explore • Logic

Have students solve the following problem:

Some of the 25 gerbils in Mrs. Carson's classroom had babies. Once their babies were grown enough to move around on their own, Mrs. Carson invited the kindergarten class down to see the new animals. However, in their excitement, the kindergarteners knocked over the cage and all the gerbils escaped.

The class did not know how many gerbils there were in all, but Mrs. Carson told them that when she counted the gerbils by 2s, 3s, 4s, or 6s, she always had one left over. When she counted by 7s, however, she came out even. What was the smallest number of gerbils that could have been in the cage? (49)

ESL/ELL STRATEGIES

When teaching the multiplication properties, review each explanation and instruction with ESL/ELL students. You may need to define and give examples of terms such as *associative, grouping, factor*, and *distributive.*

6-3 Multiplying Tens and Ones

pages 99–100

1 Getting Started

Objective

- To multiply 2-digit numbers by 1-digit numbers, no regrouping

Materials

page of 100 basic multiplication facts

Warm Up • Mental Math

Ask students to tell the number value of the following if A = 1, B = 2, C = 3, . . . , and Z = 26:

1. C × J (30)
2. Z ÷ B (13)
3. $\frac{1}{2}$ of X (12)
4. A + Z + D (31)
5. 100 × (B + C) (500)
6. sum of letters in your first name
7. Y − (D × E) (5)

Warm Up • Multiplication

Give students a timed test of the 100 basic multiplication facts. It is a good idea to repeat this test often throughout the year.

Name ____________________

Lesson 6-3

Multiplying Tens and Ones

River Road Foods processes and packs fresh fruits. How many cans of tomatoes will River Road Foods pack in 4 minutes?

Packing Rates	
Food	**Number of cans per minute**
Tomatoes	21
Apples	24
Pumpkins	19

We want to know the number of cans of tomatoes packed in __4__ minutes.

We know that __21__ cans of tomatoes are packed in 1 minute.

To find the number of cans packed in 4 minutes, we multiply __21__ by __4__.

21 × 4 = ☐	Multiply the ones. 4 × 1 = 4	Multiply the tens. 4 × 2 = 8
21 × 4	21 × 4 = 4	21 × 4 = 84

River Road Foods will pack __84__ cans of tomatoes in 4 minutes.

Getting Started

Multiply.

1. 24 × 2 = 48
2. 40 × 2 = 80
3. 31 × 3 = 93

Copy and multiply.

4. 2 × 13 = 26
5. 4 × 11 = 44
6. 4 × 22 = 88

2 Teach

Introduce the Lesson Have a student read the problem aloud. Ask students what they must find. (the number of cans of tomatoes packed in 4 minutes) Ask students if they will need all the information in the table. (no) Have students read and complete the information sentences. Guide them through the multiplication in the model, describing the regrouping to solve the problem. Tell students to round 21 cans to the nearest ten and estimate to check their solution.

Develop Skills and Concepts Write the following on the board:

$$\begin{aligned}42 \times 2 &= (40 + 2) \times 2 \\ &= (40 \times 2) + (2 \times 2) \\ &= \text{2 forties} + \text{2 twos} \\ &= 2 \times (\text{4 tens}) + 2 \times (\text{2 ones})\end{aligned}$$

Read through the problem with students. Remind students that the Multiplication-Addition Property helps us see that 42 × 2 is the same as 2 times 4 tens plus 2 times 2 ones.

- Write **42 × 2** vertically on the board. Tell students that we multiply the ones column first to find the product of 2 times 2 ones. Write **4** in the answer. Remind students that there are 4 tens in 40 and 2 times 4 tens equals 8 tens as you write 8 in the tens column. Repeat the procedure for 31 × 3 and 24 × 2.

3 Practice

Remind students to multiply the ones first. Have students complete the exercises on page 100 independently.

Practice

Multiply.

1. 13 × 2 = 26	2. 30 × 2 = 60	3. 58 × 1 = 58	4. 40 × 2 = 80	5. 21 × 3 = 63
6. 42 × 2 = 84	7. 34 × 2 = 68	8. 11 × 7 = 77	9. 24 × 2 = 48	10. 33 × 3 = 99
11. 12 × 4 = 48	12. 11 × 5 = 55	13. 44 × 2 = 88	14. 31 × 3 = 93	15. 32 × 3 = 96

Copy and multiply.

16. 10 × 5 50	17. 36 × 1 36	18. 23 × 3 69	19. 4 × 12 48
20. 3 × 13 39	21. 43 × 2 86	22. 2 × 22 44	23. 9 × 11 99
24. 41 × 2 82	25. 3 × 12 36	26. 2 × 33 66	27. 2 × 31 62

Problem Solving

Solve each problem.

28. Manuel bought 4 dozen eggs to put into the cakes he was making for the church bake sale. How many eggs did Manuel buy? 48 eggs

29. Beth bought 4 stamps. Each stamp cost 22¢. How much did Beth pay for the stamps? 88¢

Now Try This!

It's Algebra!

Fill in the blank so that both sides of the sentence are equal.

1. 145 − 12 = 143 − __10__
2. 197 − 78 = __199__ − 80
3. 41 − 13 = 48 − __20__
4. 73 − 29 = __74__ − 30
5. 187 − 65 = 182 − __60__
6. 395 − 264 = __335__ − 204
7. 359 − 126 = 363 − __130__
8. 254 − 181 = __274__ − 201

Now Try This! Ask students to first compare known minuends or subtrahends in the exercises and explain how they differ. Stress that changes made in the minuend must be repeated in the subtrahend. Discuss how changing the minuend and subtrahend through addition or subtraction can help to simplify problems such as 837 − 685 = (837 + 20) − (685 + 20) = 857 − 705.

4 Assess

Ask students to find the product of 22 × 3. (66)

For Mixed Abilities

Common Errors • Intervention

Students may get incorrect answers because they have not yet mastered basic multiplication facts. They may use flashcards to practice these facts. Have them use place-value materials to work the problems. For example, to find 2 × 24, have students first show 2 tens and 4 ones. Then, have them show 2 sets of 2 tens and 2 sets of 4 ones, or 4 tens and 8 ones. Thus, 2 × 24 = 48. At this level, understanding the concept is as important as getting the correct answer.

Enrichment • Application

Tell students to expand the table on page 99 to show the number of cans of tomatoes, apples, and pumpkins River Road Foods will pack in 1 hour and then in 8 hours.

More to Explore • Biography

Duplicate the following for students.

Directions: Fill in the blanks. The first letters, reading down, spell the name of this famous person in mathematics.

She was the first woman mathematician. She lived during the fifth century A.D. and studied medicine as well as mathematics.

(HALF)	6 is what part of 12?
(YES)	Is is true that 42 × 7 = 294?
(PRODUCT)	What is the result of a multiplication called?
(ADD)	How do you find a sum?
(TRIANGLE)	What is the name of this figure? △
(I)	What is the Roman symbol for the number 1?
(ANGLE)	What is this figure? ∠

(Hypatia. Ask a student to find additional information and report to the class.)

6-4 Multiplying, Regrouping Ones

pages 101–102

1 Getting Started

Objective
- To multiply 2-digit numbers by 1-digit numbers with regrouping

Materials
place-value materials

Warm Up • Mental Math
Dictate the following:
1. $2 \times 6 \times 3$ (36)
2. $10 \times (8 + 2)$ (100)
3. $(3 \times 6) + (4 \times 6)$ (42)
4. $(8 \times 6) - (3 \times 6)$ (30)
5. $(14 - 7) \div 7$ (1)
6. $20 \times 6 \times 1$ (120)
7. $8 \times 0 \times 10$ (0)
8. $(18 - 6) \div 3$ (4)

Warm Up • Multiplication
Have students work in pairs to write multiplication facts that have products of 10 or more. Then, have students rename each product in tens and ones.

Name ______________________

Lesson 6-4

Multiplying, Regrouping Ones

The fourth-grade science lab has 16 stations. Mr. Owens needs to make a battery-and-bulb hookup for each station. How many batteries will Mr. Owens need?

We want to find the number of batteries Mr. Owens needs.

There are __16__ stations in the science lab.

Each station requires __2__ batteries.

To find the number of batteries needed, we multiply __16__ by __2__.

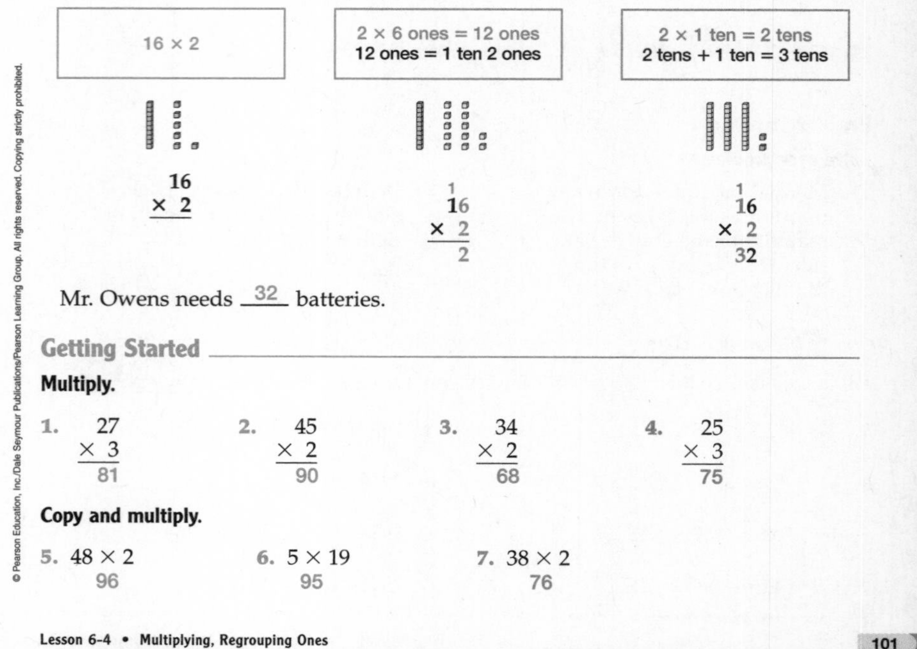

Mr. Owens needs __32__ batteries.

Getting Started

Multiply.

1. 27 × 3 = 81
2. 45 × 2 = 90
3. 34 × 2 = 68
4. 25 × 3 = 75

Copy and multiply.

5. 48 × 2 — 96
6. 5 × 19 — 95
7. 38 × 2 — 76

2 Teach

Introduce the Lesson Have a student read the problem and tell what is to be found. (how many batteries Mr. Owens will need) Ask students what information is given in the problem. (There are 16 stations.) Ask what information is needed. (number of batteries per station) Have students complete the sentences. Talk through the model multiplication with them and have students complete the solution sentence.

Develop Skills and Concepts Write the following on the board:

$$
\begin{aligned}
39 \times 2 &= (30 + 9) \times 2 \\
&= (30 \times 2) + (9 \times 2) \\
&= 2 \text{ thirties} + 9 \text{ twos} \\
&= 2 \times (3 \text{ tens}) + 9 \times (2 \text{ ones}) \\
&= 6 \text{ tens} + 18 \text{ ones} \\
&= 7 \text{ tens} + 8 \text{ ones} \\
&= 78
\end{aligned}
$$

Help students see that 10 ones are regrouped when adding 6 tens and 18 ones. Now, write **39 × 2** vertically on the board and have students tell the product of 9×2. (18) Refer students to the 6 tens + 18 ones step to see that the 10 ones are regrouped. Ask students the product of 3 tens × 2. (6 tens) Tell them now to add the 1 ten as you refer to the 7 tens + 8 ones step above. Complete the problem and then repeat the procedure for 17×3 and 43×2, where no regrouping is needed.

3 Practice

Tell students that they may not need to regroup in some of these exercises. Have students complete page 102 independently.

Practice

Multiply.

1. $\begin{array}{r} 26 \\ \times\ 3 \\ \hline 78 \end{array}$
2. $\begin{array}{r} 16 \\ \times\ 5 \\ \hline 80 \end{array}$
3. $\begin{array}{r} 34 \\ \times\ 2 \\ \hline 68 \end{array}$
4. $\begin{array}{r} 19 \\ \times\ 3 \\ \hline 57 \end{array}$
5. $\begin{array}{r} 27 \\ \times\ 3 \\ \hline 81 \end{array}$
6. $\begin{array}{r} 24 \\ \times\ 4 \\ \hline 96 \end{array}$
7. $\begin{array}{r} 17 \\ \times\ 5 \\ \hline 85 \end{array}$
8. $\begin{array}{r} 23 \\ \times\ 3 \\ \hline 69 \end{array}$
9. $\begin{array}{r} 23 \\ \times\ 4 \\ \hline 92 \end{array}$
10. $\begin{array}{r} 46 \\ \times\ 2 \\ \hline 92 \end{array}$
11. $\begin{array}{r} 12 \\ \times\ 4 \\ \hline 48 \end{array}$
12. $\begin{array}{r} 29 \\ \times\ 3 \\ \hline 87 \end{array}$
13. $\begin{array}{r} 17 \\ \times\ 4 \\ \hline 68 \end{array}$
14. $\begin{array}{r} 12 \\ \times\ 8 \\ \hline 96 \end{array}$
15. $\begin{array}{r} 13 \\ \times\ 7 \\ \hline 91 \end{array}$

Copy and multiply.

16. 2×36 72
17. 18×4 72
18. 26×3 78
19. 15×4 60
20. 28×2 56
21. 12×7 84
22. 2×43 86
23. 38×2 76
24. 24×3 72
25. 16×3 48
26. 2×19 38
27. 5×12 60
28. 18×3 54
29. 3×29 87
30. 4×19 76

Problem Solving

Solve each problem.

31. A pad of colored paper costs 29¢. How much will 3 pads cost?
 87¢
32. One pair of jeans costs $23.45. A shirt costs $16.79. The jeans cost how much more than a shirt?
 $6.66
33. A CD costs $5. How much will 16 CDs cost?
 $80
34. Mark has 47 cents. Jan has twice as much as Mark. How much does Jan have?
 94¢

4 Assess

Ask students to explain how to do the regrouping in this problem, 25×4. (Regroup the 20 ones as 2 tens. Then, add 8 tens plus 2 tens for 10 tens. Regroup the 10 tens for 1 hundred.)

5 Mixed Review

1. 28×3 (84)
2. $581.10 + $16.68 + $84.73 ($682.51)
3. 2,178 − 1,465 (713)
4. 12×3 (36)
5. 50,010 − 12,176 (37,834)
6. 21,178 + 695 (21,873)
7. 6×16 (96)
8. 82 + 1,986 (2,068)
9. 876 − 241 (635)
10. $(6 \times 8) - (4 \times 7)$ (20)

For Mixed Abilities

Common Errors • Intervention

Student may regroup correctly but then forget to add the regrouped numbers after multiplying.

Incorrect	Correct
$\begin{array}{r} {}^{2} \\ 37 \\ \times\ \ 4 \\ \hline 128 \end{array}$	$\begin{array}{r} {}^{2} \\ 37 \\ \times\ \ 4 \\ \hline 128 \end{array}$

Have students use a place-value form like the following to chart the steps.

	100s	10s	1s
4×7 ones		2	8
4×3 tens	1	2	0
	1	4	8

Enrichment • Measurement

Tell students to write and solve a multiplication and addition problem to find the perimeter of a rectangle whose sides are 29 inches and 3 inches.

More to Explore • Measurement

Put a list of customary measurements on the board. Tell students that the list of measurements matches objects that can be found somewhere in the classroom. Tell them that some of the measurements are exact, and some are estimates. Allow students time throughout the day to find and list the objects that fit the measurements, without using their rulers. Tell them to trust their eyes. At the end of the day, have students check their matches by measuring the objects with their rulers.

6-5 Multiplying 2-Digit Numbers

pages 103–104

1 Getting Started

Objective

- To multiply 2-digit numbers by 1-digit numbers with regrouping

Materials

*place-value materials

Warm Up • Mental Math

Ask students to name the number.

1. days in 5 weeks (35)
2. weeks in a decade (520)
3. milliliters in $10\frac{1}{2}$ liters (10,500)
4. hours in 3 days (72)
5. tenths in 1 whole (10)
6. years in a millennium (1,000)
7. months in 9 years (108)
8. quarts in 50 gallons (200)

Warm Up • Addition

Write on the board **16 tens + 28 ones** and have students tell how to regroup as they find the total tens and ones and the sum. (188) Repeat with more examples for sums less than 1,000.

Name ______________________________

Multiplying 2-Digit Numbers

Paul is stocking the produce bins of his grocery store. He wants to make 6 equal sections of tomatoes. How many tomatoes can Paul display?

We need to find how many tomatoes Paul can display.

There will be 24 tomatoes in each section.

Paul will pack 6 sections of tomatoes.

To find the number of tomatoes Paul can pack, we multiply 24 by 6.

Multiply the ones. Regroup if needed.	Multiply the tens. Add any extra tens.
$\begin{array}{r} {}^{2}\\ 24 \\ \times\ 6 \\ \hline 4 \end{array}$	$\begin{array}{r} {}^{2}\\ 24 \\ \times\ 6 \\ \hline 144 \end{array}$

Paul can pack 144 tomatoes.

Getting Started

Multiply.

1. $\begin{array}{r} 25 \\ \times\ 7 \\ \hline 175 \end{array}$
2. $\begin{array}{r} 32 \\ \times\ 9 \\ \hline 288 \end{array}$
3. $\begin{array}{r} 50 \\ \times\ 6 \\ \hline 300 \end{array}$
4. $\begin{array}{r} 49 \\ \times\ 8 \\ \hline 392 \end{array}$

Copy and multiply.

5. 62 × 8 496
6. 27 × 4 108
7. 8 × 36 288
8. 16 × 9 144
9. 89 × 4 356
10. 46 × 9 414
11. 73 × 7 511
12. 5 × 39 195

2 Teach

Introduce the Lesson Have students describe the picture. Have a student read the problem and tell what is to be found. (the total number of tomatoes to be displayed) Ask what information is given in the problem and the picture. (There will be 24 tomatoes in each section and there will be 6 sections.) Have students complete the sentences and guide them through the multiplication in the model. Tell students that they may want to think of the total number of pennies in 6 quarters to estimate the product of six 24s to check their solution. Have students complete the solution sentence.

Develop Skills and Concepts Write **20 × 6** vertically on the board and ask students to tell the product. (120) Write **120** and remind students that 12 tens is the same as 1 hundred 2 tens. Now, write **28 × 6** vertically on the board and ask students the product of 6 times 8. (48) Remind students that the 4 tens in 48 are extra tens and need to be added to the total tens in 6 times 2 tens. Write **16** in the answer and remind students that 16 tens equal 1 hundred 6 tens. Repeat for 40 × 8 and 49 × 8.

3 Practice

Remind students that they may not need to regroup in every problem. Before students begin the page you might have them predict the relative size of each solution using rounding. Then, have students complete page 104 independently.

Practice

Multiply.

1. 26×5 = 130	2. 64×6 = 384	3. 39×3 = 117	4. 29×2 = 58	5. 48×4 = 192
6. 53×8 = 424	7. 96×2 = 192	8. 88×3 = 264	9. 21×9 = 189	10. 32×7 = 224
11. 73×4 = 292	12. 80×9 = 720	13. 98×2 = 196	14. 81×8 = 648	15. 42×5 = 210

Copy and multiply.

16. 2×56 112	17. 3×97 291	18. 7×23 161	19. 4×59 236	20. 6×19 114
21. 5×73 365	22. 7×24 168	23. 6×28 168	24. 2×78 156	25. 9×19 171
26. 8×88 704	27. 5×68 340	28. 6×37 222	29. 3×46 138	30. 7×35 245

Problem Solving

Solve each problem.

31. The Petersons drove 317 miles the first day of their vacation. They drove 287 miles the second day. How far did the Petersons drive in 2 days?
604 miles

32. A washing machine uses 18 gallons of water for each load. How many gallons of water are used for 6 loads of wash?
108 gallons

33. A notebook contains 32 sheets of paper. How many notebooks do you need to buy to get at least 100 sheets of paper?
4 notebooks

34. A glass holds 6 ounces of juice. How many ounces of juice are needed to fill 48 glasses?
288 ounces

4 Assess

Ask students which numbers should be multiplied first when multiplying two-digit numbers. (the numbers in the ones place)

For Mixed Abilities

Common Errors • Intervention

Students may add the regrouped digit to the tens before they multiply.

Incorrect	Correct
$\begin{array}{r} {}^{4}\\ 27 \\ \times\ 6 \\ \hline 362 \end{array}$	$\begin{array}{r} {}^{4}\\ 27 \\ \times\ 6 \\ \hline 162 \end{array}$

Have students use a place-value form like the following to chart the steps.

	100s	10s	1s
6×7 ones		4	2
6×2 tens	1	2	0
	1	6	2

Enrichment • Numeration

Have students make a table showing the number of rows needed to seat 400 people if each row had 8, 10, 25, or 50 seats.

More to Explore • Statistics

Have students keep a record of how many minutes they spend eating, sleeping, playing, and watching television each day for 1 week. At the end of the week, have them convert the minutes to hours and make a bar graph showing the time spent at each activity. Display the bar graphs, and have students compare their daily routines. As an extension, have students find the average time they spent, per day, on each activity.

6-6 Multiplying Money

pages 105–106

1 Getting Started

Objective
- To multiply money

Materials
*dollars, dimes, pennies

Warm Up • Mental Math
Dictate the following:

1. 68×2 (136)
2. $68 \div 2$ (34)
3. $(68 \div 2) \times 2$ (68)
4. $(68 \times 2) \div 2$ (68)
5. $(68 \div 2) \div 2$ (17)
6. $(68 \times 2) \times 2$ (272)
7. $\frac{1}{2}$ of 68 (34)
8. $2 \times (\frac{1}{2}$ of 68) (68)

Warm Up • Numeration
Pair students at the board. Ask one student to write the number 236 in tens and ones and the second student to write the same number in dimes and pennies. Have students check their partner's work and then change roles for another problem. Continue until students can readily name 2- and 3-digit numbers in tens and ones and in dimes and pennies.

Name ______________________

Multiplying Money

Eric is buying 7 folders at the school store.

How much will this cost him?

SCHOOL SUPPLIES	
PENCILS 8¢	ERASERS 15¢
PENS 59¢	PADS 39¢
PENNANTS 79¢	FOLDERS 34¢

We want to know how much the folders will cost Eric.

Eric is buying __7__ folders.

Each folder costs __34¢__.

We find the total cost by multiplying __34¢__ by __7__.

REMEMBER $0.34 is another way of writing 34¢.

Multiply the pennies. Regroup if needed.	Multiply the dimes. Add any extra dimes. Write the dollar sign and decimal point in the product.

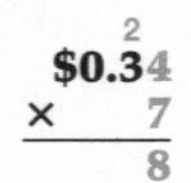

$$\begin{array}{r} \overset{2}{} \\ \$0.34 \\ \times\quad 7 \\ \hline 8 \end{array} \qquad \begin{array}{r} \overset{2}{} \\ \$0.34 \\ \times\quad 7 \\ \hline \$2.38 \end{array}$$

Seven folders will cost Eric __$2.38__.

Getting Started

Multiply.

1. $\begin{array}{r} \$0.42 \\ \times\quad 6 \\ \hline \$2.52 \end{array}$
2. $\begin{array}{r} \$0.87 \\ \times\quad 3 \\ \hline \$2.61 \end{array}$
3. $\begin{array}{r} \$0.38 \\ \times\quad 9 \\ \hline \$3.42 \end{array}$
4. $\begin{array}{r} \$0.45 \\ \times\quad 5 \\ \hline \$2.25 \end{array}$

Solve each problem using the information in the School Supplies sign.

5. Robin buys 5 pencils and 3 erasers. How much does Robin spend? 85¢
6. How much more are 3 pennants than 4 pads? 81¢

2 Teach

Introduce the Lesson Ask a student to describe the picture. Have another student read the problem and tell what needs to be found. (the cost of 7 folders) Ask students if all the necessary information is given in the problem. (no) Ask students what information is needed from the picture. (A folder costs 34¢.)

- Have students complete the sentences and guide them through the multiplication in the model. Remind them to place the dollar sign and decimal point. Ask students how they might check their solution to see if it makes sense. (Seven folders at 30¢ each would be $2.10, and 7 more 4¢ amounts would be 28¢.)

Develop Skills and Concepts Write **80 × 4** vertically on the board and have a student find the product. (320) Place dollar signs and decimal points in the problem to show money. Ask how many dollars, dimes, and pennies are in $3.20. (3, 2, 0)

- Write **$0.79 × 6** vertically on the board and have a student talk through the problem, showing the regroups, to find the product. ($4.74) Ask students to tell the number of dollars, dimes, and pennies in the answer. (4, 7, 4) Repeat for $0.89 × 7.

3 Practice

Have students work the first two rows of exercises on page 106 independently. Help students develop a plan for each of the word problems and then have students complete the DO and CHECK stages independently.

Practice

Multiply.

1. $0.63 × 3 = $1.89
2. $0.72 × 8 = $5.76
3. $0.16 × 4 = $0.64
4. $0.49 × 2 = $0.98
5. $0.84 × 3 = $2.52
6. $0.67 × 7 = $4.69
7. $0.89 × 9 = $8.01
8. $0.85 × 6 = $5.10
9. $0.53 × 6 = $3.18
10. $0.09 × 8 = $0.72
11. $0.75 × 7 = $5.25
12. $0.67 × 9 = $6.03

Problem Solving

Solve each problem. Use the School Store Supplies sign to solve Exercises 13–15.

School Store Supplies	
Pens 29¢	Pads 56¢
Pencils 6¢	Notebooks 89¢
Erasers 49¢	Highlighters 89¢

13. Adam buys an eraser and a pen. He gives the storekeeper a $1 bill. How much change should Adam receive?
22¢

14. Nancy has $4.25. She wants to buy 7 notebooks and 3 pens. How much more money does Nancy need?
$2.85

15. Todd buys 4 pencils and 5 pads. He gives the clerk a $5 bill. How much change should Todd receive?
$1.96

Use both signs to solve Exercise 16.

16. Mrs. Lopez shopped for school supplies at School Supplies Unlimited. She purchased 6 pens, 6 pencils, and 6 notebooks. How much money would she have saved if she had shopped at the School Store rather than shopping at School Supplies Unlimited?
$2.96

School Supplies Unlimited	
6 pens	$6.00
6 pencils	$0.60
6 erasers	$2.82
3 pads	$1.50
6 notebooks	$3.80
6 highlighters	$4.50

4 Assess

Ask students what they must multiply first when multiplying money. (the pennies)

5 Mixed Review

1. 18,027 − 6,510 (11,517)
2. 6 × 43 (258)
3. (3 × 7) + (4 × 9) (57)
4. 32,158 + 19,765 (51,923)
5. $728.95 − $374.27 ($354.68)
6. 27 + 338 (365)
7. $0.87 × 7 ($6.09)
8. 361 + 408 + 294 (1,063)
9. 13 × 8 (104)
10. 7,656 − 4,293 (3,363)

For Mixed Abilities

Common Errors • Intervention

Students may incorrectly place the dollar sign and decimal point in the answer because they have difficulty connecting the regrouping of dollars, dimes, and pennies to regrouping hundreds, tens, and ones. Give students the problem 3 × $0.36. Have them use real or play dimes and pennies to model $0.36 three times. Then, have them regroup 10 pennies for a dime and then 10 dimes for a dollar to demonstrate that the answer is $1.08.

Enrichment • Application

Tell students to compare two grocery ads to find the same item at different prices. Have them cut out the priced items and figure their cost savings if they bought seven at the lower price.

More to Explore • Measurement

Conduct a Class Olympics. Have students create and write rules for several skill games that could be conducted inside the classroom and whose results could be measured by length. Encourage the group to be creative in choosing appropriate events. Examples might be building a tower of milk cartons, an eraser toss, or a broom jump. Have students measure results in inches, feet, yards, and meters and record them on a classroom chart. Continue the Olympics for several days, changing the activities daily and recording all data.

ESL/ELL STRATEGIES

After completing the money exercises, call on different students to practice reading the problems and have them answer each problem aloud using dollar and cents amounts. Provide language support as needed; for example, *Forty-two cents times six equals two dollars and fifty-two cents.*

6-7 Problem Solving: Choose an Operation

pages 107–108

1 Getting Started

Objective

- To choose an operation to solve problems

Warm Up • Mental Math

Tell students to multiply by 6, then divide by 3.

1. 10 (20)
2. 15 (30)
3. 200 (400)
4. 11 (22)
5. 70 (140)
6. 22 (44)

2 Teach

Introduce the Lesson Remind students how to use the four-step problem-solving strategy for solving problems.

Develop Skills and Concepts Discuss with students how they need to know which operations to use in order to solve a problem. Ask, *Which operation is indicated by the words how many altogether?* (addition or multiplication) *Which operation is needed to find how many more, how many are left, how many fewer, and how much farther?* (subtraction) *What operation is needed to find how many apples there are in 9 rows of 8?* (multiplication or addition) *What operation is needed to find how much was earned per hour or how much each item cost?* (division)

- Have a student read the problem and tell what is to be found. (how many more cars the fourth-grade students must wash to earn $720) Ask students if the picture contains information needed to solve the problem. (yes) Ask students what needed information the picture contains. (It costs $8 to wash a car.)
- Read the SEE step with students as they fill in the information. Read the PLAN step with students and explain that in order to solve the problem, students will need to use several operations. Have students fill in the information. Finally, work through the DO and CHECK steps with students as they fill in the missing information.

It's Algebra! The concepts presented in this lesson prepare students for algebra.

Name ______________________

Problem Solving
Lesson 6-7

It's Algebra!

Problem Solving: Choose an Operation

The fourth-grade students are conducting a car wash to earn $720 to help pay for a class trip. By the end of the week, they had washed 85 cars. How many more cars must they wash to earn enough money to help pay for the trip?

★ SEE

We want to know how many more cars the fourth-grade class has to wash to earn $720.

We know the fourth-grade students washed __85__ cars. It costs __$8__ to wash 1 car.

★ PLAN

To find out how much money the fourth-grade earned this week, we __multiply__ 85 by __$8__.

To find out how much more money they need to earn, we __subtract__ what they have earned this week from $720. Then, we __divide__ by $8 to find the number of cars.

★ DO

$85 \times \$8 =$ __$680__ **Multiply.**

$\$720 - \$680 =$ __$40__ **Subtract.**

$\$40 \div \$8 =$ __5__ cars **Divide.**

The fourth-grade class needs to wash __5__ more cars.

★ CHECK

To check we add both amounts of money to see if the total is $720.

$\$680 + \$40 =$ __$720__

---

3 Apply

Solution Notes

1. Before students work on this problem, ask them to give the clue that tells which operation to use. Students will need to subtract the 32 puppies from 75.
2. Again, have students give the clue that tells which operations might be used to solve the problem. Students should realize that they can either add or multiply $0.95 by 5. Remind students to write the dollar sign and decimal point in their answer.
3. Students need to subtract to solve this problem. You may want to review how to regroup numbers when subtracting.
4. Addition is required to solve this problem. Remind students to align the numbers properly.

Apply

Write the operation(s) and solve the problem.

1. There were 75 puppies at a pet store. On Saturday, 32 of the puppies were sold. How many puppies remain at the pet store?
   subtraction; 43

2. Hamburgers at the cafeteria cost $0.95 each. How much will Eric and Graham spend together for 5 hamburgers?
   multiplication; $4.75

3. The highest point in New Mexico is Wheeler Point at 13,161 feet. The lowest point in New Mexico is Red Bluff Reservoir at 2,842 feet. How many feet higher is Wheeler Point than Red Bluff Reservoir?
   subtraction; 10,319 feet

4. Abraham Lincoln was born in 1809. He was 51 years old when he was elected to his first term as president of the United States. When was Lincoln elected for the first time?
   addition; 1860

5. Mrs. Cash is baking cookies. She has 15 rows with 7 cookies in each row. How many cookies is Mrs. Cash baking?
   multiplication; 105

Problem-Solving Strategy: Using the Four-Step Plan

★ **SEE** What do you need to find?
★ **PLAN** What do you need to do?
★ **DO** Follow the plan.
★ **CHECK** Does your answer make sense?

6. Mr. Williamson bought 6 boxes of cookies. Each box contains 45 cookies. How many cookies did Mr. Williamson buy?
   multiplication; 270

7. The area code for Hawaii is 808. The area code for Alaska is 99 greater than the area code for Hawaii. What is the area code for Alaska?
   addition; 907

8. Kelvin bought 8 stamps and 4 postcards. Stamps cost 37¢ and postcards cost 21¢. How much did Kelvin spend?
   multiplication and addition; $3.80

For Mixed Abilities

More to Explore • Numeration

Have students play "Division Charades." Distribute an index card to each student. Ask each student to write a division problem on the card. Tell students to make sure that no dividend is greater than the total number of students in the class. Collect the cards and put them in a pile. Explain to the class that in Charades, no words can be spoken. The first person selects a card and acts out the division problem. If the card reads $18 \div 3 = 6$, the student would silently tap 18 students on the shoulder, have them stand up, and then separate them into as many groups of three as possible. The other students have to guess what division fact has been acted out. If they cannot guess, the student does not earn a point. Have students prepare an additional index card with an *R* on it to represent "remainder." Problems could then be acted out in the same way with the remainder students grouped and given the *R* card to show to the class.

5. Ask students to tell what clue(s) helps them to decide which operation to use. Students can use either addition or multiplication to solve the problem.

6. This problem is similar to Exercise 5. Ask students whether it would be easier to add or multiply to find the answer.

Higher-Order Thinking Skills

7. **Analysis:** Tell students to read this problem carefully because the clue given is uncommon. Ask students to tell what words helped them to decide which operation to use.

8. **Comprehension:** Students should recognize that they need to use both multiplication and addition to solve the problem.

4 Assess

Ask students to make a chart of words or phrases that help them determine which operation(s) to use when solving word problems.

Chapter 6 Test

page 109

Items	Objectives
1–3	To recall basic multiplication facts (see page 95–96)
6–11	To solve for n using the Associative, Grouping, and Multiplying by 10 properties (see pages 97–98)
12–15	To multiply 2-digit numbers by 1-digit numbers without regrouping (see pages 99–100)
4–5, 16–19	To multiply 2-digit numbers by 1-digit numbers with regrouping (see pages 101–104)
20–23	To multiply money (see pages 105–106)

Name ______________________

CHAPTER 6 TEST

Write the first nine multiples of each number.

1. 3 0, 3, 6, 9, 12, 15, 18, 21, 24
2. 7 0, 7, 14, 21, 28, 35, 42, 49, 56
3. 9 0, 9, 18, 27, 36, 45, 54, 63, 72

Skip-count by 8.

4. 64, 72, 80, 88, 96, 104, 112, 120, 128

Skip-count by 4.

5. 12, 16, 20, 24, 28, 32, 36, 40, 44

Solve for *n*.

6. $6 \times (0 \times 3) = n$ $n = 0$
7. $8 \times 10 = n$ $n = 80$
8. $(7 \times 2) + (7 \times 3) = n$ $n = 35$
9. $9 \times (4 \times 2) = n$ $n = 72$
10. $(5 \times 0) + (5 \times 0) = n$ $n = 0$
11. $10 \times 9 = n$ $n = 90$

Multiply.

12. $21 \times 3 = 63$
13. $11 \times 7 = 77$
14. $42 \times 2 = 84$
15. $12 \times 4 = 48$
16. $36 \times 2 = 72$
17. $29 \times 3 = 87$
18. $16 \times 4 = 64$
19. $47 \times 2 = 94$
20. $\$0.37 \times 6 = \$2.22$
21. $\$0.49 \times 4 = \$1.96$
22. $\$0.73 \times 9 = \$6.57$
23. $\$0.87 \times 8 = \$6.96$

Alternate Chapter Test

You may wish to use the Alternate Chapter Test on page 356 of this book for further review and assessment.

CUMULATIVE ASSESSMENT

Circle the letter of the correct answer.

1. What is the place value of the 5 in 4,576?
a. thousands
(b.) hundreds
c. tens
d. NG

2. 5,219 ◯ 5,129
a. <
(b.) >

3. Round 735 to the nearest hundred.
(a.) 700
b. 800
c. NG

4. Round 7,295 to the nearest thousand.
a. 8,000
(b.) 7,000
c. NG

5. What is the place value of the 0 in 302,926.
a. tens
b. hundreds
c. thousands
(d.) NG

6. 49 + 73
a. 23
b. 112
c. 1,112
(d.) NG

7. 659 + 268
a. 817
b. 917
(c.) 927
d. NG

8. 3,629 + 4,381
a. 7,000
b. 8,000
(c.) 8,010
d. NG

9. 54,275 + 27,196
a. 71,371
b. 81,461
(c.) 81,471
d. NG

10. 51 − 26
(a.) 25
b. 36
c. 45
d. NG

11. 817 − 298
(a.) 519
b. 619
c. 681
d. NG

12. 61,083 − 28,596
(a.) 32,487
b. 32,587
c. 47,513
d. NG

13. 24 × 4
a. 86
(b.) 96
c. 816
d. NG

score

STOP

Cumulative Assessment

page 110

Items	Objectives
1	To identify place value through thousands (see pages 25–26)
2	To compare and order numbers through 10,000 (see pages 29–30)
3–4	To round numbers to the nearest hundred or thousand (see pages 31–32)
5	To identify place value through hundred thousands (see pages 25–26)
6–9	To add two numbers with up to 5-digits (see pages 37–42, 47–48)
10–11	To subtract two 2- or 3-digit numbers (see pages 55–58)
12	To subtract two 5-digit numbers with zero in the minuend (see pages 61–62)
13	To multiply 2-digit numbers by 1-digit numbers with two regroupings (see pages 103–104)

Alternate Cumulative Assessment

Circle the letter of the correct answer.

1. What is the place value of the 3 in 3,691?
(a) thousands
b hundreds
c tens
d NG

2. 7,364 ◯ 7,435
a >
(b) <
c =

3. Round 472 to the nearest hundred.
a 400
(b) 500
c NG

4. Round 9,327 to the nearest thousand.
a 8,000
(b) 9,000
c 10,000
d NG

5. What is the place value of the 6 in 625,430?
a hundreds
b thousands
c ten thousands
(d) NG

6. 65 + 78 =
a 1,313
(b) 143
c 133
d NG

7. 572 + 368
a 830
b 840
c 930
(d) NG

8. 6,947 + 2,164
a 8,012
(b) 9,111
c 9,011
d NG

9. 37,621 + 45,489
a 72,111
b 82,110
(c) 83,110
d NG

10. 86 − 49 =
(a) 37
b 43
c 47
d NG

11. 925 − 578
a 453
(b) 347
c 456
d NG

12. 73,071 − 34,294
a 41,223
b 39,877
c 38,877
(d) NG

13. 34 × 3 =
a 92
(b) 102
c 67
d NG

7-1 Dividing by 2 or 3

pages 111–112

1 Getting Started

Objective
- To review division facts with 2 or 3 as the divisor

Vocabulary
dividend, divisor, quotient

Warm Up • Mental Math
Have students complete the comparison: Quartet is to 4 as

1. triplet is to _____ (3)
2. unison is to _____ (1)
3. hexagonal is to _____ (6)
4. biweekly is to _____ (2)
5. baker's dozen is to _____ (13)
6. couple is to _____ (2)
7. gross is to _____ (144)

Warm Up • Numeration
Draw 9 groups of 2 Xs on the board. Ask how many groups of 2 there are. (9) Ask students how many Xs there are in all. (18) Draw a line through 1 group of Xs as you write one tally mark on the board and tell students that you have subtracted 1 group of 2. Ask how many Xs are left. (16) Continue to draw a line through 1 group at a time, record a tally mark, and ask students how many Xs are left. When all Xs have been subtracted, ask students to count the tally marks to tell how many groups of 2 Xs were subtracted in all. (9) Repeat for 8 groups of 3 Xs.

Name ____________________

CHAPTER 7

Dividing by a 1-Digit Number

Lesson 7-1

Dividing by 2 or 3

Lori is putting 2 cookies in each lunchbox. How many lunchboxes can she supply with cookies?

We want to know how many lunchboxes Lori can supply with 2 cookies each.

We know she has _12_ cookies.

Lori is putting _2_ cookies in each lunchbox.

To find how many lunchboxes can be supplied with cookies, we divide _12_ by _2_.

in all		in each		boxes		
12	÷	2	=	6	or	$2\overline{)12}$ with quotient 6
↑ dividend		↑ divisor		↑ quotient		6 ← quotient; 12 ← dividend; 2 ↑ divisor

Lori can supply 2 cookies each to _6_ lunchboxes.

Getting Started

Divide.

1. 6 ÷ 3 = _2_
2. 14 ÷ 2 = _7_
3. 8 ÷ 2 = _4_
4. 9 ÷ 3 = _3_
5. 24 ÷ 3 = _8_
6. 15 ÷ 3 = _5_
7. 6 ÷ 2 = _3_
8. 12 ÷ 2 = _6_
9. $2\overline{)16}$ 8
10. $3\overline{)27}$ 9
11. $3\overline{)18}$ 6
12. $2\overline{)4}$ 2

2 Teach

Introduce the Lesson Have a student read the problem and tell what is asked. (how many lunchboxes can have 2 cookies) Ask students if all the information they will need is given in the problem. (no) Ask what information is needed. (the number of cookies in all)

- Have students read and complete the information sentences. Point out the terms used in the model as students complete the solution sentence. Tell students to check their solution by asking what number times 2 equals 12. (6)

Develop Skills and Concepts Tell students that division is related to multiplication as you write **3 × 6 = 18** on the board. Tell students that if the product 18 is divided by 6 the answer will always be 3 because 3 × 6 = 18. Repeat for 18 divided by 3.

- Draw 18 Xs on the board and ask students to circle each group of 3. Draw 18 Xs again and repeat for groups of 6. Write **18 ÷ 3 = 6** and **18 ÷ 6 = 3** on the board and tell students that there is a name for each part of a division problem as you write the words **dividend, divisor**, and **quotient** under the respective number in each problem.
- Write $3\overline{)18}$ and $6\overline{)18}$ on the board as you tell students that this is another way to write a division problem.
- Tell students that to check a division problem using multiplication, the divisor times the quotient must equal the dividend. Provide additional division problems.

Practice

Divide.

1. $10 \div 2 =$ 5
2. $12 \div 3 =$ 4
3. $24 \div 3 =$ 8
4. $18 \div 2 =$ 9
5. $21 \div 3 =$ 7
6. $18 \div 3 =$ 6
7. $27 \div 3 =$ 9
8. $15 \div 3 =$ 5
9. $3\overline{)12}$ 4
10. $3\overline{)6}$ 2
11. $2\overline{)14}$ 7
12. $3\overline{)27}$ 9
13. $2\overline{)16}$ 8
14. $2\overline{)10}$ 5
15. $3\overline{)9}$ 3
16. $2\overline{)6}$ 3

Problem Solving

Solve each problem.

17. Joan and Angie each have 12 sweaters to store in boxes. Joan will store 3 sweaters in each box. Angie can only put 2 of her sweaters in a box. How many boxes will they need altogether for their sweaters? 10 boxes

18. Pat is putting marbles into bags. He has 24 marbles and wants to put 3 marbles into each bag. How many bags will Pat need? 8 bags

Now Try This!

Complete the tables of Arabic and Roman numerals.

hundreds	
100	C
200	CC
300	CCC
400	CD
500	D
600	DC
700	DCC
800	DCCC
900	CM

tens	
10	X
20	XX
30	XXX
40	XL
50	L
60	LX
70	LXX
80	LXXX
90	XC

ones	
1	I
2	II
3	III
4	IV
5	V
6	VI
7	VII
8	VIII
9	IX

3 Practice

Remind students to check each problem by multiplying. Tell students to make a plan to solve the word problems. Have students complete the page independently.

Now Try This! Ask students to identify the patterns that exist between the three columns such as using three letters to represent the Roman numerals for 3, 30, and 300.

Have students write expanded notations such as $23 = 20 + 3$, XXIII = XX + III, $719 = 700 + 10 + 9$, and DCCXIX = DCC + X + IX.

4 Assess

Put a division problem on the board and ask students to identify the dividend, the divisor, and the quotient.

For Mixed Abilities

Common Errors • Intervention

Some students may have difficulty finding the quotient when they are dividing by 2 or 3. For a division sentence such as $12 \div 3 = \square$, have students write a related multiplication sentence, $\square \times 3 = 12$, to help them see that 4 is the quotient. For more help, have students think $1 \times 3 = 3$, $2 \times 3 = 6$, $3 \times 3 = 9$, and $4 \times 3 = 12$, until they find the correct factor.

Enrichment • Multiplication

Tell students to write a related division problem for each multiplication fact for 2 and 3.

More to Explore • Applications

Tell students that they can find the day of the week for any given date by using these codes:

Month	Code	Month	Code
January	1	July	0
February	4	August	3
March	4	September	6
April	0	October	1
May	2	November	4
June	5	December	6

Day	Code
Sunday	1
Monday	2
Tuesday	3
Wednesday	4
Thursday	5
Friday	6
Saturday	0

Write the following directions and a month, day, and year on the board for students to use as an example:

1. Divide the last two digits of the year by 4, dropping remainders.
2. Add the code number of the month.
3. Add the day's date.
4. Divide by 7, keeping the remainder. Tell students that this remainder is the code number for the day of the week of the date they started with.

7-2 Dividing by 4 or 5

pages 113–114

1 Getting Started

Objective

- To review division facts with 4 or 5 as the divisor

Warm Up • Mental Math

Have students name a fact to check the following:

1. 7×2 ($14 \div 2$, $14 \div 7$)
2. $24 \div 6$ (4×6)
3. $15 - 7$ ($7 + 8$)
4. $36 \div 4$ (9×4)
5. $17 - 9$ ($8 + 9$)
6. $\frac{1}{2}$ of 16 (8×2, $16 \div 2$)
7. $\frac{1}{5}$ of 35 (7×5, $35 \div 5$)

Warm Up • Numeration

Divide students into pairs. Write **20** on the board and ask students to show as many different ways as they can to make 20 using addition or subtraction. (18 + 2, 25 − 5, and so on) Give students 5 minutes and then have them list their solutions in columns on the board. Ask if there are other ways not yet given. (Yes, there are an infinite number of solutions because 2,000 − 1,980 = 20, and so on.)

Name ______________________ **Lesson 7-2**

Dividing by 4 or 5

Roy is packing apples for 4 gift boxes. How many apples should he pack into each box?

We want to know how many apples Roy should pack in each box.

There are __24__ apples.

Roy is packing __4__ boxes.

To find the number of apples for each box, we divide __24__ by __4__.

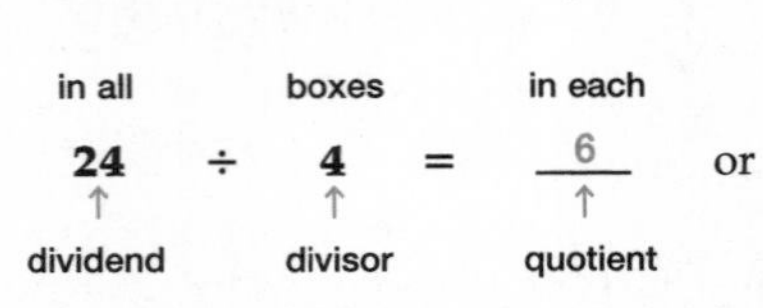

in all		boxes		in each		
24	÷	4	=	__6__	or	$4\overline{)24}$ with quotient 6
dividend		divisor		quotient		divisor; 6 ← quotient; 24 ← dividend

Roy should pack __6__ apples into each box.

Getting Started

Divide.

1. $36 \div 4 =$ __9__
2. $45 \div 5 =$ __9__
3. $16 \div 4 =$ __4__
4. $20 \div 5 =$ __4__
5. $5\overline{)30}$ __6__
6. $4\overline{)12}$ __3__
7. $4\overline{)36}$ __9__
8. $5\overline{)10}$ __2__
9. $4\overline{)32}$ __8__
10. $5\overline{)15}$ __3__
11. $4\overline{)8}$ __2__
12. $5\overline{)35}$ __7__

2 Teach

Introduce the Lesson Have students read the problem and use the picture to tell what is to be solved. (the number of apples in each box) Ask students what is known already. (24 apples are to go into 4 boxes.) Ask if there is any unnecessary information given. (no)

- Have students complete the sentences. Guide them through the division shown in the model, and have students complete the solution sentence. Ask students how they can use multiplication to check their solution. (The product of their solution and 4 should equal 24.)

Develop Skills and Concepts Have 16 students stand. Tell students that we need to find a number that multiplied by 4 will equal 16. Ask students to think of a multiplication fact in which 4 multiplied by another number equals 16. ($4 \times 4 = 16$)

- Have 4 students go to each of the 4 corners of the room and ask all students if $16 \div 4 = 4$. (yes) Now, have the 16 students sit down.
- Tell students that another way to solve this problem is to place 1 student at a time in each of the 4 areas until all 16 students are moved. Have students act this out and check to see if 4 groups of 4 equal 16 in all. (yes)
- Repeat the previous activity for $45 \div 5$ and $28 \div 4$.

3 Practice

Tell students that they will need to decide which operation to use as they make a plan to solve each word problem. Have students complete page 114 independently.

Practice

Divide.

1. $25 \div 5 =$ 5
2. $10 \div 5 =$ 2
3. $8 \div 4 =$ 2
4. $28 \div 4 =$ 7
5. $24 \div 4 =$ 6
6. $18 \div 3 =$ 6
7. $20 \div 5 =$ 4
8. $15 \div 5 =$ 3
9. $16 \div 4 =$ 4
10. $32 \div 4 =$ 8
11. $24 \div 3 =$ 8
12. $30 \div 5 =$ 6
13. $35 \div 5 =$ 7
14. $12 \div 4 =$ 3
15. $20 \div 4 =$ 5
16. $36 \div 4 =$ 9
17. $4\overline{)12}$ 3
18. $5\overline{)25}$ 5
19. $5\overline{)30}$ 6
20. $2\overline{)18}$ 9
21. $4\overline{)20}$ 5
22. $4\overline{)16}$ 4
23. $3\overline{)27}$ 9
24. $4\overline{)28}$ 7
25. $5\overline{)10}$ 2
26. $5\overline{)20}$ 4
27. $5\overline{)35}$ 7
28. $4\overline{)8}$ 2
29. $2\overline{)12}$ 6
30. $4\overline{)24}$ 6
31. $5\overline{)45}$ 9
32. $5\overline{)40}$ 8

Problem Solving

Solve each problem.

33. There are 24 students in Miss Chen's class. The students sit at 4 tables. How many students are at each table?
6 students

34. There are 28 students in Mr. Orr's class. On Tuesday, 4 students were absent. How many students were present on Tuesday?
24 students

35. There are 20 children playing soccer. There are 5 teams with the same number of children. How many children are on each team?
4 children

36. There are 5 buttons on each blouse. On Friday, Kay sewed on 30 buttons. How many blouses did Kay sew buttons on?
6 blouses

4 Assess

Have students name the multiplication fact that will help them solve $81 \div 9 = \square$. ($9 \times 9 = 81$)

For Mixed Abilities

Common Errors • Intervention

Some students may place the quotient in the wrong place when they are dividing.

Incorrect	Correct
$4\overline{)36}$ with 9 above the 3	$4\overline{)36}$ with 9 above the 6

Encourage them to first say the correct place value for the 9 and then write the digit in the ones place.

Enrichment • Numeration

Have students make a multiplication facts table for 2, 3, 4, and 5. Tell them to use their table to find how many products end in 0 or 5.

More to Explore • Numeration

Have students explain the difference between cardinal (counting) numbers and ordinal (number of position) numbers. Ask them to write out the English words for the cardinal numbers 1 through 10 and the ordinal numbers first through tenth. Ask students to go to a library, or any other source, and find the same 20 words in French or Spanish. (French: un, deux, trois, quatre, cinq, six, sept, huit, neuf, dix; premier, deuxième, troisième, quatrième, cinquième, sixième, septième, huitième, neuvième, dixième) (Spanish: uno, dos, tres, cuatro, cinco, seis, siete, ocho, nueve, diez; primero, segundo, tercero, cuarto, quinto, sexto, séptimo, octavo, noveno, décimo)

ESL/ELL STRATEGIES

To support students' comprehension and pronunciation of the words *divisor, dividend,* and *quotient,* write the words on the board, capitalizing the stressed syllables: diVISor, DIVidend, QUOtient. Review the meaning of each in simple English and have students repeat the word and the meaning.

7-3 Dividing by 6 or 7

pages 115–116

1 Getting Started

Objective

- To review division facts with 6 or 7 as the divisor

Materials

multiplication fact cards for 2 through 7

Warm Up • Mental Math

Have students name two different 1-digit numbers with a

1. sum equal to 15 plus 2 (8 and 9)
2. quotient of 4 (8 and 2, 4 and 1)
3. product equal to 60 plus 3 (7 and 9)
4. sum equal to 30 divided by 2 (7 and 8)
5. difference equal to 81 divided by 9 (9 and 0)
6. product equal to 15 times 2 (6 and 5)
7. quotient equal to 15 divided by 3 (5 and 1)
8. product > 70 (9 and 8)

Warm Up • Numeration

Write **4 × 5 = 20** on the board and have students write a related division problem. (20 ÷ 4 = 5 or 20 ÷ 5 = 4) Continue for other multiplication facts of 2, 3, 4, or 5.

It's Algebra!

Name ____________________

Dividing by 6 or 7

Mr. Lopez spent $24 on concert tickets for his family. How many tickets did he buy?

We want to know the number of tickets Mr. Lopez bought.

He spent <u>$24</u> on all the tickets.

Each ticket cost <u>$6</u>.

To find the number of tickets, we divide <u>$24</u> by <u>$6</u>.

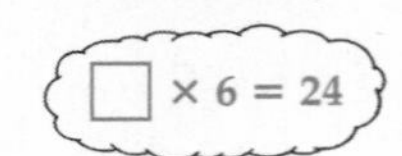

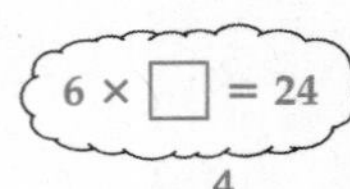

$24 ÷ $6 = <u>4</u> or **$6)$24** (4)

Mr. Lopez bought <u>4</u> tickets.

Getting Started

Divide.

1. 54 ÷ 6 = <u>9</u>
2. 28 ÷ 7 = <u>4</u>
3. 42 ÷ 6 = <u>7</u>
4. 14 ÷ 7 = <u>2</u>
5. 56 ÷ 7 = <u>8</u>
6. 18 ÷ 6 = <u>3</u>
7. 30 ÷ 6 = <u>5</u>
8. 42 ÷ 7 = <u>6</u>
9. 6)12 (2)
10. 6)24 (4)
11. 7)21 (3)
12. 7)35 (5)
13. 7)49 (7)
14. 6)36 (6)
15. 7)63 (9)
16. 6)48 (8)
17. 6)42 (7)
18. 2)14 (7)
19. 7)56 (8)
20. 6)54 (9)

2 Teach

Introduce the Lesson Have students describe the picture and read the problem. Ask students what is to be found. (the number of tickets Mr. Lopez bought) Ask what information will be used to solve the problem. (He spent $24 and 1 ticket costs $6.)

- Have students read through the plan, complete the sentences, and solve the problem. Have students check their solutions by multiplying.

Develop Skills and Concepts Write **6 × _____ = 54** on the board. Have a student complete the problem. (9)

- As you write **54 ÷ 6 =** on the board, tell students that we can also write this as a division problem. Have a student complete the problem. (9)
- Have students think of a multiplication fact for 6 or 7 and write the division problem for the fact on the board. Continue until all facts for 6 and 7 have been written.

3 Practice

Remind students that some word problems require more than one operation. Have students solve the exercises on page 116 independently.

Practice

Divide.

1. $24 \div 3 =$ 8
2. $28 \div 4 =$ 7
3. $35 \div 7 =$ 5
4. $30 \div 6 =$ 5
5. $12 \div 6 =$ 2
6. $16 \div 2 =$ 8
7. $14 \div 7 =$ 2
8. $56 \div 7 =$ 8
9. $54 \div 6 =$ 9
10. $63 \div 7 =$ 9
11. $24 \div 6 =$ 4
12. $28 \div 7 =$ 4
13. $18 \div 6 =$ 3
14. $28 \div 4 =$ 7
15. $42 \div 7 =$ 6
16. $36 \div 6 =$ 6

Copy and divide.

17. $6\overline{)30}$ 5
18. $6\overline{)36}$ 6
19. $7\overline{)21}$ 3
20. $7\overline{)49}$ 7
21. $6\overline{)24}$ 4
22. $7\overline{)42}$ 6
23. $6\overline{)18}$ 3
24. $7\overline{)28}$ 4
25. $7\overline{)63}$ 9
26. $5\overline{)40}$ 8
27. $6\overline{)54}$ 9
28. $6\overline{)48}$ 8
29. $7\overline{)35}$ 5
30. $4\overline{)32}$ 8
31. $6\overline{)42}$ 7
32. $7\overline{)56}$ 8

Problem Solving

Solve each problem.

33. Tickets for the school carnival and dinner were $7 each. The Johnson family paid $35 for tickets. How many Johnsons went to the carnival?
5 Johnsons

34. Danny bought 3 concert tickets for $5 each. Marta bought 2 tickets at $6 each. How much did they pay for all the tickets?
$27

35. Ruth bought 6 albums for $8 each. Randy bought 7 albums for $7 each. How much more did Randy spend for his albums?
$1

36. All children's books were on sale for $6 each. Rene bought books worth $30. How many books did Rene buy?
5 books

4 Assess

Ask students to write a division fact with 7 as the divisor and 6 as the quotient. ($42 \div 7 = 6$)

5 Mixed Review

1. $203.78 − $147.29 ($56.49)
2. 38×6 (228)
3. $19 + 38 + 56 + 25$ (138)
4. $27{,}265 + 4{,}070$ (31,335)
5. $28 \div 4$ (7)
6. 3,271 - 597 (2,674)
7. $(8 \times 3) + (6 \times 7)$ (66)
8. $42{,}171 - 27{,}385$ (14,786)
9. $42 \div 7$ (6)
10. $25 - 7 \times 3$ (4)

For Mixed Abilities

Common Errors • Intervention

Some students have difficulty dividing by 6 or 7. Have them write a related multiplication sentence and use that fact to find the missing factor. For example, for the problem $24 \div 6 = \square$, they should write $\square \times 6 = 24$. Then, starting with $1 \times 6 = 6$, $2 \times 6 = 12$, and so on, they will find the correct factor.

Enrichment • Numeration

Tell students to draw rectangular cookie sheets to show 2 different ways to place 28 cookies in rows.

More to Explore • Logic

Dictate the following problem for students to solve:

You are the parking attendant at a rock concert, where five of the latest and greatest rock stars are performing. You have been instructed to park their cars so that each star may dash out, get in his or her car, and drive off after the concert. You have to line up the cars in the order their owners are performing, but you don't have a program. You do overhear the following, as people are arriving: "I only came tonight to hear my favorite, Bobby Boogey! I can't stand I. M. Electric who goes on right before him, and I won't stay for Rhonda Rocker." "What a great line-up! Peter Punk followed by I. M. Electric." "I got here early because I don't want to miss The Duke!" Now, line up the cars so that these stars can get away before their fans mob them.

Ask students to draw 5 cars and label them by performer, in the order of their leaving. (The Duke, Peter Punk, I. M. Electric, Bobby Boogey, and Rhonda Rocker)

7-4 Dividing by 8 or 9

pages 117–118

1 Getting Started

Objective
- To review division facts with 8 or 9 as the divisor

Materials
counters and cups

Warm Up • Mental Math
Have students name the number referred to in the word.

1. decade (10)
2. quadrilateral (4)
3. triplets (3)
4. octagonal (8)
5. sextet (6)
6. centennial (100)
7. millionaire (1,000,000)
8. couplet (2)

Warm Up • Multiplication
Review the multiplication facts for 8 and 9 by having students make number lines on the board to show the products of 8×1 through 8×9, and similarly for 9. Have students then practice skip counting by 8s and 9s.

Name ____________________

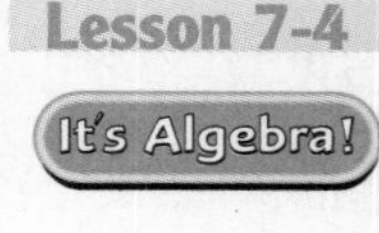

Dividing by 8 or 9

Mrs. Ferris is buying one place setting of dinnerware at a time to complete her set. How much will she pay for one place setting?

We want to find the cost of one place setting.

The sale price is $48 for 8 place settings.

To find the cost of one place setting, we divide $48 by 8.

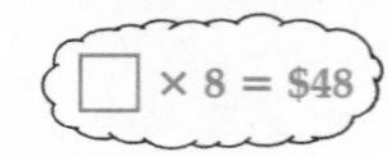

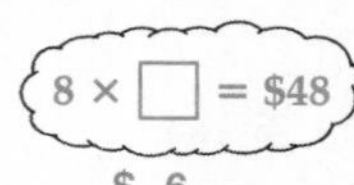

$48 ÷ 8 = $6 or **8)$48** ($6)

Mrs. Ferris will pay $6 for one place setting.

Getting Started

Divide.

1. $64 \div 8 =$ 8
2. $36 \div 9 =$ 4
3. $18 \div 9 =$ 2
4. $45 \div 9 =$ 5
5. $56 \div 8 =$ 7
6. $72 \div 8 =$ 9
7. $48 \div 8 =$ 6
8. $27 \div 9 =$ 3

Copy and divide.

9. 3)15 — 5
10. 8)32 — 4
11. 9)72 — 8
12. 9)54 — 6
13. 8)72 — 7
14. 9)63 — 7
15. 8)24 — 3
16. 8)40 — 5
17. 8)56 — 7
18. 5)45 — 9
19. 7)21 — 3
20. 4)28 — 7

2 Teach

Introduce the Lesson Have students describe the picture. Discuss the meaning of sales and what price might have originally been charged for the 8 place settings. Ask a student to read the problem and tell what is being asked. (the cost of 1 place setting) Ask students where the necessary information can be found. (in the picture)

- Have students complete the sentences and the division in the model with you. Then, have students complete the solution sentence. Ask students to check their answer by multiplying.

Develop Skills and Concepts Write **40 ÷ 5 =** _____ and **5 ×** _____ **= 40** on the board. Have students complete the problems. (8, 8) Remind students that the missing factor in the multiplication problem helps us solve the division problem.

- Write more division problems on the board and have students write missing-factor problems for each and then solve both problems.

3 Practice

Have students complete page 118 independently.

Practice

Divide.

1. $36 \div 9 =$ 4
2. $42 \div 7 =$ 6
3. $48 \div 8 =$ 6
4. $64 \div 8 =$ 8
5. $27 \div 9 =$ 3
6. $81 \div 9 =$ 9
7. $56 \div 8 =$ 7
8. $16 \div 8 =$ 2
9. $18 \div 9 =$ 2
10. $45 \div 9 =$ 5
11. $25 \div 5 =$ 5
12. $63 \div 9 =$ 7
13. $48 \div 6 =$ 8
14. $24 \div 8 =$ 3
15. $72 \div 8 =$ 9
16. $32 \div 8 =$ 4

Copy and divide.

17. $8\overline{)32}$ 4
18. $8\overline{)16}$ 2
19. $6\overline{)36}$ 6
20. $9\overline{)45}$ 5
21. $9\overline{)63}$ 7
22. $9\overline{)27}$ 3
23. $8\overline{)72}$ 9
24. $8\overline{)24}$ 3
25. $9\overline{)36}$ 4
26. $8\overline{)48}$ 6
27. $4\overline{)16}$ 4
28. $8\overline{)56}$ 7
29. $3\overline{)9}$ 3
30. $9\overline{)54}$ 6
31. $9\overline{)18}$ 2
32. $9\overline{)81}$ 9

Problem Solving

Solve each problem.

33. Katy paid $24 for 8 scarves. How much did each scarf cost?
$3

34. Phil paid 54¢ for 9 nails. How much did each nail cost?
6¢

35. Leigh paid $40 for 8 baseballs. Later, she sold the baseballs for $7 each. How much profit did Leigh make on each baseball?
$2

36. There are 9 players on a softball team. All 45 people who came to practice were put on teams. Each team paid a fee of $8 to use the field. How much was collected from the teams?
$40

4 Assess

Have students write a division fact with 9 as the divisor and 8 as the quotient. ($72 \div 9 = 8$)

For Mixed Abilities

Common Errors • Intervention

Some students will continue to have difficulty dividing by 8 and 9. Have these students work with a partner, using counters to model the problem. To model $56 \div 8$, for example, the students use 56 counters and 8 paper cups. They deal out the counters into the cups until all counters have been used. Then, they count the number of counters in each cup. The answer, 7, is the quotient. Have them repeat with other facts involving 8 and 9.

Enrichment • Division

Have students make division fact cards for dividing by 6 through 9. Tell them to work with a friend to increase their speed in recalling quotients.

More to Explore • Geometry

Illustrate a pattern on a geoboard and show students how to calculate the area by counting the number of squares enclosed. Have students look at their own geoboards. If their geoboards are marked in 1-inch squares, the area can be described in square inches. Tell them that it is also possible to use squares as a unit of measure. Ask students to make squares with the following areas: 1, 4, 9, and 16 squares. Have them complete the following chart:

Area of the Square	Number of Nails Touched	Perimeter
1	(4)	(4)
4	(8)	(8)
9	(12)	(12)
16	(16)	(16)

7-5 Understanding 1 and 0 in Division

pages 119–120

1 Getting Started

Objective
- To understand 1 and 0 in division

Materials
6 counters and 6 cups

Warm Up • Mental Math
Tell students to round the numbers to the nearest hundred and solve.

1. 2,615 − 1,398 (1,200)
2. 294 + 606 (900)
3. 719 × 2 (1,400)
4. 1,017 ÷ 2 (500)
5. 85 × 3 (300)
6. $\frac{1}{3}$ of 876 (300)
7. $\frac{1}{10}$ of 979 (100)
8. 15,982 ÷ 2 (8,000)

Warm Up • Multiplication
Have students tell the product of 1 × 0 through 1 × 9 to review that 1 times any number is that number itself. Ask students the product of zero times other numbers in order to review that the product will be zero.

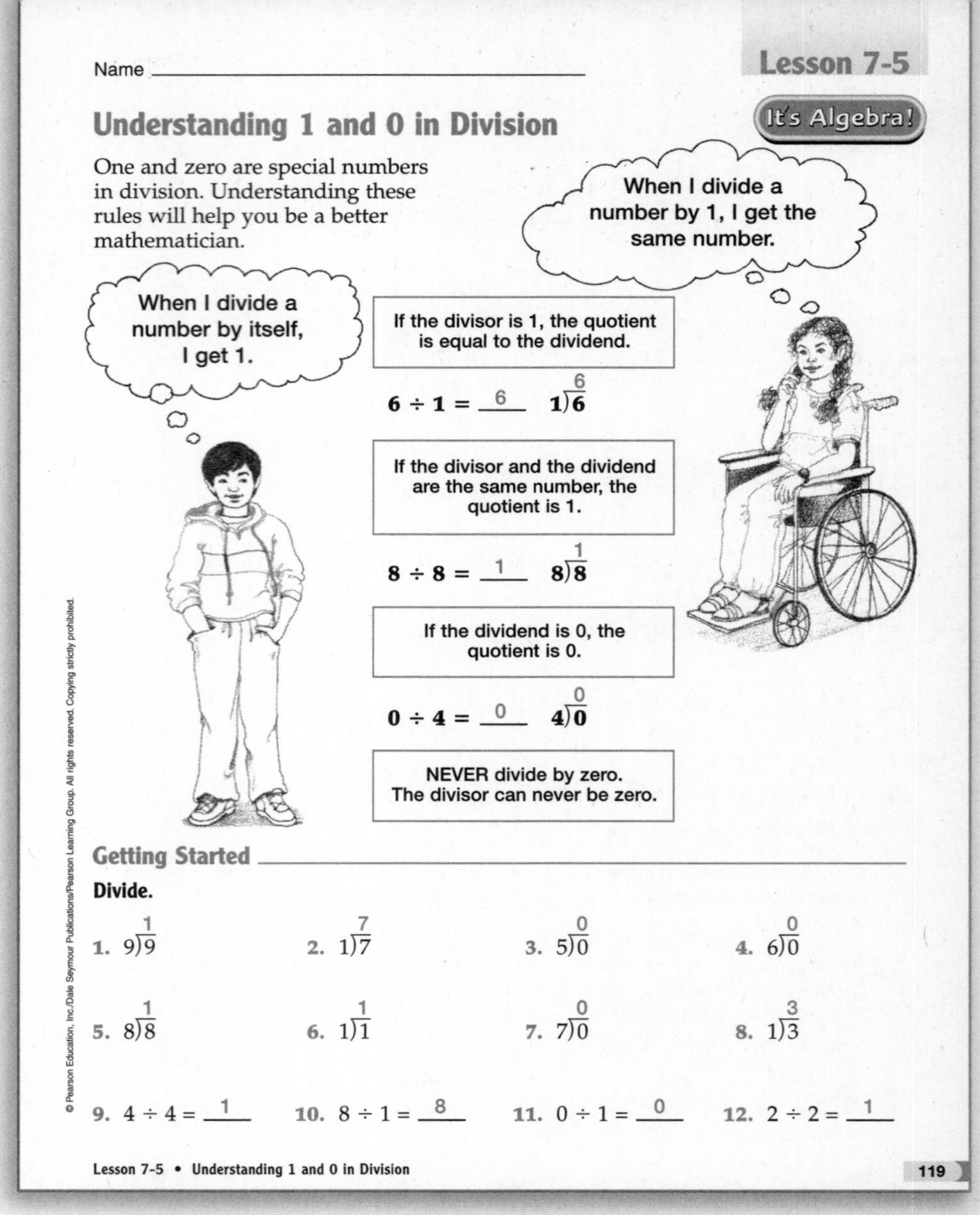

Name ______________________

Lesson 7-5

Understanding 1 and 0 in Division

It's Algebra!

One and zero are special numbers in division. Understanding these rules will help you be a better mathematician.

If the divisor is 1, the quotient is equal to the dividend.

$6 \div 1 = \underline{6}$ $\quad 1\overline{)6}$ (6)

If the divisor and the dividend are the same number, the quotient is 1.

$8 \div 8 = \underline{1}$ $\quad 8\overline{)8}$ (1)

If the dividend is 0, the quotient is 0.

$0 \div 4 = \underline{0}$ $\quad 4\overline{)0}$ (0)

NEVER divide by zero. The divisor can never be zero.

Getting Started

Divide.

1. $9\overline{)9}$ (1)
2. $1\overline{)7}$ (7)
3. $5\overline{)0}$ (0)
4. $6\overline{)0}$ (0)
5. $8\overline{)8}$ (1)
6. $1\overline{)1}$ (1)
7. $7\overline{)0}$ (0)
8. $1\overline{)3}$ (3)
9. $4 \div 4 = \underline{1}$
10. $8 \div 1 = \underline{8}$
11. $0 \div 1 = \underline{0}$
12. $2 \div 2 = \underline{1}$

2 Teach

Introduce the Lesson Have a student read the thought bubbles in the picture. Read the paragraph to the students. Have students fill in the examples as you read the rules. Emphasize that these are basic rules of division.

Develop Skills and Concepts Write **7 ÷ 1 = 7** on the board. Ask students to tell which number in this problem is the divisor. (1) Write the word *divisor* under the 1. Repeat for the dividend and quotient.

- Now, write $\mathbf{1\overline{)7}}$ on the board and have students name each number as the quotient, divisor, or dividend.
- Remind students that the quotient times the divisor must equal the dividend as you write **7 × 1 = 7** on the board.
- Help students remember that in reviewing the facts of 1, they found that 1 times any number is that number itself.
- Write **9 ÷ 1 =** on the board and ask students what number times 1 equals 9. (9) Tell students that when any dividend is divided by 1, the quotient is the same as the dividend.
- Write **9 ÷ 9 =** _____ on the board and ask students what number times 9 equals 9. (1) Continue for 8 ÷ 8, 7 ÷ 7, and so on. Tell students that a number can never be divided by zero. Tell students that 5 ÷ 0 cannot be worked because its related multiplication fact, _____ × 0 = 5, cannot be solved. Have students experiment with other problems where zero is the dividend.

3 Practice

Remind students to refer to the rules for 1 and 0 in division as they work the exercises on page 120 independently.

Practice

Divide.

1. $2\overline{)2}$ (1)	2. $5\overline{)10}$ (2)	3. $4\overline{)4}$ (1)	4. $7\overline{)14}$ (2)	5. $1\overline{)1}$ (1)
6. $2\overline{)0}$ (0)	7. $1\overline{)4}$ (4)	8. $2\overline{)6}$ (3)	9. $9\overline{)9}$ (1)	10. $3\overline{)18}$ (6)
11. $1\overline{)7}$ (7)	12. $1\overline{)8}$ (8)	13. $2\overline{)12}$ (6)	14. $4\overline{)12}$ (3)	15. $2\overline{)4}$ (2)
16. $5\overline{)5}$ (1)	17. $9\overline{)27}$ (3)	18. $1\overline{)9}$ (9)	19. $7\overline{)49}$ (7)	20. $9\overline{)0}$ (0)
21. $7\overline{)35}$ (5)	22. $3\overline{)27}$ (9)	23. $8\overline{)8}$ (1)	24. $7\overline{)63}$ (9)	25. $3\overline{)0}$ (0)
26. $7\overline{)56}$ (8)	27. $5\overline{)35}$ (7)	28. $9\overline{)72}$ (8)	29. $3\overline{)3}$ (1)	30. $2\overline{)8}$ (4)
31. $1\overline{)2}$ (2)	32. $8\overline{)40}$ (5)	33. $1\overline{)0}$ (0)	34. $6\overline{)6}$ (1)	35. $4\overline{)0}$ (0)

36. $18 \div 9 =$ 2	37. $6 \div 1 =$ 6	38. $6 \div 3 =$ 2	39. $28 \div 7 =$ 4
40. $3 \div 1 =$ 3	41. $18 \div 6 =$ 3	42. $5 \div 1 =$ 5	43. $7 \div 7 =$ 1
44. $48 \div 6 =$ 8	45. $54 \div 6 =$ 9	46. $0 \div 9 =$ 0	47. $20 \div 5 =$ 4
48. $40 \div 5 =$ 8	49. $1 \div 1 =$ 1	50. $0 \div 6 =$ 0	51. $45 \div 9 =$ 5

4 Assess

Ask, *If the divisor and the dividend are the same, what is the quotient ?* (1)

5 Mixed Review

1. $438 + 15{,}193$ (15,631)
2. $520 - 393$ (127)
3. $7 \times \$0.95$ (\$6.65)
4. $652 + 729 + 803$ (2,184)
5. $63 \div 9$ (7)
6. 78×6 (468)
7. $2{,}473 - 658$ (1,815)
8. $8 \div 1$ (8)
9. $(2 \times 5) + 18$ (28)
10. $0 \div 5$ (0)

For Mixed Abilities

Common Errors • Intervention

Some students may think the answer to both of these problems is 1.

Incorrect	Correct
$6 \div 1 = 1$	$6 \div 6 = 1$

Correct students by having them work with a partner, using counters to model the problem. Have them deal 6 counters into 6 cups to see that when the counters have been dealt, there is 1 counter in each cup. So, $6 \div 6 = 1$. Then, have them deal 6 counters into 1 cup to see that $6 \div 1 = 6$.

Enrichment • Numeration

Tell students to draw a picture that shows how sharing and $3 \div 3 = 1$ are alike. Then, have them draw a second picture that shows how selfishness and $4 \div 1 = 4$ are alike. Ask them to explain their drawings to the class.

More to Explore • Geometry

Have students research and report on geodesic domes. Ask what kinds of geometric shapes are used to construct such a structure. Have students construct their own domes as an extended project, working independently or in groups. Possible materials to use would be straws (paper straws work best), toothpicks, and paper cut into geometric shapes and glued together.

7-6 Rules of Divisibility

pages 121–122

1 Getting Started

Objective

- To test whether numbers are divisible by 2, 3, 5, 9, or 10

Materials

multiplication tables, counters

Warm Up • Mental Math

Have students find the quotient for the following:

1. 24 ÷ 3 = (8)
2. 15 ÷ 5 = (3)
3. 12 ÷ 3 = (4)
4. 32 ÷ 4 = (8)
5. 27 ÷ 9 = (3)
6. 20 ÷ 4 = (5)

Warm Up • Numeration

Ask students to tell how old the person was in 1978.

1. Eli was born in 1952. (26)
2. Selma was born in 1969. (9)
3. Grant was born in 1892. (86)
4. Corey was born in 1966. (12)
5. Evan was born in 1918. (60)
6. Katina was born in 1924. (54)
7. Cliff was born in 1971. (7)
8. Chim was born in 1953. (25)

Name ______________________

Rules of Divisibility

Stan wants to give his friends an equal number of marbles. Among how many friends can he equally divide his marbles: 2, 3, 5, 9, or 10?

We want to know among how many friends—2, 3, 5, 9, or 10—Stan can equally divide 30 marbles.

We can use rules to find out which numbers can be equally divided into 30 with no remainder.

Number of Friends	Rule	Is 30 Divisible?
2	A number is divisible by 2 if the last digit is even— 0, 2, 4, 6, or 8.	30 is divisible by 2 because the ones digit is 0 and 0 is even.
3	A number is divisible by 3 if the sum of its digits is divisible by 3.	30 is divisible by 3 because 3 + 0 = 3 and 3 is divisible by 3.
5	A number is divisible by 5 if the ones digit is 0 or 5.	30 is divisible by 5 because the ones digit is 0.
9	A number is divisible by 9 if the sum of its digits is divisible by 9.	30 is not divisible by 9 because 3 + 0 = 3 and 3 is not divisible by 9.
10	A number is divisible by 10 if the ones digit is 0.	30 is divisible by 10 because the ones digit is 0.

Stan can equally divide 30 marbles among 2, 3, 5, or 10 friends.

Getting Started

List the numbers, 2, 3, 5, 9, or 10, that the given number is divisible by.

1. 27 — 3, 9
2. 42 — 2, 3
3. 80 — 2, 5, 10
4. 35 — 5
5. 20 — 2, 5, 10

2 Teach

Introduce the Lesson Have students study the problem and tell what is being asked. (how many friends Stan can give an equal number of marbles to) Ask what information is needed. (the number of marbles and the number of friends)

- Read the plan to the class. Explain that there are several rules and that we are going to use them to solve the problem.
- Work through the model with students and have them fill in the missing information and the solution sentence. (2, 3, 5, or 10 friends)

Develop Skills and Concepts Explain to students that they can use the multiplication table to help them learn divisibility rules.

- Have students look at the 5 and 10 columns on the multiplication table and explain the pattern they see. (Any number divisible by 10 is also divisible by 5.)
- Then, have students use the multiplication table to help them skip-count by 3s and see what pattern they find. (Any number in the 6 or 9 column is divisible by 3.)

3 Practice

Have students complete all the exercises on page 122. Remind them that in Exercises 1 through 20 they are only checking for the divisibility of numbers by 2, 3, 5, 9, or 10.

Practice

List the numbers, 2, 3, 5, 9, or 10, that the given number is divisible by.

1. 32 — 2
2. 51 — 3
3. 17 — none
4. 72 — 2, 3, 9
5. 66 — 2, 3
6. 14 — 2
7. 43 — none
8. 24 — 2, 3
9. 55 — 5
10. 70 — 2, 5, 10
11. 39 — 3
12. 96 — 2, 3
13. 40 — 2, 5, 10
14. 54 — 2, 3, 9
15. 81 — 3, 9
16. 76 — 2
17. 28 — 2
18. 71 — none
19. 84 — 2, 3
20. 95 — 5

Problem Solving

Solve each problem.

21. There are 50 stars on the U.S. flag. If the stars were in equal rows, could there be 2, 3, 5, 9, or 10 rows?
2, 5, or 10 rows

22. Gina's uniform number is a 2-digit number that is divisible by 2, 3, 5, 9, and 10. Which number does Gina wear?
90

23. There are 25 members of the Evanstown Chorus. The members were to sit in rows of equal numbers for a concert. Can they sit in 2, 3, 5, 9, or 10 rows?
5 rows

24. Alyssa has 36 books in her collection. She wants to place them on shelves in equal numbers. Can she place the books in 2, 3, 5, 9, or 10 rows?
2, 3, or 9 rows

25. Tom's uniform number is between 60 and 69 and is divisible by 2 and 3 but not by 5, 9, or 10. Which number does Tom wear?
66

26. Teri says that all numbers that are divisible by 3 are divisible by 9. Neil says that all numbers that are divisible by 9 are divisible by 3. Who is correct: Teri, Neil, both, or neither?
Neil

For Mixed Abilities

Common Errors • Intervention

Some students may think that numbers divisible by 3 are automatically divisible by 9 and that numbers divisible by 5 are automatically divisible by 10. Give students two piles of counters representing numbers that are not divisible by both 3 and 9. Have students separate one pile into groups of 3 and the other into groups of 9. Repeat for 5 and 10.

Enrichment • Numeration

Tell students to expand their understanding of divisibility rules by looking at the multiplication table for 4 and 6. Ask students what patterns they see.

More to Explore • Numeration

Have students work in pairs and have them choose five 4- and 5-digit numbers of their own. Have them determine which of their numbers are divisible by 2, 3, 5, 9, or 10. Have them share their results with the class.

4 Assess

Ask students to tell which numbers—2, 3, 5, 9, or 10—90 is divisible by. (2 because the ones digit in 90, 0, is even; 3 and 9 because the sum of the digits in 90 is 9 and 9 is divisible by 3 and by 9; 5 and 10 because the ones digit in 90 is 0)

5 Mixed Review

1. $88 + 155$ (243)
2. $312 + 534$ (846)
3. $98 - 59$ (39)
4. $282 - 194$ (88)
5. 18×7 (126)
6. 38×2 (76)
7. 46×3 (138)
8. $56 \div 8$ (7)
9. $45 \div 3$ (15)
10. $92 \div 2$ (46)

7-7 Working With Remainders

pages 123–124

1 Getting Started

Objective

- To divide by 1-digit numbers with remainders

Vocabulary

remainder

Materials

*10 pencils, cups, counters

Warm Up • Mental Math

Have students tell the amount of money.

1. $(\frac{1}{10}$ of \$1$) + (\frac{1}{4}$ of \$1$)$ (35¢)
2. $2 \times \frac{1}{4}$ of \$1 (50¢)
3. $\frac{1}{20}$ of \$1 $\times$ 6 (30¢)
4. 25 dimes − 4 × 10¢ (\$2.10)
5. $\frac{1}{4}$ of \$10 (\$2.50)
6. 7 quarters and 6 dimes (\$2.35)
7. 8 dimes × 9 (\$7.20)
8. $(\frac{1}{5}$ of \$5$) \div 5$ (20¢)

Warm Up • Numeration

Write **3, 6, 9,** and **12** on the board. Review skip counting by having a student continue to write the multiples of 3 through 27. Repeat for multiples of other numbers.

Name ____________________

Working With Remainders

Keith wants to buy as many pretzels as he can with 43¢. How many can he buy? How much money will he have left?

We want to know the number of pretzels Keith can buy and how much money he will have left.

We know Keith has __43¢__ to spend.

Each pretzel costs __5¢__.

To find the number of pretzels Keith can buy and to find out how much money will be left over, we divide __43¢__ by __5¢__.

Guess the closest division fact with a product less than or equal to 43.	Multiply.	Subtract.	The remainder must be less than the divisor.
$5¢\overline{)43¢}$ quotient 8	$5¢\overline{)43¢}$ quotient 8; 40 (8 × 5)	$5¢\overline{)43¢}$ quotient 8 pretzels; − 40; 3 remainder	$5¢\overline{)43¢}$ quotient 8 R3; − 40; 3 (3¢ < 5¢)

Keith can buy __8__ pretzels. He will have __3¢__ left.

Getting Started

Divide. Show your work.

1. $6\overline{)13}$ 2 R1
2. $4\overline{)23}$ 5 R3
3. $2\overline{)19}$ 9 R1
4. $7\overline{)32}$ 4 R4
5. $8\overline{)23}$ 2 R7
6. $9\overline{)30}$ 3 R3
7. $5\overline{)37}$ 7 R2
8. $3\overline{)25}$ 8 R1
9. $8\overline{)58}$ 7 R2
10. $6\overline{)39}$ 6 R3

2 Teach

Introduce the Lesson Have a student describe the picture. Have another student read the problem and tell what two problems are to be solved. (the number of pretzels Keith can buy and the amount of money he will have left)

- Ask students what information is needed from the problem and the picture. (Keith has 43¢ and 1 pretzel costs 5¢.) Have students complete the sentences.
- Guide students through the division steps in the model to solve the problem. Tell students that they can check their answer by multiplying the quotient by the divisor and adding the remainder.

Develop Skills and Concepts Have 3 students stand. Ask another student to distribute 10 pencils, one at a time, to the 3 students so that all students have the same number. When 3 pencils are given to each of the 3 students, ask if any pencils are left over. (yes) Tell students that what is left over is called a **remainder** because it remains after dividing the pencils equally.

- Repeat the previous activity with 8 pencils and 3 students for a remainder of 2.
- Write **16 ÷ 3 =** _____ on the board and tell students that we need to think of a number that, when multiplied by 3, will be almost 16. (5) Ask students if there will be a remainder. (yes) Have a student write **5 R1** in the answer.
- Repeat the previous activity for 19 ÷ 2, 21 ÷ 4, and 37 ÷ 6.

3 Practice

Remind students to use *R* and the number to write the remainder beside the quotient. Have students complete page 124 independently.

Practice

Divide. Show your work.

1. $8\overline{)43}$ 5 R3
2. $5\overline{)39}$ 7 R4
3. $4\overline{)34}$ 8 R2
4. $9\overline{)86}$ 9 R5
5. $7\overline{)45}$ 6 R3
6. $2\overline{)13}$ 6 R1
7. $6\overline{)27}$ 4 R3
8. $3\overline{)28}$ 9 R1
9. $6\overline{)34}$ 5 R4
10. $4\overline{)30}$ 7 R2
11. $6\overline{)39}$ 6 R3
12. $9\overline{)51}$ 5 R6
13. $8\overline{)31}$ 3 R7
14. $7\overline{)29}$ 4 R1
15. $3\overline{)16}$ 5 R1

Now Try This!

It's Algebra!

Complete each table.

Multiply.	Add.	Find each missing factor.	Find each missing addend.
1 × 9 = 9	0 + 9 = 9	9 × 1 = 9	0 + 9 = 9
2 × 9 = 18	1 + 8 = 9	9 × 2 = 18	1 + 8 = 9
3 × 9 = 27	2 + 7 = 9	9 × 3 = 27	2 + 7 = 9
4 × 9 = 36	3 + 6 = 9	9 × 4 = 36	3 + 6 = 9
5 × 9 = 45	4 + 5 = 9	9 × 5 = 45	4 + 5 = 9
6 × 9 = 54	5 + 4 = 9	9 × 6 = 54	5 + 4 = 9
7 × 9 = 63	6 + 3 = 9	9 × 7 = 63	6 + 3 = 9
8 × 9 = 72	7 + 2 = 9	9 × 8 = 72	7 + 2 = 9
9 × 9 = 81	8 + 1 = 9	9 × 9 = 81	8 + 1 = 9

For Mixed Abilities

Common Errors • Intervention

Some students may think that numbers divisible by 3 are automatically divisible by 9 and that numbers divisible by 5 are automatically divisible by 10. Give students two piles of counters representing numbers that are not divisible by both 3 and 9. Have students separate one pile into groups of 3 and the other into groups of 9. Repeat for 5 and 10.

Enrichment • Numeration

Tell students to expand their understanding of divisibility rules by looking at the multiplication table for 4 and 6. Ask students what patterns they see.

More to Explore • Numeration

Have students work in pairs and have them choose five 4- and 5-digit numbers of their own. Have them determine which of their numbers are divisible by 2, 3, 5, 9, or 10. Have them share their results with the class.

Now Try This! Have students fill in the tables. When they have completed the tables, ask if they can see any patterns in the answers. (In Column 1, the sum of the digits in each product is 9.) Help them to see that the products in Column 1 can be compared to the addend pair in Column 2. (18 compared to 1 + 8) In Column 3, help them see that the missing factor is one number greater than the first digit of the product, except for 9 × 1 = 9.

4 Assess

Ask students to tell which numbers—2, 3, 5, 9, or 10—90 is divisible by. (2 because the ones digit in 90, 0, is even; 3 and 9 because the sum of the digits in 90 is 9 and 9 is divisible by 3 and by 9; 5 and 10 because the ones digit in 90 is 0)

5 Mixed Review

1. 88 + 155 (243)
2. 312 + 534 (846)
3. 98 − 59 (39)
4. 282 − 194 (88)
5. 18 × 7 (126)
6. 38 × 2 (76)
7. 46 × 3 (138)
8. 56 ÷ 8 (7)
9. 45 ÷ 3 (15)
10. 92 ÷ 2 (46)

7-8 Dividing, 2-Digit Quotients

pages 125–126

1 Getting Started

Objective

- To divide by a 1-digit number for a 2-digit quotient with no remainders

Materials

*multiplication facts; place-value blocks: tens and ones; cards; rubber bands

Warm Up • Mental Math

Tell students to start at 10 and

1. add 2, divide by 6 (2)
2. multiply by 9, add 6 (96)
3. divide by 5, subtract 2 (0)
4. add $\frac{1}{2}$ of 40 (30)
5. triple the number (30)
6. multiply by 6, divide by 5 (12)
7. quadruple the number (40)
8. add 4×9 (46)

Warm Up • Numeration

Show multiplication facts through 9 and have students write and solve the related division problems on the board.

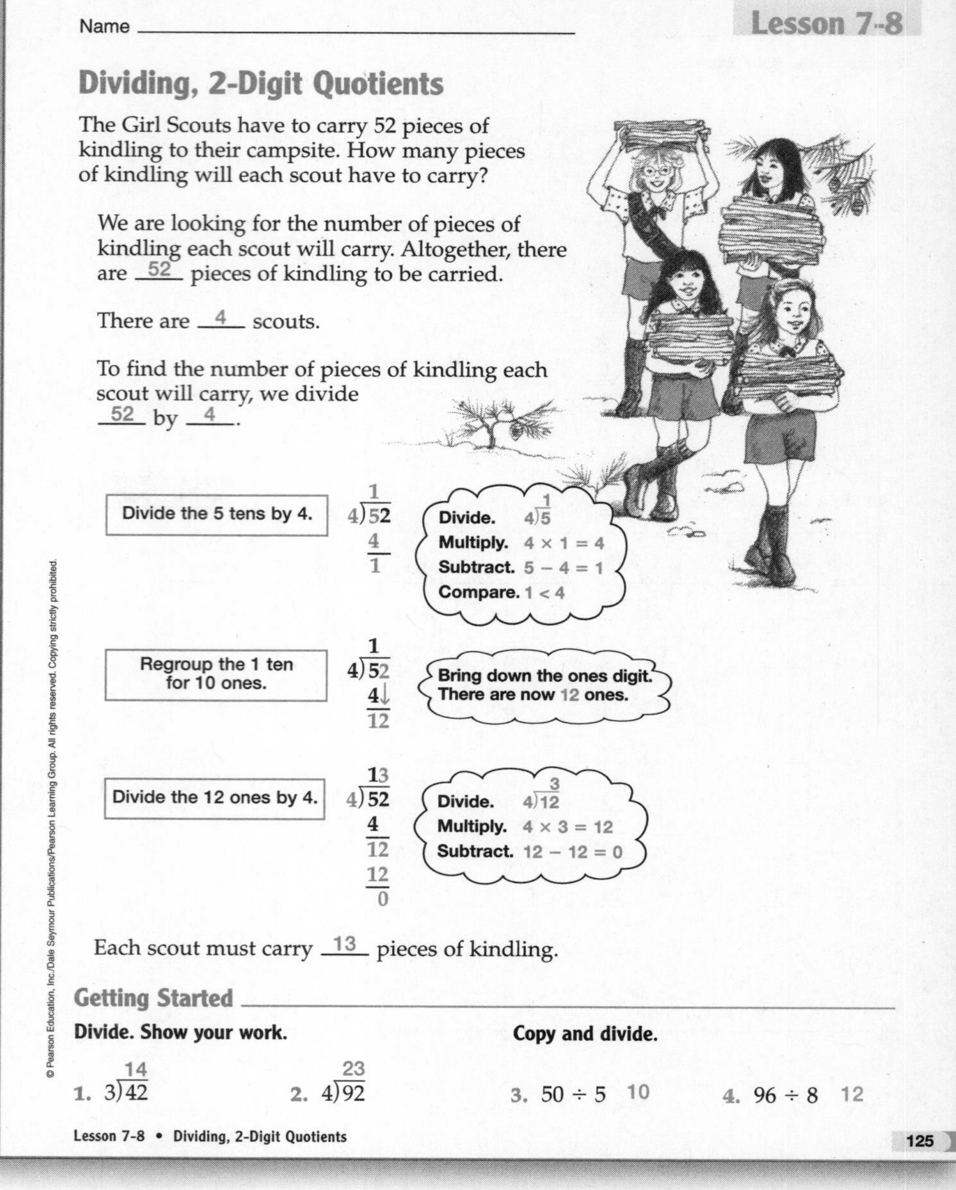

Name ______________________ Lesson 7-8

Dividing, 2-Digit Quotients

The Girl Scouts have to carry 52 pieces of kindling to their campsite. How many pieces of kindling will each scout have to carry?

We are looking for the number of pieces of kindling each scout will carry. Altogether, there are __52__ pieces of kindling to be carried.

There are __4__ scouts.

To find the number of pieces of kindling each scout will carry, we divide __52__ by __4__.

Divide the 5 tens by 4. $\begin{array}{r} 1 \\ 4\overline{)52} \\ \underline{4} \\ 1 \end{array}$

Divide. $4\overline{)5}$ (1)
Multiply. $4 \times 1 = 4$
Subtract. $5 - 4 = 1$
Compare. $1 < 4$

Regroup the 1 ten for 10 ones. $\begin{array}{r} 1 \\ 4\overline{)52} \\ \underline{4\downarrow} \\ 12 \end{array}$

Bring down the ones digit. There are now 12 ones.

Divide the 12 ones by 4. $\begin{array}{r} 13 \\ 4\overline{)52} \\ \underline{4} \\ 12 \\ \underline{12} \\ 0 \end{array}$

Divide. $4\overline{)12}$ (3)
Multiply. $4 \times 3 = 12$
Subtract. $12 - 12 = 0$

Each scout must carry __13__ pieces of kindling.

Getting Started

Divide. Show your work.

1. $3\overline{)42}$ 14
2. $4\overline{)92}$ 23

Copy and divide.

3. $50 \div 5$ 10
4. $96 \div 8$ 12

2 Teach

Introduce the Lesson Have a student read the problem and tell what is to be found. (the number of pieces of kindling each Girl Scout will carry) Ask students what information is needed. (the number of pieces of kindling in all and the number of girls) Ask if all necessary information is given in the written problem. (no) Ask students where the other information is. (in the picture)

- Have students read and complete the sentences. Guide them through the division problem in the model and have students complete the solution sentences.

Develop Skills and Concepts Write $2\overline{)56}$ on the board. Have students display the following place-value blocks: 5 tens and 6 ones. Tell students that we first divide the 5 tens by 2 by asking what number times 2 is 5, or slightly less than 5. (2) Write **2** above the 5 and tell students that we must then subtract 4 tens from the 5 to see how many tens are left. (1) Draw a line under the 4 and write **1** as you tell students that 1 ten is left. Remind students that the number of tens left must be less than the divisor.

- Tell students that to divide the ones, we regroup the 1 ten for 10 ones and add the 6 ones to make 16 ones in all. Record the 6 beside the 1 as students make the regrouping with their manipulatives. Ask students what number times 2 equals 16. (8) Record 8 in the quotient and tell students that we now subtract the 16 ones from the 16 to get zero.
- Repeat the procedure for $90 \div 5$ and $76 \div 4$.

3 Practice

Work through the first few problems on the board to provide models for students to follow if needed as they complete page 126. Have students complete the page independently.

Practice

Divide. Show your work.

1. $5\overline{)75}$ 15
2. $3\overline{)63}$ 21
3. $4\overline{)80}$ 20
4. $4\overline{)96}$ 24
5. $2\overline{)56}$ 28
6. $3\overline{)72}$ 24
7. $7\overline{)84}$ 12
8. $5\overline{)80}$ 16
9. $6\overline{)96}$ 16
10. $9\overline{)90}$ 10
11. $8\overline{)88}$ 11
12. $3\overline{)81}$ 27
13. $4\overline{)56}$ 14
14. $7\overline{)91}$ 13
15. $6\overline{)84}$ 14
16. $2\overline{)94}$ 47
17. $3\overline{)75}$ 25
18. $7\overline{)98}$ 14
19. $3\overline{)90}$ 30
20. $4\overline{)76}$ 19

Copy and divide.

21. $78 \div 6$ 13
22. $90 \div 2$ 45
23. $85 \div 5$ 17
24. $72 \div 6$ 12
25. $60 \div 5$ 12
26. $60 \div 2$ 30
27. $72 \div 4$ 18
28. $88 \div 4$ 22
29. $66 \div 3$ 22
30. $76 \div 2$ 38
31. $87 \div 3$ 29
32. $96 \div 8$ 12

Problem Solving

Solve each problem.

33. Gene bought 3 thank-you notes for 75¢. How much did each note cost?
25¢

34. Rhoda feeds her horse 3 apples a day for lunch. How many apples will the horse eat in 27 days?
81 apples

35. Lemons cost 8¢ each. Sally has 96¢. How many lemons can Sally buy?
12 lemons

36. The tennis club used 45 tennis balls in a tournament. Tennis balls are sold 3 in a can. How many cans of tennis balls did the club use? 15 cans

4 Assess

Ask students what the number left over in a division problem is called. (the remainder)

For Mixed Abilities

Common Errors • Intervention

Students may place the first digit of the quotient incorrectly when tens are involved, as in $3\overline{)48}$.

Incorrect	Correct
$\begin{array}{r} 1 \\ 3\overline{)48} \\ \underline{3} \\ 5 \end{array}$	$\begin{array}{r} 16 \\ 3\overline{)48} \\ \underline{3\downarrow} \\ 18 \\ \underline{18} \\ 0 \end{array}$

Have students work in cooperative groups of three using 48 cards. Have them group the cards into 4 tens and 8 ones, putting rubber bands around each set of ten.

Ask, *How many tens can you give to each person in your group?* (1) Have them write a **1** above the 4 tens in the dividend. Ask, *What will you do with the 1 ten left over?* (make it 10 ones to put with the others to make 18 ones) Ask, *How many ones can you give to each person in your group?* (6) Write a **6** above the ones in the dividend. Work other division problems in a similar way.

Enrichment • Numeration

Tell students to find the three 1-digit numbers that can be divided into 99 evenly. Tell them to show their work. (9, 3, 1)

More to Explore • Logic

Read or duplicate the following problem for students to solve:

A chemist rushed from his laboratory carrying a flask of steaming, purple liquid. "Look what I have invented," he said proudly to a fellow chemist. "The liquid in this bottle is such a powerful acid that it will dissolve anything it touches." "You're wrong," said the second chemist immediately. How did the second chemist know that the first was wrong? (If the liquid could dissolve anything it touched, it would have dissolved the flask in which it was being carried.)

7-9 Dividing, 2-Digit Quotients With Remainders

pages 127–128

1 Getting Started

Objective
- To divide by a 1-digit number for a 2-digit quotient with a remainder

Materials
*counters

Warm Up • Mental Math
Have students give a calculator code to find the following:

1. 7 + 2 + 10 (7 + 2 + 10 =)
2. $60 + 38.25 (60 + 38.25 =)
3. seven 8s (7 × 8 =)
4. 800 − 46 (800 − 46 =)
5. $7 + .16 − .80 (7 + .16 − .8 =)
6. $2.06 − .90 (2.06 − .9 =)
7. $4 × 3 (4 × 3 =)
8. 7 × 9 × 2¢ (7 × 9 × .02 =)

Warm Up • Division
Write 3)16 on the board and have a student work the problem to show a quotient of 5 R1. Provide more problems of 1-digit quotients with remainders.

Name ____________________

Dividing, 2-Digit Quotients With Remainders

Scott, Sam, and Tom were quizzing each other with 52 multiplication fact cards. Scott dealt all the cards so that each boy received the same number. He put the extra cards in the center. How many cards did each boy receive? How many are in the center?

We want to know the number of cards dealt to each boy and the number put in for the center.

There are 52 cards all together.

The cards were dealt to 3 boys.

To find the number of cards each boy received and the number in the center, we divide 52 by 3.

Divide the tens. Guess the closest fact with a product less than or equal to 5.	Bring down the ones. Divide the ones. Guess the closest fact with a product less than or equal to 22.	Write the remainder.
3)52 → quotient 1; 3; 2. Multiply. 3 × 1 = 3 Subtract. 5 − 3 = 2 Compare. 2 < 3	3)52 → quotient 17; 3; 22; 21. Multiply. 3 × 7 = 21	3)52 → quotient 17 R1; 3; 22; 21; 1. Subtract. 22 − 21 = 1 Compare. 1 < 3

Each boy received 17 cards. There is 1 card in the center.

Getting Started

Divide. Show your work.

1. 3)19 6 R1
2. 6)65 10 R5

Copy and divide.

3. 76 ÷ 2 38
4. 90 ÷ 4 22 R2
5. 99 ÷ 5 19 R4
6. 4)53 13 R1

---

2 Teach

Introduce the Lesson Have a student read the problem and tell what students need to find. (how many cards each boy will receive and how many cards will be left) Ask students what necessary information they will use to solve the two problems. (There are 52 cards and 3 boys.) Have students complete the sentences.

- Guide students through the division problem in the model, explaining each step through writing the remainder.
- Have students complete the solution sentence. Students may want to act out the problem to check their solution.

Develop Skills and Concepts Write **4)94** on the board. Ask students how many of the 9 tens can be given to each of 4 people. (2) Record the 2 above the 9 and ask students how many tens would be left as you subtract 8 from 9. (1) Remind students that 1 < 4.

- Ask students how 1 ten can be divided among 4 people. (change it to 10 ones) Bring down the 4 ones as you tell students that there are 14 ones to be divided by 4. Ask students to think of a multiplication fact for 4 whose product is 14 or almost 14. (3 × 4) Record the 3 above the 4 and ask students how many ones are left as you subtract 12 from 14. (2) Tell students to compare the 2 and 4 to be sure the leftover ones are less than the divisor. Record the remainder.
- Write **5)87** on the board and have a student solve the problem, talking through each step. Now, write **9)98** on the board to show students the placement of the zero in the ones place in the quotient.

3 Practice

Remind students to compare the leftover tens or ones after each subtraction to be sure the leftover number is less than the divisor. Have students complete the exercises on page 128 independently.

Practice

Divide. Show your work.

1. $4\overline{)37}$ 9 R1
2. $7\overline{)85}$ 12 R1
3. $2\overline{)57}$ 28 R1
4. $6\overline{)91}$ 15 R1
5. $5\overline{)72}$ 14 R2
6. $9\overline{)93}$ 10 R3
7. $8\overline{)67}$ 8 R3
8. $3\overline{)76}$ 25 R1
9. $4\overline{)71}$ 17 R3
10. $2\overline{)19}$ 9 R1
11. $3\overline{)97}$ 32 R1
12. $9\overline{)98}$ 10 R8
13. $8\overline{)89}$ 11 R1
14. $6\overline{)85}$ 14 R1
15. $7\overline{)96}$ 13 R5

Copy and divide.

16. 47 ÷ 3 15 R2
17. 54 ÷ 5 10 R4
18. 86 ÷ 4 21 R2
19. 62 ÷ 6 10 R2
20. 92 ÷ 7 13 R1
21. 91 ÷ 2 45 R1
22. 98 ÷ 8 12 R2
23. 59 ÷ 3 19 R2
24. 71 ÷ 4 17 R3
25. 87 ÷ 5 17 R2
26. 76 ÷ 7 10 R6
27. 80 ÷ 6 13 R2

Problem Solving

Solve each problem.

28. Margaret has 76¢ to buy stickers. Stickers cost 5¢ each. How many stickers can Margaret buy? How much money will she have left?
15 stickers; 1¢ left

29. John has 65 flowers that he wants to plant in 6 rows. How many flowers will go into each row. How many flowers will be left?
10 flowers; 5 flowers left

30. Yoko has divided her 51 books equally on 4 shelves. How many books go on each shelf? How many books are left?
12 books; 3 books left

31. Tyrone is putting 5 flowers into each vase. He is filling 16 vases. How many flowers altogether does Tyrone need to put into the vases?
80 flowers

4 Assess

Ask, *When a division problem has a remainder, what must you compare it to and why?* (the divisor; to make sure it is less than the divisor)

5 Mixed Review

1. 28,127 + 34,392 (62,519)
2. 57 × 9 (513)
3. 7 + 4 × 8 (39)
4. 51 ÷ 3 (17)
5. $70.00 − $27.68 ($42.32)
6. 5 × $0.25 ($1.25)
7. 396 + 275 + 427 (1,098)
8. 95 ÷ 6 (15 R5)
8. 964,326 − 19,610 (944,716)
10. 84 ÷ 8 (10 R4)

For Mixed Abilities

Common Errors • Intervention

Students may repeatedly multiply or subtract incorrectly when working a division problem because they have not yet mastered their facts. Have these students work with partners on sets of subtraction or multiplication facts in order to firm up these skills.

Enrichment • Logic

Read or duplicate the following for students to solve: A man was in his car headed due north. The road he was on was perfectly straight.

After driving one mile and making no turns at all, the driver discovered he was exactly one mile south of his starting point. How could this happen? (The man drove his car backward.)

More to Explore • Applications

Have students make a dot grid on their papers, made up of 10 rows of dots with 10 dots in each row. Have students use the dot grid to show division problems. Give students a word problem such as the following:

Jason has 24 cents. The stickers he wants to buy cost 6¢ each. How many stickers can he buy?

Ask students to circle sets of dots showing how many stickers Jason can buy. (4 groups of 6 dots) Have students write a division fact for the problem. (24 ÷ 6 = 4) Then, have students create their own division problems. Ask them to trade problems and solve each other's by using the dot grid.

ESL/ELL STRATEGIES

Students may need extra help with phrases such as *guess the closest fact, less than or equal to,* and *bring down the ones.* Write problems like those in the book on the board and use these phrases as you solve them. Ask students to repeat each phrase as you use it.

7-10 Problem Solving: Draw a Diagram

pages 129–130

1 Getting Started

Objective

- To draw a picture or diagram to solve a problem

Warm Up • Mental Math

Tell students to divide the sum or product by 60.

1. 80 + 220 (5)
2. 100 + 1,700 (30)
3. 60 × 40 (40)
4. 24 × 200 (80)

Tell students to divide the sum or product by 6.

5. 80 + 220 (50)
6. 100 + 1,700 (300)
7. 60 × 40 (400)
8. 24 × 200 (800)

Name ____________________

Problem Solving: Draw a Diagram

Five girls are running a race. Elaine is 50 yards behind Diane. Cassie is 10 yards behind Elaine. Barb is 50 yards ahead of Cassie. Ann is 20 yards behind Barb. Ann is 30 yards behind Diane.

What is the order of the girls at this time?

★ SEE

We want to know the order of the five girls in the race.

We know Elaine is 50 yards behind Diane. Cassie is 10 yards behind Elaine. Cassie is 50 yards behind Barb. Ann is 20 yards behind Barb. Ann is 30 yards behind Diane.

★ PLAN

You can draw a diagram of the race, showing positions 10 yards apart.

★ DO

Using the facts, place the initial of each girl on the diagram to show her position.

START •—•—•—•—•—•—•—•—•—• FINISH
C E A B D

The order of the girls from first to last is Diane, Barb, Ann, Elaine, and Cassie.

★ CHECK

Compare your diagram of the race with the facts in the original problem to see if each girl is in the correct position.

2 Teach

Introduce the Lesson Remind students of the four-step method for solving problems.

Develop Skills and Concepts Ask students if they have ever tried to describe something to someone and found the best way was to draw a picture or diagram to show what they meant. Discuss with students how we draw a map to tell someone how to get to a specific place. Tell students that we often draw a picture or diagram to solve such problems because it can make a problem easier for us to understand.

- Have a student read the problem. Tell students that in order to make sense of all this information, we need to organize it by drawing the position of each runner.
- Have a student read the SEE step and fill in the blanks. Tell students that sometimes we need to reword the information in the SEE step to make it more clear. Have students note that some of the sentences have been reworded so that each tells who is behind whom.
- Have a student read the PLAN step. Tell students that although the problem does not ask the distance between each runner, it will help to draw the spaces marked off at 10-yard intervals.
- Have students locate each runner to complete the DO step and then see if each statement in the CHECK step can be answered yes.
- Ask students if any of the facts are unnecessary to solve the problem. (Yes, Ann is 30 yards behind Diane.) Now, have students use their diagrams to find the distance between every 2 runners.

Apply

Draw a diagram to help you solve each problem. Remember to use the four-step plan.

1. Sue is waiting to buy tickets for a movie. There are 5 people waiting ahead of her and 7 people waiting behind her. How many people are in line for tickets?

   13 people

2. If it takes 3 minutes to make a cut in a log, how long will it take to cut a 20-foot log into 10 equal lengths?

   27 minutes

3. If you count the number of pickets and spaces along a fence, you will find that the number of spaces is always one less than the number of pickets. How many spaces will a bicycle wheel with 15 spokes contain?

   15 spaces

4. Valleyview is south of Columbus but north of Lexington. Trenton is south of Valleyview but north of Lexington. Mintown is north of Valleyview but south of Columbus. Which town is the second most southerly town of the five?

   Trenton

Problem-Solving Strategy: Using the Four-Step Plan

- ★ **SEE** What do you need to find?
- ★ **PLAN** What do you need to do?
- ★ **DO** Follow the plan.
- ★ **CHECK** Does your answer make sense?

5. One dozen eggs and 2 half-gallons of milk cost $3.35. Two dozen eggs and 2 half-gallons of milk cost $4.30. How can you use subtraction to find which costs more, 1 dozen eggs or 1 half-gallon of milk?

   See Solution Notes.

6. Five dogs are chasing a cat. The dogs are 3 yards behind the cat and running hard. The cat is 12 feet from a tree. The cat and the dogs are running at the same rate. Will the cat reach the tree and safety before the dogs catch the cat?

   yes

7. A brown horse, a gray horse, and a white horse were in a race on a rainy day. Draw a picture to show how many different ways the horses can finish if the gray horse never wins on rainy days.

   See Solution Notes.

For Mixed Abilities

More to Explore • Geometry

Have students create tangram designs to share with others in the class. A tangram is a Chinese puzzle made by cutting a square of paper into five triangles, a square, and a rhomboid, and arranging them into many different figures. Have students trace their design on paper and ask them to exchange it with another student. Then, have students calculate the perimeter of each of the geometric shapes in the tangram design.

3 Apply

Solution Notes

1. Remind students to draw a diagram.
2. Have students draw a board cut into 10 pieces to help them see that they only need 9 cuts, which will take 27 minutes.
3. Have students draw a picket fence and count the spaces. Then, have students draw a wheel with 15 spokes and count the 15 spaces. Help students see that in counting the spaces, they can avoid confusion if they mark the space where they begin. Have them discuss the difference in the number of spaces on the fence and the wheel. You may want to have students extend this activity by finding other examples.
4. The complete order of the towns is Columbus, Mintown, Valleyview, Trenton, and Lexington.

Higher-Order Thinking Skills

5. **Analysis and Evaluation:** Because the second price involves only 1 more dozen of eggs, a dozen costs the difference between $4.30 and $3.35, or $0.95. Two half-gallons of milk costs the difference between $3.35 and $0.95, or $2.40, which means that 1 half-gallon of milk costs more than $0.95; so milk costs more.
6. **Synthesis:** Students should deduce that if both the cat and the dogs are running at about the same rate, the cat will reach and run up the tree before the dogs reach the cat, regardless of the distance.
7. **Analysis:** Any picture will do that shows the following arrangements:

   brown, gray, white
   brown, white, gray
   white, gray, brown
   white, brown, gray

7-11 Calculator: Use the Multiplication and Division Keys

pages 131–132

1 Getting Started

Objective
- To use a calculator to multiply and divide

Materials
calculators; multiplication fact cards

Warm Up • Mental Math
Have students name the fact to check the division problem.

1. $16 \div 4$ (4×4)
2. $32 \div 8$ (8×4)
3. $12 \div 6$ (2×6)
4. $56 \div 7$ (8×7)
5. $81 \div 9$ (9×9)
6. $36 \div 6$ (6×6)
7. $54 \div 9$ (9×6)
8. $42 \div 7$ (6×7)

Warm Up • Calculator
Have students find the product of 86 and 7 on their calculators. (602) Give other multiplication problems for students to practice multiplying a 2-digit number by a 1-digit number.

Name ______________________

Calculator: Use the Multiplication and Division Keys

Natalie is packing lunches for a picnic. She needs to buy 5 apples. How much will Natalie pay for the 5 apples?

We want to know the price for 5 apples.

We know that __3__ apples cost __51¢__.

To find the cost of 5 apples, we first find the cost of 1 by dividing __\$0.51__ by __3__. Then, we multiply the cost of 1 apple by __5__.

This can be done on the calculator in one code.

[.] 51 [÷] 3 [×] 5 [=] [0.85]

Natalie will pay __\$0.85__ for 5 apples.

Complete each calculator code.

1. 42 [÷] 7 [=] [6]
2. 76 [÷] 2 [=] [38]
3. 96 [÷] 4 [=] [24]
4. 52 [÷] 4 [=] [13]
5. 36 [÷] 9 [×] 7 [=] [28]
6. 84 [÷] 4 [×] 3 [=] [63]
7. 72 [÷] 6 [×] 9 [=] [108]
8. 75 [÷] 5 [×] 9 [=] [135]

2 Teach

Introduce the Lesson Ask students to tell what is shown in the picture. (A girl is looking at apples at a fruit stand.) Have a student read the problem aloud and tell what students are to find. (the cost of 5 apples)

- Ask students what information is needed to solve this problem. (the cost of 1 apple) Ask if this information is given. (no) Ask what is given that may be helpful. (The cost of 3 apples is 51¢.)
- Have students complete the sentences and the calculator code with you to solve the problem. Have students decide if their solution makes sense.

Develop Skills and Concepts Write **$6 \div 2 \times 3 =$** on the board. Have students work the problem on their calculators. Ask students why this calculator code is easier than $6 \div 2 =$ and $3 \times 3 =$. (There are fewer steps or fewer keys to press.)

- Give more practice with $20 \div 4 \times 7 =$, $6 \times 5 \div 4 =$, and $72 \div 6 - 3 =$. Have students complete the codes at the bottom of the page.

3 Practice

Remind students to clear their calculators before each new problem. Tell students to make a plan of one calculator code to work each story problem. Have students complete page 132 independently.

Now Try This! Tell students that it will be helpful for them to set up each problem using a solve for *n* format. Have them use their calculators and remind students to read the operation words carefully.

Practice

Complete each calculator code.

1. 85 ÷ 5 = 17
2. 57 ÷ 3 = 19
3. 91 ÷ 7 = 13
4. 96 ÷ 6 = 16
5. 88 ÷ 2 = 44
6. 90 ÷ 9 = 10
7. 48 ÷ 4 × 9 = 108
8. 63 ÷ 7 × 8 = 72
9. 63 ÷ 9 × 7 = 49
10. 75 ÷ 5 × 6 = 90
11. 62 + 26 ÷ 4 = 22
12. 216 − 158 ÷ 2 = 29

Problem Solving

Solve each problem. Use a calculator.

13. Jeff can clean 4 rugs every 76 minutes. How long will it take Jeff to clean 9 rugs?
171 minutes
14. Sandi earns $32 every 4 days selling flowers. How much will Sandi earn in 7 days?
$56
15. Nathan can jog 5 miles in 65 minutes. How long will it take Nathan to jog 8 miles?
104 minutes
16. Bananas are on sale at 6 for 96¢. How much do 8 bananas cost?
$1.28

Now Try This!

It's Algebra!

Find each missing number.

1. The product of 2 numbers is 45. Their sum is 14. What are the numbers?
5, 9
2. The sum of 2 numbers is 60. Their difference is 12. What are the numbers?
24, 36
3. The quotient of two numbers is 15. Their sum is 144. What are the numbers?
135, 9
4. Five times one number is three more than six times another number. The difference between the numbers is 1. What are the numbers? 3, 2

4 Assess

Have students use their calculators to solve the following exercise: 1293 − 13 ÷ 5. (256)

For Mixed Abilities

Common Errors • Intervention

Some students may not understand the calculator code for finding the cost of a number of objects. Write the following calculator code on the board:

84 ÷ 3 × 5 = □

Ask, *If 3 whistles cost 84¢, how do you find the cost of 1 whistle?* (divide 84 by 3) Point to the calculator code for this operation. Have students find the cost of 1 whistle. Ask, *How can you find the cost of 5 whistles?* (multiply 28¢ by 5) Point to the calculator code for this operation. Then, have students use their calculators to find the cost ($1.40)

Enrichment • Calculator

Tell students to use their calculators to find 9 times the sum of the digits 0 through 9.

More to Explore • Applications

Incorporate the students' interest in sports to give them practice in division. Have students list their favorite team sports. Next to each sport, have them list the number of players on a team. Now, have students create mathematical problems, involving any combination of operations, using this list. Have them solve the following examples to get them started:

1. How many soccer teams can be formed from 36 players? (36 players ÷ 6 per team = 6 teams)
2. How many players would you need for 4 baseball teams? (4 teams × 9 players = 36 players)
3. How many players are on a team if you have 45 players and 9 teams? (45 players ÷ 9 teams = 5 players per team)

Chapter 7 Test

page 133

Items	Objectives
1–18	To master division facts using 2 through 9 as divisors (see pages 111–118)
3–8	To master understanding 1 and 0 in division (see pages 119–120)
19–22	To solve basic division problems with remainders (see pages 123–124)
23–26	To divide 2-digit numbers by 1-digit numbers to get 2-digit quotients without a remainder (see pages 125–126)
27–34	To divide 2-digit numbers by 1-digit numbers to get 2-digit quotients with remainders (see pages 127–128)

Name ______________________

CHAPTER 7 TEST

Divide.

1. $5\overline{)30}$ 6
2. $7\overline{)49}$ 7
3. $1\overline{)6}$ 6
4. $8\overline{)56}$ 7
5. $4\overline{)4}$ 1
6. $3\overline{)9}$ 3
7. $6\overline{)42}$ 7
8. $9\overline{)0}$ 0
9. $2\overline{)18}$ 9
10. $6\overline{)36}$ 6

11. $45 \div 5 =$ 9
12. $21 \div 7 =$ 3
13. $36 \div 4 =$ 9
14. $18 \div 9 =$ 2
15. $24 \div 8 =$ 3
16. $24 \div 6 =$ 4
17. $36 \div 3 =$ 12
18. $25 \div 5 =$ 5

Divide. Show your work.

19. $4\overline{)17}$ 4 R1
20. $7\overline{)50}$ 7 R1
21. $3\overline{)26}$ 8 R2
22. $9\overline{)46}$ 5 R1
23. $3\overline{)36}$ 12
24. $5\overline{)50}$ 10
25. $2\overline{)84}$ 42
26. $4\overline{)76}$ 19
27. $6\overline{)73}$ 12 R1
28. $8\overline{)94}$ 11 R6
29. $2\overline{)93}$ 46 R1
30. $7\overline{)70}$ 10
31. $4\overline{)82}$ 20 R2
32. $5\overline{)79}$ 15 R4
33. $9\overline{)97}$ 10 R7
34. $3\overline{)88}$ 29 R1

Alternate Chapter Test

You may wish to use the Alternate Chapter Test on page 358 of this book for further review and assessment.

CUMULATIVE ASSESSMENT

Circle the letter of the correct answer.

1. Round 809 to the nearest hundred.
 a. 700
 (b.) 800
 c. NG

2. Round 8,575 to the nearest thousand.
 a. 7,000
 b. 8,000
 (c.) NG

3. What is the place value of the 3 in 423,916?
 a. tens
 b. hundreds
 (c.) thousands
 d. NG

4. 587 + 274
 a. 751
 b. 761
 (c.) 861
 d. NG

5. 4,297 + 3,486
 a. 7,683
 (b.) 7,783
 c. 8,783
 d. NG

6. 57,198 + 23,488
 a. 70,686
 (b.) 80,686
 c. 81,686
 d. NG

7. 73 − 29
 a. 52
 b. 56
 c. 102
 (d.) NG

8. 603 − 256
 (a.) 347
 b. 357
 c. 453
 d. NG

9. 42,823 − 29,647
 (a.) 13,176
 b. 13,276
 c. 27,224
 d. NG

10. 8 × 3
 (a.) 24
 b. 25
 c. 32
 d. NG

11. 37 × 2
 a. 64
 (b.) 74
 c. 614
 d. NG

12. 59 × 6
 a. 304
 (b.) 354
 c. 356
 d. NG

13. 4)93
 a. 2 R1
 b. 23
 (c.) 23 R1
 d. NG

score

STOP

Cumulative Assessment

page 134

Items	Objectives
1–2	To round numbers to the nearest hundred or thousand (see pages 31–32)
3	To identify place value through hundred thousands (see pages 25–26)
4–6	To add two numbers with up to 5 digits (see pages 37–48)
7	To subtract two 2-digit numbers (see pages 55–56)
8	To subtract two 3-digit numbers with zero in the minuend (see pages 63–64)
9	To subtract two 5-digit numbers (see pages 61–62)
10	To multiply using 8 as a factor (see pages 85–86)
11–12	To multiply 2-digit numbers by 1-digit numbers with two regroupings (see pages 103–104)
13	To divide by a 1-digit number for a 2-digit quotient with a remainder (see pages 127–128)

Alternate Cumulative Assessment

Circle the letter of the correct answer.

1. Round 659 to the nearest hundred.
 a 500
 b 600
 (c) 700
 d NG

2. Round 4,306 to the nearest thousand.
 a 3,000
 (b) 4,000
 c 5,000
 d NG

3. What is the place value of the 9 in 659,243?
 a hundreds
 (b) thousands
 c ten thousands
 d NG

4. 376 + 547
 a 813
 b 814
 c 913
 (d) NG

5. 3,163 + 2,498
 a 5,552
 b 5,551
 (c) 5,661
 d NG

6. 65,376 + 16,596
 a 71,973
 b 71,862
 (c) 81,972
 d NG

7. 85 − 48
 a 43
 b 47
 c 33
 (d) NG

8. 808 − 249
 (a) 559
 b 569
 c 641
 d NG

9. 36,715 − 17,426
 (a) 19,289
 b 19,299
 c 21,311
 d NG

10. 7 × 6 =
 a 35
 b 36
 (c) 42
 d NG

11. 26 × 4 =
 (a) 104
 b 824
 c 84
 d NG

12. 75 × 5 =
 a 104
 (b) 375
 c 130
 d NG

13. 6)85 =
 a 14
 (b) 14 R1
 c 10
 d NG

8-1 Time

pages 135–136

1 Getting Started

Objectives

- To tell time on a standard or digital clock
- To tell time when hours and minutes are added or subtracted

Materials

*demonstration analog clock

Warm Up • Mental Math

Ask students which operation(s) can be used to get from

1. 16 to 10 (−)
2. 200 to 4 (−, ÷)
3. $\frac{1}{5}$ to $\frac{4}{5}$ (+, ×)
4. 20 to 5 (−, ÷)
5. 5 to 20 (+, ×)
6. 1,000 to 10,000 (×, +)
7. 12 to 2 and then to 3 (− or ÷ and then +)
8. 45 to 9 and then to 27 (− or ÷ and then × or +)

Warm Up • Measurement

Have students arrange the hands on the demonstration clock to show the times they rise and go to bed on school and weekend days, eat lunch, and so on. Have other students read the times shown. Include some times on the quarter-hour.

Name ______________________

Measurement

CHAPTER 8

Lesson 8-1

Time

Mr. Jameson can drive home from work in 25 minutes. Draw on a standard clock and a digital clock the time when Mr. Jameson will arrive.

We want to know what time Mr. Jameson will get home.

The time now is 4:28.

His trip home will take 25 minutes.

Arrival Time

Arrival Time

Mr. Jameson will get home at 4:53 or 7 minutes to 5.

Getting Started

Write each time as you would see it.

1. five after twelve

2. quarter to four

3:45

3. seven thirty-five

Write each time as you would say it.

4. 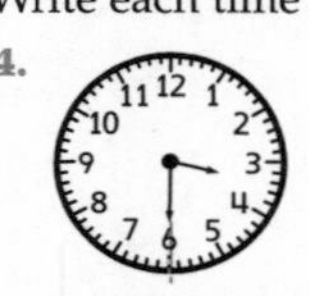

three thirty or thirty minutes after three

5. seven forty-five or quarter to eight

6. ten forty-one or nineteen minutes to eleven

---

2 Teach

Introduce the Lesson Have a student read the problem aloud and tell what is to be solved. (the time when Mr. Jameson will arrive home from work) Ask students what information is needed to solve this problem. (the time he leaves work and the amount of time it takes him to get home) Tell students to write this information in the sentences.

- On the demonstration clock, show students how they can determine the arrival time. Have students draw the hands on the analog clock to show this time. Then, have them write the answer in the digital clock and supply the answers in the solution sentence. Tell students that they can check their solution by adding or by counting on 25 minutes.

Develop Skills and Concepts Show 3:15 on the demonstration clock. Ask how many minutes after 3:00 are shown. (15) Ask a student to move the minute hand to show 5 minutes later than 3:15. Ask students to tell and write ways to say this time. (3:20, twenty minutes after 3)

- Show 2:25 on the demonstration clock and slowly move the minute hand 40 minutes forward as you tell students that 40 minutes later than 2:25 is 3:05 or 5 minutes after 3. Write **12:55** on the board and tell students that we can find the time that is 42 minutes later than 12:55 by adding 55 and 42. (97 minutes) Remind students that 60 of the 97 minutes equals 1 hour, so the time would be 1 hour and 37 minutes later than 12:00 or 1:37. Repeat for subtracting minutes from a specific time to show the regrouping of 1 hour for 60 minutes.
- Give more examples. Do not use times that require the use of A.M. and P.M. notations at this time.

Practice

Show each time.

1. 8:20

2. 3:50

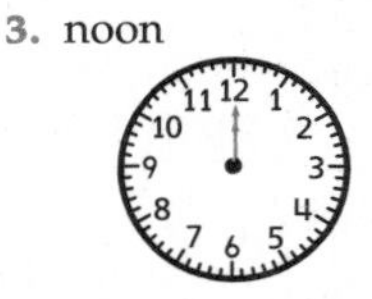

3. noon

4. quarter to eleven

10:45

5. ten minutes after six

6. three twenty-five

Write each time as you would say it.

7.

two sixteen or sixteen minutes after two

8.

twenty to ten or nine forty

9.

four oh two or two minutes after four

10.

twelve thirty or thirty minutes after twelve

11.

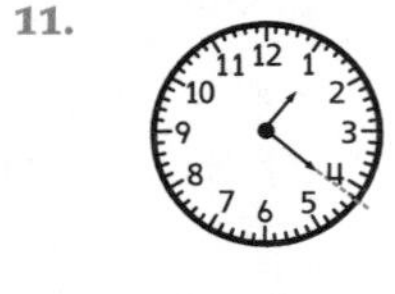

one twenty-one or twenty-one minutes after one

12.

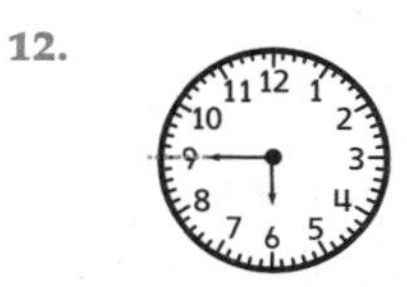

six forty-five or quarter to seven

Problem Solving

Solve each problem.

13. The donut shop opens at 8:30 A.M. The baker needs 2 hours and 25 minutes to prepare the bakery for his first customers. At what time should he start baking? 6:05 A.M.

14. At 11:50 A.M. Mrs. Miller dismisses her morning kindergarten class. She has 1 hour and 50 minutes until the afternoon class arrives. At what time will she greet her afternoon students? 1:40 P.M.

3 Practice

Have students complete page 136 independently.

4 Assess

Have students look at the current time and write it two different ways. (Answers will vary.)

5 Mixed Review

1. 8,020 − 5,372 (2,648)
2. 73 × 6 (438)
3. \$253.49 + \$128.92 (\$382.41)
4. 28 × 8 (324)
5. \$100.00 − \$64.38 (\$35.62)
6. 75 ÷ 4 (18 R3)
7. 4 × \$0.68 (\$2.72)
8. 363 + 257 + 1,806 (2,426)
9. 45 ÷ 6 (7 R3)
10. (6 × 8) + (7 × 7) (97)

For Mixed Abilities

Common Errors • Intervention

Some students may read the minutes incorrectly on a standard clock. They might write a time such as 2:30 as 2:20 because they are counting the minutes by starting at 2 instead of 12. Have each student work with a partner to use a model of a clock. One student sets a time on the clock and the partner writes down the time. Both students should check the answer to make sure it is correct. Then, they trade roles to set another time on the clock and write the time.

Enrichment • Applications

Have students make a time line to include a minimum of three activities they perform during a normal morning in their lives. Have them illustrate the activity under each time.

More to Explore • Numeration

Tell students that they will be doing an experiment to find out if it is easier to count objects in a row or objects grouped in a pattern. Have each student draw several arrays of countable marks or letters on a sheet of paper. Tell students to put some of the marks or letters in a straight row and others forming patterns of some kind.

Have students work in pairs. One student covers the worksheet with paper, revealing only one set of marks or letters at a time for a brief moment. The second student writes an estimate of the number of marks or letters shown. Have students reverse roles. At the end of the activity, have them compare their estimates with the actual numbers. Ask students to conclude whether it is easier to count objects in a row or grouped in a pattern.

8-2 Time A.M. and P.M.

pages 137–138

1 Getting Started

Objective
- To use A.M. and P.M. notations

Materials
*demonstration analog clock

Warm Up • Mental Math
Tell students to start at 400 and
1. subtract their age
2. multiply by their age
3. divide by 4 (100)
4. find $\frac{1}{2}$ that number (200)
5. multiply by 5 (2,000)
6. subtract 3.9 (396.1)
7. add 2 times the product of 10×6 (520)

Warm Up • Measurement
Have students use the demonstration clock to show and compare their rising times, bedtimes, supper times, and so on.

Then, have students figure the amount of time between breakfast and lunch, lunch and supper, rising time and breakfast, and so forth. Discuss how different homes have different time schedules.

Name ______________________

Time A.M. and P.M.

It takes 20 minutes to load the buses after school. What time did school end?

We want to find the time that school ends.

The buses leave the school at 12:10 P.M..

It takes 20 minutes to load the bus.

To find the time that school ends, we can move the minute hand back 20 minutes.

Units of Time
1 minute (min) = 60 seconds (s)
1 hour (h) = 60 minutes
1 day (d) = 24 hours

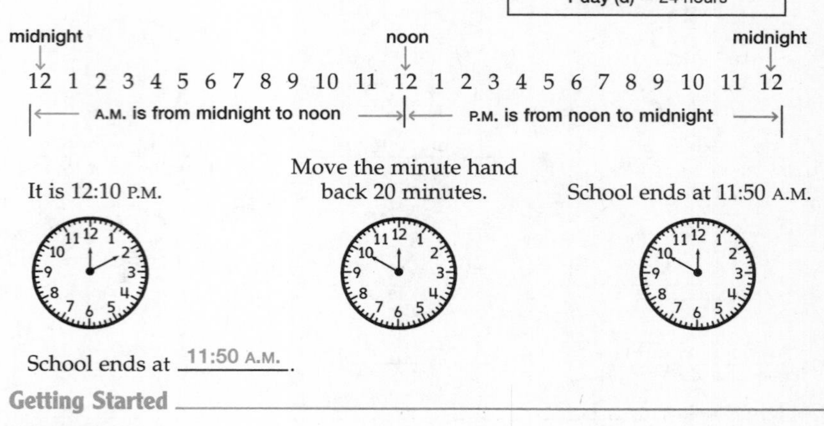

School ends at 11:50 A.M..

Getting Started

Write each time.

1. 45 minutes later than 9:15 A.M. 10:00 A.M.
2. 4 hours before 12:00 midnight 8:00 P.M.
3. 15 minutes earlier than 3:04 P.M. 2:49 P.M.
4. 1 hour and 20 minutes before 6:45 A.M. 5:25 A.M.
5. 3 hours after 10:00 A.M. 1:00 P.M.
6. 25 minutes after 12:15 A.M. 12:40 A.M.

2 Teach

Introduce the Lesson Have a student read the problem aloud and tell what is to be solved. (the time school ends) Ask students what information is given. (Loading the buses takes 20 minutes. Buses leave at 12:10 P.M.)

- Discuss A.M. and P.M. and work through the model of moving the clock back 20 minutes from P.M. to A.M. Then, have students complete the solution sentence.

Develop Skills and Concepts Write **7:00** on the board and ask students what meal they might eat at this time. (breakfast or supper)

- Tell students that there are two 7:00 times in each day because there are 24 hours in a day and a clock can only show 12 hours. Tell students that 12:00 midnight to 12:00 noon is noted by the use of **A.M.** Write **6:15 A.M.** on the board and tell students that this time occurs between midnight and noon.
- Tell students that the time from 12:00 noon to 12:00 midnight is noted by the use of **P.M.** Write **10:00** on the board. Ask students to write A.M. or P.M. for class time (A.M.) or sleeping time (P.M.). Now, write **7:45** A.M. and **2:15 P.M.** on the board and tell students that we find the time between these times by counting to noon and beyond noon. (6 hours 30 minutes)
- Repeat for the number of hours and minutes between breakfast at 6:30 and lunch at 12:09. (5 hours 39 minutes) Give other examples for students to practice.

3 Practice

Remind students that time is divided into two 12-hour periods, with A.M. meaning morning and P.M. meaning afternoon and evening. Have students complete page 138 independently.

Practice

Write each time.

1. 30 minutes later than 4:15 A.M. 4:45 A.M.
2. 20 minutes earlier than 12:45 P.M. 12:25 P.M.
3. 25 minutes before 2:15 P.M. 1:50 P.M.
4. 40 minutes after 6:22 A.M. 7:02 A.M.
5. 2 hours after noon 2:00 P.M.
6. 1 hour and 15 minutes before 5:00 P.M. 3:45 P.M.

Problem Solving

Solve each problem.

7. What time is 2 hours and 25 minutes before 8:30 A.M.? 6:05 A.M.
8. What time is 1 hour and 52 minutes after 11:50 A.M.? 1:42 P.M.
9. School begins at 8:30 A.M. Gina wakes up at 7:15 A.M. How much time does Gina have before school? 1 hour and 15 minutes
10. Ralph ran a marathon in 3 hours and 56 minutes. He began the race at 9:25 A.M. When did Ralph finish the race? 1:21 P.M.

Now Try This!

Help this school bus pick up its students in the correct order. Draw a line from the bus garage to the school in the time order of earliest to latest.

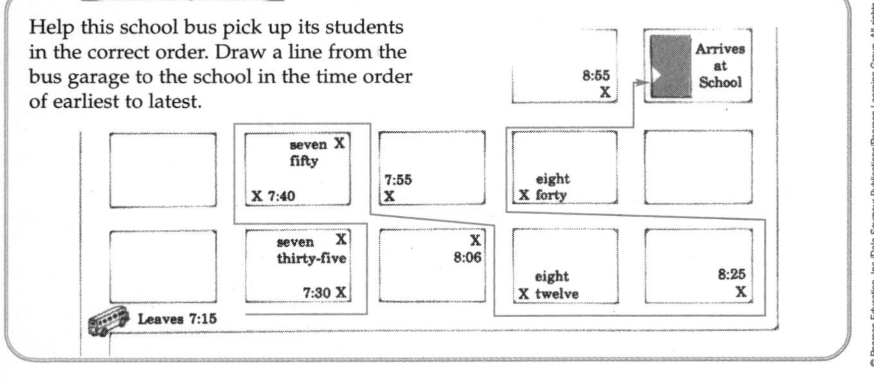

Now Try This! Tell students to first read all the times on the map. Make sure they see that the times are written in three different ways. It may help students to first number the times in order from earliest to latest. They can then easily draw the bus route.

4 Assess

Give students the following problem: It takes Roger 45 minutes to get from his house to Sam's house. If Roger arrived at Sam's house at 4:15 P.M., what time did Roger leave his house? (3:30 P.M.)

For Mixed Abilities

Common Errors • Intervention

Many students write the time before or after a given time incorrectly when first learning how to tell time. Have each student work with a partner, making a list of times each one gets up, goes to school, goes to bed, and so forth. Have pairs use a model of a clock to set these times and then move the hands to show 30 minutes before and 45 minutes after the time. If students need more practice, have them repeat the activity with other lists of times.

Enrichment • Measurement

Have students make a time line to include a minimum of six activities during a normal day in their lives. Then, have them illustrate an activity under each time and find the number of hours and minutes between each two consecutive activities.

More to Explore • Measurement

Have pairs of students make cards to play measurement Concentration. Each card should have a measurement on it that will pair up with an equivalent measure on another card. Tell them to use pairs, such as 8 ounces and 1 cup, 12 inches and 1 foot, 2 pints and 1 quart, 365 days and 1 year, and so on. Tell students to shuffle the cards and arrange them facedown. Have them take turns turning the cards over to find matching pairs, keeping the cards that match, and replacing cards that do not match. The player with the most cards after all are matched wins.

8-3 Elapsed Time

pages 139–140

1 Getting Started

Objective
- To find elapsed time

Vocabulary
elapsed time

Materials
*demonstration analog clock

Warm Up • Mental Math
Have students find the quotient for the following:

1. 18 ÷ 9 = (2)
2. 54 ÷ 6 = (9)
3. 63 ÷ 9 = (7)
4. 32 ÷ 4 = (8)
5. 40 ÷ 5 = (8)
6. 36 ÷ 9 = (4)

Warm Up • Pencil and Paper
Tell students to find the product for the following:

1. (35) = 5 × 7
2. (21) = 7 × 3
3. (72) = 8 × 9
4. (15) = 3 × 5
5. (28) = 7 × 4
6. (24) = 4 × 6

Name ______________________

Elapsed Time

Senya leaves for dance class at 12:45 P.M. She returns home at 3:30 P.M. How long is Senya away from home?

We want to find how much time **elapsed**, or passed, from 12:45 P.M. to 3:30 P.M.

Count the hours.
There are __2__ hours from 12:45 P.M. to 2:45 P.M.

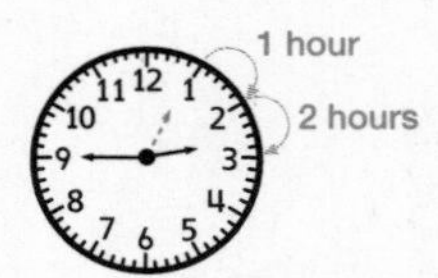

Count the minutes.
There are __45__ minutes from 2:45 P.M. to 3:30 P.M.

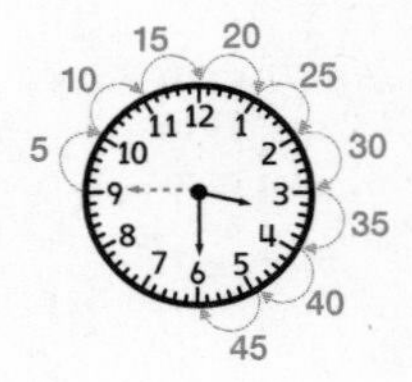

Senya was away from home for __2 hours__ and __45 minutes__.

Getting Started

Write each elapsed time.

1. Begin: 7:30 A.M.
   End: 11:15 A.M.
   __3 hours and 45 minutes__
2. Begin: 12:40 P.M.
   End: 6:35 P.M.
   __5 hours and 55 minutes__
3. Begin: 8:27 A.M.
   End: 11:00 A.M.
   __2 hours and 33 minutes__
4. Begin: 5:15 P.M.
   End: 10:03 P.M.
   __4 hours and 48 minutes__
5. Begin: 2:12 P.M.
   End: 3:35 P.M.
   __1 hour and 23 minutes__
6. Begin: 1:59 P.M.
   End: 9:00 P.M.
   __7 hours and 1 minute__

2 Teach

Introduce the Lesson Have students study the problem and tell what is being asked. (how long Senya is away from home) Ask students what facts are given in the problem. (Senya leaves for dance class at 12:45 P.M. and returns home at 3:30 P.M.)

- Read the next sentence to the class and explain that **elapsed time** is the amount of time that has passed from one event to another.
- Read through each step in the model with students, making sure they understand that in the first step they are counting the hours and in the second they are counting the minutes. Have students fill in the information in the model and the solution sentence. (2; 45; 2 hours and 45 minutes)

Develop Skills and Concepts Write the following on the board:

begin: 12:16 P.M.
end: 5:39 P.M.

- Using a demonstration analog clock, move the minute hand as students count the minutes from 16 to 39 with you. Write the answer on the board. (23 minutes) Then, move the hour hand from 12 to 5 and have students count the hours with you. Write **5 hours** in front of the 23 minutes.
- Repeat with other examples of elapsed time.

3 Practice

Have students complete all the exercises on page 140.

Practice

Write each elapsed time.

1. Begin: 3:15 P.M.
   End: 5:12 P.M.
   1 hour 57 minutes
2. Begin: 9:05 A.M.
   End: 11:59 A.M.
   2 hours 54 minutes
3. Begin: 4:32 P.M.
   End: 5:19 P.M.
   47 minutes
4. Begin: 1:48 P.M.
   End: 10:22 P.M.
   8 hours 34 minutes
5. Begin: 12:36 P.M.
   End: 5:35 P.M
   4 hours 59 minutes
6. Begin: 7:06 A.M.
   End: 10:20 A.M.
   3 hours 14 minutes

7. 2 hours and 25 minutes later

5:40

8. 7 hours and 32 minutes later

5:40

9. 5 hours and 9 minutes later

10:07

10. 4 hours and 18 minutes later

4:50

Problem Solving

Solve each problem.

11. Rhonda boarded a bus that left at 9:35 A.M. It arrived at its destination 1 hour and 45 minutes later. At what time did Rhonda's bus arrive?
    11:20 A.M.
12. The program that Belinda attended ended at 7:45 P.M. If it lasted 1 hour and 50 minutes, at what time did it start?
    5:55 P.M.
13. Lonnie left for the movies at 5:27 P.M. and returned home at 8:25 P.M. How long was Lonnie away from home?
    2 hours and 58 minutes
14. Michele took a nap at 2:30 P.M. She woke up at 4:45 P.M. For how long was Michele asleep?
    2 hours and 15 minutes

4 Assess

Ask students to tell how much time has elapsed if they wake up at 5:55 A.M. and arrive at school at 8:45 A.M.
(2 hours and 50 minutes)

5 Mixed Review

1. 32 + 702 (734)
2. 44 + 56 (100)
3. 62 − 59 (3)
4. 348 − 179 (169)
5. 522 − 506 (16)
6. 24 × 5 (120)
7. 80 × 3 (240)
8. 34 ÷ 2 (17)
9. 88 ÷ 8 (11)
10. 32 ÷ 2 (16)

For Mixed Abilities

Common Errors • Intervention

Some students may confuse the hour hand and the minute hand. Tell them that even though 1 hour is longer than 1 minute, the hour hand is shorter than the minute hand. Have these students work in pairs finding elapsed times. Have one student find the number of elapsed minutes and the other find the number of elapsed hours.

Enrichment • Measurement

Have students imagine that they are going to take a bus ride that leaves at 10:45 A.M. and is due to arrive at its final destination 2 hours and 20 minutes later. Have them figure out what time the bus will arrive. (1:05 P.M.) Remind them to think about A.M. and P.M. Ask students to explain how they determined whether the time was A.M. or P.M. (If it is A.M. before the time hits 12 o'clock, it becomes the opposite or P.M.)

More to Explore • Measurement

Have students expand their understanding of elapsed time by showing them a map of the United States. Point out the different time zones to them. Ask students to name a state in another time zone and give the current time in that time zone.

8-4 Calendars

pages 141–142

1 Getting Started

Objective

- To use a calendar

Warm Up • Mental Math

Have students find the product of the following:

1. $7 \times 8 =$ (56)
2. $9 \times 2 =$ (18)
3. $3 \times 12 =$ (36)
4. $10 \times 4 =$ (40)
5. $12 \times 7 =$ (84)
6. $10 \times 9 =$ (90)

Warm Up • Division

Have students copy and find the quotient.

1. $15 \div 3 =$ (5)
2. $60 \div 12 =$ (5)
3. $52 \div 7 =$ (7 R3)
4. $72 \div 8 =$ (9)
5. $48 \div 12 =$ (4)
6. $52 \div 4 =$ (13)

Name ______________________

Calendars

Dontrelle has a piano recital 15 days after Labor Day. Use the calendar to determine the day of the week and the date of the recital.

Units of Time
1 week = 7 days
1 year = 12 months
1 year = 52 weeks
1 year = 365 days
1 leap year = 366 days
1 decade = 10 years
1 century = 100 years

We need to find the day and date of Dontrelle's piano recital.

Labor Day is on a Monday, September 5.

Use the day after Labor Day as the first day and count 15 days.

The recital is on a Tuesday, September 20.

Getting Started

Use the calendar above to write each date or day of the week.

1. What date is the fourth Wednesday in September? September 28
2. On what day of the week is September 17? Saturday
3. Ron has his piano recital one week after Labor Day. What is the date of Ron's piano recital? September 12
4. Tamika has a soccer match on the third Sunday in September. What date is Tamika's soccer match? September 18
5. How many days are in September? 30
6. On what day of the week is October 1? Saturday

2 Teach

Introduce the Lesson Have students study the problem and tell what is being asked. (the day of the week and the date of Dontrelle's piano recital) Ask students what fact is given in the problem. (Dontrelle's piano recital will take place 15 days after Labor Day.) Ask students where they can find the date for Labor Day. (on the calendar)

- Read the plan to the class and have a volunteer state the required steps. (Use the day after Labor Day as the first day and count 15 days.) Have a student tell the day and the date as the remaining students fill in the solution sentence. (Tuesday, September 20)

Develop Skills and Concepts Explain to students that a date, for example September 1, is really just a number used to tell order. Numbers used to tell order are called ordinal numbers. When you say a date you can use an ordinal number, such as first, sixth, or eighteenth.

- Write the following dates on the board: **September 30, January 31, May 2,** and **August 23**. Have students say each date using ordinal numbers. (September thirtieth, January thirty-first, May second, and August twenty-third) Then, ask students to say their date of birth. Remind them that birthdays can be ordinal numbers, too.

3 Practice

Have students complete all the exercises on page 142. Tell students for Exercises 1 through 7 they should use the calendars at the top of the page.

Practice

Use the calendar to answer each question.

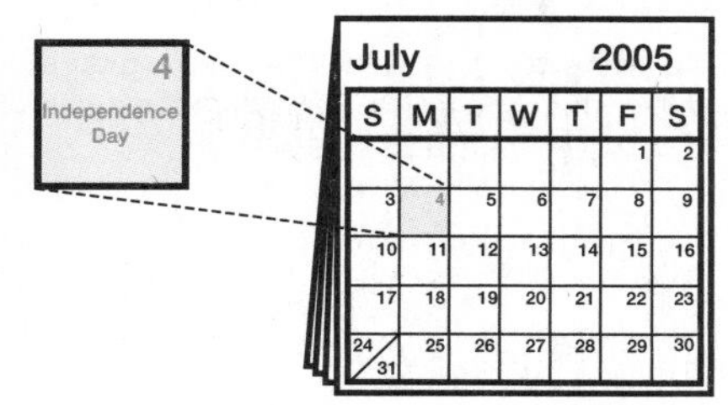

1. Tony begins a 21-day vacation on June 25. On what day does he return home?
   July 16

2. On what day of the week is Flag Day?
   Tuesday

3. Counting the day after Flag Day as the first day, how many days is it until Independence Day?
   20 days

4. Montel has a baseball playoff game exactly 2 weeks after Independence Day. When is Montel's playoff game?
   July 18

5. The town swimming pool opened on Memorial Day. Memorial Day is the last Monday in May. What date was Memorial Day?
   May 30

6. Marta returns home from her vacation on the fourth Friday of July. What day of the month does Marta return home?
   Friday, July 22

Problem Solving

Solve each problem.

7. A decade is a period of 10 years. What year is two decades from 2005?
   2025

8. Suppose it is Monday, October 3. On what day of the week is October 30?
   Sunday

4 Assess

Ask students to figure out what day is the last day of school if final exams are three days long and the first day of testing is Monday, June 28. (Wednesday, June 30)

5 Mixed Review

1. 365 + 12 (377)
2. 30 + 255 (285)
3. 200 − 185 (15)
4. 366 − 179 (187)
5. 12 × 5 (60)
6. 31 × 8 (248)
7. 120 × 3 (360)
8. 52 ÷ 4 (13)
9. 72 ÷ 6 (12)
10. 46 ÷ 8 (5 R6)

For Mixed Abilities

Common Errors • Intervention

Some students may have difficulty calculating dates, forgetting to count the start date. Have them draw circles on their calendars to mark the start dates. Remind students to always include the circled start date when they are calculating dates.

Enrichment • Biography

Tell students to figure out how many full decades Sir Isaac Newton lived. He was born in 1642 in Great Britain and died in 1727. (8 full decades) Newton discovered the laws of gravity and is credited with founding calculus.

More to Explore • Estimation

Have students find out which day of the week they were born on. Have them use the day of the week of their birthday this year to start a table that will display each year and the day of the week of their birthday for each year of their life. Have them count back 1 day from the day of the week of their birthdate for each year and 2 days for each leap year. Remind students that 1996, 2000, and 2004 are leap years.

8-5 Inches, Feet, Yards, and Miles

pages 143–144

1 Getting Started

Objective
- To determine and use the appropriate unit of measure: inch, foot, yard, and mile

Vocabulary
inch, foot, yard, mile

Materials
*pictures of a newborn baby, a sitting or crawling baby, and a teenager; a book measuring about 9 inches on one side

Warm Up • Mental Math
Tell students that a gross is 12 dozen. Ask how many dozens are in the following:

1. 2 gross (24 dozen)
2. $\frac{1}{2}$ gross (6 dozen)
3. 6 gross (72 dozen)
4. $\frac{1}{3}$ gross (4 dozen)
5. $\frac{1}{6}$ gross (2 dozen)
6. 5 gross (60 dozen)
7. 10 gross (120 dozen)
8. $1\frac{1}{2}$ gross (18 dozen)

Name ____________________

Lesson 8-5

Inches, Feet, Yards, and Miles

The **customary system** is a measurement system we frequently use. The **inch**, **foot**, **yard**, and **mile** are units of length in this system. Measure this pencil to the nearest eighth of an inch.

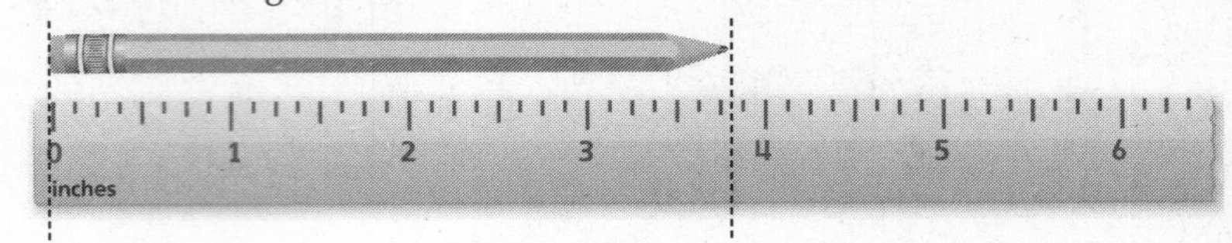

This inch ruler is divided into eighths of an inch. The small lines between inch markings help you know to which eighth of an inch your measurement is closer.

The measurement of the pencil to the nearest eighth of an inch is $3\frac{7}{8}$ inches.

We know the following:

1 foot = 12 inches	1 yard = 3 feet	1 mile = 5,280 feet
1 ft = 12 in.	1 yd = 3 ft	1 mi = 5,280 ft
	1 yd = 36 in.	1 mi = 1,760 yd

Getting Started

Measure the length of the paintbrush to the nearest eighth of an inch.

1.  $4\frac{5}{8}$ inches

Choose inches, feet, yards, or miles to measure each.

2. height of a tall tree feet
3. length of a sheet of paper inches

Compare each pair of lengths. Write <, >, or = in the circle.

4. 3 ft (>) 30 in.
5. 55 in. (<) 5 ft

Warm Up • Numeration
Review skip counting by 3s, 5s, and 10s. Have students use addition to skip-count by 12s and then use multiplication to check their answers. Repeat for skip counting by 24s and 36s.

2 Teach

Introduce the Lesson Have a student read the opening paragraph and tell what they need to do. (measure the pencil to the nearest eighth of an inch)

- Work through the instructions with students, pointing out the location of the eighth markings on a ruler. Then, have students fill in the solution sentence.
- Next, have students fill in the equivalents. Have students tell the abbreviation for each of the four units of measure. (in., ft, yd, mi)

Develop Skills and Concepts Show the pictures of the children. Ask students which child might be 6 or 9 months old, 15 years old, and 6 or 7 days old.

- Discuss with students how a baby's age is generally given in days until the baby is a week or two old and then in weeks until the baby is a month and then in months. Discuss how ages are given in months until the age of 2 or so and then in years because days, weeks, and months are cumbersome once years can be used.
- Now, have a student draw on the board and label a line that is about 1 **inch** long. Have other students draw and label lines that are about 1 **foot** long and 1 **yard** long. Ask students if a line that is 1 **mile** long can be drawn on the board. (no) Tell students to imagine 1,760 lines the length of the 1-yard line and that would be 1 mile.
- Hold up a book and tell students that the book is about 9 units of measure in length. Ask if it would be 9 inches, 9 feet, 9 yards, or 9 miles. (9 inches) Name some objects in and outside of the classroom and have students choose the best unit of measure to describe each.

Practice

Measure the length of each item to the nearest eighth of an inch.

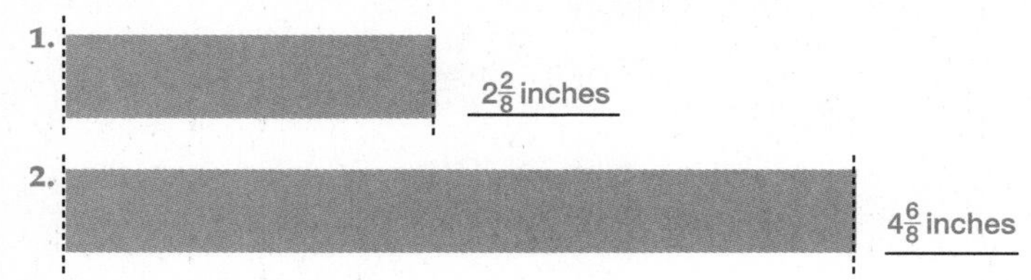

1. $2\frac{2}{8}$ inches

2. $4\frac{6}{8}$ inches

Choose inches, feet, yards, or miles to measure each.

3. length of a city miles
4. length of a football field yards
5. height of a giraffe feet
6. length of Texas miles

Compare each pair of lengths. Write <, >, or = in the circle.

7. 4 ft (<) 50 in.
8. 4 yd (=) 12 ft
9. 2,500 ft (>) 800 yd
10. 200 in. (<) 6 yd

Write the equivalent.

11. 6 ft = 72 in.
12. 2 mi = 3,520 yd
13. 3 yd = 108 in.
14. 10,560 ft = 2 mi
15. 27 ft = 9 yd
16. 12 ft = 144 in.

Problem Solving

Solve each problem.

17. Choose the most appropriate tool Martha should use to measure the length of her bedroom?

ruler (yardstick) tape measure

18. Ben is replacing the front door to his house. He wants to find the measure of the door. Should he measure to the nearest $\frac{1}{8}$ inch, inch, foot, or yard?

$\frac{1}{8}$ inch

3 Practice

Have students complete the page independently.

Assess

Ask students to name three things they would measure in miles. (Answers will vary.)

For Mixed Abilities

Common Errors • Intervention

Some students may have difficulty choosing the appropriate unit of measure. Have them work in groups to find and list items in the classroom that measure approximately an inch, a foot, and a yard. Have them use these items as models as they estimate the measures of other objects in the classroom. This activity provides students with common objects that can be used as models for estimating and comparing.

More to Explore • Measurement

Have students trace their hand on a sheet of paper and cut out the figure. Tell them that everyday objects can be used to measure things. Explain that when objects such as a hand are used as units of measure, they are called nonstandard units of measure.

Have students measure the length of their desk using their hand cutouts. Remind them that their answer will be in hands, just like their answer is in inches when they measure with a ruler. Then ask, *What type of unit of measure is demonstrated using the hand cutouts?* (nonstandard) Finally, have students explain why a hand cutout is a nonstandard unit of measure. (because people have different hand sizes)

Enrichment • Measurement

Have students record the odometer reading of four different cars. Tell students to estimate how many thousands of miles the cars have traveled altogether. They should then estimate in thousands how much farther the most-traveled car has been driven than each of the others.

8-6 Cups, Pints, Quarts, and Gallons

pages 145–146

1 Getting Started

Objective
- To determine and use the appropriate unit of measure: cup, pint, quart, and gallon

Vocabulary
cup, pint, quart, gallon

Materials
*measuring cup; *pint, quart, $\frac{1}{2}$ gallon, and gallon containers

Warm Up • Mental Math
Ask students to explain these figurative phrases:

1. to sum it up (a summary)
2. six of one and a half dozen of the other (all the same)
3. two steps forward and one step back (slow progress)
4. just a second (tiny bit of time)
5. the second Tuesday of next week (never)
6. An ounce of prevention is worth a pound of cure. (better and easier to prevent)

Warm Up • Numeration
Have students skip-count by 2s, 4s, and 6s and use multiplication or addition to check their answers.

Name ______________________

Lesson 8-6

Cups, Pints, Quarts, and Gallons

Antonio wants to have enough fruit punch for 32 people to have 1 cup each. How many quarts of fruit punch should Antonio make?

We want to know the number of quarts of fruit punch that Antonio should make.

We know the following:

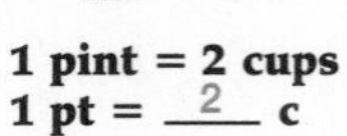

1 pint = 2 cups
1 pt = _2_ c

1 quart = 2 pints
1 qt = _2_ pt
1 qt = _4_ c

1 gallon = 4 quarts
1 gal = _4_ qt
1 gal = _8_ pt
1 gal = _16_ c

Antonio wants to have _32_ cups of fruit punch.
There are _4_ cups in a quart.

We find the number of quarts by dividing _32_ cups by _4_ cups.

32 ÷ 4 = 8 **32 cups = 8 quarts**

Antonio should make _8_ quarts of fruit punch.

Getting Started

Circle the better estimate.

1.  (2 cups) 2 quarts

2. 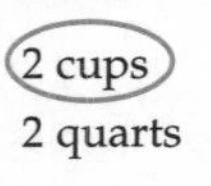  (6 pints) 6 gallons

Rename each measurement.

3. 16 pints = _8_ quarts
4. 2 gallons = _8_ quarts

Compare each pair of capacities. Write <, >, or = in the circle.

5. 5 cups (<) 2 quarts
6. 6 pt (>) 3 cups

2 Teach

Introduce the Lesson Have a student read the problem aloud and tell what is to be solved. (how many quarts of fruit punch Antonio has to make for 32 people)

- Work through the model with students as they fill in the equivalents. Have students tell the abbreviation for each of the four units of measure. (c, pt, qt, gal) Then, have them complete the plan and solution sentences.

Develop Skills and Concepts Tell students to imagine being sent to the store to buy milk. Ask students what they would need to know. (the size of the container wanted) Students may also question the kind of milk wanted, for example, 1%, 2%, whole, skim, and so on.

- Tell students that milk is packaged in half-pint, pint, quart, half-gallon, and gallon containers. Tell students that these units are measures of capacity and that a cup is another measure of capacity.
- Fill a gallon container with water and have students use the water to fill quart containers. Ask, *How many quart containers of water are needed to fill a gallon?* (4) Have a student record the finding on the board. (4 quarts = 1 gallon)
- Continue to find the number of pints in a gallon, cups in a pint, pints in a quart, cups in a gallon, and so on. Have students record all findings. Ask students to name items and products that can be purchased in these various units. (Answers may vary.)

3 Practice

Remind students to think of their experiences to help them select the best estimate of capacity. Have students complete page 146.

Practice

Circle the better estimate.

1. 
(1 cup) 1 quart

2. 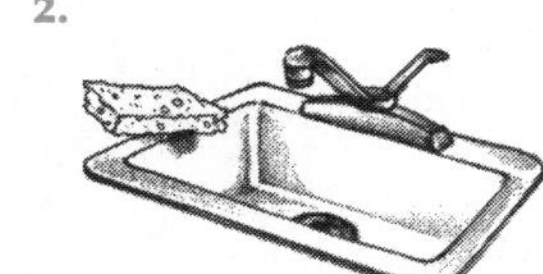
20 pints (20 gallons)

3. 

1 quart (1 gallon)

Rename each measurement.

4. 4 cups = 2 pints
5. 8 quarts = 16 pints
6. 3 gallons = 24 pints
7. 16 quarts = 4 gallons
8. 6 quarts = 24 cups
9. 16 pints = 2 gallons
10. 4 gallons = 64 cups
11. 10 quarts = 2 gallons 2 quarts

Compare each pair of capacities. Write <, >, or = in the circle.

12. 4 gallons (<) 20 quarts
13. 12 cups (>) 2 quarts
14. 8 pints (=) 16 cups
15. 18 quarts (>) 4 gallons
16. 10 pints (<) 6 quarts
17. 3 gallons (=) 48 cups

Problem Solving

Solve each problem.

18. Fran is making a recipe for chocolate pudding. Should she measure the milk needed for the recipe to the nearest cup, pint, quart, or gallon?
cup

19. Another recipe calls for 6 pints of water. Fran only has a 1-cup measuring cup. How many times does she need to fill it to measure 6 pints?
12 times

4 Assess

Ask, *How many times can you fill a gallon container with water from a quart container?* (4)

5 Mixed Review

1. 17,243 + 878 (18,121)
2. 5 × 86 (430)
3. 16,029 − 9,726 (6,303)
4. 56 ÷ 9 (6 R2)
5. $0.37 × 8 ($2.96)
6. 27 − 4 × 4 (11)
7. 372 + 478 + 708 (1,558)
8. 83 ÷ 3 (27 R2)
9. 8 + 9 + 3 + 9 + 7 (36)
10. $503.75 − $378.36 ($125.39)

For Mixed Abilities

Common Errors • Intervention

Some students will have difficulty choosing the appropriate unit of capacity. Have them work in groups with models of a cup, pint, quart, and gallon. Have them use water or sand to see how many of each unit it takes to fill the next larger unit. This activity provides students with experience in the relative values of the measures.

Enrichment • Applications

Tell students to cut from 12 product containers the unit measurement words that tell the container size. They should include at least one of each size—cup, pint, quart, and gallon— in the selection of items.

More to Explore • Measurement

Explain to students that the more commonly used smaller units for measuring capacity are teaspoons and tablespoons. Have students use measuring cups and water to determine how many teaspoons there are in 1 tablespoon. (3)

8-7 Ounces, Pounds, and Tons

pages 147–148

1 Getting Started

Objective

- To determine the appropriate unit of measure: ounce, pound, and ton

Vocabulary

ounce, pound, ton

Materials

*product containers having weight information on the labels

Warm Up • Mental Math

Dictate the following problems:

1. $20 \times \frac{1}{2}$ of 8 (80)
2. $16 \div 2 \times 0$ (0)
3. $\$1.10 \times 4$ ($4.40)
4. $6\frac{1}{2} - 4\frac{1}{2}$ (2)
5. $56 \div (2 + 5)$ (8)
6. $60 \times (36 \div 6)$ (360)
7. 50¢ × 10 ($5)
8. $\$1 \times \frac{1}{2}$ of 6 ($3)

Warm Up • Numeration

Have students work the following problems on the board:

16 × 4	64 ÷ 4	67 ÷ 4	65 ÷ 4
16 × 6	96 ÷ 6	98 ÷ 6	99 ÷ 6
16 × 5	80 ÷ 5	84 ÷ 5	82 ÷ 5

Have students compare the three division problems to each multiplication problem to see that the number of 16s remains the same in each problem, but there are different remainders.

Name ______________________

Ounces, Pounds, and Tons

Some customary units of weight are **ounces**, **pounds**, and **tons**. A scale is used to measure weight.

Would the weight of a stick of butter more likely be 4 ounces, 4 pounds, or 4 tons?

We know the following:

1 pound = 16 ounces — <u>1</u> **lb =** <u>16</u> **oz**

1 ton = 2,000 pounds — <u>1</u> **T =** <u>2,000</u> **lb**

Study these comparisons:

about 1 ounce

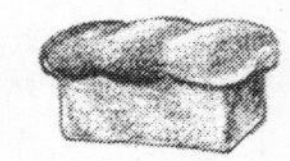
about 1 pound

about 1 ton

The weight of a stick of butter would probably be about <u>4 ounces</u>.

Getting Started

Would these objects be weighed in ounces, pounds, or tons?

1. 

<u>ounces</u>

2. 
<u>tons</u>

3. 
<u>pounds</u>

Compare each pair of capacities. Write <, >, or = in the circle.

4. 4 lb (>) 60 oz
5. 6,000 lb (<) 4 T

Use the table to help complete the measurements.

Pounds	1	2	3	4	5	6
Ounces	16	32	48	64	80	96

6. 50 oz = 3 lb <u>2</u> oz
7. 2 lb 5 oz = <u>37</u> oz
8. 4 lb 7 oz = <u>71</u> oz
9. 72 oz = <u>4</u> lb <u>8</u> oz

2 Teach

Introduce the Lesson Ask a student to describe the picture. Have a student read the problem aloud and tell what is to be solved. (if a stick of butter weighs 4 ounces, 4 pounds, or 4 tons)

- Work through the model with students, discussing the equivalents and comparisons. Have students tell the abbreviations for each of the three units of measure. (oz, lb, T) Then, have them complete the solution sentence.

Develop Skills and Concepts Write the following on the board: **16 oz = 1 lb; 2,000 lb = 1 T**. Ask students how many **tons** equal 4,000 pounds. (2) Repeat for other multiples of 2,000. Ask students how many pounds equal 32 ounces. (2) Repeat for other multiples of 16 through 96. Tell students that 2,100 pounds would equal 1 ton with 100 pounds left over. Ask students how many tons and leftover pounds there are in 2,300 pounds. (1 ton; 300 pounds)

- Provide more examples for students to figure the tons and pounds left over. Repeat for pounds and ounces left over for 18 ounces, 37 ounces, and 39 ounces. Have students list on the board the words *ton, ounce,* and *pound* in order from least to greatest weight and then from greatest to least weight. (ounce, pound, ton; ton, pound, ounce)

Practice

Would these objects be weighed in ounces, pounds, or tons?

1.  ounces

2.  tons

3. 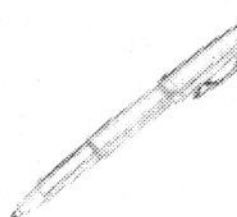 ounces

4. pounds

Rename each measurement.

5. 53 ounces = 3 pounds 5 ounces

6. 18 ounces = 1 pound 2 ounces

7. 4 pounds 6 ounces = 70 ounces

8. 3 pounds 4 ounces = 52 ounces

9. 85 ounces = 5 pounds 5 ounces

10. 100 ounces = 6 pounds 4 ounces

11. 3 tons 1,500 pounds = 7,500 pounds

12. 4 tons 725 pounds = 8,725 pounds

Compare each pair of weights. Write <, >, or = in the circle.

13. 65 ounces (>) 4 pounds

14. 3 pounds 12 ounces (>) 50 ounces

15. 5 tons 1,000 pounds (>) 9,000 pounds

16. 6 pounds 14 ounces (=) 110 ounces

17. 2,400 pounds (<) 1 ton 1,400 pounds

18. 42 ounces (<) 3 pounds

Problem Solving

Solve each problem.

19. The baseball bat that Tony uses weighs 1 ounce less than 2 pounds. How many ounces does Tony's bat weigh?
31 ounces

20. Mr. Robinson's new car weighs 1 ton 500 pounds. How many pounds does Mr. Robinson's new car weigh?
2,500 pounds

3 Practice

Tell students that they are to write the most appropriate unit of measure for each object in Exercises 1 to 4. Remind students that the equivalencies on page 147 will help them with Exercises 5 to 18. Have them complete the page independently.

4 Assess

Ask students to list at least ten items that they would weigh in ounces. (Answers will vary.)

For Mixed Abilities

Common Errors • Intervention

Some students may have difficulty converting from ounces to pounds and vice versa. To give students more practice, have them work in groups with products from the grocery store. Have them make a list of all the containers with weights in pounds and use the information on page 147 to help convert those weights to ounces. Then, have students convert the weights of containers expressed in ounces as pounds and ounces.

Enrichment • Measurement

Write the following on the board:

C = 32 ounces
2 = 1 ton
U = 3 pounds
I = 1 pound 8 ounces
R = 1,500 pounds
4 = 3,000 pounds
__ __ __ __ __ SMART __ ME.

Have students copy the incomplete sentence. Explain that each blank line represents a word. Have students arrange the weights in order from least to greatest. Then, ask them to write each letter or number in that order, from left to right, in the blanks. (I C U R 2 SMART 4 ME.) Ask a volunteer to read the sentence. (I see you are too smart for me.)

More to Explore • Statistics

Tell students that the daily weather report provides many opportunities for the gathering of data and the use of graphs. Have students design a calendar for the current month. Tell them to record the daily high and low temperatures, precipitation, and other weather information on this calendar.

Have students make bar graphs showing the high and low temperatures for the month. Tell them to make a line graph showing the amount of precipitation recorded. Ask students to notice if any relationship exists between temperature and precipitation and to explain it.

8-8 Centimeters

pages 149–150

1 Getting Started

Objective
- To measure length in centimeters

Vocabulary
metric system, centimeter

Materials
rulers marked in centimeters; cards with lengths and widths varying in tenths of centimeters through 15.0 centimeters

Warm Up • Mental Math
Have students solve for n.

1. $48 - n = 32$ (16)
2. $5 \div n = 1$ (5)
3. $(n \times 7) + 16 = 65$ (7)
4. n is 1 more than $32 \div 2$ (17)
5. $n \times n = 60 + 4$ (8)
6. $\frac{1}{n}$ of $100 = 10$ (10)
7. $n \div 3 = 63 \div 7$ (27)
8. $n + n + n = 18$ (6)

Warm Up • Numeration
Review rounding by drawing a number line from 0 to 100 with multiples of 10 noted. Ask students to name the multiple of 10 that is closest to 87. (90) Have students tell the multiple of 10 that is closest to other numbers you name.

Name ______________________

Centimeters

The **metric system** is another measurement system that we frequently use. The **centimeter** is a unit of length in this system. Measure this toothbrush to the nearest centimeter.

The toothbrush bristles are about 1 cm.

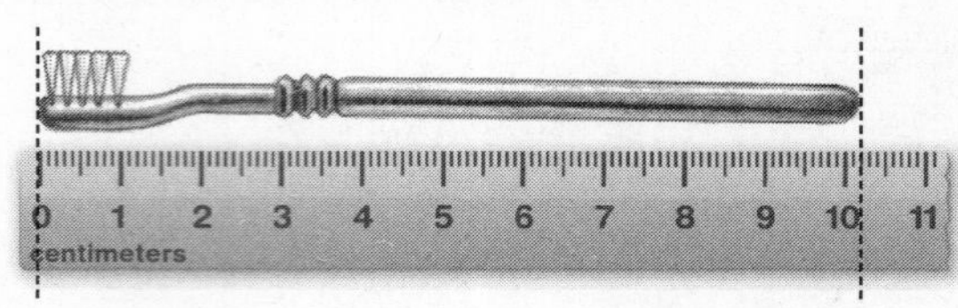

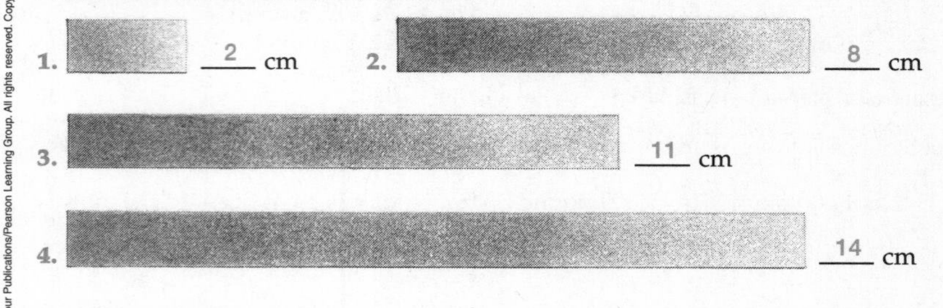

This metric ruler is divided into centimeters. The abbreviation for centimeter is cm. The small lines between centimeter markings help you know to which centimeter your measurement is closer.

Measured to the nearest centimeter, the toothbrush is about 10 centimeters long.

Getting Started

Measure each bar to the nearest centimeter.

1. 2 cm
2. 8 cm
3. 11 cm
4. 14 cm

Measure each object to the nearest centimeter.

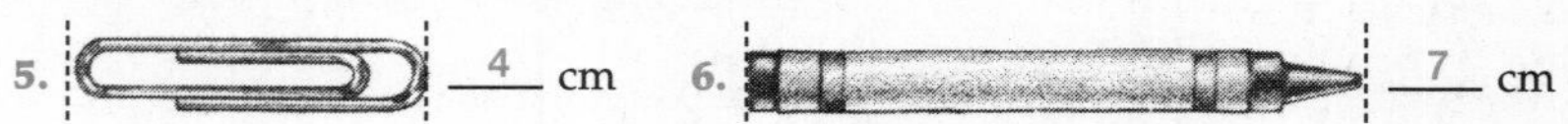

5. 4 cm
6. 7 cm
7. Estimate the length of your arm in centimeters. ____ cm Answers will vary.
8. Measure your arm to the nearest centimeter. ____ cm Answers will vary.

2 Teach

Introduce the Lesson Have a student describe the picture and read the thought bubble. Read the problem to the students and ask what is to be found. (the length of the toothbrush to the nearest centimeter)

- Read and discuss the paragraph about centimeters with students. Then, have students complete the solution sentence.

Develop Skills and Concepts Remind students that there are two systems of measure used in the United States. Tell students that they are going to work with the **metric** system now.

- Have students lay their **centimeter** rulers along the width of a card. Remind students to line the left edge of the card with zero on the ruler.
- Suppose the card is 3.7 centimeters by 8.2 centimeters. Ask students if the card is exactly 3 centimeters in width. (no) Ask if it is closer to the 3- or the 4-centimeter mark. (4-centimeter mark) Tell students that we would say the card's width is about 4 centimeters.
- Repeat the procedure for the length of the card to find it is about 8 centimeters. Have students check the measurements of other cards.

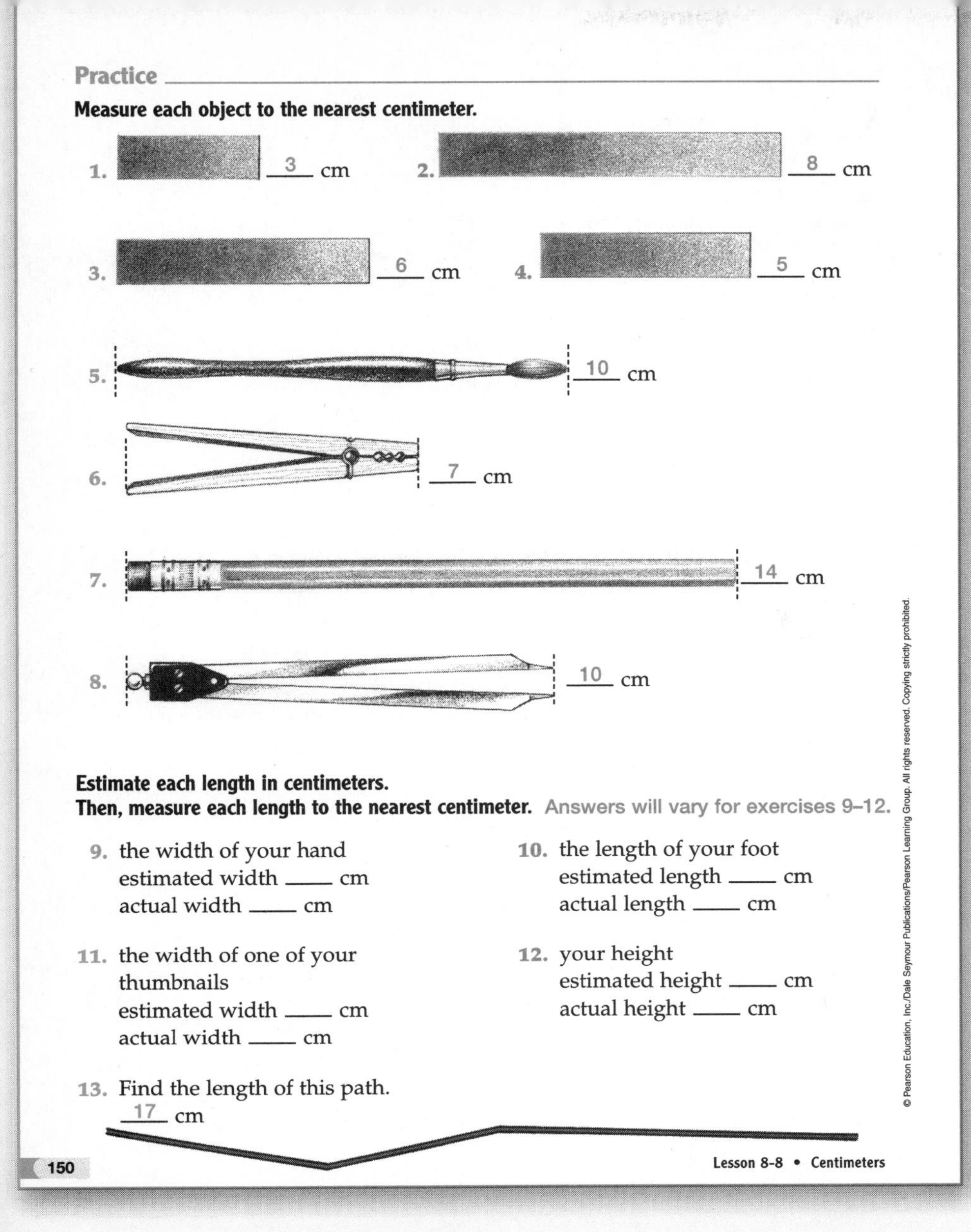

Practice

Measure each object to the nearest centimeter.

1. 3 cm
2. 8 cm
3. 6 cm
4. 5 cm
5. 10 cm
6. 7 cm
7. 14 cm
8. 10 cm

Estimate each length in centimeters.
Then, measure each length to the nearest centimeter. Answers will vary for exercises 9–12.

9. the width of your hand
estimated width ____ cm
actual width ____ cm

10. the length of your foot
estimated length ____ cm
actual length ____ cm

11. the width of one of your thumbnails
estimated width ____ cm
actual width ____ cm

12. your height
estimated height ____ cm
actual height ____ cm

13. Find the length of this path.
17 cm

3 Practice

Tell students that for Exercises 1 to 8 on page 150 they are to measure each object and write its length to the nearest centimeter. Tell students that for Exercises 9 to 13 on page 150 they are to write the estimated and actual lengths of each fact about themselves and then find the total length of the path. Have students complete the page independently.

4 Assess

Ask students to list five objects in the classroom that they could measure in centimeters. (Answers will vary.)

For Mixed Abilities

Common Errors • Intervention

Some students measure incorrectly because they do not properly line up the end of their ruler or the mark for 0 with the end of the line segment. Have these students work in pairs to measure line segments, checking each other to make sure that the ruler is correctly placed.

Enrichment • Measurement

Have students draw a treasure map whose total distance from the starting point to the treasure is about 72 centimeters.

More to Explore • Measurement

Help the class construct a bulletin board entitled "How Would You Measure This?" Tell students to bring in pictures of common consumer products, for example, milk cartons, potatoes, and so on. Have students mount the pictures on the board and place blank answer strips under each. Have individuals or groups of students write the correct metric units that would be used to measure the item on the answer strip under each picture. This activity could be displayed indefinitely, with pictures changed weekly.

8-9 Decimeters, Meters, and Kilometers

pages 151–152

1 Getting Started

Objective
- To determine and use the appropriate measure of length: decimeter, meter, and kilometer

Vocabulary
decimeter, meter, kilometer

Materials
metersticks marked in centimeters and decimeters

Warm Up • Mental Math
Ask students which customary unit(s) of measure they would use for the following:
1. their height (foot, inch)
2. oil for car motor (quart)
3. newborn baby's length (inch)
4. buying fabric (yard)
5. speed limit (mile)
6. small amount of candy (ounce)
7. weight of tractor (ton)

Warm Up • Numeration
Have students count by 10s through 2,000 in a relay, with each student counting the next 10 numbers from where the previous student left off. Remind students that when multiplying by 10, the product is that number with a zero to the right of it. Have students write the products of various 2-digit numbers multiplied by 10.

Name ______________________

Lesson 8-9

Decimeters, Meters, and Kilometers

Some other units of length in the metric system are **decimeters**, **meters**, and **kilometers**. How do these compare with centimeters?

We know the following:

1 decimeter = 10 centimeters

1 dm = 10 cm

1 meter = 10 decimeters

1 m = 10 dm

1 m = 100 cm

1 kilometer = 1,000 meters

1 km = 1,000 m

Study these comparisons:

about 1 decimeter

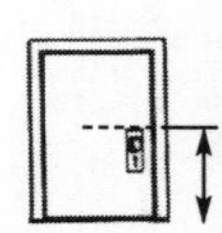

about 1 meter

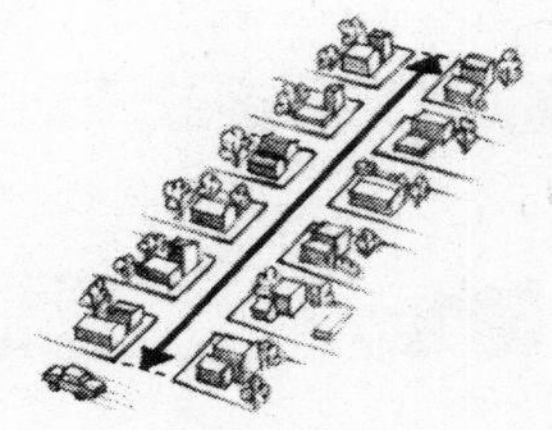

about 1 kilometer

It takes 10 centimeters to make a decimeter, 100 centimeters to make a meter, and 1,000 meters to make a kilometer.

Getting Started

Circle the best estimate.

1. the width of this page

(21 cm) 21 m 21 km

2. the distance from Chicago to Miami

2,000 dm 2,000 m (2,000 km)

Compare each pair of lengths. Write <, >, or = in the circle.

3. 800 m (<) 1 km

4. 30 dm (=) 3 m

2 Teach

Introduce Lesson Have a student describe the picture and read the thought bubbles. Read the problem to students and ask what question is to be answered. (how decimeters, meters, and kilometers compare with centimeters)

- Work through the equivalents with students and discuss the comparisons. Then, have them complete the solution sentence.

Develop Skills and Concepts Ask students why we would not use centimeters to measure a street's length. (The numbers would be large and awkward to use.)

- Tell students that the metric system's units to measure longer lengths are the **decimeter, meter**, and **kilometer.** Discuss the abbreviations for each.
- Tell students that the metric system is based on units of 10. Have a student draw a 1-centimeter line on the board. Tell students that 10 centimeters is equal to 1 decimeter. Have a student draw a 10-centimeter line under the 1-centimeter line. Tell students that 10 decimeters is equal to 1 meter.
- Next, have a student draw a 1-meter line under the other two lines. Show a meterstick marked in centimeters and ask how many centimeters are in 1 meter. (100) Have a student draw a 100-centimeter line. Ask students if 100 centimeters is equal to 10 decimeters. (yes)
- Tell students that a much longer line of 1,000 meters is equal to 1 kilometer. Remind students that a centimeter ruler is used to measure smaller lengths and a meterstick is used to measure longer lengths. Tell students that the width of their fingernail gives an idea of 1 centimeter and a giant step gives an idea of the length of 1 meter.

Practice

Circle the best estimate.

1. the length of 6 city blocks

   1 dm (1 km) 1 m

2. the height of your classroom

   4 cm (4 m) 4 km

3. the length of a paper clip

   (3 cm) 3 dm 3 m

4. the length of a school playground

   (50 m) 50 dm 50 km

Compare each pair of lengths. Write <, >, or = in the circle.

5. 2 km (>) 2,000 dm
6. 4 m (=) 400 cm
7. 5 cm (<) 1 dm
8. 3 km (>) 2,750 m
9. 80 m (<) 850 dm
10. 55 cm (<) 6 dm
11. 3 km = 3,000 m
12. 55 dm = 550 cm
13. 175 m = 1,750 dm
14. 1,500 cm = 15 m
15. 38 m = 3,800 cm
16. 20,000 dm = 2 km

Complete the table. Then, write the answers for Exercises 17 and 18.

17. Approximately how far can an athlete run in one hour? 18 km
18. What would be the approximate time for a 15-kilometer run? 50 min

Athlete's Pace						
Minutes	10	20	30	40	50	60
Kilometers	3	6	9	12	15	18

Problem Solving

Solve each problem.

19. A fence is made of 25 sections that are each 6 meters long. How many decimeters long is the fence?
    1,500 dm

20. Fred's sister is 80 centimeters tall. How many decimeters does she need to grow to be 1 meter tall?
    2 dm

3 Practice

Tell students that they are to skip-count by 3s to complete the table and then use that table to solve Exercises 17 and 18. Have students complete the page independently.

4 Assess

Ask students if a person would measure the distance from New York to California in meters or kilometers. Have them explain their answer. (kilometers; because the distance is very long)

For Mixed Abilities

Common Errors • Intervention

Some students may have difficulty choosing the appropriate unit of measure. Have them find and list items in the classroom that measure approximately a centimeter, a decimeter, and a meter. Have them use these items as models as they estimate the measures of other objects in the classroom. This activity provides students with common models to use for comparisons and estimates.

Enrichment • Measurement

Have students make a list that shows the number of inches in a foot, yard, and mile and the number of centimeters in a decimeter, meter, and kilometer.

More to Explore • Measurement

Have students imagine that they are shopping for a new desk. The desk must fit into a space that is 128 centimeters long. Explain to them that they must find a desk that is as close in length to 128 centimeters as possible because they will need enough space for a computer and worktable. Ask students whether they would use a standard, metric, or nonstandard unit to measure the desks and have them name the unit they would use. (metric; centimeter) Have them explain why it is important to use a metric unit of measure. (because the measurement has to be accurate)

8-10 Milliliters and Liters

pages 153–154

1 Getting Started

Objective

- To determine and use the appropriate measure of capacity: milliliters and liters

Materials

*various small containers;
*transparent liter containers;
*4 drinking glasses; *eyedroppers with 1 milliliter marked; *teaspoons

Warm Up • Mental Math

Ask students to find how many hours and minutes it is from the following:

1. noon to 1:30 A.M. (13, 30)
2. 2:16 A.M. to 4:26 A.M. (2, 10)
3. 7:00 P.M. to 3:30 A.M. (8, 30)
4. 4:28 A.M. to 7:30 A.M. (3, 2)
5. midnight to 1:46 P.M. (13, 46)
6. 8:00 A.M. to 4:45 P.M. (8, 45)
7. 9:15 P.M. to 6:30 A.M. (9, 15)
8. 11:49 A.M. to 3:30 P.M. (3, 41)

Warm Up • Fractions

Write the following on the board:

$\frac{1}{2}$ **of 10 =** _____.

Have a student complete the problem. (5) Repeat for $\frac{1}{5}$ of 10, $\frac{1}{2}$ of 100, $\frac{1}{4}$ of 100, $\frac{1}{5}$ of 100, $\frac{1}{2}$ of 1,000, $\frac{1}{4}$ of 1,000, and $\frac{1}{5}$ of 1,000.

Name ______________________

Milliliters and Liters

Two of the most often used metric units of capacity are the **liter** and the **milliliter**. Would the glass Kaitlyn is using hold about 250 liters or 250 milliliters of milk?

We know the following:

1 liter = 1,000 milliliters

__1__ L = __1,000__ mL

Study these comparisons:

about 1 liter | about 250 milliliters | about 10 milliliters

The glass probably holds about 250 __milliliters__ of milk.

Getting Started

Circle the better estimate.

1. (150 mL) 150 L
2. 1 mL (1 L)

Compare each pair of capacities. Write <, >, or = in the circle.

3. 3 mL (<) 3 L
4. 655 mL (<) 6 L

2 Teach

Introduce the Lesson Have a student describe the picture. Have another student read the problem and tell what question is to be answered. (if a person would drink 250 liters or 250 milliliters of milk at one meal)

- Work through the equivalents with students and discuss the comparisons. Have students tell the abbreviations for liter and milliliter. (L, mL) Then, have students complete the solution sentence.

Develop Skills and Concepts Fill a liter container with water. Tell students that the **liter** is a metric unit of capacity. Have a student pour the liter of water into the 4 eight-ounce glasses to show that a liter is about 4 glassfuls.

- Now, fill the teaspoon with water and tell students that the spoon holds about 4 milliliters of water. Tell students that 1 **milliliter** is a very small amount and an eyedropper works well to measure that quantity. Have students work with the eyedroppers and teaspoons to realize the comparable capacities of 1 milliliter and a teaspoon.

3 Practice

Tell students that they are to find the better estimate of capacity for each container shown in the first group of exercises. Have students complete page 154 independently.

Practice

Circle the better estimate.

1\. 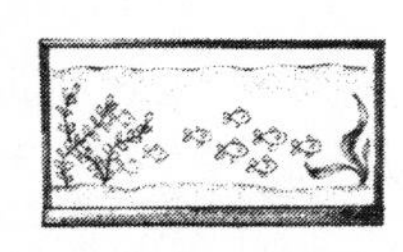

25 mL (25 L)

2\. 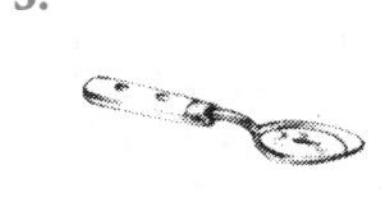

(200 mL) 200 L

3\.

(5 mL) 5 L

Compare each pair of capacities. Write <, >, or = in the circle.

4\. 1,500 mL (<) 2 L

5\. 3,200 mL (>) 3 L

6\. 15 L (>) 1,500 mL

7\. 42 L (=) 42,000 mL

Problem Solving

Solve each problem.

8\. Mark is making strawberry jello. Should he try to measure the water needed for the recipe to the nearest milliliter or the nearest liter?
milliliter

9\. Patti has to use 2 liters of water for a recipe. She only has a 250-milliliter measuring cup. How many times will she need to fill the 250-milliliter measuring cup to measure 2 liters? 8 times

Now Try This!

Circle the smaller containers that can be filled by the larger container in each exercise.

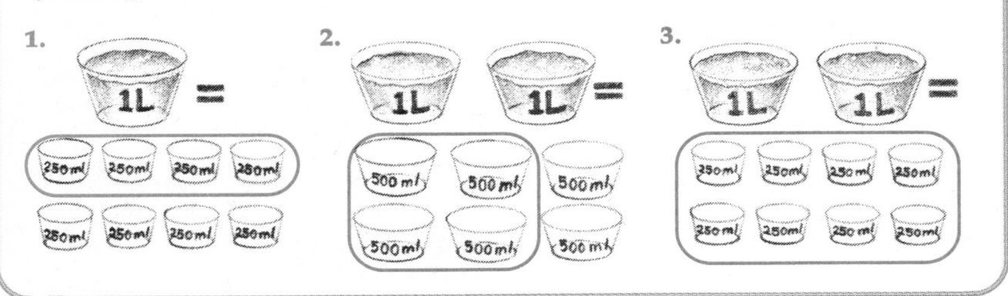

Now Try This! Remind students that there are 1,000 milliliters in 1 liter. There are two times 500 mL and four times 250 mL in each liter.

4 Assess

Ask students to list five items that they would measure in milliliters. (Answers will vary.)

For Mixed Abilities

Common Errors • Intervention

Some students will have difficulty choosing the appropriate unit of capacity. Have them work in groups to find the capacity of containers using the eyedroppers and liter measures as standards, recording their measurements. Rotate the containers among the groups and have students compare their findings.

Enrichment • Measurement

Have students find the number of milliliters in different-size drinking glasses in their kitchen cupboard at home. Ask them to make a chart that pictures each different-size glass and to record its capacity. Tell students to find the capacity of a smaller glass and use that measurement to find the capacity of larger glasses.

More to Explore • Measurement

Provide students with five shoeboxes, books, a balance scale, a spring scale, and a yardstick or ruler. Put the following formula on the board and tell students that work can be measured using it: **work = force × distance**. The amount of force is measured in pounds and the distance is measured in feet. Putting the two together, the amount of work is measured in **foot-pounds**.

Ask students to experiment with the five shoeboxes loaded with different weights of books. Have them measure the weight of each shoebox. Next, have them attach a spring scale to the box and slide it along the floor. They must measure the distance the box is moved and note on the spring scale the amount of force used. Have them use the formula to find the number of foot-pounds of work for each box. For example, if the box is pulled 3 feet and takes 9 pounds of force, 27 foot-pounds of work have been done. Help students to see the correlation between the weight of each box and the foot-pounds it generates.

8-11 Grams and Kilograms

pages 155–156

1 Getting Started

Objective

- To determine and use the appropriate measure of mass: gram and kilogram

Vocabulary

gram, kilogram

Materials

*items weighing about 1 gram;
*items weighing about 1 kilogram;
*scale that weighs in kilograms;
*eyedroppers

Warm Up • Mental Math

Tell students to find the following if $n = 4$ and $p = 6$:

1. $2 * n + 2 * p$ (20)
2. $\frac{1}{n}$ of 40 (10)
3. $(2 \times p) - n$ (8)
4. $6\,(n + p)$ (60)
5. $(2 \times p) \div n$ (3)
6. $5 \times p - 2 \times n$ (22)

Warm Up • Measurement

Write the following on the board:

Capacity	
Customary	**Metric**
cup (c)	**milliliter** (mL)
pint (pt)	**liter** (L)
quart (qt)	**gallon** (gal)

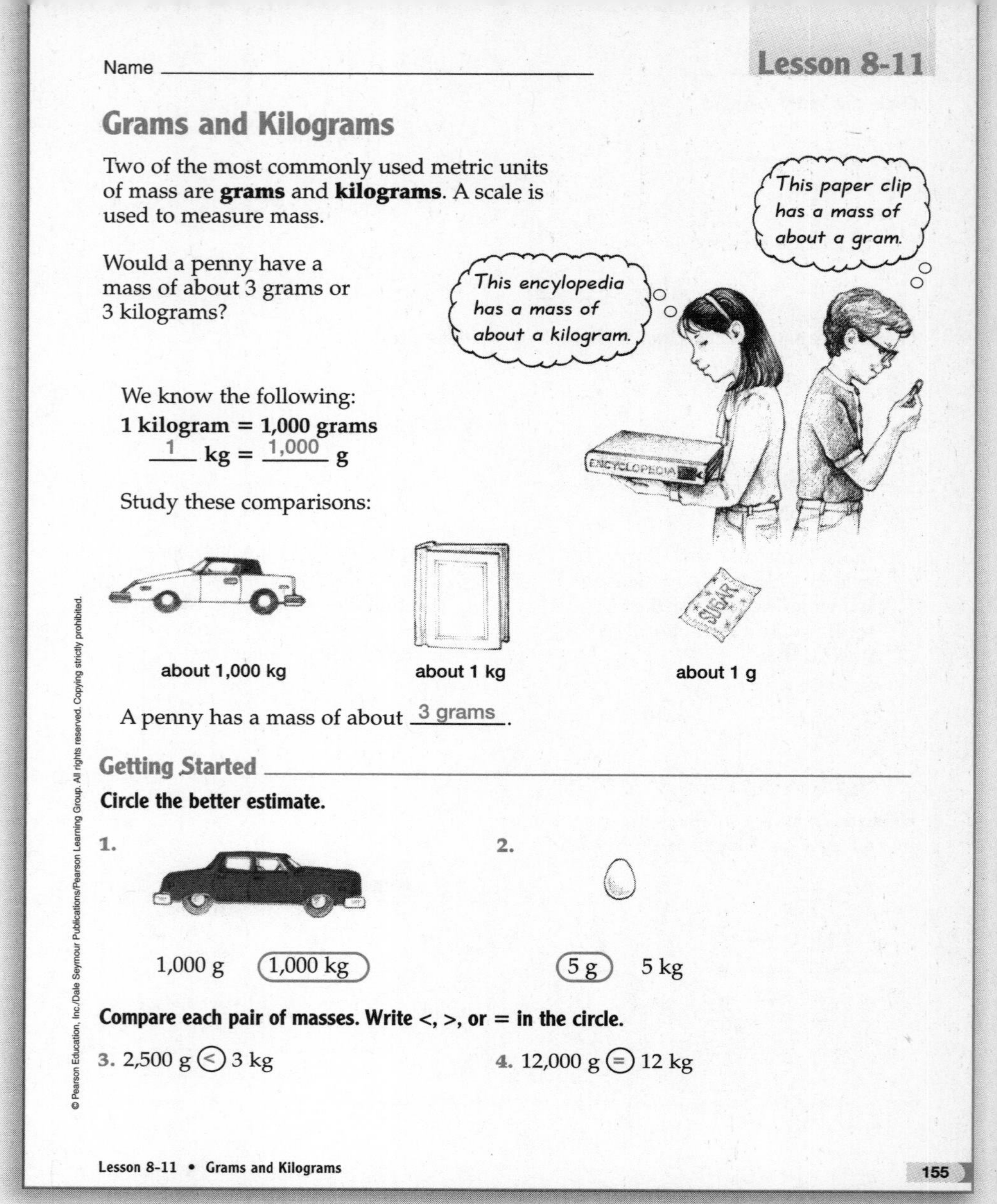

Name ____________________

Lesson 8-11

Grams and Kilograms

Two of the most commonly used metric units of mass are **grams** and **kilograms**. A scale is used to measure mass.

Would a penny have a mass of about 3 grams or 3 kilograms?

We know the following:
1 kilogram = 1,000 grams
__1__ **kg =** __1,000__ **g**

Study these comparisons:

about 1,000 kg | about 1 kg | about 1 g

A penny has a mass of about __3 grams__.

Getting Started

Circle the better estimate.

1. 1,000 g (1,000 kg)
2. (5 g) 5 kg

Compare each pair of masses. Write <, >, or = in the circle.

3. 2,500 g (<) 3 kg
4. 12,000 g (=) 12 kg

Have students write the abbreviations for each unit. Repeat for both customary and metric units of length.

2 Teach

Introduce the Lesson Have a student describe the picture and read the thought bubbles. Read the problem to students and ask what question is to be answered. (if a penny has a mass of about 3 grams or 3 kilograms)

- Work through the equivalents with students and discuss the comparisons. Have students tell the abbreviations for gram and kilogram. (g, kg) Then, have students complete the solution sentence.

Develop Skills and Concepts Show students an array of items weighing about 1 gram or 1 kilogram each.

- Tell students that 1 **gram** is a very small amount of weight and that 1,000 grams equals 1 **kilogram**.
- Tell students that they are to use this information to make an educated guess of the weight of each item and sort them into groups weighing about 1 gram or 1 kilogram.
- Have students use the scale to find the actual weight of each item to verify their guesses.

Practice

Circle the better estimate.

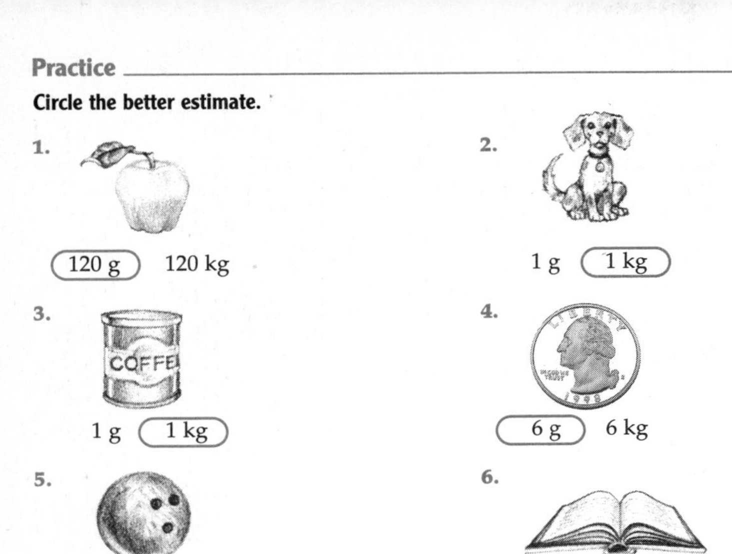

1. (120 g) 120 kg
2. 1 g (1 kg)
3. 1 g (1 kg)
4. (6 g) 6 kg
5. 6 g (6 kg)
6. 1 g (1 kg)

Compare each pair of masses. Write <, >, or = in the circle.

7. 600 g (<) 6 kg
8. 5,300 g (>) 5 kg
9. 20 kg (=) 20,000 g
10. 3 kg (<) 3,300 g

Problem Solving

Solve each problem.

11. Don is trying to find out how much mass he can lift. Should he measure the mass to the nearest gram or to the nearest kilogram? kilogram
12. Jill took 750 grams of potato salad to a picnic. She also took 1 kilogram of macaroni salad. How many more grams of macaroni salad than potato salad did Jill take? 250 grams
13. Sandra is going to bake cookies for her cousins. The recipe calls for a measure of flour. Should she try to estimate her measure to the nearest gram or kilogram? gram
14. A large paper clip has a mass of 3 grams. Sean has a box of paper clips that has a mass of 87 grams. How many paper clips are in the box? 29

3 Practice

Tell students that they are to choose the better estimate for the weight of each object in Exercises 1 to 6. Have students complete the page independently.

4 Assess

Ask students to name five objects that have a mass of about 1 gram. (Answers will vary.)

For Mixed Abilities

Common Errors • Intervention

Some students will have difficulty choosing the appropriate unit of weight. Have them work in groups to weigh 1 liter of water (1 kilogram) and 1 milliliter of water (1 gram). They can then experiment with other amounts and weights of water.

Enrichment • Measurement

Have students cut pictures from magazines or newspapers of five items for each of the following approximate weights: 10 grams or less, 10 ounces or less, 1 or more kilograms, and $\frac{1}{2}$ ton or more.

More to Explore • Measurement

Draw a picture of a book on the board and label its mass as 1,000 grams. Tell students that a book has a mass of about 1,000 grams. Remind them that 1,000 grams is equal to 1 kilogram. Explain to them that it is easy to convert objects with a mass of 1,000 or more grams into kilograms by dividing.

Have students determine what numbers they need to divide to convert the mass of the book into kilograms. (1,000 ÷ 1,000) Ask them to calculate the mass of the book in kilograms. (1 kg) Encourage students to select other objects around the classroom that have a mass of 1,000 grams or more. Have them estimate each object's mass in grams and convert the mass to kilograms as a class.

8-12 Celsius and Fahrenheit Temperatures

pages 155–156

1 Getting Started

Objective

- To read temperature in Celsius and Fahrenheit

Warm Up • Mental Math

Have students find the product of the following.

1. 6 × 7 = (42)
2. 4 × 2 = (8)
3. 5 × 12 = (60)
4. 14 × 4 = (56)
5. 9 × 7 = (63)
6. 10 × 9 = (90)

Warm Up • Numeration

Have students copy the sentence and determine the correct quotient.

1. 45 ÷ 3 = (15)
2. 72 ÷ 12 = (6)
3. 94 ÷ 2 = (47)
4. 56 ÷ 8 = (7)
5. 33 ÷ 11 = (3)
6. 15 ÷ 7 = (2 R1)

Name ______________________

Lesson 8-12

Celsius and Fahrenheit Temperatures

You can use a **Celsius** scale or a **Fahrenheit** scale to measure temperature. A Celsius scale measures metric units of temperature written as °C. A Fahrenheit scales measures customary units of temperature written as °F. What Celsius temperature and Fahrenheit temperature would be good for swimming?

°F °C
Boiling point of water
Normal human body temperature
Hot day
Room temperature
Freezing point of water
Cold day

Study these comparisons:

0°C 32°F 10°C 50°F

It would be good for swimming at 32°C on the Celsius scale and 90°F on the Fahrenheit scale.

Getting Started

Circle the better estimate.

1. You are wearing shorts. (90°F) 90°C
2. You are skiing. −15°F (−15°C)

Write the temperature reading for each thermometer.

3. 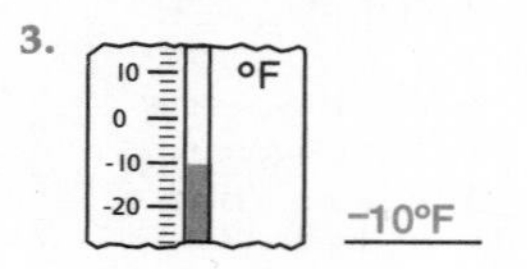 −10°F

4. °C 5°C

5. 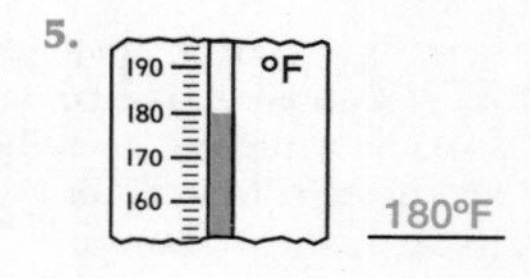 180°F

2 Teach

Introduce Lesson Have students study the problem and tell what is being asked. (what temperature on each scale Celsius and Fahrenheit would be good for swimming)

- Ask students what facts are given in the illustration. (the boiling point of water, normal body temperature, hot day temperature, room temperature, the freezing point of water, and cold day temperature in both Celsius and Fahrenheit)
- Read the thermometer with the class. Ask students to find the Celsius and Fahrenheit temperatures for swimming. (32°C, 90°F) Then, study the comparisons with them.

Develop Skills and Concepts Explain to students that temperatures can be below 0°F or 0°C. When temperatures are below 0 they are negative temperatures.

- Draw a number line showing −10 to 10 on the board. Explain how a number line is similar to a thermometer. Add degree symbols to the numbers on the number line. Ask students the difference between a number line and a thermometer. (A number line is horizontal and a thermometer is vertical.)

3 Practice

Have students complete the exercises on page 158.

Practice

Circle the better estimate for each.

1. You are ice skating.

   (20°F) 20°C

2. You are in a rain forest.

   (100°F) 100°C

3. What is the temperature inside a refrigerator?

   (40°F) 40°C

4. You are in Hawaii.

   30°F (30°C)

Write the temperature reading for each thermometer.

5. 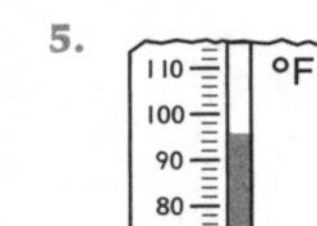

96°F

6. 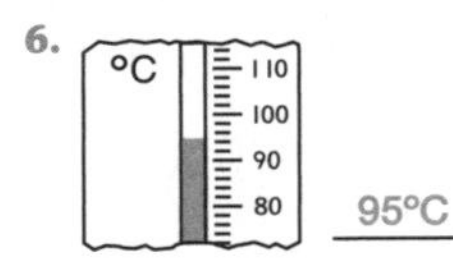

95°C

7. 
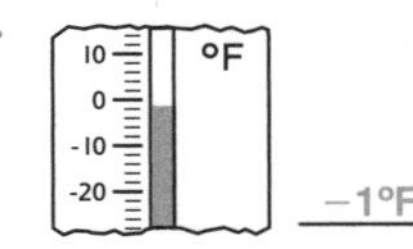

−1°F

8. 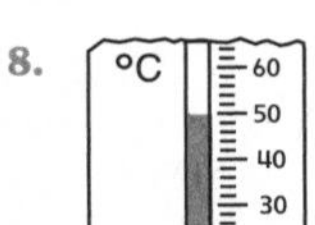

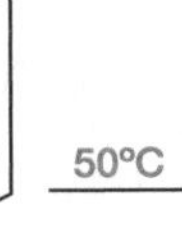
50°C

9. 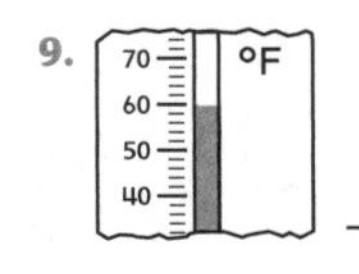

60°F

10. 
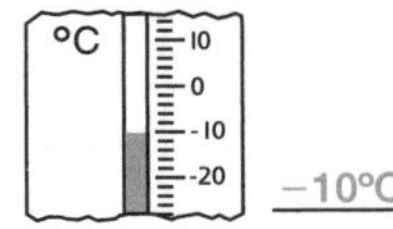

−10°C

Problem Solving

Solve each problem.

11. The temperature outside is 34°F. How much does the temperature have to drop for water to freeze?
   2°F

12. Tami said, "It is 32°, let's go swimming." Is the temperature in degrees Fahrenheit or degrees Celsius?
   Celsius

13. John is boiling water. The temperature of the water is 95°C. How much does the temperature need to rise for the water to boil?
   5°C

14. Mr. Watson turned the thermostat in his house up to 72°. Is the temperature in degrees Fahrenheit or degrees Celsius?
   Fahrenheit

For Mixed Abilities

Common Errors • Intervention

Some students may have difficulty reading the Celsius thermometer because it has 2-degree intervals. Have students skip-count by 2s to help them read the Celsius thermometer.

Enrichment • Statistics

Have students create their own classroom weather station, equipping it with a thermometer, barometer, and student-made rain gauge. Have them record weather statistics for a month, using their station equipment.

Have students keep records by making graphs of the high and low temperatures for a week, make a similar graph for the temperatures for the same week a year ago and compare. Also have them research the record high and low for each day's date, and so on. See if students can think of other graphing activities.

More to Explore • Measurement

Tell students to expand their ability to use a thermometer by calculating the temperature difference between 20°F and 45°F. (25°F) Remind them to be sure to include °F in their answer.

4 Assess

Ask students to draw their own Celsius and Fahrenheit thermometers. Have them shade each thermometer to show the temperature on a warm summer day. (about 80°F or 30°C)

5 Mixed Review

1. 102 + 526 (628)
2. 72 + 315 (387)
3. 58 − 39 (19)
4. 455 − 250 (205)
5. 9 × 6 (54)
6. 55 × 2 (110)
7. 9 × 12 (108)
8. 76 ÷ 2 (38)
9. 58 ÷ 4 (14 R2)
10. 38 ÷ 8 (4 R6)

8-13 Problem Solving: Make a Model

pages 159–160

1 Getting Started

Objective
- To make a model to solve problems

Materials
items such as pennies or similar circles, blocks, straws or toothpicks, and so forth; graph paper

Warm Up • Mental Math
Have students tell the cost.

1. 20 at 14¢ each ($2.80)
2. 1 if 60 costs $1.80 (3¢)
3. 40 if 2 costs $1 ($20)
4. 9 if 5 costs $400 ($720)
5. 60 if 5 costs $1 ($12)
6. 800 at 10¢ each ($80)
7. 4 if 8 costs $10 ($5)
8. 15 if 5 costs 75¢ ($2.25)

Name ______________________________

Problem Solving: Make a Model

It's Algebra!

How many ways can you tear off 4 attached stamps from a sheet of postage stamps?

SEE
We want to know how many patterns of 4 attached stamps we can make.

PLAN
We can use a sheet of graph paper to represent the stamps and cut out different patterns of four squares.

DO

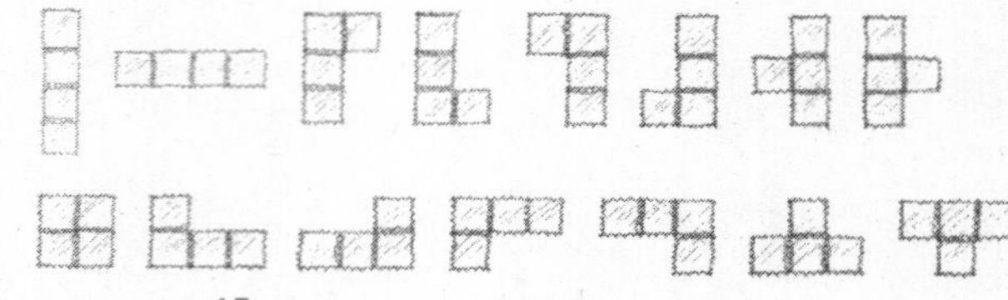

There are 15 different ways to tear off 4 attached stamps.

CHECK

$4 + 0 = 4$ There are 2 ways we can tear off stamps with 4 in a row or column.

$3 + 1 = 4$ There are 12 ways we can tear off stamps with 3 in one row or column and 1 in another row or column.

$2 + 2 = 4$ There is 1 way we can tear off stamps with 2 stamps in one row or column and 2 stamps in another row or column.

2 Teach

Introduce the Lesson Read the problem aloud. Remind students of the four-step method for solving problems.

Develop Skills and Concepts Remind students that they have learned to solve problems by making drawings. Tell students that some problems are easier for us to understand if we use actual items that can be moved around and looked at from different sides. Tell students that we call this a model and it represents data in a problem like a model airplane represents the real vehicle.

- Tell students that it is important when making a model to choose objects that accurately represent the real thing. Have a student read the problem and its SEE and PLAN steps.
- Have students use graph paper to make a model if stamps are not available. Tell them that they can make the graph paper more like stamps by drawing pictures on each square. Have students experiment by drawing each 4-stamp pattern.
- Help students develop strategies to be sure they have included all possibilities and then fill in the solution sentences. Have students complete the CHECK step.

3 Apply

Solution Notes
1. Help students understand that they should move one block at a time until they find that the 9 block needs to be moved to the far left column.
2. Remind students that a cube has equal edges. Help students search for a number that when multiplied by itself and again by itself, or cubed, is equal to 27. (3) The solution of 26 cubes is then found by thinking of a cube with 3 across, 3 up, and 3 down so that only the 1 cube in the middle does not touch the sides, top, or bottom.

Apply

Make a model to help you solve each problem.

1. Move one block to another stack so that the sums of all the stacks are equal.

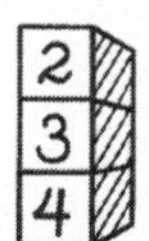

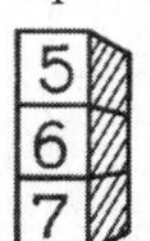

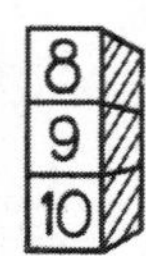

Move the 9 block to the first column.

2. It takes 27 sugar cubes to fill a cubical box. How many of the cubes will be touching the sides, top, or bottom of the box?
26 cubes

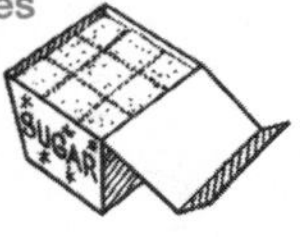

3. Change the arrangement of the circles on the left to the arrangement on the right by moving only three of the circles.

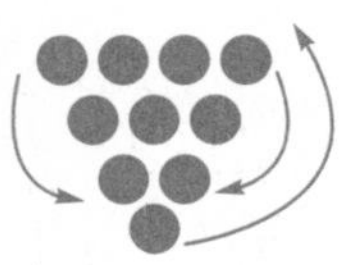
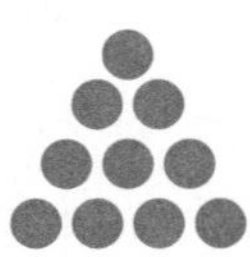

4. Form five equilateral triangles using nine sticks of equal length.
See Solution Notes.

Problem-Solving Strategy: Using the Four-Step Plan

- ★ **SEE** What do you need to find?
- ★ **PLAN** What do you need to do?
- ★ **DO** Follow the plan.
- ★ **CHECK** Does your answer make sense?

5. Kevin makes a triangle with cubic blocks. He puts 1 block in the first row, 2 blocks in the second row, 3 blocks in the third row, and so on. Predict whether he will need more or fewer than 100 blocks to make 10 rows. Then, prove that your prediction was correct.
See Solution Notes.

6. A ladybug walks around the edges of the top of a box of crackers. The top is shaped like a rectangle. When the bug gets back to where she started, she has walked 48 cm. If the top of the box is 6 cm wide, explain how you could find the length of the top of the box.
Answers will vary.

7. There are 12 cubic blocks on the bottom layer of a box. The box is completely filled with blocks. Why could there not be 10 layers if there are fewer than 100 blocks in the box?
Answers will vary.

For Mixed Abilities

More to Explore • Biography

Mary Somerville was born in Scotland in 1780. Like other women mathematicians, Mary had to struggle against her parents' opposition to her studies. Intelligent women of the time were urged to hide their learning to avoid drawing envy and losing friends. Mary was sent to boarding school at age 9 to learn to read and write.

As a teenager, Mary discovered algebra and began to study it intensely. She read far into the night, using up the family's supply of candles. Her parents hid the candles from her, but she persisted, rising at dawn to study. Mary became a science writer, a person who could make complex technical subjects understandable. She surpassed her male colleagues by the success of her work. Mary was one of the first women to become a member of the Royal Astronomical Society. Mary continued her studies and writing until she died at age 92.

3. Have students model solutions until they find one that works.
4. Remind students that an equilateral triangle has three equal sides. Ask if all the triangles must be the same size. (no) Remind students to keep this in mind as they form their model because the fifth triangle is the outline formed by the four smaller triangles.

Higher-Order Thinking Skills

5. **Analysis:** Each row has the same number of blocks as the number of the row; the tenth row has 10 blocks and the greatest number of blocks in any of the 10 rows. So, there are fewer than ten 10-block rows and less than 100 blocks. Students can draw a picture or use blocks to prove their answer.

6. **Synthesis:** Answers will vary but should include that there are two edges of 6 cm; so, the other 2 edges must have a total length of 48 cm minus 12 cm, or 36 cm. Being equal in length, each edge is therefore 18 cm.

7. **Synthesis:** A possible answer is that because there are 12 blocks in each layer, $10 \times 12 = 120$, which is more than 100.

page 161

Items	Objectives
1–2	To use A.M. and P.M. notation; find elapsed time (see pages 137–140)
3–4	To determine the appropriate customary unit of length (see pages 143–144)
5–6	To determine the appropriate customary unit of capacity (see pages 145–146)
7–8	To determine the appropriate customary unit of weight (see pages 147-148)
9–10	To determine the appropriate metric unit of length (see pages 149–152)
11	To determine the appropriate metric unit of capacity (seé pages 153–154)
12	To determine the appropriate metric unit of mass (see pages 155–156)

Name ______________________

Write the time. Include A.M. or P.M.

1. 45 minutes before midnight

2. 3 hours and 15 minutes after 10:30 A.M.

Circle the best estimate.

3. 6 in. (6 ft) 6 mi

4. 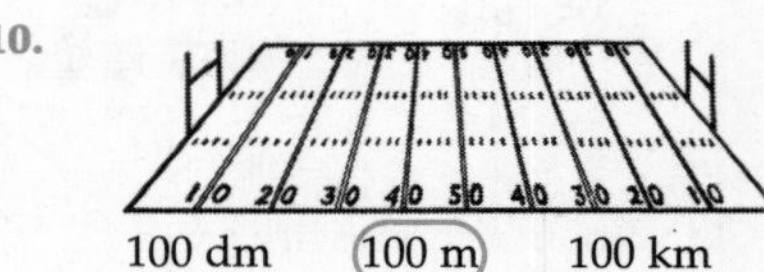

(7 in.) 7 ft 7 yd

5. 

1 pt (1 qt) 1 gal

6. (2 c) 2 pt 2 qt

7. 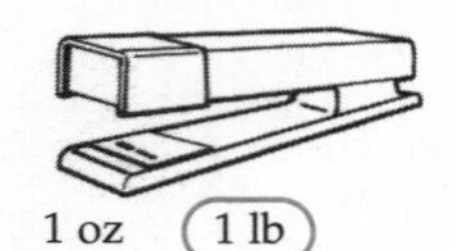

1 oz (1 lb)

8. 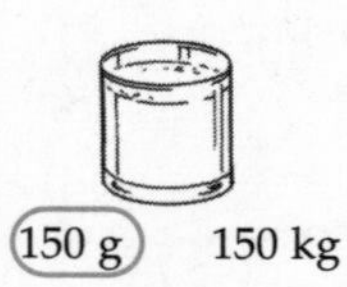

(15 oz) 15 lb

9. (3 cm) 3 dm 3 m

10. 100 dm (100 m) 100 km

11. 2 mL (2 L)

12. (150 g) 150 kg

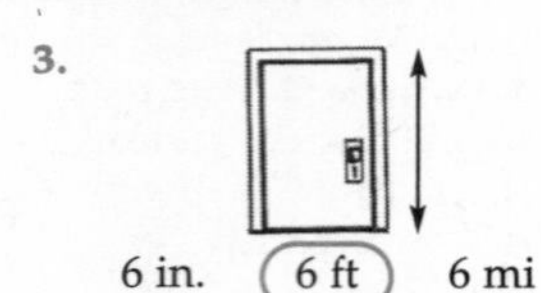

Alternate Chapter Test

You may wish to use the Alternate Chapter Test on page 360 of this book for further review and assessment.

Circle the letter of the correct answer.

1. Round 650 to the nearest hundred.
   a. 60
   (b.) 700
   c. NG

2. Round 4,296 to the nearest thousand.
   (a.) 4,000
   b. 5,000
   c. NG

3. What is the place value of the 0 in 625,309?
   a. ten thousands
   b. thousands
   c. hundreds
   (d.) NG

4. 3,475 + 2,686
   a. 5,061
   b. 5,161
   (c.) 6,161
   d. NG

5. 36,439 + 17,806
   a. 43,235
   b. 54,215
   (c.) 54,245
   d. NG

6. 822 − 387
   (a.) 435
   b. 535
   c. 662
   d. NG

7. 86,042 − 41,385
   a. 44,357
   b. 45,343
   c. 54,357
   (d.) NG

8. 6×8
   (a.) 48
   b. 54
   c. 63
   d. NG

9. 26 × 3
   a. 68
   (b.) 78
   c. 618
   d. NG

10. 65×7
   a. 425
   b. 435
   c. 635
   (d.) NG

11. $42 \div 7$
   a. 5
   b. 7
   c. 9
   (d.) NG

12. 4)48
   a. 1
   b. 2
   (c.) 12
   d. NG

13. $86 \div 6$
   (a.) 14 R2
   b. 14 R24
   c. 14
   d. NG

score

STOP

Cumulative Assessment

page 162

Items	Objectives
1–2	To round numbers to the nearest hundred and thousand (see pages 31–32)
3	To identify place value through hundred thousands (see pages 17–18, 25–26)
4–5	To add two numbers up to 5 digits (see pages 41–48)
6–7	To subtract two numbers up to 5 digits (see pages 55–66)
8	To multiply using the factor 8 (see pages 85–86)
9	To multiply 2-digit numbers by 1-digit numbers with two regroupings (see pages 103–104)
10–11	To divide using 7 as the divisor (see pages 115–116)
12	To divide 2-digit numbers by 1-digit numbers without remainder (see pages 125–126)
13	To divide 2-digit numbers by 1-digit numbers with remainders (see pages 127–128)

Alternate Cumulative Assessment

Circle the letter of the correct answer.

1. Round 836 to the nearest hundred.
   a 700
   (b) 800
   c 900
   d NG

2. Round 5,687 to the nearest thousand.
   a 5,000
   (b) 6,000
   c 7,000
   d NG

3. What is the place value of the 5 in 723,586?
   a ten thousands
   b thousands
   c hundred thousands
   (d) NG

4. 4,953 + 2,268
   a 6,111
   b 6,122
   c 7,111
   (d) NG

5. 27,266 + 24,817
   a 41,073
   (b) 52,083
   c 52,183
   d NG

6. 718 − 469
   (a) 249
   b 351
   c 349
   d NG

7. 92,031 − 51,654
   a 40,387
   b 41,487
   c 41,623
   (d) NG

8. $7 \times 9 =$
   a 49
   b 56
   (c) 63
   d NG

9. $19 \times 6 =$
   a 6,543
   (b) 114
   c 75
   d NG

10. $37 \times 6 =$
   (a) 222
   b 272
   c 1,842
   d NG

11. $81 \div 9 =$
   a 8
   (b) 9
   c 10
   d NG

12. 2)46
   a 2
   b 20 R6
   (c) 23
   d NG

13. 8)96
   a 1
   (b) 12
   c 11 R8
   d NG

9-1 Multiplying by Powers of 10

pages 163–164

1 Getting Started

Objectives

- To multiply 1-digit numbers by a power of 10

Warm Up • Mental Math

Have students name the unit of measure.

1. L (liter)
2. ft (foot)
3. lb (pound)
4. kg (kilogram)
5. mL (milliliter)
6. c (cup)
7. dm (decimeter)
8. qt (quart)
9. oz (ounce)

Warm Up • Multiplication

Write the following on the board: **6 × 10** and **9 × 10**. Have students tell the products. (60, 90) Ask students to recall the rule for multiplying a number by 10. (The product is that number followed by a zero.) Have students find the products of more 1-digit numbers and 10.

Name ______________________

CHAPTER 9

Multiplying Whole Numbers

Lesson 9-1

It's Algebra!

Multiplying by Powers of 10

Ricky's job is shoveling snow from his front walk. The walk is 9 meters long. Find the length of Ricky's shoveling job in centimeters.

We want to know how many centimeters of walk Ricky has to shovel.

The walk is __9__ meters long.

Each meter contains __100__ centimeters. To find the length of the walk in centimeters, we multiply __9__ by __100__.

Study these multiplications:

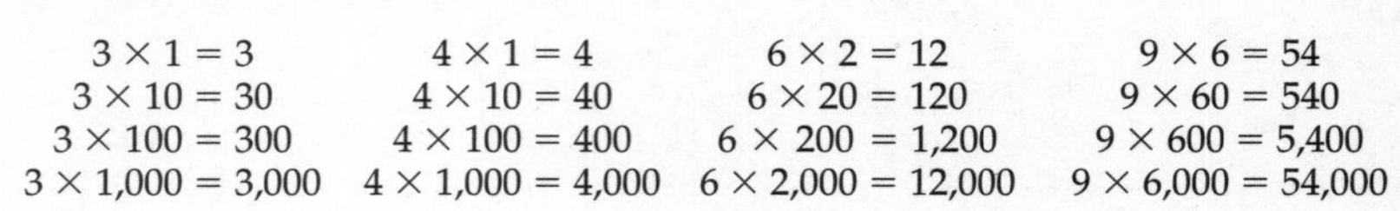

3 × 1 = 3	4 × 1 = 4	6 × 2 = 12	9 × 6 = 54
3 × 10 = 30	4 × 10 = 40	6 × 20 = 120	9 × 60 = 540
3 × 100 = 300	4 × 100 = 400	6 × 200 = 1,200	9 × 600 = 5,400
3 × 1,000 = 3,000	4 × 1,000 = 4,000	6 × 2,000 = 12,000	9 × 6,000 = 54,000

Multiply the digits that are not zeros. The product has the same number of zeros as there are zeros in the factors.

9 × 100 = __900__

Ricky's front walk is __900__ centimeters long.

Getting Started

Multiply.

1. 6 × 100 = __600__
2. 5 × 100 = __500__
3. 7 × 1,000 = __7,000__
4. 9 × 10 = __90__
5. 7 × 6,000 = __42,000__
6. 6 × 9,000 = __54,000__

---

2 Teach

Introduce the Lesson Have a student describe the picture. Ask students to tell the units of measure that are abbreviated on the sign. (centimeter, decimeter, meter) Ask a student to read the problem and tell what is to be found. (the length of the walk in centimeters)

- Have students complete the information sentences as they read with you. Point out the multiplication progression shown in the model. Have students complete the equation and the solution sentence.

Develop Skills and Concepts Write **100 × 6 = 600** on the board and tell students that a number times 100 is that number followed by two 0s. Write **2 × 200** on the board and ask students to apply the rule for multiplying by 100 to find the product. (2 × 1 = 2 and 2 followed by two 0s is 200.) Have students work more problems of 1-digit numbers times 100.

- Write **1,000 × 6** on the board and help students develop a rule for multiplying by 1,000. (The product is that number followed by three 0s.) Tell students that when multiplying by multiples of 10, the product is the product of the two numbers that are not zero, followed by the number of zeros in the factors.
- Now, write **200 × 9** on the board and tell students to multiply the 2 by the 9 for a product of 18. Ask students how many zeros are in the factors. (2) Write **00** after the 18 and have students read the number. (1,800) Repeat for more numbers.

3 Practice

Remind students to refer to the model on page 163 if necessary as they work to find the products in Exercises 1 through 39 and 41 through 43. Have students complete page 164 independently.

Practice

Multiply.

1. 5 × 100 = 500
2. 6 × 3,000 = 18,000
3. 9 × 20 = 180
4. 7 × 10 = 70
5. 7 × 1,000 = 7,000
6. 3 × 80 = 240
7. 5 × 400 = 2,000
8. 2 × 9,000 = 18,000
9. 9 × 700 = 6,300
10. 8 × 800 = 6,400
11. 3 × 5,000 = 15,000
12. 8 × 7,000 = 56,000
13. 4 × 40 = 160
14. 7 × 300 = 2,100
15. 2 × 6,000 = 12,000
16. 6 × 70 = 420
17. 5 × 2,000 = 10,000
18. 9 × 80 = 720
19. 5 × 600 = 3,000
20. 8 × 4,000 = 32,000
21. 9 × 30 = 270
22. 7 × 700 = 4,900
23. 6 × 5,000 = 30,000
24. 4 × 300 = 1,200
25. 7 × 80 = 560
26. 7 × 6,000 = 42,000
27. 4 × 9,000 = 36,000
28. 5 × 800 = 4,000
29. 3 × 700 = 2,100
30. 8 × 5,000 = 40,000
31. 2 × 8,000 = 16,000
32. 9 × 9,000 = 81,000
33. 6 × 40 = 240
34. 2 × 7,000 = 14,000
35. 4 × 500 = 2,000
36. 9 × 60 = 540
37. 6 × 6,000 = 36,000
38. 4 × 8,000 = 32,000
39. 9 × 400 = 3,600

Problem Solving

Solve each problem.

40. How many centimeters long is a table that is 9 decimeters in length?
90 centimeters

41. A carton of drinking straws contains 800 straws. How many straws are in 7 cartons?
5,600 straws

42. Computers cost $2,000 each. How much will a school pay for 7 computers?
$14,000

43. A small car weighs 3,000 pounds. How much do 8 small cars weigh?
24,000 pounds

4 Assess

Ask students to tell how to multiply by multiples of 10. (When multiplying by multiples of 10, the product is the product of the two numbers that are not zero, followed by the number of zeros in the factors.)

5 Mixed Review

1. 38 × 5 (190)
2. 350 − 209 (141)
3. 17 ÷ 8 (2 R1)
4. (9 ÷ 9) × 9 (9)
5. 72,156 + 27,408 (99,564)
6. 6 × $0.73 ($4.38)
7. $743.20 − $270.94 ($472.26)
8. 64 ÷ 5 (12 R4)
9. (8 × 7) + 4 (60)
10. 462 + 703 + 297 (1,462)

For Mixed Abilities

Common Errors • Intervention

Some students will write an incorrect number of zeros when multiplying. Have them complete the table shown below. For each answer, ask them to compare the number of zeros in the multiples of 10, 100, and 1,000 with the number in the product.

	6	3	8	7
10	60	30	80	70
100	(600)	(300)	(800)	(700)
500	(3,000)	(1,500)	(4,000)	(3,500)
1,000	(6,000)	(3,000)	(8,000)	(7,000)
6,000	(36,000)	(18,000)	(48,000)	(42,000)
9,000	(54,000)	(27,000)	(72,000)	(63,000)

Enrichment • Applications

Tell students to make a 5-year table to show the total mileage on a car at the end of each year if it is driven an average of 6,000 miles each year. Then, have students show the mileage if the car is driven 9,000 or 7,000 miles each year.

More to Explore • Numeration

Have students look up Egyptian hieroglyphics in an encyclopedia. Ask them to list the symbols for our numbers 1 through 10, for 20, 30, 40, 100, 1,000, 10,000, 100,000, and 1,000,000.

ESL/ELL STRATEGIES

Explain the meaning and use of powers of 10. Powers of 10 are numbers that have 10 as a factor. Powers of 10 allow us to quickly estimate answers to problems containing very large numbers.

9-2 Multiplying 3-Digit Numbers

pages 165–166

1 Getting Started

Objective

- To multiply 3-digit numbers by 1-digit numbers without regrouping

Warm Up • Mental Math

Have students complete the comparison: 6 is to 36 as 3 is to 18 as

1. 4 is to ____ (24)
2. 8 is to ____ (48)
3. 5 is to ____ (30)
4. 9 is to ____ (54)
5. 7 is to ____ (42)
6. 2 is to ____ (12)
7. 1 is to ____ (6)
8. 10 is to ____ (60)

Warm Up • Multiplication

Write problems such as **36 × 8** and **43 × 3** on the board for review of multiplying 2-digit numbers by 1-digit numbers, with or without regrouping. Have students talk through each problem as it is worked.

Name ______________________

Lesson 9-2

Multiplying 3-Digit Numbers

Mach 1 is the unit used to measure the speed of sound. Major Joseph Rogers was the first person to fly a plane faster than Mach 2. How fast was Major Rogers flying when he reached Mach 2?

Mach 1 is 742 miles per hour.
Mach 2 is twice as fast as Mach 1.

We want to know the speed for Mach 2.

We know Mach 1 is 742 miles per hour.

Mach 2 is 2 times as fast.

To find Mach 2, we multiply 742 by 2.

Multiply the ones.	Multiply the tens.	Multiply the hundreds.
742 × 2 = 4	742 × 2 = 84	742 × 2 = 1,484

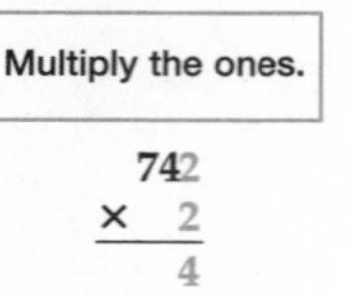

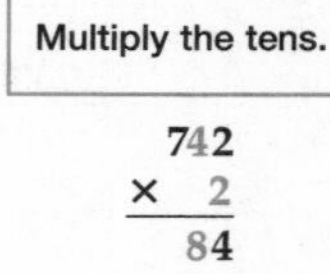

Major Rogers was flying 1,484 miles per hour when he reached Mach 2.

Getting Started

Multiply.

1. 221 × 4 = 884
2. 611 × 5 = 3,055
3. 730 × 3 = 2,190
4. 411 × 5 = 2,055

Copy and multiply.

5. 304 × 2 608
6. 801 × 5 4,005
7. 612 × 4 2,448
8. 512 × 3 1,536

2 Teach

Introduce the Lesson Have a student tell what is shown in the picture. Tell students that a Mach is a unit of measure that indicates the speed of sound. The unit was named for the physicist who studied sound. Discuss how sound is produced as air vibrates.

- Have a student read the problem and tell what is to be found. (the speed of Mach 2) Ask what information is needed to solve the problem. (the speed of Mach 1, which is 742 miles per hour)
- Have students complete the sentences and guide them through the multiplication shown in the model. Have students round the speed of Mach 1 to the nearest hundred and multiply by 2 to check their solution.

Develop Skills and Concepts Draw the following place-value chart on the board:

hundreds	tens	ones
1	3	2
×		3

Cover the 1 in the hundreds place and remind students that they already know how to multiply 32 × 3. Have a student multiply the ones and tens columns. Uncover the 1 and tell students that they now need to multiply the hundreds by 3. Have a student tell what 3 × 100 equals. (300) Show students that the number of hundreds goes in the hundreds place.

- Repeat for 423 × 4 and 642 × 4.

Practice

Multiply.

1. 823 × 3 = 2,469
2. 510 × 5 = 2,550
3. 302 × 4 = 1,208
4. 721 × 4 = 2,884
5. 634 × 2 = 1,268
6. 823 × 3 = 2,469
7. 704 × 2 = 1,408
8. 911 × 5 = 4,555

Copy and multiply.

9. 2 × 734 — 1,468
10. 3 × 931 — 2,793
11. 620 × 4 — 2,480
12. 814 × 2 — 1,628
13. 433 × 3 — 1,299
14. 311 × 5 — 1,555
15. 4 × 320 — 1,280
16. 5 × 901 — 4,505

Problem Solving

Solve each problem.

17. There are 212 rubber bands in a box. How many rubber bands are in 4 boxes?
848 rubber bands

18. Marty ran 425 meters in 2 minutes. Todd ran 512 meters in the same time. How much farther did Todd run?
87 meters

Now Try This!

What is a Roman emperor's favorite food? Find the correct letters by matching the numeral under each blank to the Roman numeral in the chart.

S XXII	L XXXVII	A XI
E IV	C I	S LXVII
D LVI	A XLVI	A XXIX
A II	R XVI	S VII

C	A	E	S	A	R
1	2	4	7	11	16

S	A	L	A	D	S
22	29	37	46	56	67

3 Practice

Remind students to place a comma after the digit in the thousands place. Have students complete the page independently.

Now Try This! Discuss the Roman system of numeration. Study the sequence of numbers under the blanks and have students write the next three numbers in Roman numerals also.

4 Assess

Ask students to explain how to multiply a 3-digit number by a 1-digit number. (multiply the one-digit number by the digits in the ones place, the tens place, and then the hundreds place)

For Mixed Abilities

Common Errors • Intervention

Some students may begin multiplying with the hundreds rather than with the ones. Remind them to start in the same place as they do in addition and subtraction. Students also can benefit conceptually if they work some problems using a place-value approach as shown below.

```
  231
×   3
    3  ← 3 × 1
   90  ← 3 × 30
  600  ← 3 × 200
  693
```

Enrichment • Applications

Tell students to estimate Major Rogers's speed at Mach 2 to the nearest hundred and to find out how many times faster he went than they do traveling in a car going 10 miles per hour.

More to Explore • Measurement

Have students look at their handspans, the distance between the tip of the little finger and the tip of the thumb at its widest stretch. Explain that this is a commonly used, nonstandard unit of measure.

Tell students to measure their handspans as accurately as possible to the nearest $\frac{1}{2}$ inch. Have students write their hand measurements on the board. Then, tell students to each make a bar graph showing the handspan measurements of all the students. Have each student interpret his or her graph for the class. Display completed graphs on a bulletin board.

9-3 Multiplying Money

pages 167–168

1 Getting Started

Objective

- To multiply money through $9.99 by 1-digit numbers with regrouping

Warm Up • Mental Math

Have students name the equivalent.

1. 32 oz = (2) lb
2. 8,000 lb = (4) t
3. 2,000 m = (2) km
4. 88 qt = (22) gal
5. 4 dm = (40) cm
6. 6 L = (6,000) mL
7. 300 cm = (3) m
8. 3 pt = (6) c

Warm Up • Multiplication

Write the following problems vertically on the board to have students review multiplication of money: **$0.76 × 3, $0.19 × 6, $0.47 × 6**, and **$0.59 × 2.**

Have students work the problems and then check them by estimating the amounts to the nearest dime. ($2.28, $1.14, $2.82, $1.18)

Name ______________________

Lesson 9-3

Multiplying Money

Mrs. Juarez is planning to cook dinner on the grill. She needs 4 pounds of steak. How much will Mrs. Juarez spend if she buys porterhouse steaks?

We want to know how much money Mrs. Juarez will spend on steaks.

She needs __4__ pounds of steak.

She wants to buy __porterhouse__ steak, which sells for __$5.19__ a pound.

To find the total cost of the steak, we multiply __$5.19__ by __4__.

Multiply the pennies. Regroup if needed.	Multiply the dimes. Add the extra dimes. Write the decimal point in the product.	Multiply the dollars. Write the dollar sign in the product.
$5.1^{3}9 × 4 = 6	$5.^{3}19 × 4 = .76	$5.19 × 4 = $20.76

Mrs. Juarez will spend __$20.76__ on steak.

Getting Started

Multiply.

1. $6.13 × 4 = $24.52
2. $8.26 × 3 = $24.78
3. $7.18 × 5 = $35.90
4. $4.37 × 2 = $8.74

Copy and multiply.

5. 6 × $4.09 = $24.54
6. $3.29 × 2 = $6.58
7. 5 × $3.19 = $15.95
8. 2 × $6.01 = $12.02

2 Teach

Introduce the Lesson Have a student describe the picture. Ask students how much each kind of steak costs per pound. (sirloin is $3.29, T-bone is $4.09, and porterhouse is $5.19 per pound.) Have a student read the problem and tell what is to be solved. (the cost of 4 pounds of porterhouse steak) Ask students if there is unnecessary information given. (yes)

- Have students read and complete the sentences to solve the problem. Guide students through the multiplication steps in the model and have them complete the solution sentence. Have students estimate the cost of 1 steak to the nearest dime to check their solution.

Develop Skills and Concepts Write **326 × 4** and **$3.26 × 4** vertically on the board. Ask students how the two problems are similar. (Both have the same numbers.) Ask students how these problems differ. (One has a dollar sign and a decimal point.)

- Have students talk through the problems as they work. Remind students that the decimal point goes before the dimes digit. Have a student place the dollar sign and decimal point. Repeat for $7.86 × 2 and $3.99 × 4.
- Have students read the products to practice reading money amounts. Remind students that they say the word *and* for the decimal in dollar amounts.

3 Practice

Remind students to place the dollar sign and decimal in each answer. Tell students to use the sign on page 167 to solve the last two word problems. Have students complete page 168 independently.

Practice

Multiply.

1. $6.23 × 4 = $24.92
2. $7.17 × 5 = $35.85
3. $5.38 × 2 = $10.76
4. $9.15 × 6 = $54.90
5. $4.08 × 7 = $28.56
6. $8.14 × 6 = $48.84
7. $3.49 × 2 = $6.98
8. $6.26 × 3 = $18.78
9. $2.25 × 3 = $6.75
10. $9.09 × 8 = $72.72
11. $7.29 × 3 = $21.87
12. $3.17 × 4 = $12.68

Copy and multiply.

13. 3 × $7.24 $21.72
14. $4.12 × 8 $32.96
15. $7.19 × 5 $35.95
16. 2 × $8.37 $16.74
17. $6.09 × 7 $42.63
18. $8.13 × 6 $48.78
19. 4 × $9.23 $36.92
20. 3 × $4.29 $12.87

Problem Solving

Solve each problem.

21. Hal bought 5 gallons of paint. Each gallon cost $7.15. How much did Hal pay for the paint? $35.75
22. A fishing rod costs $8.97. A spinning reel costs $7.97. What is the total cost of the rod and reel? $16.94
23. Radial tires are $49.35 each. Regular tires are $34.89 each. How much more are the radial tires? $14.46
24. Oil filters are packed 6 to a box. One oil filter costs $1.09. How much does one box cost? $6.54

Use the data on page 167 to solve Exercises 25 and 26.

25. How much more will 3 pounds of T-bone steak cost than 3 pounds of sirloin steak? $2.40
26. Mr. Kelly bought 2 pounds of T-bone steak and 2 pounds of porterhouse steak. How much change did he receive from a $20 bill? $1.44

4 Assess

Ask students to explain the difference between multiplying with money and multiplying with numbers.
(For money, you need to put the dollar sign and decimal point in the product.)

5 Mixed Review

1. 86 ÷ 2 (43)
2. 28,155 + 36,345 (64,500)
3. 500 × 3 (1,500)
4. 62,176 − 24,259 (37,917)
5. 72 − 8 × 8 (8)
6. 8 × 6,000 (48,000)
7. 376 + 1,204 + 314 (1,894)
8. $0.85 × 4 ($3.40)
9. $50.70 − $38.29 ($12.41)
10. 72 ÷ 5 (14 R2)

For Mixed Abilities

Common Errors • Intervention

Some students may add the renamed number before they multiply.

Incorrect	Correct
^{1} $6.24 × 3 = $18.92	^{1} $6.24 × 3 = $18.72

Have students work with partners to use the following form to compute:

```
  $3.19
×     4
   0.36 ← 4 × 9 pennies
   0.40 ← 4 × 1 dime
  12.00 ← 4 × 3 dollars
 $12.76
```

Enrichment • Application

Tell students to cut from newspapers or catalogs two priced items they might buy in quantities of 2 or more. Then, have them find the total cost if they bought 3 of one item and 4 of the other.

More to Explore • Logic

Tell students that you are going to write the numerals 0 through 9 on the board in a definite order. Ask them if they can figure out the logic represented by this order of numbers. Then, write the following number sequence on the board for students to identify the pattern:

8, 5, 4, 9, 1, 7, 6, 3, 2, 0

(They are in alphabetical order according to the number word for each numeral.)

9-4 Multiplying 3-Digit Numbers With Regrouping

pages 169–170

1 Getting Started

Objective

- To multiply 3-digit numbers by 1-digit numbers with regrouping

Warm Up • Mental Math

Ask students which is greater.

1. 8×9 or $\frac{1}{2}$ of 140 (8×9)
2. $42 \div 6$ or $42 \div 7$ ($42 \div 6$)
3. $35 \div 7$ or $54 \div 9$ ($54 \div 9$)
4. $\frac{1}{4}$ of 40 or $\frac{1}{4}$ of 36 ($\frac{1}{4}$ of 40)
5. $26 - 18$ or $18 \div 3$ ($26 - 18$)
6. 64×1 or 20×4 (20×4)
7. $80 \div 8$ or $99 \div 9$ ($99 \div 9$)
8. $\frac{1}{10}$ or $\frac{1}{9}$ ($\frac{1}{9}$)

Warm Up • Numeration

Write **7 × 2 tens** on the board. Have students tell how many tens there are in all (14 tens). Then, ask how many hundreds, tens, and ones are in the product. (1, 4, 0) Repeat for **7 ones × 9**. (6, 3) Present more problems for students to review the regroupings.

Name ______________________________

Lesson 9-4

Multiplying 3-Digit Numbers With Regrouping

The distance from Denver to Seattle is approximately 3 times as far as the distance from Chicago to Pittsburgh. About how far is it from Seattle to Denver?

We want to know the approximate distance between Seattle and Denver.

The distance from Chicago to Pittsburgh is __457__ miles.

The distance from Seattle to Denver is approximately __3__ times that number.

To find the distance between Seattle and Denver, we multiply __457__ by __3__.

Multiply the ones. Regroup if needed.	Multiply the tens. Add any extra tens. Regroup if needed.	Multiply the hundreds. Add any extra hundreds.
$\begin{array}{r} {}^{2} \\ 457 \\ \times\ 3 \\ \hline 1 \end{array}$	$\begin{array}{r} {}^{1}{}^{2} \\ 457 \\ \times\ 3 \\ \hline 71 \end{array}$	$\begin{array}{r} {}^{1} \\ 457 \\ \times\ 3 \\ \hline 1{,}371 \end{array}$

It is about __1,371__ miles from Seattle to Denver.

Getting Started

Multiply.

1. $525 \times 7 = 3{,}675$
2. $289 \times 4 = 1{,}156$
3. $328 \times 8 = 2{,}624$
4. $\$2.48 \times 6 = \$14.88$

Copy and multiply.

5. $\$5.96 \times 9$ $53.64
6. 709×5 3,545
7. 492×3 1,476
8. 587×2 1,174

2 Teach

Introduce the Lesson Ask students what the map shows. (4 cities and the mileage from Chicago to Pittsburgh as 457 miles) Have a student read the problem and tell what is to be found. (distance from Seattle to Denver) Ask students what information is given to help solve this problem. (It is 457 miles from Pittsburgh to Chicago and it is approximately 3 times that distance from Seattle to Denver.)

- Have students read and complete the sentences. Guide them through the multiplication steps in the model. Have students complete the solution sentence.

Develop Skills and Concepts Remind students that they already know how to work this problem as you write **78 × 4** vertically on the board. Have students talk through the work, as you record their responses, to find the product. (312) Ask students why there are 31 tens instead of 28. (added the 3 tens that were regrouped from the ones to 28)

- Now, write **278 × 4** on the board and have students talk through the work to find the ones and then the tens. Ask students if some of the 31 tens can be regrouped for hundreds. (yes) Record the regrouping and ask students to find the total number of hundreds. (11 hundreds) Ask students if some hundreds can be regrouped for thousands. (yes) Record the regrouping and ask students to read the product. (1,112)
- Repeat for 69×3 and 469×3.

Practice

Multiply.

1. 357 × 4 = 1,428
2. 292 × 8 = 2,336
3. 537 × 5 = 2,685
4. 673 × 2 = 1,346
5. 896 × 6 = 5,376
6. 383 × 7 = 2,681
7. $4.77 × 3 = $14.31
8. 709 × 9 = 6,381
9. 372 × 8 = 2,976
10. 628 × 3 = 1,884
11. 548 × 6 = 3,288
12. 419 × 5 = 2,095

Copy and multiply.

13. 5 × 386 — 1,930
14. 457 × 9 — 4,113
15. $6.75 × 2 — $13.50
16. 3 × 628 — 1,884
17. 727 × 8 — 5,816
18. $2.94 × 5 — $14.70
19. 7 × 929 — 6,503
20. 6 × 848 — 5,088

Problem Solving

Solve each problem.

21. The Cammero family drove 346 miles on their vacation. The Johnsons drove 4 times as far as the Cammeros. How far did the Johnsons drive? 1,384 miles

22. The distance from San Francisco to Los Angeles is 403 miles. Mr. Harris left San Francisco at 10:00 A.M. and drove 115 miles. How far does Mr. Harris still have to drive? 288 miles

23. A jet airliner can hold 186 people. How many people can 7 jet airliners carry? 1,302 people

24. A china platter costs $6.75 and cereal bowls cost $3.95 each. How much would Sharon pay for a platter and 4 bowls? $22.55

3 Practice

Remind students to work the ones column first and then the tens and then the hundreds. Have students complete the page independently.

4 Assess

Ask students to find the product of $7.85 × 6. ($47.10)

For Mixed Abilities

Common Errors • Intervention

Some students may regroup but forget to record and add the regrouped number when multiplying.

Incorrect	Correct
357 × 4 = 1,208	(regrouped 2 2) 357 × 4 = 1,428

Have students use a place-value chart to show the steps in the problem.

	100s	10s	1s
4 × 7 ones		2	8
4 × 5 tens	2	0	0
4 × 3 hundreds 1	2	0	0
1,	4	2	8

Enrichment • Measurement

Tell students to find out how many weeks they have lived since their birth. Then, have them find out how many weeks they will have lived when they are 3 times as old as they are now.

More to Explore • Money

Divide the class into small groups. Provide each group with 2 quarters, 5 dimes, 2 nickels, and 5 pennies. Have each group find as many combinations of coins as it can to total a value of 50¢. Ask groups to record their combinations on a chart and compare the results with other groups to make sure all possible combinations are found.

To extend this activity, have students make as many sets as possible for other amounts of money.

9-5 Multiplying 4-Digit Numbers

pages 171–172

1 Getting Started

Objective

- To multiply 4-digit numbers by 1-digit numbers with regrouping

Warm Up • Mental Math

Have students name the century for the following years:

1. 1977 (twentieth)
2. 2001 (twenty-first)
3. 1776 (eighteenth)
4. 1492 (fifteenth)
5. 961 (tenth)
6. 1382 (fourteenth)
7. 2164 (twenty-second)

Warm Up • Numeration

Ask students to tell the total number of hundreds in $(9 \times 700) + (4 \times 100)$. (67)
Have students tell the number of thousands in 1,300. (1)
Repeat for $(200 \times 6) + (9 \times 100)$, $(900 \times 9) + (1 \times 100)$, and other similar problems.

Name ______________________

Multiplying 4-Digit Numbers

The course of the North Coast Cross Country Ski Race is 8 miles long. How many yards long is the race course?

We want to know how many yards long the race course is.

The race course is 8 miles long.

There are 1,760 yards in 1 mile.

To find the total number of yards in the race course, we multiply 1,760 by 8.

Multiply the ones. Regroup if needed.	Multiply the tens. Add any extra tens. Regroup if needed.	Multiply the hundreds. Add any extra hundreds. Regroup if needed.	Multiply the thousands. Add any extra thousands.
$1{,}760 \times 8 = 0$	(4) $1{,}760 \times 8 = 80$	(6 4) $1{,}760 \times 8 = 080$	(6) $1{,}760 \times 8 = 14{,}080$

The course of the North Coast Cross Country Ski Race is 14,080 yards long.

Getting Started

Multiply.

1. $6{,}243 \times 3$ = 18,729
2. $4{,}086 \times 3$ = 12,258
3. $5{,}248 \times 7$ = 36,736
4. $\$75.76 \times 5$ = $378.80

Copy and multiply.

5. $4{,}273 \times 2$ = 8,546
6. $\$16.59 \times 9$ = $149.31
7. $\$90.06 \times 8$ = $720.48
8. $3{,}876 \times 6$ = 23,256

2 Teach

Introduce the Lesson Have a student read the problem and tell what is to be found. (the race course length in yards) Ask students what information is given in the problem. (The race course is 8 miles long.) Ask students what helpful information is given in the picture. (1 mile = 1,760 yards) Ask students why the number of feet in 1 mile is not useful. (do not need to answer the question in feet)

- Have students complete the sentences. Guide students through the multiplication steps in the model and have them complete the solution. Have students round the number of yards in 1 mile to the nearest hundred and estimate to check their answer.

Develop Skills and Concepts Write **789 × 4** on the board and have a student work the problem to review multiplying a 3-digit number. (3,156)

- Now, write **4,789 × 4** on the board and tell students that they will multiply the ones, tens, and hundreds as in the first problem, but they will regroup the hundreds for thousands and then multiply the thousands and add the regrouped thousands. Have a student work the problem. (19,156)
- Repeat for 816×9 and $3{,}816 \times 9$. (7,344; 34,344)

Practice

Multiply.

1. 3,216 × 5 = 16,080
2. 7,926 × 8 = 63,408
3. 2,079 × 6 = 12,474
4. 8,273 × 2 = 16,546
5. $37.86 × 7 = $265.02
6. 9,376 × 3 = 28,128
7. 5,163 × 4 = 20,652
8. $48.06 × 9 = $432.54
9. 1,673 × 8 = 13,384
10. $32.85 × 5 = $164.25
11. 8,269 × 2 = 16,538
12. 4,675 × 4 = 18,700

Copy and multiply.

13. 8 × 4,271 34,168
14. 3 × 6,129 18,387
15. $63.38 × 9 $570.42
16. 4,350 × 2 8,700
17. 4,925 × 9 44,325
18. 4 × $57.83 $231.32
19. 5 × 6,256 31,280
20. 9,816 × 6 58,896

Problem Solving

Solve each problem.

21. The highest mountain in Russia is 18,510 feet tall. The highest mountain in the United States is 1,810 feet higher than that. How tall is the highest mountain in the United States?
20,320 feet

22. A computer with 1K memory can store 1,024 bits of information. How many bits of information can an 8K computer store?
8,192 bits

23. The race cars drove a total of 4 miles in 9 laps. How many feet did they travel?
21,120 feet

24. A dump truck can hold 1,426 pounds of dirt. How many pounds can the dump truck haul in 7 trips?
9,982 pounds

25. Rebecca ran 450 yards less than 6 miles. How many yards did Rebecca run?
10,110 yards

26. It costs $23.43 for a pair of jeans and $14.26 for a shirt. Pablo bought 2 pairs of jeans and 3 shirts. How much did Pablo spend?
$89.64

3 Practice

Remind students to place a dollar sign and decimal point in the product for money problems. Tell students that they will need to use the information on page 171 when solving Exercise 25 on page 172. Remind students to watch for word problems that require two operations. Have students complete the page independently.

4 Assess

Ask students what they must remember to do after multiplying the tens, hundreds, and thousands in a number. (add any extra tens, hundreds, or thousands)

For Mixed Abilities

Common Errors • Intervention

Some students may make errors because they have difficulty writing numbers for the partial products in the correct place of the product. Have students work with partners and use the following form to compute:

```
    4,329
  ×     5
       45   ← 5 × 9
      100   ← 5 × 20
    1 500   ← 5 × 300
   20 000   ← 5 × 4,000
   21,645
```

Enrichment • Application

Tell students to use a map or an almanac to find the distance they would travel if they drove 2 round trips from New York City to San Francisco.

More to Explore • Statistics

Using the newspaper weather page, have students make a bar graph showing the daily high temperatures of various cities around the world. Have them check a world map to make sure they have included cities in both hemispheres and in the temperate zones, tropic zones, and polar zones.

After 2 weeks of graphing, tell students to write five questions based on information shown on their graphs, for example, Which city has the lowest temperature? Which city has the most consistent temperature? Have them exchange graphs with a partner who will answer the questions.

9-6 Multiplying by Multiples of 10

pages 173–174

1 Getting Started

Objective

- To multiply 2- or 3-digit numbers by a 2-digit multiple of 10

Warm Up • Mental Math

Have students complete the following equivalents:

1. (60) min = 1 h
2. 1 gal = (8) pt
3. 3 h = (180) min
4. (8) oz = $\frac{1}{2}$ lb
5. 49 d = (7) wk
6. 40 cm = (4) dm
7. (60) s = 1 min
8. 2 L = (2,000) mL

Warm Up • Multiplication

Ask students what 8 × 0 equals. (0) Repeat for 0 × 2 and 4 × 0. Write **24 × 0** vertically on the board. Ask students the product. (0) Ask students what any number times zero equals. (zero) Write **464 × 0** on the board and have a student write the product. (0) Repeat for more problems up to 4-digit numbers times zero.

Name ______________________

Multiplying by Multiples of 10

Ronald is learning health skills in his CPR class. He is taking his pulse to find his heart rate for 1 minute. How many times will Ronald's heart beat in 1 hour?

...70, 71, 72

We want to know how often Ronald's heart beats hourly.

His heart beats 72 times in 1 minute.

We know that there are 60 minutes in 1 hour.

To find Ronald's hourly heart rate, we multiply 72 by 60.

Multiply by the digit in the ones place.	Multiply by the digit in the tens place. Add any extra tens.
72 × 60 = 0	72 × 60 = 4,320 (regroup 1)

Ronald's heart beats 4,320 times in 1 hour.

Getting Started

Multiply.

1. 36 × 20 = 720
2. 35 × 40 = 1,400
3. 50 × 60 = 3,000
4. 125 × 30 = 3,750

Copy and multiply.

5. 625 × 70 = 43,750
6. 820 × 80 = 65,600
7. 635 × 50 = 31,750
8. 708 × 90 = 63,720

2 Teach

Introduce the Lesson Have a student describe the picture and read the thought bubble. Ask a student to read the problem aloud and tell what is to be solved. (the number of times Ronald's heart will beat in 1 hour) Ask students what information is given. (Ronald's heart beats 72 times in 1 minute.) Ask students what additional information is needed. (the number of minutes in 1 hour)

- Have students read and complete the sentences and work through the model with them to solve the problem. Tell students to round 72 to the nearest ten to check their answer.

Develop Skills and Concepts Remind students that the rule for multiplying by a multiple of 10 is to multiply the numbers that are not zero and write that product followed by the number of zeros in the factors.

- Write **84 × 90** on the board and have a student multiply 84 times 9. (756) Ask how many zeros are in the two factors. (1) Have the student write one 0 after 756 and read the number. (7,560)
- Now, write **84 × 90** vertically on the board and ask students what zero times 84 equals. (0) Write a zero in the ones place. Tell students that we now multiply the numbers that are not zero. Have a student multiply 84 × 9. (756) Write **756** in front of the zero.
- Repeat for 96 × 40 and 38 × 70. (3,840; 2,660)

3 Practice

Tell students that there is unnecessary information in one word problem and another problem requires them to supply some of the information. Have students complete page 174.

Practice

Multiply.

1. 52 × 30 = 1,560	2. 76 × 40 = 3,040	3. 27 × 70 = 1,890	4. 80 × 90 = 7,200
5. 48 × 20 = 960	6. 63 × 60 = 3,780	7. 400 × 50 = 20,000	8. 88 × 80 = 7,040
9. 153 × 30 = 4,590	10. 400 × 70 = 28,000	11. 94 × 50 = 4,700	12. 785 × 20 = 15,700

Copy and multiply.

13. 40 × 253 (10,120)	14. 70 × 36 (2,520)	15. 90 × 573 (51,570)	16. 426 × 30 (12,780)
17. 651 × 60 (39,060)	18. 20 × 879 (17,580)	19. 89 × 50 (4,450)	20. 80 × 600 (48,000)

Problem Solving

Solve each problem.

21. How many minutes are there in 36 hours?
2,160 minutes

22. How many seconds are there in 15 minutes?
900 seconds

23. Charlotte can walk 1 kilometer in 9 minutes. One day, Charlotte walked for 90 minutes. How many kilometers did she walk?
10 kilometers

24. A container holds 245 milliliters of juice. The cafeteria used 80 containers. How many milliliters of juice did the cafeteria use?
19,600 milliliters

25. The school photographer took 875 pictures. Each picture takes 40 seconds to develop. How many seconds will it take to develop the pictures?
35,000 seconds

26. Bobby read 16 chapters of a book. Each chapter was 30 pages long. Bobby took 50 minutes to read each chapter. How long did it take him to read the book?
800 minutes

4 Assess

Ask students to multiply 836 by 90. (75,240)

5 Mixed Review

1. 9 + 5 + 6 + 3 + 1 (24)
2. 4 × 309 (1,236)
3. $8.27 × 2 ($16.54)
4. 6,210 − 3,008 (3,202)
5. 67 ÷ 4 (16 R3)
6. $563.17 + $327.93 ($891.10)
7. 47 × 6 (282)
8. 75 + 1,027 + 472 (1,574)
9. 39 ÷ 8 (4 R7)
10. 376 − 256 (120)

For Mixed Abilities

Common Errors • Intervention

When multiplying by a multiple of 10, students may add the renamed number before they multiply.

Incorrect	Correct
¹72 × 60 = 4,820	¹72 × 60 = 4,320

Correct students by having them review renaming when multiplying 6 × 72 and apply it to multiplying by 60.

Enrichment • Patterns

Tell students to make a table to show the total number of eggs if they had 10 dozen, 20 dozen, 40 dozen, and 70 dozen.

More to Explore • Probability

Give each pair of students a bag containing 25 assorted colored cubes. Ask them to pick out 5 of the cubes. Explain that there are a total of 25 cubes in the bag and ask them to guess at the contents based on their sample of 5. Tell students to put the cubes back in the bag and draw another 5. Ask them again to estimate the contents after each sample of 5. See if any group is able to guess the total contents after they have seen 20 of the 25 cubes.

9-7 Multiplying by 2-Digit Numbers

pages 175–176

1 Getting Started

Objective
- To multiply 2-digit numbers by 2-digit numbers, without regrouping

Warm Up • Mental Math
Dictate the following:
1. 5 × 50 (250)
2. 84 ÷ 2 (42)
3. 6 × 10 × 2 (120)
4. (7 × 5) ÷ 5 (7)
5. 238 − 3 tens (208)
6. 4 hundreds plus 28 ones (428)
7. 6,000 × 9 (54,000)
8. 43 × 0 × 8 (0)

Warm Up • Numeration
Write **(30 × 84) + (4 × 84)** on the board and have students find the sum of the two multiplications. (2,520 + 336 = 2,856) Repeat for **(26 × 70) + (26 × 9)** and **(40 × 18) + (18 × 7)**. (1,820 + 234 = 2,054; 720 + 126 = 846)

Name ______________________

Multiplying by 2-Digit Numbers

January, usually the coldest month of the year, is named for the Roman god Janus. How many hours are there in the month of January?

We want to know the total number of hours in January.

January has __31__ days.

We know that there are __24__ hours in 1 day.

To find the total number of hours, we multiply __31__ by __24__.

Multiply by the ones.	Multiply by the tens.	Add the products.
31 × 24 124 ← 4 × 31	31 × 24 124 620 ← 20 × 31	31 × 24 124 + 620 744 ← 24 × 31

There are __744__ hours in January.

Getting Started

Multiply.

1. 23 × 32 = 736
2. 42 × 24 = 1,008
3. 50 × 35 = 1,750
4. 64 × 22 = 1,408

Copy and multiply.

5. 12 × 14 168
6. 21 × 28 588
7. 53 × 23 1,219
8. 96 × 11 1,056

2 Teach

Introduce the Lesson Ask students what tool of measurement is shown in the picture. (calendar) Ask what a calendar measures. (the time in days, weeks, and months)

- Have a student read the problem and tell what is to be found. (the number of hours in the month of January) Ask students what information will be needed. (the number of days in January and the number of hours in a day) Ask which of that information is not given in the problem or picture. (the number of hours in a day)
- Have students complete the sentences as you read with them. Guide students through the multiplication in the model to solve the problem. Tell students to check their solution by rounding 31 to 30 and multiplying.

Develop Skills and Concepts Write the following on the board:

61 × 46 = (61 × 40) + (61 × 6)

Tell students that the problem of 61 × 46 can be rewritten as forty 61s plus six 61s.

- Now, write the following on the board:

```
   61
 × 46
  366  (6 × 61)
2,440  (40 × 61)
2,806  (46 × 61)
```

- Talk through multiplying the ones, then multiplying the tens, and finally adding the products.
- Repeat the exercise for 23 × 71 and 32 × 24. (1,633; 768)

Practice

Multiply.

1. $\begin{array}{r} 32 \\ \times 43 \\ \hline \end{array}$ 1,376
2. $\begin{array}{r} 71 \\ \times 56 \\ \hline \end{array}$ 3,976
3. $\begin{array}{r} 23 \\ \times 23 \\ \hline \end{array}$ 529
4. $\begin{array}{r} 75 \\ \times 11 \\ \hline \end{array}$ 825
5. $\begin{array}{r} 32 \\ \times 32 \\ \hline \end{array}$ 1,024
6. $\begin{array}{r} 22 \\ \times 43 \\ \hline \end{array}$ 946
7. $\begin{array}{r} 61 \\ \times 38 \\ \hline \end{array}$ 2,318
8. $\begin{array}{r} 80 \\ \times 49 \\ \hline \end{array}$ 3,920
9. $\begin{array}{r} 54 \\ \times 12 \\ \hline \end{array}$ 648
10. $\begin{array}{r} 62 \\ \times 33 \\ \hline \end{array}$ 2,046
11. $\begin{array}{r} 84 \\ \times 21 \\ \hline \end{array}$ 1,764
12. $\begin{array}{r} 32 \\ \times 42 \\ \hline \end{array}$ 1,344

Copy and multiply.

13. 41×38 1,558
14. 24×62 1,488
15. 63×33 2,079
16. 60×57 3,420
17. 76×81 6,156
18. 79×51 4,029
19. 96×11 1,056
20. 23×52 1,196

Problem Solving

Solve each problem.

21. How many hours are there in 2 weeks and 5 days? 456 hours
22. How many inches are there in 25 feet and 11 inches? 311 inches
23. How many ounces are there in 3 pounds 9 ounces? 57 ounces
24. How many feet are there in 15 yards 2 feet? 47 feet

3 Practice

Remind students to think of the multiplier or bottom factor as a multiple of 10 plus some ones. Tell students that they must supply some missing information to work each word problem in Exercises 21 to 24. Have students complete page 176.

4 Assess

Ask students to solve 21×14. (294)

For Mixed Abilities

Common Errors • Intervention

Students may forget to write a zero in the ones place of the second partial product when multiplying by the tens. Correct students by having them rewrite the problems as shown below.

$21 \times 43 = (20 \times 43) + (1 \times 43)$
$860 + 43 = 903$

Enrichment • Measurement

Have students find the combined weight, in ounces, of 2 babies if one weighs 11 pounds 6 ounces and the other weighs 13 pounds 10 ounces. (25 pounds)

More to Explore • Measurement

Have students estimate the weight, in grams, of various familiar objects. First, give examples of items that weigh about a gram, such as a safety pin or a paper clip. Next, name items that weigh about 1 kilogram, for example, a baseball bat or two footballs. Have students gather six to eight objects such as a dime, a comb, a classroom globe, a pair of tennis shoes, or a math book.

Have students estimate the weight of each item and gather the items into groups according to the following categories: less than 10 grams, 11–100 grams, 101–500 grams, 501–999 grams, 1–5 kilograms, 5–10 kilograms, and more than 10 kilograms. Tell students to check their estimations for accuracy by weighing the lighter objects on a balance scale, using a set of metric weights, or by weighing the heavier items on a metric bathroom scale.

9-8 Multiplying by 2-Digit Numbers With Regrouping

pages 177–178

1 Getting Started

Objective

- To multiply 2-digit numbers by 2-digit numbers with regrouping

Warm Up • Mental Math

Have students tell the profit if they buy at the following price:

1. 5/$1 and sell at 25¢ each (5¢)
2. 40¢ each and sell at 6/$3 (10¢)
3. $1 each and sell at 2/$3 (50¢)
4. $35 each and sell at 2/$90 ($10)
5. 8/$20 and sell at $3.50 each ($1)
6. $3.19 and sell at $4.50 ($1.31)
7. 4/88¢ and sell at 3/99¢ (11¢)
8. $4.80 for 2 and sell at $5.00 each ($2.60)

Warm Up • Numeration

Write **69 × 3** on the board and have a student talk through the multiplication and the regrouping. Repeat for more problems of multiplying 2-digit numbers by 1-digit numbers with regrouping.

Name ______________________

Multiplying by 2-Digit Numbers With Regrouping

After Mike fills all the shelves, how many pounds of sugar will there be?

We want to find how many pounds of sugar will be in the display.

Each bag of sugar weighs __25__ pounds.

There will be __56__ bags of sugar.

To find the total number of pounds of sugar, we multiply __56__ by __25__.

Multiply by the ones. Add any extra tens.	Multiply by the tens. Add any extra tens.	Add the products.
3 56 × 25 280 ← 5 × 56	1 56 × 25 280 1120 ← 20 × 56	56 × 25 280 + 1,120 1,400 ← 25 × 56

There are __1,400__ pounds of sugar in the display.

Getting Started

Multiply.

1. 36 × 24 = 864
2. 23 × 42 = 966
3. 39 × 56 = 2,184
4. 74 × 38 = 2,812

Copy and multiply.

5. 86 × 58 — 4,988
6. 73 × 66 — 4,818
7. 48 × 93 — 4,464
8. 47 × 47 — 2,209

2 Teach

Introduce the Lesson Have a student read the problem aloud and tell what is being asked. (to find the total number of pounds of sugar)

- Ask students what information is needed to answer the problem. (the number of bags and the number of pounds per bag) Ask students where this information is given. (The sign says fifty-six 25-pound bags is the capacity.)
- Have students read with you as they complete the sentences. Guide students through the multiplication steps in the model to solve the problem. Tell students that they can estimate the correctness of their answer by finding the total number of pounds in 60 bags of sugar.

Develop Skills and Concepts Write **43 × 6** on the board and have a student talk through the regrouping and find the product. (258)

- Now, write **43 × 56** on the board and have the same student multiply the six 43s, make the regrouping, and write the product. (258) Remind students that the 5 in the tens place means 5 tens or 50, and fifty 43s equals 2,150 as you write 2,150 under 258. Tell students that we add the product of six 43s and the product of fifty 43s to find that fifty-six 43s equals 2,408 as you write the answer.
- Repeat for 32 × 9 and 32 × 69. (288; 2,208)

3 Practice

Remind students to think of the bottom factor as a multiple of 10 plus ones. Have students complete page 178 independently.

Practice

Multiply.

1. 26 × 34 = 884
2. 42 × 29 = 1,218
3. 85 × 37 = 3,145
4. 67 × 26 = 1,742
5. 73 × 48 = 3,504
6. 96 × 53 = 5,088
7. 59 × 74 = 4,366
8. 77 × 23 = 1,771
9. 89 × 76 = 6,764
10. 45 × 67 = 3,015
11. 28 × 98 = 2,744
12. 56 × 39 = 2,184
13. 35 × 35 = 1,225
14. 82 × 64 = 5,248
15. 19 × 56 = 1,064
16. 97 × 48 = 4,656

Copy and multiply.

17. 29 × 46 1,334
18. 85 × 37 3,145
19. 52 × 74 3,848
20. 65 × 98 6,370
21. 57 × 58 3,306
22. 26 × 76 1,976
23. 87 × 38 3,306
24. 79 × 89 7,031

Problem Solving

Solve each problem.

25. Each package of trivia cards contains 75 questions. There are 25 packages in the game. How many trivia questions are there in the game?
1,875 questions

26. There are 28 chairs in each classroom in school. The school has 15 classrooms. How many classroom chairs are in the school?
420 chairs

27. Eric rode his bicycle 26 miles. He still has 17 miles to go. How far will Eric ride his bike?
43 miles

28. There are 96 oranges in each crate. Casey put 38 crates in inventory. How many oranges are in inventory?
3,648 oranges

4 Assess

Ask students to solve 88 × 66 and show the regrouping.
(5,808)

5 Mixed Review

1. 39 ÷ 7 (5 R4)
2. $468.27 + $693.84 ($1,162.11)
3. 362 + 507 + 221 (1,090)
4. 576 × 8 (4,608)
5. 1,875 − 658 (1,217)
6. 76 ÷ 6 (12 R4)
7. $19.03 − $7.98 ($11.05)
8. $8.42 × 7 ($58.94)
9. 46 − 3 × 9 (19)
10. 3 × 958 (2,874)

For Mixed Abilities

Common Errors • Intervention

Some students may confuse the regrouped digits from finding the first partial product with the regrouped digits from finding the second partial product. To avoid such errors, have them cross out the first set of regrouped digits when they have finished multiplying by the ones digit.

Enrichment • Measurement

Tell students to look up the world record for the long jump, round the number to the nearest foot, and convert that measurement to inches.

More to Explore • Numeration

Have students use their table of Egyptian hieroglyphics made for the More to Explore on page T164 or put a chart of the hieroglyphics for the numbers 1 through 10, 100, 1,000, and 10,000 on the board. Explain that ancient Egyptians used hieroglyphics as a tallying system.

Have students write out the hieroglyphics for 59; 882; 4,205; and 552.

9-9 Multiplying 3-Digit by 2-Digit Numbers With Regrouping

pages 179–180

1 Getting Started

Objective
- To multiply 3-digit numbers by 2-digit numbers with regrouping

Warm Up • Mental Math
Have students complete the comparison: 49 is to 7 as 14 is to 2 as

1. 140 is to (20)
2. 7 is to (1)
3. \$3.50 is to (\$0.50)
4. (42) is to 6
5. 210 is to (30)
6. (84) is to 12
7. \$1.75 is to (\$0.25)
8. 56 is to (8)

Warm Up • Multiplication
Write several 2-digit-by-2-digit multiplication problems on the board. Have students work in pairs at the board to solve the problems and check their solutions.

Name ______________________

Multiplying 3-Digit Numbers by 2-Digit Numbers With Regrouping

Marcia is the hostess at the Village Lunch Shop. Her regular work week is 35 hours. How much does Marcia earn each week?

Village Lunch Shop Wages	
Job	Pay
Host/hostess	\$5.85/hour
Cook	\$6.27/hour
Busperson	\$4.85/hour
Dishwasher	\$5.25/hour

We want to know Marcia's weekly wage.

We know Marcia works 35 hours each week and earns \$5.85 each hour.

To find her total weekly wages, we multiply \$5.85 by 35.

Multiply by the ones. Add any extra dimes and dollars.	Multiply by the tens. Add any extra dimes and dollars.	Add the products. Place the dollar sign and decimal point.
4 2 \$5.85 × 35 2925 ← 5 × 585	2 1 \$5.85 × 35 2925 17550 ← 30 × 585	\$5.85 × 35 29 25 + 175 50 \$204.75

Marcia earns \$204.75 each week.

Getting Started

Multiply.

1. \$3.65 × 28 = \$102.20
2. 575 × 47 = 27,025
3. \$8.09 × 53 = \$428.77
4. 639 × 82 = 52,398

Copy and multiply.

5. \$0.85 × 39 = \$33.15
6. \$7.00 × 56 = \$392.00
7. 825 × 78 = 64,350
8. 960 × 62 = 59,520

2 Teach

Introduce the Lesson Have a student read the sign. Ask another student to read the problem and tell what is being asked. (how much Marcia earns each week) Ask students what information is needed to solve the problem. (the number of hours she works per week and her pay per hour) Ask what information will be used from the problem and the sign. (35 hours per week at \$5.85 per hour)

- Have students read and complete the sentences. Guide students through the multiplication steps in the model to solve the problem. Ask students how they might check the solution. (round the pay to \$6.00 or the hours to 40 and multiply)

Develop Skills and Concepts Write **286 × 2, 286 × 23,** and **\$2.86 × 23** in vertical form on the board. Have a student work the first problem, talking through each regrouping. Review with students how the two 200s are multiplied.

- Have a student work the second problem and talk through each step as 3 times 286 is found and 20 times 286 is found.
- Ask students how the third problem differs from the second. (A dollar sign and decimal point are added.) Have a student talk through each step, work the problem, and place the dollar sign and decimal point in the product.
- Repeat for 168 × 7, 168 × 97, and \$1.68 × 97. (1,176; 16,296; \$162.96)

3 Practice

Remind students to place a decimal point and a dollar sign in the product of each money problem. Tell students to use the information on page 179 to solve the word problems. Have students complete page 180.

Practice

Multiply.

1. 216 × 27 = 5,832	2. $9.45 × 46 = $434.70	3. $8.73 × 39 = $340.47	4. $0.46 × 78 = $35.88
5. 915 × 67 = 61,305	6. 707 × 54 = 38,178	7. $6.38 × 85 = $542.30	8. 784 × 96 = 75,264
9. $3.90 × 59 = $230.10	10. 458 × 66 = 30,228	11. 823 × 92 = 75,716	12. $5.85 × 78 = $456.30
13. 867 × 84 = 72,828	14. $6.81 × 49 = $333.69	15. $9.06 × 37 = $335.22	16. 738 × 29 = 21,402

Copy and multiply.

17. 47 × 368 17,296
18. $5.47 × 34 $185.98
19. 296 × 88 26,048
20. 56 × $8.28 $463.68
21. $9.25 × 67 $619.75
22. 96 × 428 41,088
23. 72 × $7.93 $570.96
24. 408 × 27 11,016

Problem Solving

Solve each problem. Use the chart on page 179.

25. Winston cooked for the Village Lunch Shop for 4 weeks. He worked 8 hours a day for 5 days each week. What did he earn during that time?
$1,003.20

26. Richard worked as a dishwasher for 40 hours. How much more did he earn than the busperson who worked 40 hours?
$16

27. Penny washed dishes at the Village Lunch Shop for 35 hours. How much did she earn?
$183.75

28. Lisa works 32 hours each week as a cook and 10 hours as a hostess. How much will Lisa earn in 3 weeks?
$777.42

4 Assess

Ask students to solve the following problem and show the multiplication: $4.84 × 34. ($164.56)

5 Mixed Review

1. 72,176 + 38,934 (111,110)
2. $781.27 − $350.89 ($430.38)
3. 473 × 40 (18,920)
4. _____ × 5 = 45 (9)
5. 795 − 432 (363)
6. (8 × 9) + (5 × 7) (107)
7. 2,567 × 7 (17,969)
8. 342 + 178 + 456 (976)
9. 91 ÷ 7 (13)
10. 80 × 500 (40,000)

For Mixed Abilities

Common Errors • Intervention

Some students may have difficulty aligning the numbers properly and thereby make careless errors. To avoid such errors, have these students work their problems on grid paper.

Enrichment • Numeration

Ask students if (80 × 436) + (436 × 7) = (6 × 87) + (30 × 87) + (87 × 400). Have them explain why.

More to Explore • Probability

Explain that probabilities are frequently expressed as fractions. If there is one chance in two that something will happen, the probability is $\frac{1}{2}$. Point out that this fraction is the number of specific ways the event can happen over all possible ways.

Show students a coin. Explain that there is one way of getting a specific result, for example, flipping a head, and there are two possible ways the flip can turn out, heads or tails. The probability of flipping a head is $\frac{1}{2}$. Have students express the following probabilities as fractions:

1. drawing an ace from a deck of 52 cards ($\frac{4}{52}$ or $\frac{1}{13}$)
2. getting a six on one roll of the dice ($\frac{1}{6}$)
3. drawing a red cube out of a bag containing one red, one blue, and one yellow cube ($\frac{1}{3}$)

9-10 Using Estimation

pages 181–182

1 Getting Started

Objective
- To use estimation in multiplication

Vocabulary
leap year

Warm Up • Mental Math
Have students tell their cumulative bank balance if they first do the following:

1. deposit $26 ($26)
2. deposit $24.50 ($50.50)
3. write a check for $12.50 ($38)
4. write a check for $14.50 ($23.50)
5. earn interest of 7¢ ($23.57)
6. deposit $3.43 ($27)
7. write a check for $26.93 (7¢)
8. deposit $15.09 ($15.16)

Warm Up • Rounding
Write on the board random numbers up to 100,000. Have students round each number to the nearest ten, hundred, thousand, and ten thousand.

Name ______________________

Using Estimation

Estimation is used to see if an answer seems reasonable. How many days are in the 9 years shown here? The years 2004 and 2008 each have an extra day because they are leap years.

We want to find the number of days in the years 2001 through 2009.

There are __9__ years, and each year has __365__ days.

2004 and 2008 have __1__ extra day each because they are __leap years__.

To find the total number of days in these years, we multiply __365__ by __9__ and add __2__.

$$\begin{array}{r} 365 \\ \times \quad 9 \\ \hline 3,285 \end{array}$$

__3,285__ + __2__ = __3,287__

There are __3,287__ days in 2001 through 2009.

Use estimation to check if your answer seems reasonable.

365 is about 370.

$$\begin{array}{r} 370 \\ \times \quad 9 \\ \hline 3,330 \end{array}$$

The answer seems reasonable.

Getting Started

Copy and multiply. Use estimation to check your answers.

1. 68×39 = 2,652
2. 94×8 = 752
3. 27×58 = 1,566
4. 186 × 7 — 1,302
5. $4.37 × 55 — $240.35
6. 749 × 89 — 66,661

2 Teach

Introduce the Lesson Have students read the problem with you and tell what is to be answered. (the total number of days in the years 2001 through 2009)

- Tell students that a year is the time it takes Earth to revolve around the Sun and is slightly more than 365 days. Tell students that every 4 years, the extra bit of time over 365 days is made up by adding a day called a **leap day**. Thus, the year is called a **leap year**.
- Ask students what information is given. (There are 9 years all together and 2 of the years have 1 extra day each.) Remind students that when they round a number, they are estimating. Have students complete the sentences to solve the problem with you.

Develop Skills and Concepts Write **29 × 71** on the board. Have students find the product. (2,059) Tell students that we can check this answer by rounding one of the numbers to the nearest ten before multiplying.

- Have students tell the nearest multiple of 10 for 29. (30) Have a student write the problem on the board and find its product. (2,100) Ask students if the product of 2,059 is reasonable. (yes)
- Repeat for 182 × 64 and $1.49 × 18. (11,648; $26.82)

Practice

Multiply. Use estimation to check your answers.

1. 77×32 = 2,464
2. 57×89 = 5,073
3. 92×8 = 736
4. 476×67 = 31,892
5. $\$5.09 \times 62$ = $315.58
6. 895×82 = 73,390

Copy and multiply.

7. 83×7 581
8. 6×247 1,482
9. 39×81 3,159
10. 73×24 1,752
11. 75×87 6,525
12. 13×98 1,274
13. 9×576 5,184
14. $37 \times \$6.58$ $243.46

Problem Solving

Solve each problem. Use estimation to check your answers.

15. Brenda runs up the stadium steps each day. There are 89 steps in the stadium. How many steps does Brenda run in 38 days? 3,382 steps

16. It takes 4 cups of whole wheat flour to make 1 loaf of bread. The City Bakery bakes 673 loaves of bread each day. How many cups of flour are used each day? 2,692 cups

17. Juan has saved $136.46 to buy a color television set. The television set sells for $249.19. How much more does Juan need to save? $112.73

18. Paint sells for $7.79 a gallon. Mr. Jameson used 18 gallons of paint to paint his farm buildings. How much did the paint cost Mr. Jameson? $140.22

19. Annette gets 3¢ for each can she saves. One week Annette received 87¢. How many cans did she save? 29 cans

20. Pete and his sister have the same birthdate. When Pete is 12, his sister is 23. How many days older is Pete's sister? Do not add extra days for leap year. 4,015 days

3 Practice

Remind students to round each number to the nearest ten before multiplying to estimate the correctness of each product. Have students circle each problem in the first two groups where a dollar sign and decimal point will be needed in the answer. Tell students to disregard leap years in Exercise 20 because it asks about how many days rather than an exact number. Have students complete page 182.

4 Assess

Ask students to find the product for 78×29 and to use estimation to check their answer (2,262; $78 \times 30 = 2,340$).

For Mixed Abilities

Common Errors • Intervention

Some students may give estimates that are not reasonable because they round incorrectly. Have these students work in pairs with different 2- and 3-digit numbers. Have them round first to the nearest ten and then, when possible, to the nearest hundred. They should make and use a number line to show they have rounded correctly.

Enrichment • Measurement

Tell students to draw a number line to show why 1986 is in the twentieth century rather than the nineteenth. Then, tell them to find all the leap years in the last half of the twentieth century.

More to Explore • Measurement

Fill various small paper cups with different substances such as rice, salt, sugar, flour, and so on. Tell students to estimate the weight of each cup in ounces, record the estimates, and arrange the cups in order from heaviest to lightest. Then, have students find the actual weight of each cup to the nearest $\frac{1}{2}$ ounce and compare their estimate to the actual weight.

Have them make two bar graphs, one that shows their weight estimates and one that shows the actual weight for each cup. Discuss with students when estimating weight would be a valuable tool and when it would not.

ESL/ELL STRATEGIES

Discuss the meaning of the terms *estimate, estimation*, and *reasonable*. Explain that an estimate is a good guess, and that estimation is the process of finding an answer that you know is very close to the correct answer, a reasonable answer that makes sense.

9-11 Problem Solving: Try, Check, and Revise

pages 183–184

1 Getting Started

Objective

- To solve problems by trying, checking, and revising

Warm Up • Mental Math

Have students count 2s, 3s, 5s, 10s, or 25s to tell the total.

1. 11 rows of 10 seats (110)
2. 4 groups of 5 girls (20)
3. 9 quarters (225)
4. 7 boxes of 25 pencils (175)
5. socks in 12 pairs (24)
6. 9 groups of 10 boys (90)

Warm Up • Algebra

Have students copy and find the missing addend.

1. $n + 12 = 30$ (18)
2. $n + 8 = 29$ (21)
3. $n + 5 = 47$ (42)
4. $n + 3 = 92$ (89)
5. $n + 15 = 35$ (20)
6. $n + 25 = 78$ (53)

It's Algebra!

Name ______________________

Problem Solving: Try, Check, and Revise

Marcy is paying for lunch with only quarters and dimes. She has 28 coins for a total of $5.20. How many quarters does Marcy have? How many dimes does Marcy have?

SEE

Marcy has __28__ quarters and dimes. She has __$5.20__ to pay for lunch.

We need to find how many __quarters__ and __dimes__ Marcy has.

We know a quarter is worth __$0.25__ and a dime is worth __$0.10__.

PLAN

Try to guess the number of quarters and dimes.

Check your guess by finding the value of a number of coins and adding these values to see if the total is $5.20.

Revise your guess if it does not equal __$5.20__. Make a reasonable second guess. You can revise your guesses several times until you find the correct answer.

DO

First Try 14 quarters and 14 dimes

$14 \times \$0.25 =$ __$3.50__

$14 \times \$0.10 =$ __$1.40__

$\$3.50 + \$1.40 =$ __$4.90__ *Too low!*

$\$4.90 \; (<) \; \$5.20$

Second Try 16 quarters and 12 dimes

$16 \times \$0.25 =$ __$4.00__

$12 \times \$0.10 =$ __$1.20__

$\$4.00 + \$1.20 =$ __$5.20__ *Correct!*

$\$5.20 \; (=) \; \$5.20$

Marcy has __16__ quarters and __12__ dimes.

CHECK

__16__ quarters + __12__ dimes = __28__ coins

2 Teach

Introduce the Lesson Remind students of the four-step problem-solving strategy for solving problems. Explain that in this lesson they will make a reasonable guess first in order to solve the problem as quickly as possible. Explain that a good first guess can help them refine their guesses until they find the answer.

Develop Skills and Concepts Have students read the problem and tell what they are to find. (how many quarters and dimes Marcy has)

- Read each sentence under the SEE step, asking for volunteers to give the missing information as students fill in the answers.
- Read the Try and Check statements in the PLAN step. Ask students what the word *revise* means (to change) and then read the Revise section. Have students fill in the missing information.
- Put the exercise on the board as it appears in the DO section of the text. Have a student fill in the numbers as others suggest possible answers. Help students find more clues and complete the exercise together. Have students complete the solution sentence. Then, have students work through and complete the CHECK step.

3 Apply

Solution Notes

1. Suggest that students make a chart and start with calculated guesses.

$20	$50	Total
30	30	$2,100
40	20	$1,800
42	18	$1,740

Apply

Try, check, and revise to solve each problem.

1. An amusement park cashier counts only $20 bills and $50 bills. She has counted 60 bills so far. The bills have a value of $1,740. How many $20 bills has she counted? How many $50 bills has she counted?
forty-two $20 dollar bills and eighteen $50 dollar bills
2. There are 1,005 students who are bused to Carr Memorial Elementary School. The school rents 25 buses. The buses are all filled to capacity. Some hold 45 students and some hold 35 students. How many buses hold 45 students? How many buses hold 35 students?
13 buses hold 45 students and 12 buses hold 35 students.
3. A total of 1,290 students go to Carr Memorial Elementary School. Carr Memorial has classrooms that hold 25 students and 30 students. The school has 48 classrooms. How many classrooms hold 30 students? How many classrooms hold 25 people?
Eighteen classrooms hold 30 people and 30 classrooms hold 25 people.
4. A museum charges $18 for adults and $12 for children. Yesterday, the museum earned $1,362 from ticket sales. A total of 94 people bought tickets. How many children paid to visit the museum yesterday? How many adults paid to visit the museum?
55 children and 39 adults

Problem-Solving Strategy: Using the Four-Step Plan

* SEE What do you need to find?
* PLAN What do you need to do?
* DO Follow the plan.
* CHECK Does your answer make sense?

5. A tour group has rented 38 rooms at a motel on a lake. Some rooms cost $85 a night. Other rooms cost $65 a night. If the tour group spent $2,990, how many rooms were rented for $85 each? How many rooms were rented for $65 each?
Twenty-six rooms were rented for $85 and 12 rooms were rented for $65.
6. In June, Mary ran every day of the month. On some days she ran 15 miles and on other days she ran 12 miles. She ran a total of 417 miles in June. How many days did Mary run 15 miles? How many days did Mary run 12 miles?
Mary ran 15 miles on 19 days and 12 miles on 11 days.
7. Mickey was playing a game in which he gets 75 points for every game that he wins and 25 points for every game he loses. He has played 90 games and has a total of 4,750 points. How many games has he won? How many games has he lost?
Mickey has won 50 games and lost 40 games.

For Mixed Abilities

More to Explore • Application

Have students look through newspapers and magazines to find the prices of two different items rounded to the nearest dollar. Have students pick a number of total items and determine the cost of those items. Students may write their own try, check, and revise problem to share with other students.

2. Have students write an expression that they can use for their first guess. For example, $(15 \times 45) + (10 \times 35)$.
3. Have students write an expression such as the following to use for their first guess: $(24 \times 25) + (24 \times 30)$.
4. Suggest that students begin by writing different addition sentences with a sum of 94 and multiplying by $18 and $12.

Higher-Order Thinking Skills

5. **Analysis:** Have students divide $2,990 by 38 and determine if the quotient is closer to 65 or 85 (the quotient is 79 rounded to the nearest whole number, which is closer to 85). Then, ask them what that will indicate. (The number that it is closer to will have more of the rooms. There are more $85 rooms than $65 rooms.)

6. **Analysis:** Ask students what information they need to infer in order to make their first guess. (June has 30 days.)

7. **Evaluation:** Have students write the sentence they used to solve this problem. $(50 \times 75) + (40 \times 25) = 4{,}750$

4 Assess

Ask students to explain how to use the try, check, and revise strategy for solving problems. Have them give an example to show how to use it. (Answers will vary.)

Chapter 9 Test

page 185

Items	Objectives
1, 3, 4	To multiply 3-digit numbers by 1-digit numbers with regrouping (see pages 169–170)
2	To multiply 3-digit amounts of money by 1-digit numbers (see pages 167–168)
5–8, 22–23	To multiply 4-digit numbers or money by 1-digit numbers with or without regrouping (see pages 171–172)
9–12	To multiply two 2-digit numbers without regrouping (see pages 175–176)
13–16	To multiply two 2-digit numbers with regrouping (see pages 177–178)
17–21, 24–28	To multiply 3-digit numbers or money by 2-digit numbers (see pages 179–180)

Name ______________________

CHAPTER 9 TEST

Multiply.

1. $224 \times 3 = 672$
2. $\$3.43 \times 4 = \$13.72$
3. $679 \times 7 = 4{,}753$
4. $425 \times 9 = 3{,}825$
5. $3{,}208 \times 5 = 16{,}040$
6. $5{,}728 \times 2 = 11{,}456$
7. $\$38.51 \times 6 = \$231.06$
8. $7{,}286 \times 8 = 58{,}288$
9. $21 \times 32 = 672$
10. $43 \times 23 = 989$
11. $57 \times 11 = 627$
12. $72 \times 43 = 3{,}096$
13. $76 \times 29 = 2{,}204$
14. $59 \times 75 = 4{,}425$
15. $83 \times 46 = 3{,}818$
16. $67 \times 96 = 6{,}432$
17. $245 \times 36 = 8{,}820$
18. $\$7.93 \times 18 = \$142.74$
19. $425 \times 86 = 36{,}550$
20. $729 \times 79 = 57{,}591$
21. $\$6.37 \times 45 = \$286.65$
22. $4{,}759 \times 8 = 38{,}072$
23. $6{,}904 \times 9 = 62{,}136$
24. $\$5.94 \times 63 = \$374.22$
25. $748 \times 59 = 44{,}132$
26. $286 \times 43 = 12{,}298$
27. $\$8.99 \times 74 = \$665.26$
28. $987 \times 39 = 38{,}493$

Alternate Chapter Test

You may wish to use the Alternate Chapter Test on page 362 of this book for further review and assessment.

Circle the letter of the correct answer.

1. Round 723 to the nearest hundred.
   (a.) 700
   b. 800
   c. NG

2. Round 5,329 to the nearest thousand.
   (a.) 5,000
   b. 6,000
   c. NG

3. What is the place value of the 5 in 315,603?
   a. tens
   b. hundreds
   (c.) thousands
   d. NG

4. 4,279 + 3,651
   a. 7,820
   (b.) 7,930
   c. 8,930
   d. NG

5. 20,926 + 50,285
   a. 70,211
   b. 71,101
   (c.) 71,211
   d. NG

6. 703 − 246
   a. 447
   b. 543
   c. 557
   (d.) NG

7. 52,186 − 19,429
   a. 32,657
   (b.) 32,757
   c. 47,363
   d. NG

8. 63 ÷ 9
   (a.) 7
   b. 8
   c. 9
   d. NG

9. 72 ÷ 3
   a. 20
   b. 22
   (c.) 24
   d. NG

10. 4)93
   a. 2 R1
   b. 23
   (c.) 23 R1
   d. NG

11. Choose the best estimate of length.
   (a.) 7 cm
   b. 7 dm
   c. 7 m

12. Choose the best estimate of weight.
   a. 2 oz
   (b.) 2 lb

13. 4,032 × 3
   a. 1,296
   (b.) 12,096
   c. 12,396
   d. NG

14. 27 × 46
   a. 1,222
   b. 1,322
   c. 1,442
   (d.) NG

score

STOP

Cumulative Assessment

page 186

Items	Objectives
1–2	To round numbers to the nearest hundred or thousand (see pages 31–32)
3	To identify place value through hundred thousands (see pages 25–26)
4–5	To add two numbers up to 5 digits (see pages 41–48)
6–7	To subtract two numbers up to 5 digits (see pages 55–65)
8–10	To divide up to 2-digit numbers by 1-digit numbers with and without remainders (see pages 121–128)
11	To determine the appropriate metric unit of length (see pages 149–152)
12	To determine the appropriate customary unit of weight (see pages 147–148)
13–14	To multiply up to two 4-digit numbers with and without regrouping (see pages 165–180)

Alternate Cumulative Assessment

Circle the letter of the correct answer.

1. Round 621 to the nearest hundred.
   a 500
   (b) 600
   c 700
   d NG

2. Round 6,497 to the nearest thousand.
   a 5,000
   (b) 6,000
   c 7,000
   d NG

3. What is the place value of the 7 in 279,654?
   a tens
   b thousands
   (c) ten thousands
   d NG

4. 6,372 + 2,348
   a 8,610
   (b) 8,720
   c 9,720
   d NG

5. 62,438 + 33,675
   (a) 96,113
   b 95,014
   c 95,003
   d NG

6. 907 − 329
   (a) 578
   b 622
   c 588
   d NG

7. 37,243 − 18,376
   a 18,877
   (b) 18,867
   c 28,867
   d NG

8. 48 ÷ 6 =
   a 5
   b 7
   c 9
   (d) NG

9. 4)96
   a 2 R1
   b 21 R2
   (c) 24
   d NG

10. 3)85
   a 20 R5
   (b) 28 R1
   c 28
   d NG

11. What would you use to measure a candle?
   a mm
   (b) cm
   c m
   d NG

12. What would you use to measure a person's weight?
   a oz
   (b) lb
   c t
   d NG

13. 5,344 × 2
   a 7,566
   (b) 10,688
   c 11,688
   d NG

14. 36 × 46
   a 320
   b 2,656
   (c) 1,656
   d NG

10-1 Dividing, 3-Digit Quotients

pages 187–188

1 Getting Started

Objective
- To divide 3-digit numbers by 1-digit numbers, no remainders

Materials
*place-value blocks: hundreds, tens, ones

Warm Up • Mental Math
Dictate the following:
1. 20×12 (240)
2. 13×30 (390)
3. 100×50 (5,000)
4. 70×8 (560)
5. 30×40 (1,200)
6. $\$4 \times 50$ (\$200)
7. $\$10 \times 40$ (\$400)
8. 16×60 (960)

Warm Up • Division
Review 2-digit division by writing the following exercises on the board:

3)36 (12) 4)96 (24) 6)72 (12) 5)55 (11) 7)84 (12)

Have students work the exercises and then check their answers by multiplying the quotient by the divisor.

Name ______________________

Division of Whole Numbers

CHAPTER 10

Lesson 10-1

Dividing, 3-Digit Quotients

Sydney and Alicia have 254 invitations to address for a PTA dance. How many invitations will each girl have to address?

We want to know the number of invitations each girl will address.

Altogether, there are <u>254</u> invitations.

There are <u>2</u> girls doing the addressing.

To find the number of invitations each girl will address, we divide <u>254</u> by <u>2</u>.

Divide the hundreds.	Divide the tens.	Regroup the extra ten for 10 ones. Divide the ones.
2)254 → 1; 2; 0	2)254 → 12; 2↓; 5; 4; 1	2)254 → 127; 2; 5; 4↓; 14; 14; 0
Each girl will address at least 100 invitations.	Each girl will address at least 120 invitations.	Each girl will address 127 invitations.

Sydney and Alicia will each have to address <u>127</u> invitations.

Getting Started

Divide. Show your work.

1. 3)636 (212)
2. 4)456 (114)

Copy and divide. Show your work.

3. $392 \div 2$ (196)
4. $714 \div 6$ (119)

---

2 Teach

Introduce the Lesson Have students describe the picture. Have a student read the problem and tell what is to be found. (the number of invitations each girl will need to address) Ask students what information is given. (There are 254 invitations and 2 girls.) Ask what operation should be used. (division)

- Have students read and complete the sentences through each stage of the plan. Guide students through the division steps in the model. Tell students to multiply their solution by 2 to check their work.

Develop Skills and Concepts Write **426** on the board. Ask students how many hundreds, tens, and ones are in the number. (4, 2, 6)

- Display place-value blocks—hundreds, tens, and ones—to represent 426 and ask students to help you put the blocks in two equal groups by starting with the hundreds, then the tens, and finally the ones. Tell students that they are dividing 426 by 2. Ask students to name the number in each group. (213)
- Repeat for $316 \div 2$, where the extra hundred is regrouped for 10 tens and the extra ten is regrouped for 120 ones. (158) Continue for $816 \div 6$ and $645 \div 5$. (136; 129)

Practice

Divide. Show your work.

1. $3\overline{)366}$ 122
2. $2\overline{)648}$ 324
3. $5\overline{)555}$ 111
4. $4\overline{)844}$ 211
5. $3\overline{)651}$ 217
6. $4\overline{)856}$ 214
7. $2\overline{)658}$ 329
8. $6\overline{)672}$ 112
9. $4\overline{)968}$ 242
10. $7\overline{)791}$ 113
11. $3\overline{)759}$ 253
12. $2\overline{)938}$ 469

Copy and divide. Show your work.

13. $575 \div 5$ 115
14. $736 \div 4$ 184
15. $832 \div 2$ 416
16. $864 \div 3$ 288
17. $996 \div 2$ 498
18. $579 \div 3$ 193
19. $912 \div 4$ 228
20. $597 \div 3$ 199

Problem Solving

Solve each problem.

21. A computer printer printed 924 lines in 4 minutes. How many lines were printed each minute? 231 lines

22. Ivan helped his grandfather box his collection of 462 books that he will donate to 3 school libraries. How many books will each library receive? 154 books

23. There are 3 feet in 1 yard. Change 486 feet to yards. 162 yards

24. The pond is 165 yards wide. How wide is the pond in feet? 495 feet

3 Practice

Remind students that all exercises on the page require division. Ask students what information they will need to solve Exercise 24. (the number of feet in 1 yard) Tell students to check their answers by multiplying. Have students complete the page independently.

4 Assess

Ask students what place to start dividing a 3-digit number by a 1-digit number. (hundreds)

For Mixed Abilities

Common Errors • Intervention

Because they do not align digits carefully, some students often miss a digit in the dividend that should be brought down to get the correct quotient.

Incorrect	Correct
$2\overline{)352}$ = 16 2 12 12 0	$2\overline{)352}$ = 176 2↓ 15 14↓ 12 12 0

Have students work in pairs to model the problem using place-value blocks. In order to separate 3 hundreds, 5 tens, and 2 ones into two equal groups, students will have to regroup 1 hundred for 10 tens, giving them 15 ones. Then, they will regroup 1 ten, giving them 12 ones. Each equal group contains 1 hundred, 7 tens, and 6 ones, showing the quotient is 176.

Enrichment • Numeration

Tell students to show their work and explain why $\frac{1}{3}$ of 834 is equal to 2×139.

More to Explore • Statistics

Have students make two horizontal bar graphs to illustrate the difference between random numbers and nonrandom numbers.

Write the year of birth and the last four digits of the phone number of six students on the board. Ask students to graph the year each of the six students was born on one graph and the last four digits of the six students' telephone numbers on the other graph. When students have completed their graphs, display them and ask which graphs show random numbers (the graphs with telephone numbers) and which show nonrandom numbers (the graphs with birthdates).

Have students name two more sets of random and nonrandom numbers to graph and repeat the activity.

10-2 Dividing, Remainders

pages 189–190

1 Getting Started

Objective

- To divide 3-digit numbers by 1-digit numbers with remainders

Warm Up • Mental Math

Dictate the following:

1. $\frac{2}{6} + \frac{2}{6} + \frac{1}{6}$ $(\frac{5}{6})$
2. 0.1 + 0.4 + 0.3 (0.8)
3. 20 × 400 (8,000)
4. 11 + 22 + 33 (66)
5. (8 × 6) + (10 × 6) (108)
6. 7 × (48 ÷ 6) (56)
7. 5 lb 2 oz × 8 (41 lb)
8. 25¢ × 6 ($1.50)

Warm Up • Division

Review 2-digit division with remainders by writing the following exercises on the board:

$3\overline{)37}$ $7\overline{)93}$ $6\overline{)85}$ $7\overline{)83}$

(12 R1; 13 R2; 14 R1; 11 R6)

Have students work the exercises and then check their answers by multiplying the quotient by the divisor and adding the remainder.

Name ____________________

Lesson 10-2

Dividing, Remainders

The Quick Computer Company packs 6 CDs in a package. In one day, the company made 748 CDs. How many packages did the company fill? How many CDs were left over?

We want to find the number of packages of CDs produced and the number of CDs left over.

Quick Computer made __748__ CDs in 1 day.

There are __6__ CDs in a package.

To find the number of packages, we divide __748__ by __6__. The remainder is the number of CDs left over.

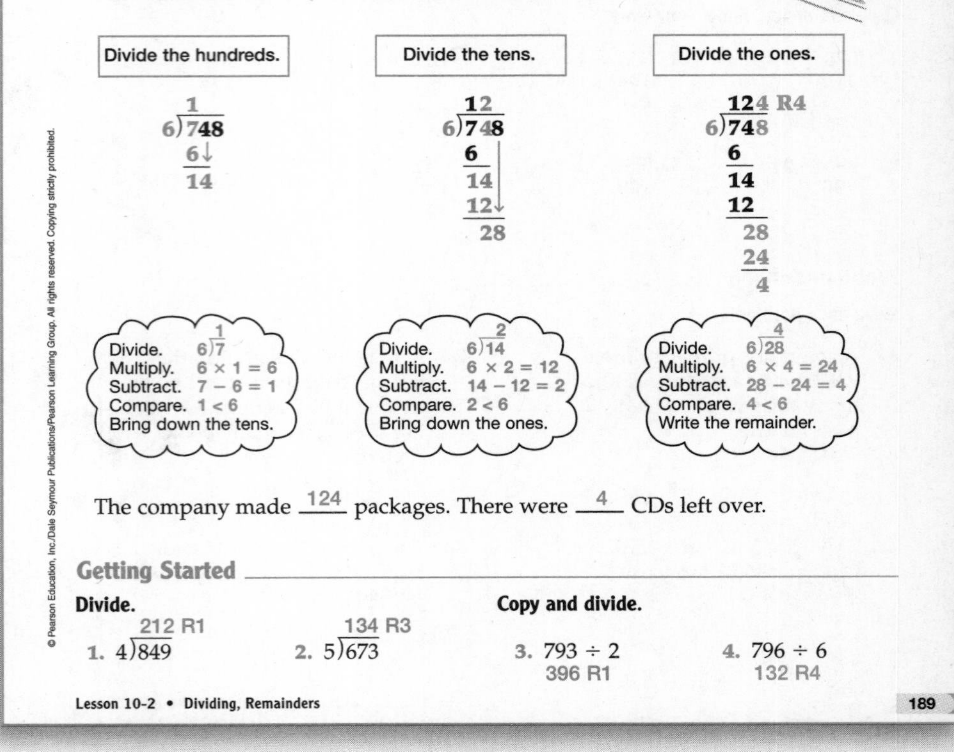

The company made __124__ packages. There were __4__ CDs left over.

Getting Started

Divide.

1. $4\overline{)849}$ 212 R1
2. $5\overline{)673}$ 134 R3

Copy and divide.

3. 793 ÷ 2 396 R1
4. 796 ÷ 6 132 R4

2 Teach

Introduce the Lesson Tell students that the picture shows a machine that sorts CDs. Have a student read the problem. Ask students what two questions are to be answered. (the number of packages of CDs and the number of CDs left over) Ask what information is needed. (the number of CDs in all and the number of CDs per package) Ask what information is given. (748 CDs in all and 6 per package)

- Have students read and complete the sentences. Guide students through the division steps in the model, pointing out the remainder. Have students check their answer by multiplying the divisor and the quotient and adding the remainder.

Develop Skills and Concepts Write $6\overline{)732}$ on the board and have a student work the problem, talking through each step. (122) Ask students if there is a remainder. (no)

- Now, write $6\overline{)735}$ on the board. Have students talk through each step as you work the problem. (122) Ask students if there are ones left over. (yes) Ask how many ones are left over. (3) Remind students that leftover ones in division are called the remainder as you write **R3** after the 122. Tell students that the leftover number must be less than the divisor.
- Repeat the procedure for $2\overline{)268}$, $2\overline{)269}$, $8\overline{)992}$, and $8\overline{)999}$. (134, 134 R1, 124, 124 R7)

Practice

Divide. Show your work.

1. $4\overline{)449}$ 112 R1
2. $6\overline{)682}$ 113 R4
3. $3\overline{)635}$ 211 R2
4. $2\overline{)869}$ 434 R1
5. $5\overline{)594}$ 118 R4
6. $4\overline{)925}$ 231 R1
7. $7\overline{)784}$ 112
8. $6\overline{)826}$ 137 R4
9. $8\overline{)916}$ 114 R4
10. $5\overline{)727}$ 145 R2
11. $2\overline{)916}$ 458
12. $7\overline{)919}$ 131 R2
13. $4\overline{)857}$ 214 R1
14. $6\overline{)885}$ 147 R3
15. $3\overline{)558}$ 186
16. $5\overline{)962}$ 192 R2

Copy and divide.

17. $437 \div 2$ 218 R1
18. $789 \div 4$ 197 R1
19. $896 \div 8$ 112
20. $416 \div 3$ 138 R2
21. $775 \div 6$ 129 R1
22. $815 \div 5$ 163
23. $593 \div 2$ 296 R1
24. $779 \div 7$ 111 R2
25. $956 \div 7$ 136 R4
26. $651 \div 3$ 217
27. $912 \div 5$ 182 R2
28. $852 \div 4$ 213

Problem Solving

Solve each problem.

29. Li is packing tomato plants into boxes that hold 5 plants each. Li has 598 plants. How many boxes will she need? How many plants are left over?
119 boxes; 3 plants left over

30. Mr. Hawthorne is putting 4 chairs at each table in his cafe. If Mr. Hawthorne has 462 chairs, how many tables can he supply? How many chairs are left over?
115 tables; 2 chairs left over

3 Practice

Tell students that some exercises on page 190 will not have remainders. For the word problems remind students to write how many plants and chairs are left over. Have students complete the page independently.

4 Assess

Ask students whether the remainder must be more or less than the divisor. (less than the divisor)

For Mixed Abilities

Common Errors • Intervention

Some students may give incorrect answers by forgetting to give the remainder as part of the answer. Show them how to check all answers using multiplication and addition. For example,

PROBLEM: $873 \div 6 = 145$ R3

CHECK: $(6 \times 145) + 3 = 873$

Students who continue to have difficulty with long division should work in small groups with place-value blocks, such as hundreds, tens, and ones, to model the problems.

Enrichment • Division

Have students write a 3-digit division problem that has a divisor of 6 and a remainder of 4. Suggest that they start with a 3-digit multiple of 6.

More to Explore • Applications

Ask students to calculate what their age would be if they lived on an imaginary planet that orbits the Sun once every 60 days. Then, ask how old they would be if the planet had an orbit of 1,095 days.

10-3 Dividing, Zeros in Quotients

pages 191–192

1 Getting Started

Objective

- To divide 3-digit numbers by 1-digit numbers for quotients with zeros

Warm Up • Mental Math

Ask if the number is divisible by 3.

1. 102 (yes)
2. 391 (no)
3. 475 (no)
4. 261 (yes)
5. 592 (no)
6. 375 (yes)
7. 87 (yes)
8. 799 (no)

Warm Up • Division

Review dividing a 3-digit number by a 1-digit number by having students work the following exercises on the board:

(212)	(239 R1)	(114 R3)
4)848	3)718	5)573
(122 R5)	(118 R3)	(326)
6)737	7)829	2)652

Name ______________________

Dividing, Zeros in Quotients

How long will it take for 424 quarts of blood to be pumped through the human heart?

We want to know the number of minutes it will take for the heart to pump 424 quarts of blood.

We know that 4 quarts are pumped each minute.

To find the length of time it takes to pump the blood, we divide 424 by 4.

Divide the hundreds.	Divide the tens. 2 < 4 so we write a 0 in the tens place.	Bring down the ones. Divide the ones.
$\begin{array}{r} 1 \\ 4\overline{)424} \\ 4 \\ \hline 0 \end{array}$	$\begin{array}{r} 10 \\ 4\overline{)424} \\ 4\downarrow \\ \hline 2 \end{array}$	$\begin{array}{r} 106 \\ 4\overline{)424} \\ 4\ \downarrow \\ \hline 24 \\ 24 \\ \hline 0 \end{array}$

It will take 106 minutes to pump 424 quarts of blood through the heart.

Getting Started

Divide. Show your work.

1. 7)735 — 105
2. 5)534 — 106 R4
3. 3)309 — 103
4. 2)803 — 401 R1

Copy and divide. Show your work.

5. 826 ÷ 8 — 103 R2
6. 816 ÷ 4 — 204
7. 600 ÷ 6 — 100
8. 948 ÷ 9 — 105 R3

2 Teach

Introduce Lesson Have a student read the problem aloud and tell what is to be solved. (the number of minutes it will take for 424 quarts of blood to be pumped through the human heart) Ask students if the information in the picture is needed to solve the problem. (yes) Have a student read the sign.

- Have students complete the sentences to solve the problem. Guide students through each step of the division. Tell students to multiply their answer by the number of quarts per minute to see if they get the total number of quarts.

Develop Skills and Concepts Write 3)627 on the board. Have students divide the hundreds and then tell if there are enough tens to be divided by 3. (no) Tell students that we then place a zero in the tens place in the quotient to show that each of the 3 groups gets no tens. Write **0** in the quotient.

- Tell students that we must regroup the 2 tens for 20 ones and then see how many ones there are in all. (27) Ask students to divide the 27 ones by 3 and tell the number. (9)
- Now, have students multiply the quotient by the divisor to check their answer.
- Repeat for 9)997 and 4)830, where the quotients have zero and a remainder. (110 R7, 207 R2)

Practice

Divide. Show your work.

1. $6\overline{)618}$ 103
2. $4\overline{)832}$ 208
3. $7\overline{)749}$ 107
4. $2\overline{)816}$ 408
5. $9\overline{)927}$ 103
6. $5\overline{)545}$ 109
7. $8\overline{)854}$ 106 R6
8. $9\overline{)906}$ 100 R6

Copy and divide. Show your work.

9. 436 ÷ 4 — 109
10. 200 ÷ 2 — 100
11. 709 ÷ 7 — 101 R2
12. 651 ÷ 6 — 108 R3
13. 837 ÷ 8 — 104 R5
14. 320 ÷ 3 — 106 R2
15. 529 ÷ 5 — 105 R4
16. 973 ÷ 9 — 108 R1

Problem Solving

Solve each problem.

17. How many gallons of water are there in 408 quarts? 102 gallons
18. Change 318 feet to yards. 106 yards
19. Change 36 feet to inches. 432 inches
20. How many pints are there in 216 quarts? 432 pints

Now Try This!

It's Algebra!

Study the grid. Complete each statement.

The gym is at (A, 2).

The office is at (B, 1).

The cafeteria is at (C, 2).

The library is at (B, 2).

The art room is at (C, 1).

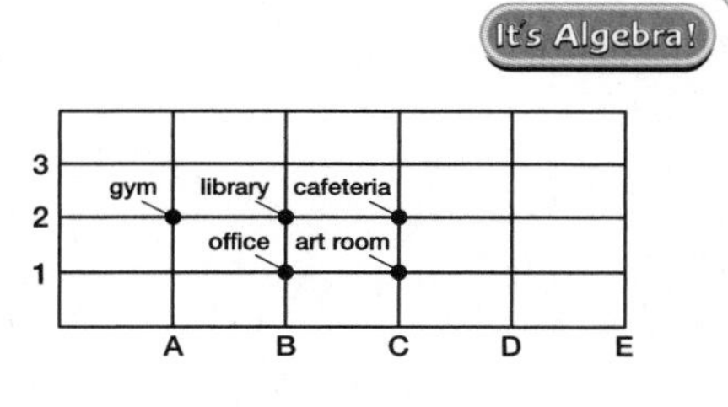

3 Practice

Remind students to record a zero in the quotient when there are not enough tens to divide. Tell students that they will need to supply some information for each of the word problems. Have students complete the page independently.

Now Try This! Review naming points by coordinates. Demonstrate proper written form.

4 Assess

Ask students what number needs to be written in the tens place of a quotient when there are not enough tens to divide. (0)

For Mixed Abilities

Common Errors • Intervention

When dividing with zeros in the quotient, some students may omit the zero or write it in an incorrect place.

Incorrect

$8\overline{)873}$ 19 R1 $8\overline{)873}$ 190 R1

Correct

$8\overline{)873}$ 109 R1

Have students use place-value blocks to model the problem. They should recognize that the 8 hundreds can be divided into 8 groups of 1 hundred each, but 7 tens cannot be divided into 8 groups of any number; thus a zero should be written in the tens place of the quotient to show this.

Enrichment • Application

Tell students to find the number of hours and minutes it would take for the human heart to pump 824 quarts of blood if 4 quarts are pumped each minute.

More to Explore • Probability

Have students calculate the probabilities for drawing a club from a pack of cards ($\frac{1}{4}$), for throwing a two on a single die ($\frac{1}{6}$), and for flipping a penny and getting a tail ($\frac{1}{2}$). Then, give each student a sheet of paper with three circles on it. Remind them that each circle can be divided into fractional parts like a pie. Have them show each of the following probabilities as a fraction of a circle.

of drawing a club

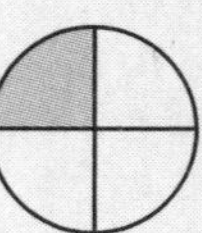

of throwing a two

of getting a tail

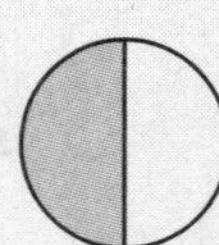

10-4 Dividing Larger Dividends

pages 193–194

1 Getting Started

Objective

- To divide 3-digit numbers by 1-digit numbers for 2-digit quotients

Warm Up • Mental Math

Have students name the decade number that is closest to the answer.

1. 82 ÷ 2 (40)
2. 27 × 3 (80)
3. (6 × 9) + 12 (70)
4. 69 ÷ 3 (20)
5. 25 × 3 (80)
6. (32 ÷ 8) + 7 (10)
7. $\frac{1}{2}$ of 96 (50)
8. 10 + 60 − 43 (30)

Warm Up • Division

Ask students to tell the numbers from 1 to 9 that can be divided by 3 evenly and those that have a remainder. Continue for numbers from 1 to 9 when divided by 2, 4, 6, 7, and 9.

Name ____________________

Dividing Larger Dividends

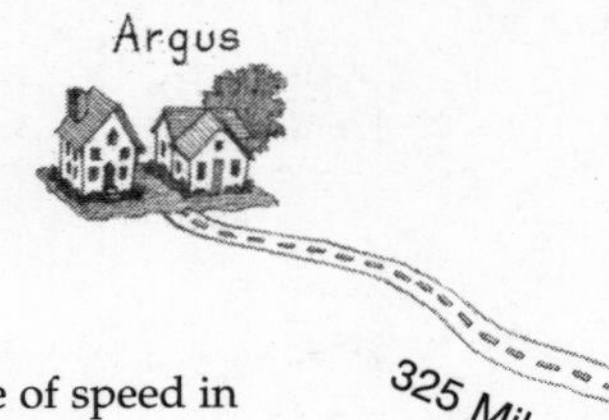

It took the Johnson family 6 hours to drive from Argus to Chester. What was their rate of speed in miles per hour?

We are looking for the Johnson's rate of speed in miles per hour.

The Johnsons traveled __325__ miles from Argus to Chester.

It took them __6__ hours to make this trip.

To find the rate of speed, we divide the total miles by the number of hours. We divide __325__ by __6__.

Divide the hundreds. 3 < 6 Regroup the 3 hundreds as 30 tens.	Divide the tens. Write the quotient digit above the tens.	Regroup the 2 tens as 20 ones. Divide the ones.	Check by multiplying. Add any remainders.
6)325	5 6)325 30 2	54 R1 6)325 30↓ 25 24 1	54 × 6 324 + 1 325

The Johnson family drove at a rate of speed of __54__ miles per hour.

Getting Started

Divide. Check your work.

1. 4)276 (69)
2. 8)433 (54 R1)

Copy and divide. Check your work.

3. 850 ÷ 9 (94 R4)
4. 249 ÷ 3 (83)

2 Teach

Introduce the Lesson Have students tell what can be learned from the pictured map. (Argus to Chester is 325 miles.) Have a student read the problem and tell what is to be solved. (the speed at which the Johnsons traveled in miles per hour) Ask what information would be used to find the speed. (the distance traveled and the number of hours they drove)

- Have students complete the sentences and work the division steps in the model with you. Have them complete the solution sentence.

Develop Skills and Concepts Write **4)424** on the board. Ask students if the 4 hundreds can be divided by 4. (yes) Erase the 4 hundreds and replace it with a 1. Ask students if the 1 hundred can be divided by 4. (no)

- Tell students that since 1 hundred cannot be divided by 4, they must regroup 1 hundred for 10 tens. Have students tell how many tens in all. (12) Ask students to divide the 12 tens by 4. (3) Write **3** in the tens place in the quotient and ask students if any tens are left over and need to be regrouped for ones. (no) Have students divide the ones (1) and tell if there is a remainder. (no) Have students check the solution by multiplying.
- Repeat the procedure for 9)549 and 8)709. (61; 88 R5)

Practice

Divide. Check your work.

1. $7\overline{)511}$ 73
2. $4\overline{)268}$ 67
3. $2\overline{)108}$ 54
4. $6\overline{)354}$ 59
5. $9\overline{)468}$ 52
6. $3\overline{)159}$ 53
7. $5\overline{)262}$ 52 R2
8. $8\overline{)290}$ 36 R2
9. $5\overline{)476}$ 95 R1
10. $2\overline{)137}$ 68 R1
11. $9\overline{)755}$ 83 R8
12. $4\overline{)317}$ 79 R1
13. $8\overline{)363}$ 45 R3
14. $3\overline{)209}$ 69 R2
15. $7\overline{)539}$ 77
16. $6\overline{)540}$ 90

Copy and divide. Check your work.

17. 627 ÷ 8 78 R3
18. 248 ÷ 5 49 R3
19. 615 ÷ 9 68 R3
20. 137 ÷ 2 68 R1
21. 316 ÷ 6 52 R4
22. 209 ÷ 7 29 R6
23. 312 ÷ 4 78
24. 115 ÷ 3 38 R1
25. 423 ÷ 7 60 R3
26. 196 ÷ 5 39 R1
27. 120 ÷ 3 40
28. 517 ÷ 6 86 R1

Problem Solving

Solve each problem.

29. It is 385 miles from Al's house to his grandparents' house. It takes Al's family 7 hours to drive there. What is their rate of speed in miles per hour? 55 miles per hour
30. The cafeteria serves 6-ounce glasses of milk. How many full glasses of milk can be poured from 225 ounces of milk? 37 glasses
31. How many pints are there in 10 gallons? 80 pints
32. It is 137 miles from Argus to Lincoln and 188 miles from Lincoln to Chester. How much farther is it from Lincoln to Chester than from Lincoln to Argus? 52 miles

3 Practice

Remind students that they will need to provide additional information in the word problems and that some may require operations other than division. Have students complete the page independently.

4 Assess

Ask students to describe how to solve $7\overline{)642}$. (Divide the hundreds. Regroup the 6 hundreds as 60 tens. Divide the tens. Write the quotient above the tens. [9] Regroup 1 ten as 10 ones. Divide the ones. [1] Write the remainder. [5])

For Mixed Abilities

Common Errors • Intervention

When the first digit in the dividend cannot be divided by the divisor, some students simply ignore it.

Incorrect

$$\begin{array}{r} 23 \\ 3\overline{)169} \end{array}$$

Have students use place-value blocks to model the problem. The first step is to regroup 1 hundred for 10 tens, giving a total of 16 tens. With such experiences, students soon will see the 16 tens right away without having to regroup or work with manipulatives.

Enrichment • Division

Have students tell how they know, without working the problem, that the quotient of $6\overline{)542}$ will be 2 digits. Tell them to write four such problems for a friend to solve and check.

More to Explore • Biography

Galileo Galilei, an Italian scientist of the sixteenth and seventeenth centuries, is best remembered for his work in astronomy. He invented the telescope and within one year had observed and written about sunspots, mountains on the Moon, Jupiter's moons, the phases of Venus, and the rings of Saturn.

However, Galileo also went against many popular views of the time. He theorized that objects fall at the same rate, no matter how much they weigh. Galileo also supported Copernicus's view that the Sun was the center of the solar system. The church officials of the time said this view went against the Holy Scriptures, and they threatened Galileo with torture if he refused to take back what he had said. Today, we know that Galileo was right about our solar system.

ESL/ELL STRATEGIES

Review the meaning of the word *dividend*. It is the number to be divided. When you divide 50 by 5, the dividend is 50.

10-5 Dividing Money

pages 195–196

1 Getting Started

Objective
- To divide money through $9.99 by 1-digit numbers

Warm Up • Mental Math
Have students complete the comparison:

A century is to 25 as
1. 12 is to (3)
2. 60 is to (15)
3. 40 is to (10)
4. 16 oz is to (4 oz)
5. 36 is to (9)
6. $1 is to (25¢)
7. 48 is to (12)
8. 56 is to (14)

Warm Up • Division
Write the following exercises on the board to have students review dividing a 3-digit number by a 1-digit number for a 2-digit quotient:

(78 R2)	(69 R4)	(93 R8)	(55)
8)626	6)418	9)845	5)275

Name ______________________

Dividing Money

At the print shop, Fred had 5 copies made of his science report. What was the cost of each copy?

We want to know the cost of making one copy of Fred's report.

Fred paid $2.45 to get 5 copies made.

To find the cost of one copy, we divide the cost of all the copies by the number of copies made.

We divide $2.45 by 5.

Divide the dollars. 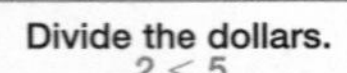 2 < 5 Write a zero above the 2. Write the decimal point above the decimal in the dividend.	Regroup the 2 dollars for 20 dimes. Divide the dimes.	Regroup the 4 leftover dimes for pennies. Divide the pennies. Write the dollar sign.
0. 5)$2.45	0.4 5)$2.45 2 0 4	$0.49 5)$2.45 2 0 45 45 0

One copy cost Fred $0.49.

Getting Started

Divide. Check your work.

1. 7)$3.36 ($0.48)
2. 3)$6.81 ($2.27)
3. 4)$1.80 ($0.45)

Copy and divide. Check your work.

4. $7.45 ÷ 5 ($1.49)
5. $4.74 ÷ 6 ($0.79)
6. $9.09 ÷ 9 ($1.01)

2 Teach

Introduce the Lesson Have a student read the problem aloud and tell what is to be found. (the cost of 1 copy) Ask students what information is given in the picture. (The bill reads 5 copies for $2.45.) Ask how many reports Fred had printed. (5)

- Have students read and complete the sentences to solve the problem. Guide them through the division steps in the model. Have students multiply the divisor and the quotient to check their answer.

Develop Skills and Concepts Write 4)$3.24 on the board. Tell students that this problem is like those they have been doing, but there is a dollar sign and a decimal point.

- Ask if there are enough dollars to divide by 4. (no) Tell students that in dollar amounts, we place a zero in the dollar place to show no dollars. Tell students that we then regroup $3 for dimes.
- Ask how many dimes are in $1. (10) Then, ask how many dimes there are in all. (32) Have students divide the dimes by 4, record the 8 in the tens place, and subtract the 32 from 32. (0)
- Ask students to divide the ones (1) and subtract the product of 4 from 4. (0) Tell students that the quotient of a money problem must have a dollar sign and decimal point. Have students read the quotient of $0.81. Ask a student to multiply the divisor and the quotient to check the answer.
- Repeat the procedure for 7)$3.05 and 9)$8.01. ($0.05; $0.09)

Practice

Divide. Check your work.

1. $4\overline{)\$5.16}$ $1.29
2. $8\overline{)\$3.36}$ $0.42
3. $7\overline{)\$8.26}$ $1.18
4. $3\overline{)\$3.09}$ $1.03
5. $9\overline{)\$5.76}$ $0.64
6. $2\overline{)\$7.58}$ $3.79
7. $5\overline{)\$9.35}$ $1.87
8. $6\overline{)\$2.88}$ $0.48
9. $7\overline{)\$7.14}$ $1.02
10. $6\overline{)\$4.62}$ $0.77
11. $8\overline{)\$9.44}$ $1.18
12. $3\overline{)\$9.18}$ $3.06

Copy and divide. Check your work.

13. $6.25 ÷ 5 $1.25
14. $4.24 ÷ 8 $0.53
15. $1.38 ÷ 2 $0.69
16. $8.13 ÷ 3 $2.71
17. $9.52 ÷ 7 $1.36
18. $1.56 ÷ 6 $0.26
19. $9.20 ÷ 4 $2.30
20. $7.65 ÷ 9 $0.85

Problem Solving

Solve each problem.

21. Rob bought 5 paperback books for $7.95. Each book cost the same amount. What was the cost of each book? $1.59
22. Two hundred eight quarts of milk are to be poured into gallon jugs. How many jugs will be needed? 52 jugs
23. Vince and Don purchased 3 birthday cards, each costing $1.20. They decided to split the cost evenly. How much did each boy pay? $1.80
24. A 6-pound package of hamburger costs $8.34. How much would a 4-pound package cost? $5.56

3 Practice

Remind students that when using a dollar sign to write an amount of money less than $1, we place a zero in the dollar place. Remind students that the word problems will have answers with dollar signs and decimal points and may require more than one operation. Have students complete the page independently.

4 Assess

Ask students what they must remember to write in the quotient when dividing money. (a dollar sign and a decimal point)

For Mixed Abilities

Common Errors • Intervention

Students may divide money and give an incorrect answer because they forget to include the dollar sign and decimal point. Once these students recognize that a problem involves amounts of money, have them write the dollar sign and decimal point in the quotient before they begin dividing.

Enrichment • Numeration

Have students cut from a newspaper an ad that offers several items for one price. Tell them to find the cost of one item to the nearest penny.

More to Explore • Statistics

Have students work in groups of two. Give each group a pair of number cubes and graph paper. Have them list all possible sums of the pair of number cubes (2 through 12) across the bottom of the graph paper. Ask them to roll the number cubes 50 times and indicate the sum of each roll on the graph paper by placing an X in the column above that sum.

When they have finished, post their graphs and have other students examine them. Ask someone to describe the shape of the graphs. Explain that this is called a **bell-shaped curve**. Ask students to explain why rolling number cubes produces a graph with this shape. If students have trouble explaining, have them list all the ways each sum, 2 through 12, can be formed with two number cubes.

10-6 Dividing by Multiples of 10

pages 197–198

1 Getting Started

Objective

- To divide 2- or 3-digit multiples of 10 by 2-digit multiples of 10

Warm Up • Mental Math

Ask students if there is a remainder.

1. $25 \div 3$ (yes)
2. $84 \div 7$ (no)
3. $666 \div 2$ (no)
4. $35 \div 6$ (yes)
5. $100 \div 4$ (no)
6. $68 \div 6$ (yes)
7. $38 \div 3$ (yes)
8. $72 \div 2$ (no)

Warm Up • Numeration

Remind students that when we multiply a number by a multiple of 10, we multiply the numbers that are not zero and follow that product with the number of zeros in the factors. Ask students to write and solve the following exercises as you dictate them:

$26 \times 20 =$ (520), $2 \times 400 =$ (800), $40 \times 50 =$ (2,000), $70 \times 9 =$ (630), $75 \times 20 =$ (1,500)

Name ______________________

Dividing by Multiples of 10

Michelle is making bows that use 20 centimeters of ribbon each. How many bows can Michelle make with one spool of ribbon?

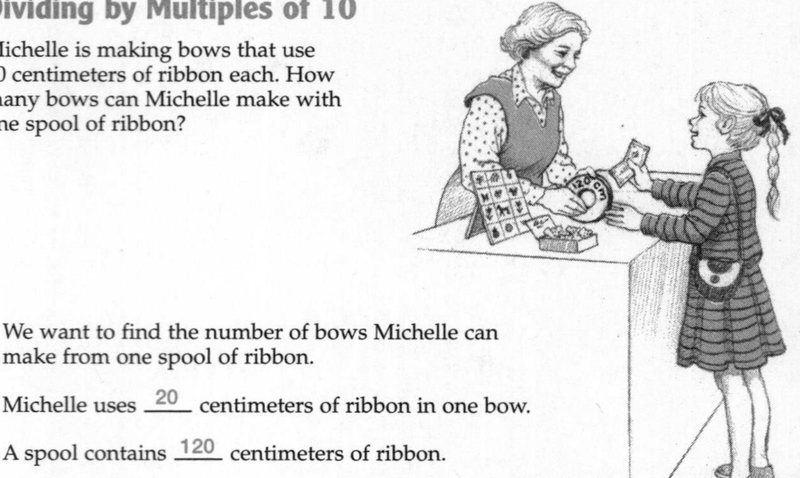

We want to find the number of bows Michelle can make from one spool of ribbon.

Michelle uses __20__ centimeters of ribbon in one bow.

A spool contains __120__ centimeters of ribbon.

To find the number of bows, we divide __120__ by __20__.

We know that $12 \div 2 = 6$. **We use this fact to find $120 \div 20$.**

$$\begin{array}{r} 6 \\ 20\overline{)120} \\ \underline{120} \\ 0 \end{array}$$

Michelle can make __6__ bows from one spool of ribbon.

Getting Started

Divide. Check your work.

1. $30\overline{)60}$ 2
2. $20\overline{)80}$ 4
3. $80\overline{)640}$ 8
4. $40\overline{)240}$ 6

Copy and divide. Check your work.

5. $60 \div 20$ 3
6. $90 \div 30$ 3
7. $320 \div 40$ 8
8. $360 \div 60$ 6

2 Teach

Introduce the Lesson Have a student read the problem aloud and describe the picture. Ask students what question is to be answered. (the number of bows that can be made from one spool of ribbon) Ask what information is given. (One spool holds 120 centimeters of ribbon and one bow uses 20 centimeters of ribbon.)

- Have students complete the sentences and work the division with you. Tell students to check their solution by multiplying the divisor and the quotient after filling in the solution.

Develop Skills and Concepts Write $80\overline{)240}$ on the board. Tell students that we know 2 is too small to divide by 80, and 24 is also too small, so we need to find out how many 80s are in 240.

- Tell students that we look at the 8 and think of how many 8s are in 24. (3) Remind students that they learned that $8 \times 3 = 24$, so $80 \times 3 = 240$. Therefore, there are three 8s in 24 and there are three 80s in 240. Write **3** in the ones place of the quotient. Ask students to multiply 80 by 3 to see if the answer is correct.
- Write $90\overline{)810}$ on the board and remind students that we ask ourselves how many 9s are in 81 to estimate the number of 90s in 810. Help students complete and check the exercise.
- Continue having students use partial divisors to work $80\overline{)720}$ and $90\overline{)180}$. (9; 2)

Practice

Divide. Check your work.

1. $20\overline{)60}$ 3
2. $30\overline{)60}$ 2
3. $40\overline{)80}$ 2
4. $30\overline{)90}$ 3
5. $60\overline{)60}$ 1
6. $50\overline{)150}$ 3
7. $70\overline{)490}$ 7
8. $60\overline{)300}$ 5
9. $80\overline{)240}$ 3
10. $90\overline{)270}$ 3
11. $40\overline{)360}$ 9
12. $20\overline{)160}$ 8
13. $70\overline{)490}$ 7
14. $90\overline{)270}$ 3
15. $80\overline{)400}$ 5
16. $30\overline{)150}$ 5
17. $60\overline{)480}$ 8
18. $50\overline{)200}$ 4
19. $60\overline{)240}$ 4
20. $90\overline{)540}$ 6
21. $20\overline{)100}$ 5
22. $40\overline{)320}$ 8
23. $30\overline{)210}$ 7
24. $70\overline{)140}$ 2
25. $80\overline{)240}$ 3

Copy and divide. Check your work.

26. $270 \div 30$ 9
27. $450 \div 50$ 9
28. $420 \div 60$ 7
29. $560 \div 80$ 7
30. $200 \div 40$ 5
31. $810 \div 90$ 9

Problem Solving

Solve each problem.

32. There are 30 pounds of potatoes in each sack. How many sacks of potatoes will be needed to fill an order of 180 pounds? 6 sacks

33. There are 20 small glasses of juice in a quart. How many glasses can be made from 80 quarts? 1,600 glasses

3 Practice

Remind students to use partial dividends to estimate the quotients. Have students complete the page independently.

4 Assess

Ask students what is helpful to know when dividing by multiples of 10. (basic division facts)

For Mixed Abilities

Common Errors • Intervention

Students may have difficulty visually identifying the basic fact involved when they are dividing multiples of 10. Before they divide, ask students to circle the two numbers, or digits, in the problem that are part of a basic fact. For example, the 3 and the 6 would be circled in $30\overline{)60}$ as part of the basic fact $6 \div 3 = 2$.

Enrichment • Multiples

Tell students to make a table showing the products of multiples of 10, through 90 times the multiples of 10.

More to Explore • Applications

Have students use an almanac to complete the following activity.

Divide students into groups. Tell each group to write any 5-digit number on paper. Tell students that they are the navigators of a large passenger plane that will travel to five cities and then return home.

Tell students that the number they wrote is their present odometer reading. Tell them that they must plan a route to include any five cities in the world, outside of the United States. Then, have them look up the distance from their starting point to the first destination, chart the mileage, and add it to their odometer. Have them continue this as they "travel" to each city, adding each mileage. After they have completed their trip by returning home, have them display their data charts.

10-7 Dividing by Multiples of 10, Remainders

pages 199–200

1 Getting Started

Objective

- To divide 2- or 3-digit numbers by 2-digit multiples of 10 with remainders

Warm Up • Mental Math

Have students name a unit of measure for the following:

1. metric length (cm, dm, m, km)
2. customary capacity (c, pt, qt, gal)
3. metric capacity (L, mL)
4. customary length (in., ft, yd, mi)
5. customary weight (oz, lb, T)
6. metric weight (g, kg)
7. metric length less than dm (cm)

Warm Up • Division

Ask students to write and solve the division problem that tells the number of 40s in 280. ($280 \div 40 = 7$)

Dictate the following for more practice in writing and solving problems: 50s in 850, 70s in 840, 20s in 900, 60s in 720, and 40s in 960. (17, 12, 45, 12, 24)

Name ______________________ **Lesson 10-7**

Dividing by Multiples of 10, Remainders

Pam counted 625 pennies in her savings. How many penny wrappers will she use to package them for the bank? How many pennies will she have left over?

We want to find the number of penny wrappers that Pam needs.

She has <u>625</u> pennies in her savings.

A penny wrapper will hold <u>50</u> pennies.

To find the number of penny wrappers needed, we divide the total number of pennies by the number of pennies in one wrapper.

We divide <u>625</u> by <u>50</u>.

How many 50s are in 6? How many 50s are in 62?

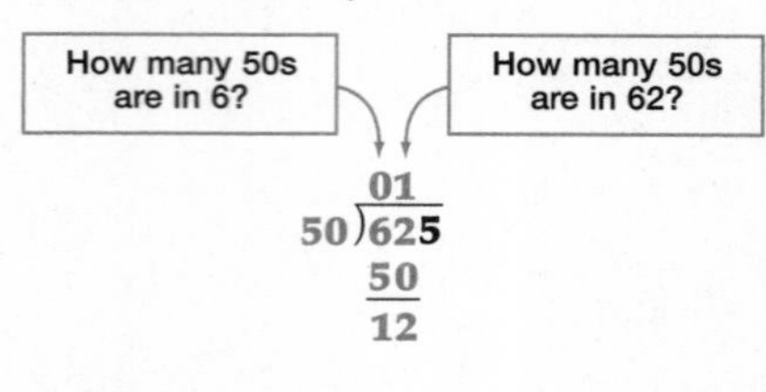

How many 50s are in 125?

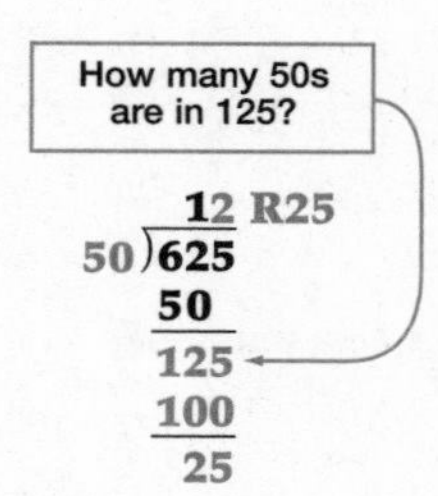

Pam needs <u>12</u> wrappers. She will have <u>25</u> pennies left over.

Getting Started

Divide. Check your work.

1. $10\overline{)38}$ 3 R8
2. $40\overline{)725}$ 18 R5
3. $70\overline{)90}$ 1 R20
4. $20\overline{)738}$ 36 R18

Copy and divide. Check your work.

5. $87 \div 40$ 2 R7
6. $906 \div 70$ 12 R66
7. $67 \div 50$ 1 R17
8. $959 \div 90$ 10 R59

2 Teach

Introduce the Lesson Have students tell about the picture. Discuss coin wrappers and reasons for using them. (ease in handling, counting, and shipping quantities) Have a student read the problem and tell what is to be solved. (the number of penny wrappers Pam will use and the number of pennies left over) Ask students what information is given. (Pam has 625 pennies and one wrapper holds 50 pennies.)

- Have students read and complete the sentences to solve the problem. Guide them through the division steps in the model. Tell students to multiply the divisor and the quotient and add the remainder to check their solution.

Develop Skills and Concepts Write $40\overline{)715}$ on the board. Ask students how many 40s are in 7. (0) Tell students that we can think of zero in the hundreds place in the quotient.

- Ask how many 40s are in 71. (1) Write 1 in the tens place and tell students to subtract the product of 1×40 from 71 to see how many tens are left. (31) Tell students to think of 31 tens as 310 ones and to tell how many ones there are in all. (315)
- Ask how many 40s are in 315. (7) Have students tell the product of 7×40 as you write a 7 in the ones place. (280) Ask the number of ones left over. (35) Have a student write the remainder in the answer. Have a student check by multiplying the divisor and the quotient and then adding the remainder.
- Repeat for $50\overline{)726}$ and $30\overline{)504}$. (14 R26; 16 R24)

Practice

Divide. Check your work.

1. $30\overline{)68}$ 2 R8
2. $40\overline{)75}$ 1 R35
3. $10\overline{)83}$ 8 R3
4. $20\overline{)96}$ 4 R16
5. $50\overline{)600}$ 12
6. $80\overline{)960}$ 12
7. $70\overline{)824}$ 11 R54
8. $60\overline{)730}$ 12 R10
9. $80\overline{)325}$ 4 R5
10. $40\overline{)765}$ 19 R5
11. $20\overline{)708}$ 35 R8
12. $50\overline{)436}$ 8 R36
13. $70\overline{)842}$ 12 R2
14. $90\overline{)425}$ 4 R65
15. $60\overline{)375}$ 6 R15
16. $30\overline{)852}$ 28 R12

Copy and divide. Check your work.

17. 800 ÷ 40
20
18. 735 ÷ 60
12 R15
19. 867 ÷ 30
28 R27
20. 428 ÷ 70
6 R8
21. 650 ÷ 50
13
22. 329 ÷ 80
4 R9

Problem Solving

Solve each problem.

23. There are 40 paper clips to a box. How many boxes will you buy if you need 360 paper clips for your project?
9 boxes
24. A container holds 80 milliliters of juice. How many containers are needed to hold 480 milliliters of juice?
6 containers
25. Robbie worked 40 hours a week on his summer job. His paycheck for 1 week amounted to $240. How much did Robbie make per hour?
$6
26. There are 50 clothespins to a box. Each box costs $3.75. How much will 350 clothespins cost?
$26.25

3 Practice

Tell students to watch for remainders. Remind them that they may need more than one operation in the word problems. Have students complete the page independently.

4 Assess

Ask students to find the quotient for $80\overline{)874}$ (10 R74)

For Mixed Abilities

Common Errors • Intervention

Some students have difficulty estimating the quotient when dividing by multiples of 10. Have those students write a basic-fact form to help them estimate the quotient as shown below.

PROBLEM: $60\overline{)251}$

BASIC FACT: 6 tens$\overline{)24 \text{ tens}}$

The quotient is about 4.

Enrichment • Money

Tell students to find the total amount of money they would have if they had one full wrapper each of pennies, nickels, dimes, and quarters.

More to Explore • Probability

Explain that the probabilities students have talked about to this point have been even probabilities. For example, the chance of getting heads when they flip a coin is the same as the chance of getting tails, one-half.

Give each pair of students a flat-headed thumbtack in a small cup. Tell them to shake the cup and check each time whether the tack lands point up or point down. Have them shake the cup 50 times and record their results. Ask the class if it was more likely for the tack to land point up or point down. (point up) Ask students to offer explanations. (Point up was most likely because the head of the tack is heavier.) Tell them that this is an example of an unequal probability.

10-8 Dividing, 2-Digit Divisors

pages 201–202

1 Getting Started

Objective

- To divide 3-digit numbers by 2-digit numbers for 1-digit quotients with and without remainders

Warm Up • Mental Math

Ask students to tell which is more.

1. 40 × 40 or 180 ÷ 1 (40 × 40)
2. 8 × 700 or 9 × 600 (8 × 700)
3. $\frac{1}{2}$ of 320 or 244 ÷ 2 ($\frac{1}{2}$ of 320)
4. 8 lb or 9 × 16 oz (9 × 16 oz)
5. 42 dimes or 16 quarters (42 dimes)
6. 8 gal or 33 qt (33 qt)
7. (6 × 30) + 18 or 3 × 70 (3 × 70)

Warm Up • Numeration

To help students see more complicated division problems as multiples of 10 divided by multiples of 10, have students rewrite each of the following exercises by rounding both numbers to the nearest ten: 94)789 (90, 790), 66)118 (70, 120), 56)478 (60, 480), 61)127 (60, 130), and 78)326 (80, 330).

Name ____________________

Lesson 10-8

Dividing, 2-Digit Divisors

The Valdez family will be traveling 289 miles to visit relatives. About how many gallons of gas will the Valdez family use on the trip?

We want to find the number of gallons of gas the Valdez family will use on the trip.

The Valdez family is driving 289 miles.

The family car gets 32 miles on each gallon of gas. To find the number of gallons needed, we divide the distance by the number of miles per gallon. We divide 289 by 32.

Round the divisor to the nearest ten. Estimate how many 30s are in 289.

32)289 (30)289 → 9)

Because 289 ÷ 30 is about 9, use 9 as the digit in the quotient.

```
   9 R1
32)289
   288   ← 9 × 32
     1
```

Check a division problem by multiplying and then adding if there is a remainder.

```
  32    divisor
×  9    × quotient
 288
+  1    + remainder
 289    dividend
```

The Valdez family will use about 9 gallons of gas on the trip.

Getting Started

Divide. Check your work.

1. 22)176 (8)
2. 49)325 (6 R31)

Copy and divide. Check your work.

3. 225 ÷ 31 (7 R8)
4. 142 ÷ 18 (7 R16)

2 Teach

Introduce the Lesson Have students tell about the picture. Have students tell what information is given in the operator's manual. Ask a student to read the problem and tell what is to be solved. (the approximate number of gallons of gas to be used on a trip)

- Help students decide which information will be used. (The trip is 289 miles and the car gets about 32 miles per gallon.) Have students read and complete the sentences to solve the problem. Guide them through the rounding and division steps in the model.

Develop Skills and Concepts Write **59)486** on the board. Tell students that we want to see how many 59s are in 486 and we know neither 4 nor 48 is divisible by 59. Tell students that we then must find a number that multiplied by 59 will be about 486.

- Tell students that we can think of 59 as rounded to 60 and think of how many 60s are in 486. Remind students that they know that eight 60s are 480 and 480 is almost 486, so 8 would be a good estimate.
- Write **8** in the quotient of 59)486 and have a student multiply 8 × 59 to see if the estimate is about right. (472) Have a student subtract to see how many are left. (14) Tell students that because 14 is less than 59, it is the remainder.
- Repeat for 92)786 and 46)187. (8 R50; 4 R3)

3 Practice

Remind students to think of the divisors as rounded to the nearest ten. Tell students to be sure to record any remainders. Have students complete page 202 independently.

Practice

Divide. Check your work.

1. $31\overline{)162}$ 5 R7
2. $38\overline{)250}$ 6 R22
3. $42\overline{)298}$ 7 R4
4. $72\overline{)598}$ 8 R22
5. $37\overline{)285}$ 7 R26
6. $28\overline{)186}$ 6 R18
7. $51\overline{)268}$ 5 R13
8. $19\overline{)165}$ 8 R13
9. $89\overline{)812}$ 9 R11

Copy and divide. Check your work.

10. 268 ÷ 43 — 6 R10
11. 208 ÷ 45 — 4 R28
12. 846 ÷ 92 — 9 R18
13. 198 ÷ 32 — 6 R6
14. 658 ÷ 73 — 9 R1
15. 575 ÷ 67 — 8 R39
16. 322 ÷ 46 — 7
17. 585 ÷ 68 — 8 R41
18. 148 ÷ 21 — 7 R1
19. 561 ÷ 58 — 9 R39
20. 389 ÷ 53 — 7 R18
21. 233 ÷ 32 — 7 R9
22. 248 ÷ 28 — 8 R24
23. 616 ÷ 83 — 7 R35
24. 321 ÷ 47 — 6 R39

Problem Solving

Solve each problem.

25. Miss Douglas worked 71 hours during one pay period. She made $568. How much does Miss Douglas earn in 1 hour? $8

26. Mr. Kowalski put 14 gallons of gas into his car. His car averages 37 miles to a gallon. About how far can Mr. Kowalski drive his car on this tank of gas? about 518 miles

4 Assess

Ask students how to check a division problem. (multiply and add if there is a remainder)

5 Mixed Review

1. 6,307 × 9 (56,763)
2. 837 ÷ 6 (139 R3)
3. 27,190 + 356 (27,546)
4. 85 × 408 (34,680)
5. 6,719 − 4,285 (2,434)
6. 471 ÷ 4 (117 R3)
7. 17 + 23 + 48 + 19 (107)
8. 4,818 − 865 (3,953)
9. 16 + 7 × 5 (51)
10. 2,000 × 5 (10,000)

For Mixed Abilities

Common Errors • Intervention

Some students may record the remainder as the last digit of the quotient.

Incorrect	Correct
$32\overline{)258}$ 82	$32\overline{)258}$ 8 R2

Have students use multiplication and addition to check their answers. (8 × 32) + 2 = 258

Enrichment • Division

Tell students to write three division problems of 3-digit dividends and 1-digit divisors where the quotient of each problem is 8 R17. Suggest that they start with some 3-digit multiples of 8.

More to Explore • Applications

Ask students to bring in several ads from newspapers and flyers from local grocery stores to help them find the unit prices of grocery items.

Using a calculator, show them how to divide the total price of an item by the number of units in the item to find the unit price.

Have students make up a shopping list for supplies for a class luncheon. One student can write it on the board as the others dictate.

Divide the class into small groups. Each group will need some ads, a calculator, and the shopping list. Ask groups to find the lowest unit prices of the items on the list and where they could be bought. Remind them that it is important that they use the correct unit of measure when making their calculations.

Ask students what else they learned about pricing, besides the unit price applications.

10-9 Estimating Quotients

pages 203–204

1 Getting Started

Objective
- To estimate quotients

Warm Up • Mental Math
Tell students to round to the nearest ten to estimate the following:

1. 683 + 124 (800)
2. 16 × 31 (600)
3. 89 ÷ 13 (9)
4. 67 × 22 (1,400)
5. 443 ÷ 16 (22)
6. 78 ÷ 19 (4)
7. 1,251 − 703 (550)
8. 346 × 21 (7,000)

Warm Up • Multiplication
Review multiplying 2-digit numbers by 1-digit numbers.

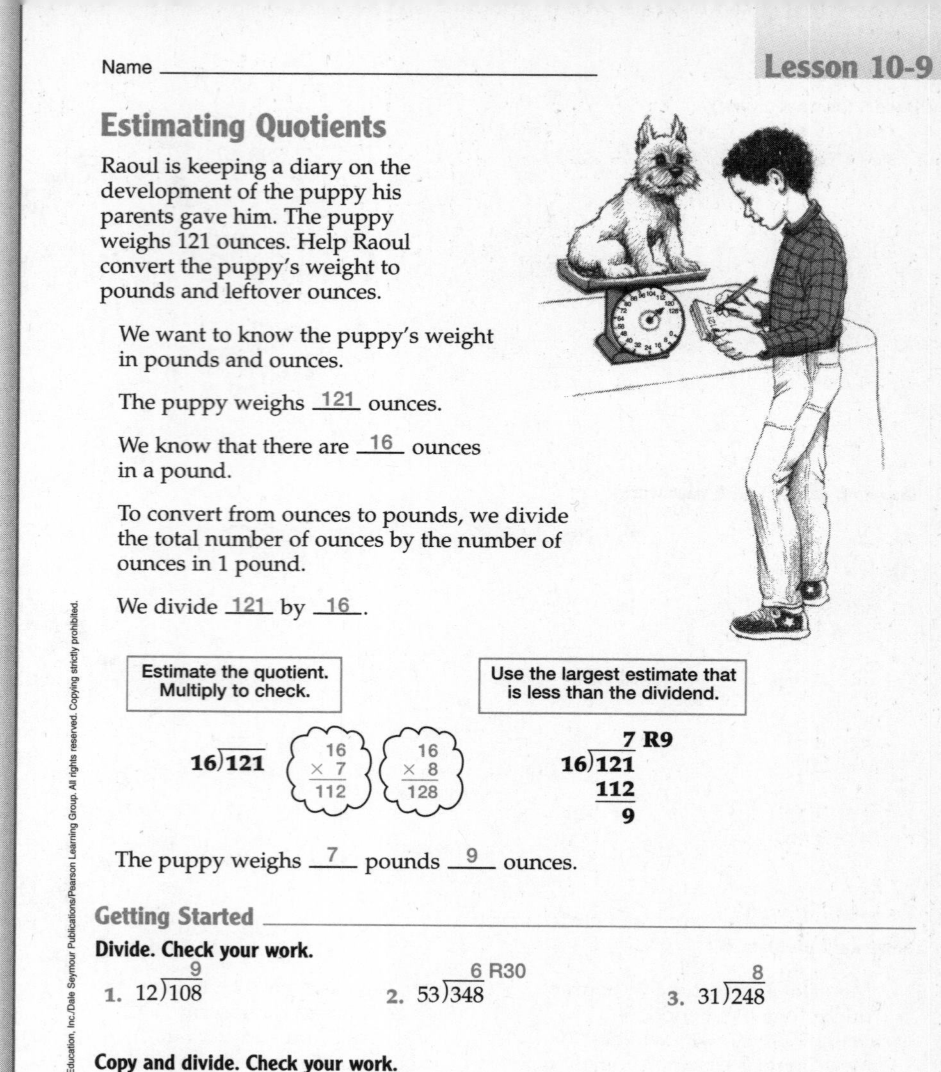

Name ________________________ **Lesson 10-9**

Estimating Quotients

Raoul is keeping a diary on the development of the puppy his parents gave him. The puppy weighs 121 ounces. Help Raoul convert the puppy's weight to pounds and leftover ounces.

We want to know the puppy's weight in pounds and ounces.

The puppy weighs 121 ounces.

We know that there are 16 ounces in a pound.

To convert from ounces to pounds, we divide the total number of ounces by the number of ounces in 1 pound.

We divide 121 by 16.

Estimate the quotient. Multiply to check.

16)121 — 16 × 7 = 112; 16 × 8 = 128

Use the largest estimate that is less than the dividend.

16)121 = 7 R9; 121 − 112 = 9

The puppy weighs 7 pounds 9 ounces.

Getting Started

Divide. Check your work.

1. 12)108 (9)
2. 53)348 (6 R30)
3. 31)248 (8)

Copy and divide. Check your work.

4. 268 ÷ 27 (9 R25)
5. 144 ÷ 24 (6)
6. 340 ÷ 84 (4 R4)

2 Teach

Introduce the Lesson Ask a student to read the problem aloud and tell what it asks. (the puppy's weight in pounds and ounces) Ask students the weight of the puppy. (121 ounces) Ask what information is still needed. (16 ounces equal 1 pound)

- Have students read and complete the sentences. Guide them through the conversion in the model. Ask students how they can check their answer. (multiply 7 by 16 and add 9)

Develop Skills and Concepts Write 47)368 on the board. Tell students to look for a number that, when multiplied by 47, will be slightly less than 368.

- Tell students that another way to estimate the quotient is to guess and check. Tell students that we think of the number of 47s in 368 and know it will be less than 10 because ten 47s would be 470 and way too much. Tell students that we then think of how much less than 10 and could guess there may be eight 47s in 368.
- Have a student find the product of 47 × 8 to check this guess. (376) Ask if 8 was a good guess. (no, too high)
- Tell students to check if seven 47s would be a better estimate by multiplying 47 × 7. (329) Tell students that the 7 is the better estimate because it is close to but less than 368. Have students find the remainder. (39)
- Repeat to estimate quotients for 54)476 and 39)263. (8 R44; 6 R29)

Practice

Divide. Check your work.

1. $43\overline{)344}$ 8
2. $56\overline{)180}$ 3 R12
3. $72\overline{)608}$ 8 R32
4. $37\overline{)163}$ 4 R15
5. $94\overline{)715}$ 7 R57
6. $86\overline{)816}$ 9 R42
7. $16\overline{)100}$ 6 R4
8. $48\overline{)312}$ 6 R24
9. $55\overline{)476}$ 8 R36

Copy and divide. Check your work.

10. $250 \div 39$ 6 R16
11. $512 \div 62$ 8 R16
12. $273 \div 57$ 4 R45
13. $288 \div 48$ 6
14. $743 \div 85$ 8 R63
15. $214 \div 96$ 2 R22
16. $536 \div 75$ 7 R11
17. $160 \div 23$ 6 R22
18. $308 \div 68$ 4 R36

Problem Solving

Solve each problem.

19. How many feet and inches are in 105 inches? 8 feet 9 inches
20. How many days are in 168 hours? 7 days

Now Try This!

It's Algebra!

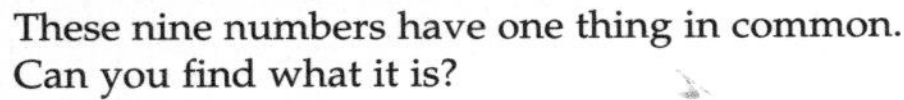

These nine numbers have one thing in common. Can you find what it is?

222	315	2,124
405	999	6,111
210	1,905	1,158

They are all multiples of 3.

3 Practice

Remind students to estimate a quotient by thinking of each divisor as rounded to the nearest ten. Tell students that some exercises on this page have no remainders and the word problems need additional information. Have students complete the page independently.

Now Try This! Help students recognize that those numbers that can be divided evenly by 3 are also the numbers whose digits have a sum that is evenly divisible by 3.

4 Assess

Ask students what they must remember when estimating quotients. (use the largest estimate that is less than the dividend)

For Mixed Abilities

Common Errors • Intervention

Watch for students who have difficulty finding the largest estimate that is less than the dividend. Have students work in pairs, guessing and checking to write related multiplication problems.

$$76\overline{)423} \rightarrow \begin{array}{l} 4 \times 76 = 304 \\ 5 \times 76 = 380 \\ 6 \times 76 = 456 \end{array}$$

Enrichment • Division

Tell students to write and solve division problems that have the same 3-digit dividend and the same remainder but have different 2-digit divisors.

More to Explore • Statistics

Have each student ask five friends or neighbors what brand of toothpaste and hand soap they use. Then, have one student tally these results on the board while the others copy the results on paper.

To have students compare the frequency of use for both types of products, have them make a bar graph for each, using the tally information. Display the graphs and ask students what conclusions they can draw from them.

10-10 Partial Dividends

pages 205–206

1 Getting Started

Objective

- To divide 3-digit numbers by 2-digit numbers using a partial dividend

Warm Up • Mental Math

Ask students to name a multiple of 3.

1. > 100 (102, 105, and so on)
2. equal to 174 ÷ 2 (87)
3. with 4 digits (1,002, and so on)
4. with 9 hundreds (900, 903, and so on)
5. < 212 (3, 6, 9, . . . , 210)
6. whose digits have a sum of 27 (999, 9,891, and so on)
7. with one 0 (102, 120, and so on)
8. that is divisible by 27 (81, 108, and so on)

Warm Up • Division

Review problems such as 80)480 and 30)540 for students to practice dividing multiples of 10 using partial divisors and dividends.

Name ______________________

Partial Dividends

Quan's job at the dairy is to run the machine that attaches the labels to the bottles. Quan figures that 330 bottles pass through the machine in 1 hour. How many cases of milk are labeled in 1 hour? How many bottles are left over?

We want to know the number of cases of milk the machine labels in 1 hour and how many bottles are left over.

Quan's machine labels <u>330</u> bottles in an hour.

There are <u>12</u> bottles in a case.

To find the number of cases, we divide the total number of bottles by the number in one case. We divide <u>330</u> by <u>12</u>.

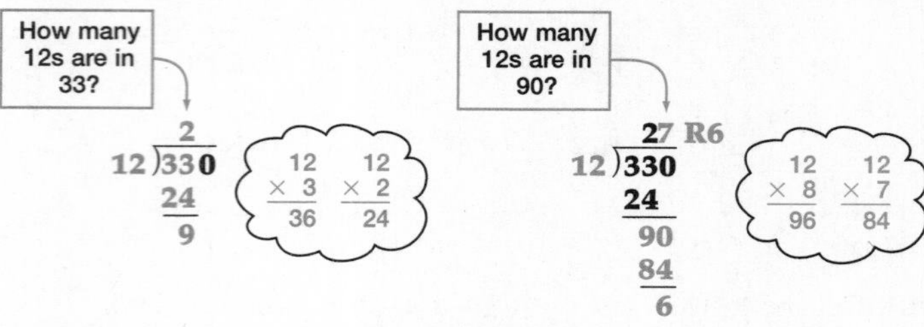

Quan's machine labels <u>27</u> cases an hour with <u>6</u> bottles labeled for the next case.

Getting Started

Divide. Check your answer.

1. 26)475   18 R7
2. 36)829   23 R1
3. 63)245   3 R56

Copy and divide. Check your answer.

4. 930 ÷ 24   38 R18
5. 856 ÷ 35   24 R16
6. 758 ÷ 48   15 R38

2 Teach

Introduce the Lesson Have a student read the problem aloud and describe the picture. Ask students what question is to be answered. (the number of cases of milk labeled in 1 hour and how may bottles are left over) Ask students what information is given in the problem. (330 bottles are labeled in 1 hour.) Ask students what information is needed from the picture. (12 bottles are in each case.)

- Have students complete the sentences and solve the problem. Guide students through the steps in the model. Have students multiply the divisor and the quotient and add the remainder to check their work.

Develop Skills and Concepts Write **26)436** on the board. Tell students that we want to know the number of 26s in 436. Because we know that 4 is too small, we look at the number of 26s in 43.

- Write 1 above the 3 and tell students that we subtract the product of 1 × 26 from 43 to see how many tens are left. (17) Tell students that we then regroup the 17 tens for 170 ones and add the 6 ones for 176 ones.
- Tell students that we want to find the number of 26s in 176 and we can estimate that there are 6 because 6 × 26 is 156. Record the 6, subtract to find the remainder, and record the remainder of 20. Have students multiply the divisor and the quotient and add the remainder to check the problem.
- Repeat for 8)732 and 21)265. (91 R4; 12 R13)

Practice

Divide.

1. $24\overline{)536}$ 22 R8
2. $18\overline{)627}$ 34 R15
3. $39\overline{)846}$ 21 R27
4. $21\overline{)756}$ 36
5. $58\overline{)427}$ 7 R21
6. $63\overline{)915}$ 14 R33
7. $37\overline{)854}$ 23 R3
8. $16\overline{)898}$ 56 R2
9. $84\overline{)963}$ 11 R39

Copy and divide.

10. 945 ÷ 45 — 21
11. 763 ÷ 17 — 44 R15
12. 812 ÷ 26 — 31 R6
13. 518 ÷ 63 — 8 R14
14. 838 ÷ 28 — 29 R26
15. 946 ÷ 37 — 25 R21
16. 683 ÷ 12 — 56 R11
17. 915 ÷ 48 — 19 R3
18. 709 ÷ 29 — 24 R13

Problem Solving

Solve each problem.

19. Mr. Wallace drove 385 miles at 55 miles per hour. How many hours did Mr. Wallace drive? 7 hours

20. How many pounds are in 950 ounces? 59 pounds 6 ounces

21. Sandy ran 720 meters running wind sprints. She ran 45 meters each time. How many wind sprints did Sandy run? 16 wind sprints

22. Richard used 864 grams of batter to make cookies for the bake sale. Each cookie weighed 24 grams and was sold for 15¢. How much did the bake sale make from Richard's cookies? $5.40

3 Practice

Remind students that they know several ways to divide by a 2-digit number. Tell students that one of the word problems requires more than one operation. Have students complete the page independently.

4 Assess

Ask students to divide $14\overline{)830}$. (59 R4)

For Mixed Abilities

Common Errors • Intervention

Watch for students who fail to notice that their first estimated partial quotient is not enough.

Incorrect	Correct
310 R14	40 R14
$23\overline{)934}$	$23\overline{)934}$
69	92
244	14
230	0
14	14

Each time they subtract, students should check to make sure the difference is less than the divisor. If it is not, then the quotient figure should be increased.

Enrichment • Numeration

Tell students to show their work and tell why 31 is not a good quotient estimate to find how many 26s are in 832.

More to Explore • Logic

Read the following story to students and tell them to use logic to answer the question:

A salesman was talking to a customer in a pet store. "This parrot is a fantastic bird," said the pet store salesman. "He can give the correct answer to absolutely every mathematical problem he hears." The customer was so impressed that he bought the parrot. But when he got it home, the bird wouldn't answer a single problem. If what the salesman said was true, why wouldn't the bird answer the problems? (The parrot was deaf.)

10-11 Finding Averages

pages 207–208

1 Getting Started

Objective
- To find averages

Vocabulary
average

Warm Up • Mental Math
Have students name the remainder.

1. 73 ÷ 9 (1)
2. 67 ÷ 7 (4)
3. 50 ÷ 8 (2)
4. 70 ÷ 9 (7)
5. 84 ÷ 9 (3)
6. 49 ÷ 5 (4)
7. 29 ÷ 3 (2)
8. 53 ÷ 6 (5)

Warm Up • Addition
Review column addition through 4-digit numbers. Have students read the sum of each problem and tell how many numbers were added to reach that sum.

Name ______________________

Finding Averages

Andy, Bart, and Larry are the top collectors of autographs in their class. Andy has 14 autographs. Bart has 9 and Larry has 13. What is the average number of autographs each boy has?

We want to find the **average** number of autographs each boy has.

Andy has <u>14</u> autographs, Bart has <u>9</u>, and Larry has <u>13</u>.

To find the average, we add all the individual numbers of autographs and divide that sum by the number of boys. We add <u>14</u>, <u>9</u>, and <u>13</u> and divide by <u>3</u>.

$$\begin{array}{r} 14 \\ 9 \\ +\ 13 \\ \hline 36 \end{array} \qquad 3\overline{)36}\ \text{(1)} \qquad 3\overline{)36}\ \text{(12)}$$

The three boys have an average of <u>12</u> autographs each.

Getting Started

Find the average.

1. 16, 54 <u>35</u>
2. 34, 27, 48, 51, 35 <u>39</u>
3. 158, 196, 243, 203 <u>200</u>

2 Teach

Introduce the Lesson Have a student read the problem aloud and tell what is to be solved. (the average number of the 3 boys' autographs) Ask what useful information is given in the problem. (Andy has 14, Bart has 9, and Larry has 13 autographs.)

- Have students complete the sentences with you. Guide them through the addition and division steps in the model to find the average number of autographs.

Develop Skills and Concepts Tell students that an **average** gives us an idea of how a group of numbers can be represented by one number. Write the following on the board:

10 13 8 7 12

Tell students that we want to find the average of these numbers, so we add them together and divide by 5.

- Have students add the numbers. (50) Ask students why we would divide by 5. (There are 5 numbers in all.) Have students divide 50 by 5. (10) Tell students that 10 is the average of these 5 numbers.
- Tell students that if the group of numbers told us how many students were absent with the flu each day for a week, then we would say an average of 10 students were absent each day that week.
- Have students find the average of 14, 18, and 100 and then discuss why the average is much larger than the two smaller numbers and much smaller than 100. (The smaller numbers pull the average down and the larger number pulls it up.)
- Give students more groups of numbers that average to a whole number.

Practice

Find the average.

1. 17, 28, 36 <u>27</u>
2. 146, 254 <u>200</u>
3. 86, 58, 37, 49, 45 <u>55</u>
4. 624, 534, 810, 712 <u>670</u>
5. 1,496, 4,868 <u>3,182</u>
6. 3,467, 2,948, 4,511 <u>3,642</u>

Problem Solving

Solve each problem.

7. Maria bowled games of 125, 136, and 138. What was Maria's average score? 133
8. Mrs. Jordan drove 243 miles on Monday and 485 miles on Tuesday. What was her average driving distance for 1 day? 364 miles
9. Art earned $243 in May, $316 in June, $375 in July, and $286 in August. What was Art's average monthly earnings? $305
10. Arnold took part in a free-throw contest. His scores were 13, 19, 17, 18, and 18. Find his average score and how many times he scored below it. average, 17; scored once below average

Use the graph of Martina's Free-Throw Record to solve Exercises 11 to 15.

11. In which round did Martina get her highest score? <u>Tuesday</u>
12. What was her highest score? <u>23</u>
13. In which round did Martina get her lowest score? <u>Friday</u>
14. What was her lowest score? <u>12</u>
15. What was Martina's average? <u>18</u>

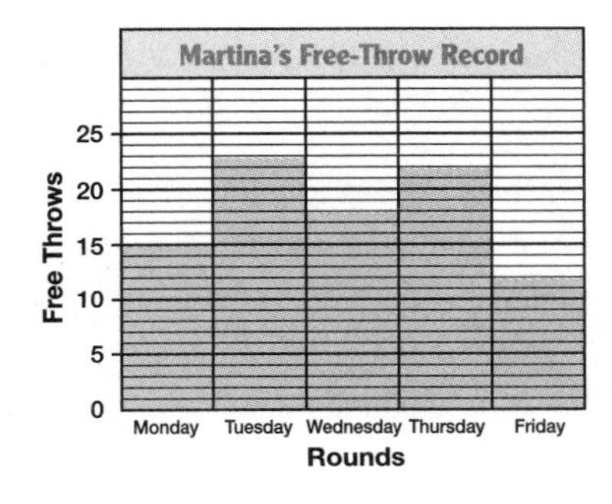

3 Practice

Remind students that when finding an average, the divisor is the number of items in all. Tell students to check to be sure their answers are reasonable. Ask students some questions about the graph at the bottom of the page to review graph skills. Have students complete the page independently.

4 Assess

Ask students how to determine the divisor when finding an average. (count the number of items in all)

For Mixed Abilities

Common Errors • Intervention

Some students may always divide by 2 or some other incorrect number when they are finding averages. Have them work in small cooperative groups to discuss why the divisor must be the number of addends. They should recognize that if the average represents each addend, then the number of addends times the average should have the same total as the sum of all the addends. Then, have students make up sets of 3, 4, and 5 addends and give them to other groups to find the averages.

Enrichment • Application

Tell students to use an almanac to find the average population of the three largest cities in your state.

More to Explore • Measurement

Have students make a floor plan of their classroom. Have them make a list of all the measurements they will need, such as the length of each wall; the width of doors and windows, and where they are placed in the walls; and so on. For this project, also have them measure the furniture. Discuss why they will not need to make height measurements. Showing them a sample floor plan will help. Complete the measurements with students, using a meterstick. Review the meaning of scale.

Give each student a sheet of graph paper and a metric ruler, and tell them to use the following scale: 1 cm = 1 m. Have students draw the floor plan to scale on the graph paper. On another sheet of graph paper, have them draw the furniture to scale. Tell them to cut the furniture out and place it on the first graph paper in the room where it belongs. Let them have fun arranging their classroom in many different ways.

10-12 Problem Solving: Multistep Problems

pages 209–210

1 Getting Started

Objective
- To solve multistep problems

Warm Up • Mental Math
Have students get to 246 if they start at the following:

1. 200 (add 46)
2. 400 ÷ 5 (add 166)
3. 123 (multiply by 2)
4. 4 × 40 (add 86)
5. 1,000 (subtract 754)
6. 492 (divide by 2)

Warm Up • Pencil and Paper
Have students copy the exercise and determine the sum.

1. 8 + (2 + 0) (10)
2. 12 + (7 + 2) (21)
3. 9 + (12 + 9) (30)
4. 15 + (22 + 2) (39)
5. 4 + (11 + 0) (15)
6. 25 + (12 + 9) (46)

Name ______________________

Problem Solving: Multistep Problems

Tamika, a photographer, has taken 456 photographs. She wants to display them in two photo albums. How many pages will she fill in the second album?

60 pages 6 photos on each page

30 pages 4 photos on each page

SEE

We want to know how many pages Tamika will fill in the second album.

We know the first album has 60 pages with 6 photos on each page. The second album has 30 pages and holds 4 photos on each page.

PLAN

We multiply 60 by 6 to find out the number of photos in the first album. We subtract that number from 456. Then, we divide the number of photos that are left by 4 to find the number of pages Tamika will fill in the second album.

DO

60 pages × 6 photos = 360 photos

456 − 360 = 96 photos

96 ÷ 4 = 24 pages

Tamika will fill 24 pages in the second photo album.

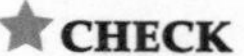

CHECK

360 ÷ 6 = 60

96 + 360 = 456

24 × 4 = 96

2 Teach

Introduce the Lesson Point out the four-step problem-solving strategy and have students describe how to use each step.

Develop Skills and Concepts Have students read the problem and tell what they are to find. (how many pages 456 photographs can fill if one album has 60 pages and holds 6 photographs on each page and the other album has 30 pages and holds 4 photographs on each page) Explain to students that this problem will require using several different operations to find the answer.

- Read each sentence under the SEE and PLAN steps, asking for volunteers to give the missing information as students fill in the answers.
- Work through the DO and CHECK steps with students. Make sure students understand that the order of the operations can change the answer.

3 Apply

Solution Notes

1. Ask students to explain which two operations they need to use to solve the problem. (division, to find the miles per hour; subtraction, to compare the speeds)
2. Ask students how they can check their answer. (multiply 58 by 15 and 62 by 12; the products should be the number of each type of card that Hideki has.)
3. Have students read the problem carefully. Ask them what the word *average* means. (the number obtained by adding two or more quantities and dividing by the number of quantities added)
4. Make sure students understand that Derek flew a total of 10,820 miles and therefore the interisland flights must be less than 10,820 miles.

Apply

Solve each problem. Show your work.

1. Mr. Tompkins drove 427 miles in 7 hours on Monday. He drove 318 miles in 6 hours on Tuesday. How many miles per hour faster did he drive on Monday?
   8 miles per hour faster

2. Hideki has 744 baseball cards in 12 boxes. He also has 870 football cards in 15 boxes. How many more baseball cards are in each box than football cards?
   4 more baseball cards in each box

3. Marta is reading a book that is 425 pages long. She has read 15 pages each day for 7 days. If she wants to finish the book in 5 more days, how many pages does she have to average per day?
   64 pages per day

4. Derek flew from Newark, New Jersey, to Honolulu, Hawaii. He received 4,910 frequent flyer miles each way. He also took 2 interisland flights while in Hawaii. Derek received 10,820 frequent flyer miles for the entire trip. How many frequent flyer miles did he get for each interisland flight?
   500 frequent flyer miles

Problem-Solving Strategy: Using the Four-Step Plan

- ★ **SEE** What do you need to find?
- ★ **PLAN** What do you need to do?
- ★ **DO** Follow the plan.
- ★ **CHECK** Does your answer make sense?

5. The Titans have played 12 games this season and have scored 504 points. Donna has played in each of the 12 games. She is the team's leading scorer with an average of 15 points per game. How many points per game have the rest of the Titans scored?
   27 points per game

6. Anjali bowled scores of 175, 122, and 129. Ramona bowled 3 games and averaged 125. How many pins per game better did Anjali bowl than Ramona?
   17 pins per game

7. Alex wants to average a 95 on his math tests. So far, he has scored 92, 98, 95, and 92. What is the lowest score he can get on his next test to have an average of 95?
   98

For Mixed Abilities

More To Explore • Estimation

Have students compile a list of times when the skill of estimating is helpful. Then, have students collect an assortment of containers in many shapes: cottage cheese carton, oatmeal box, juice can, milk carton, and so on. Fill one of the containers with lima beans. Ask students to estimate how many beans it takes to fill the container. Discuss how they arrived at their estimations. Repeat for other containers. Extend the activity by having students make a list of situations in which estimating would not be applicable.

Higher-Order Thinking Skills

5. **Analysis:** Students can find Donna's point total first (180 points) and then find how many points per game the rest of the Titans scored. (subtract Donna's points from 504 and divide by 12) Instead, some students might find the average Titan score per game ($504 \div 12 = 42$) and then subtract Donna's average points. ($42 - 15 = 27$)

6. **Comprehension:** Some students may add 125 to Anjali's total. Help these students understand that Ramona's average is separate from Anjali's.

7. **Comprehension:** Have students read the problem carefully. Ask them what the key phrase is in the question. (lowest score) Have them explain what that means.

4 Assess

Ask students to explain two different ways to find the answer to Exercise 6. (add, multiply, subtract, and divide; add, divide, and subtract)

10-13 Calculator: Comparison Shopping

pages 211–212

1 Getting Started

Objective
- To use a calculator to do comparison shopping

Vocabulary
unit cost

Materials
calculators; priced product containers in 2 sizes

Warm Up • Mental Math
Ask students if each is evenly divisible by 5.

1. 1,986 (no)
2. 2,000,000 (yes)
3. 4,695 (yes)
4. 10 × 46 (yes)
5. 10 + 15 + 4 (no)
6. 5 × 45 (yes)
7. 38 ÷ 2 (no)
8. $\frac{1}{2}$ of 460 (yes)

Warm Up • Calculator
To review calculator codes for money amounts, dictate amounts of money for students to enter into their calculators. Now, have students write calculator codes on the board for problems of 1- and 2-digit numbers times money amounts. Have students complete the codes.

Name ____________________

Lesson 10-13

Calculator: Comparison Shopping

Mrs. Anderson is shopping for the best buy in baby diapers. Should Mrs. Anderson buy the package of 17 diapers or the package of 42 diapers?

We are looking for the better buy in diapers.

The package of 17 diapers costs $10.20.

The package of 42 diapers costs $16.80.

To decide which is the better buy, we must find the **unit cost** of each diaper. To find the unit cost, we divide the total cost of the package by the number of units in it. We divide $10.20 by 17 and $16.80 by 42.

Complete each code.

10 [.] 20 [÷] 17 [=] [0.6] 0.6 means $0.60 or 60¢.

16 [.] 80 [÷] 42 [=] [0.4] 0.4 means $0.40 or 40¢.

60¢ > 40¢

The package of 42 diapers is the better buy.

Getting Started

Find the cost per ounce. Circle the better buy.

1. 8 ounces $6.48 $0.81 per oz
   (18 ounces) $12.96 $0.72 per oz

2. 6 ounces $87.60 $14.60 per oz
   (8 ounces) $105.68 $13.21 per oz
   12 ounces $170.88 $14.24 per oz

2 Teach

Introduce the Lesson Have a student read the problem aloud and tell what is to be solved. (which of the 2 packages of diapers is the best buy) Ask students what information is given. (The package of 17 diapers costs $10.20 and the package of 42 diapers costs $16.80.)

- Have students read and complete the sentences. Help them complete the calculator codes in the model to find the best buy.

Develop Skills and Concepts Tell students that we often need to know which packages in a store give us the most for our money. Tell students that we have to find the **unit cost**, or the cost per cookie in each bag of cookies, in order to know which package is the best buy.

- Tell students to suppose that there are 2 bags of cookies, one costing $2.80 for 40 cookies and the other costing $3.00 for 50 cookies. Tell students that to find the cost of 1 cookie in each bag, we divide $2.80 by 40 and $3.00 by 50.
- Have students write and solve the calculator codes. Remind them to enter the decimal point and that no dollar sign will appear on the screen.
- Ask students which bag offers a cookie at a lower price. ($3.00 bag) Repeat for 22 tea bags for $2.64 and 35 tea bags for $3.85.

Practice

Use your calculator to find the answer.

1. 7)$8.61 — $1.23
2. 6)$15.72 — $2.62
3. 15)$31.05 — $2.07
4. 26)$6.76 — $0.26
5. 34)$816.34 — $24.01
6. 59)$35,577 — $603

Find the cost per unit. Circle the better buy.

7. (6 pounds $23.94) — $3.99 per pound
   15 pounds $60.45 — $4.03 per pound
8. 12 hats $187.80 — $15.65 per hat
   (18 hats $267.48) — $14.86 per hat
9. 8 pounds $75.92 — $9.49 per pound
   (14 pounds $124.32) — $8.88 per pound
   18 pounds $164.34 — $9.13 per pound
10. 15 tickets $297.90 — $19.86 per ticket
    25 tickets $482.25 — $19.29 per ticket
    (35 tickets $638.75) — $18.25 per ticket

Problem Solving

Solve each problem.

11. A store sells one brand of hot chocolate at 3 pounds for $2.43. It sells another brand at 5 pounds for $3.95. Which is the better buy? How much do you save?
    5 pounds for $3.95; 2¢ per pound

12. The school store sells binders for $2.49 each. If you buy 6 binders of the same color, the cost is $13.26. How much is saved if you buy 6 red binders?
    $1.68

13. Mrs. Crowe wants to buy new plates for her kitchen. She can buy a set of 8 plates for $18.25 each or a set of 12 for $208.80. Which set should she buy to get the better buy? How much will Mrs. Crowe save per plate?
    set of 12; $0.85 per plate

14. Just CDs is going out of business! Lashanta can buy 24 CDs for $190.80 or 15 CDs for $133.35. Which number of CDs is the better buy? How much will Lashanta save per CD?
    set of 24; $0.94 per CD

3 Practice

Remind students to place a dollar sign and decimal point in each answer of a money amount. Tell them that the word problems involve more than one operation. Tell students to complete the page independently.

4 Assess

Ask students how to find the better buy. (Find the unit cost of an item by dividing the total cost of the package by the number of units in it.)

For Mixed Abilities

Common Errors • Intervention

Some students may try to determine the best buy without finding the unit cost. Have them work in small groups to discuss why they are not necessarily getting a better buy because they are paying less for a group of items. Students should recognize that the cost of one unit in one grouping may be more than it would be if you bought another grouping of a like item. Students should recognize also that "better buy" assumes two offerings of equal quality. Have students visit grocery stores as a group project to find examples.

Enrichment • Application

Have students make a table of unit costs of two different-sized containers for each of four different products used often in their homes.

More to Explore • Numeration

Tell students that they will be playing tic-tac-toe with numbers, following some very special rules. Have students draw the standard game grid of 3 rows of 3 squares.

Working in pairs, tell one student to use only odd numbers, 1 through 9. The other player will use only even numbers, 2 through 10. To play, each player writes any of his or her allowed numbers in any square, on each turn. The winning player completes a row, either horizontal, diagonal, or vertical, whose sum is 15.

Students will see plans of strategy after they play a few times. Each time, they must reach a sum of 15 to win. Have students reverse roles so that they have a chance to work with both even and odd numbers.

Chapter 10 Test

page 213

Items	Objectives
1–2	To divide 3-digit numbers by 1-digit numbers to get a 3-digit quotient without remainder (see pages 187–188)
3–4	To divide 3-digit numbers by 1-digit numbers to get a 3-digit quotient with remainder (see pages 189–190)
5–8	To divide 3-digit numbers by 1-digit numbers to get a 3-digit quotient with zero (see pages 191–192)
9–12	To divide 3-digit numbers by 1-digit numbers to get a 2-digit quotient with or without a remainder (see pages 193–194)
13–16	To divide money by a 1-digit number (see pages 195–196)
17–20	To divide 3-digit numbers by 2-digit numbers to get a 1-digit quotient with or without a remainder (see pages 201–202)
21–24	To divide 3-digit numbers by 2-digit numbers to get a 2-digit quotient with or without a remainder (see pages 205–206)
25–28	To find the average of a set of numbers (see pages 207–208)

Alternate Chapter Test

You may wish to use the Alternate Chapter Test on page 364 of this book for further review and assessment.

Name ____________________

CHAPTER 10 TEST

Divide. Check your work.

1. 3)696 — 232
2. 5)555 — 111
3. 2)843 — 421 R1
4. 4)725 — 181 R1
5. 2)408 — 204
6. 3)900 — 300
7. 4)813 — 203 R1
8. 3)625 — 208 R1
9. 6)276 — 46
10. 7)399 — 57
11. 8)623 — 77 R7
12. 9)738 — 82
13. 4)$8.24 — $2.06
14. 8)$5.76 — $0.72
15. 2)$9.38 — $4.69
16. 7)$4.97 — $0.71
17. 27)162 — 6
18. 39)247 — 6 R13
19. 53)325 — 6 R7
20. 44)428 — 9 R32
21. 18)468 — 26
22. 25)$9.00 — $0.36
23. 39)756 — 19 R15
24. 57)912 — 16

Find the average.

25. 6, 9, 4, 5, 1 — 5
26. 92, 46, 75, 55 — 67
27. 47, 69, 101, 45, 48 — 62
28. 155, 245, 125 — 175

Cumulative Assessment

CUMULATIVE ASSESSMENT

Circle the letter of the correct answer.

1. Round 639 to the nearest hundred.
   (a.) 600
   b. 700
   c. NG

2. Round 7,350 to the nearest thousand.
   (a.) 7,000
   b. 8,000
   c. NG

3. What is the place value of the 1 in 517,296?
   a. thousands
   (b.) ten thousands
   c. hundred thousands
   d. NG

4. 3,249 + 1,816
   a. 4,055
   b. 4,065
   (c.) 5,065
   d. NG

5. 26,795 + 27,586
   a. 43,271
   b. 44,381
   (c.) 54,381
   d. NG

6. 829 − 136
   (a.) 693
   b. 713
   c. 793
   d. NG

7. 13,053 − 12,875
   (a.) 178
   b. 1,178
   c. 1,822
   d. NG

8. Choose the better estimate of weight.
   (a.) 1.5 g
   b. 1.5 kg

9. Choose the better estimate of volume.
   (a.) 350 mL
   b. 350 L

10. $2.79 × 6
    a. $12.24
    (b.) $16.74
    c. $16.76
    d. NG

11. 39 × 23
    a. 117
    (b.) 897
    c. 7,107
    d. NG

12. 607 × 58
    a. 34,204
    b. 35,786
    c. 36,206
    (d.) NG

13. 624 ÷ 3
    a. 28
    (b.) 208
    c. 280
    d. NG

14. 42)226
    a. 5
    b. 5 R26
    c. 526
    (d.) NG

STOP

score

page 214

Items	Objectives
1–2	To round numbers (see pages 31–32)
3	To identify place value through hundred thousands (see pages 17–18, 27–28)
4–5	To add two numbers up to 5 digits (see pages 47–48)
6	To subtract two 3-digit numbers (see pages 57–58)
7	To subtract two 5-digit numbers (see pages 61–62)
8–9	To determine the appropriate metric unit of weight or capacity (see pages 153–156)
10–12	To multiply up to 3-digit numbers by up to 2-digit numbers including money with and without regrouping (see pages 167–168, 175–180)
13	To divide 3-digit numbers by up to 1-digit numbers to get a 3-digit quotient with zero (see pages 191–192)
14	To divide 3-digit numbers by 2-digit numbers to get a 1-digit quotient with a remainder (see pages 199–200)

Alternate Cumulative Assessment

Circle the letter of the correct answer.

1. Round 756 to the nearest hundred.
   a 700
   (b) 800
   c NG

2. Round 9,227 to the nearest thousand.
   a 8,000
   (b) 9,000
   c 10,000
   d NG

3. What is the place value of the 2 in 213,654?
   a thousands
   b ten thousands
   (c) hundred thousands
   d NG

4. 6,157 +2,937
   a 8,084
   (b) 9,094
   c 9,194
   d NG

5. 16,975 + 37,145
   (a) 54,120
   b 53,010
   c 43,121
   d NG

6. 348 − 162
   a 286
   b 226
   (c) 186
   d NG

7. 18,062 − 16,473
   a 2,411
   (b) 1,589
   c 2,699
   d NG

8. What unit would you use to weigh a loaf of bread?
   a mg
   b kg
   (c) g
   d NG

9. What unit of capacity would you use to measure a small pond?
   a mL
   (b) L

10. $4.86 × 4
    a 13.24
    b $17.64
    (c) $19.44
    d NG

11. 27 × 34
    a 189
    b 698
    (c) 918
    d NG

12. 409 × 36
    a 3,681
    b 14,728
    c 14,984
    (d) NG

13. 6)36
    a 16
    b 160
    c 106
    (d) NG

14. 67)423
    a 60 R21
    (b) 6 R21
    c 621
    d NG

11-1 Points, Lines, Rays, and Line Segments

pages 215–216

1 Getting Started

Objective

- To identify and draw points, lines, rays, and line segments

Vocabulary

line segment, endpoint, line, ray, intersect, perpendicular, parallel

Materials

*straightedge; rulers marked in centimeters; cards measuring less than 15 cm; graph paper

Warm Up • Mental Math

Have students give the perimeter of the following:

1. square with 1 side of 14 cm (56 cm)
2. hexagon whose sides are each 22 in. (132 in.)
3. triangle whose sides are each 38 ft (114 ft)
4. 60 dm × 150 dm rectangle (420 dm)

Warm Up • Measurement

Have students work in pairs to find the length and width of various-sized cards. Have students write the measurements on the cards, trade cards with another pair of students, and check each other's work.

Name ____________________

Geometric Figures

CHAPTER 11

Lesson 11-1

Points, Lines, Rays, and Line Segments

A straight path from point A to point B is called a **line segment**. Points A and B are called **endpoints**.

We say **line segment** AB or **line segment** BA.
We write $\overline{AB}$ or $\overline{BA}$.

A line segment is part of a **line**. A line extends indefinitely in both directions.

We say **line** XY or **line** YX.
We write $\overleftrightarrow{XY}$ or $\overleftrightarrow{YX}$.

A **ray** is also part of a line. It has one endpoint and extends forever in one direction.

We say **ray** MN.
We write $\overrightarrow{MN}$.

Some lines **intersect**. Line CD intersects line EF at point P.

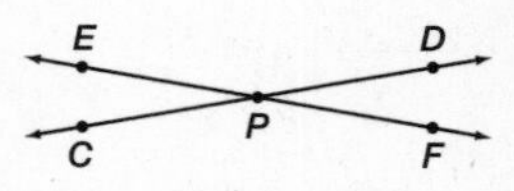

Lines that intersect at right angles are called **perpendicular** lines. Line GH is perpendicular to line JK.

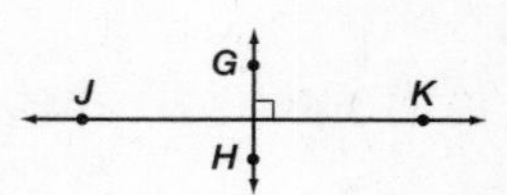

Some lines do not intersect. Line LM is **parallel** to line ST.

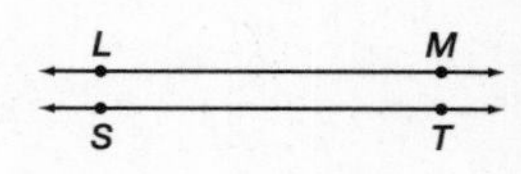

Getting Started

Write the name of the figure.

1. • C

point C

Draw and label the figure.

2. $\overline{TV}$ T V

Check students' drawings.

2 Teach

Introduce the Lesson Have students read with you to learn about points, endpoints, lines, line segments, rays, intersecting lines, perpendicular lines, and parallel lines.

- Point out that letters that describe points are italicized. Help students find the points of each figure on the right as they are described in the model.

Develop Skills and Concepts Use the straightedge to draw line segment XY on the board and have students measure it. Explain that the figure is called **line segment** XY or line segment YX. Tell students that X and Y are **endpoints**.

- Extend the segment beyond both endpoints and draw an arrow at each end. Tell students that the figure is now called **line** XY or line YX because its ends do not have definite points.
- Have a student use a straightedge to draw a line from point G. Tell the student to stop at some point and draw an arrow at that place. Explain that this is an example of a **ray**.
- Draw lines BC and DE on the board so that they intersect and tell students that these lines cross each other and are called **intersecting lines**.
- Draw lines QR and ST on the board so that they intersect and form a right angle. Tell students that these lines are called **perpendicular lines**.
- Draw lines MN and OP on the board as parallel lines and tell students that **parallel lines** would never meet.
- Have students use graph paper to practice drawing and labeling perpendicular, intersecting, and parallel lines.

Practice

Write the name of the figure.

1. line XY or $\overleftrightarrow{XY}$

2. line segment RS or $\overline{RS}$

3. ray ST or $\overrightarrow{ST}$

4. line segment MN or $\overline{MN}$

5. line LM or $\overleftrightarrow{LM}$

6. point G

Draw and label the figure.

7. point R

•R

8. $\overleftrightarrow{UV}$

U V

9. $\overrightarrow{GH}$

G H

10. $\overleftrightarrow{MN}$ parallel to $\overleftrightarrow{RS}$

M N / R S

11. $\overleftrightarrow{CD}$

C D

12. $\overleftrightarrow{OP}$ intersecting $\overleftrightarrow{ST}$

O T / S P

Write *intersecting*, *parallel*, or *perpendicular*.

13. 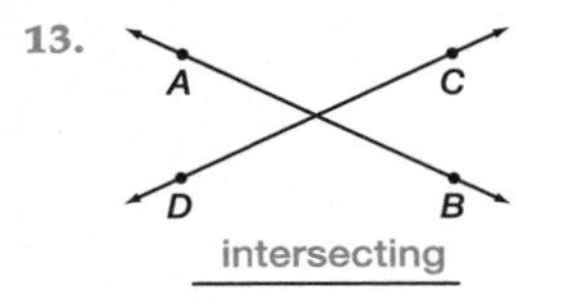

intersecting

14. 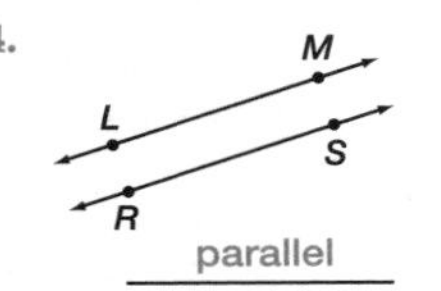

parallel

15. 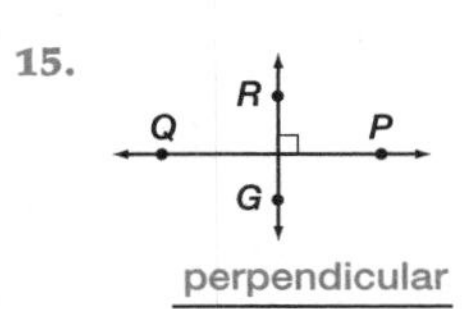

perpendicular

Name six different line segments in the line.

16. 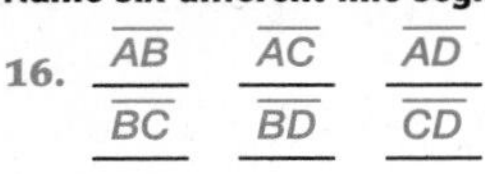

$\overline{AB}$ $\overline{AC}$ $\overline{AD}$

$\overline{BC}$ $\overline{BD}$ $\overline{CD}$

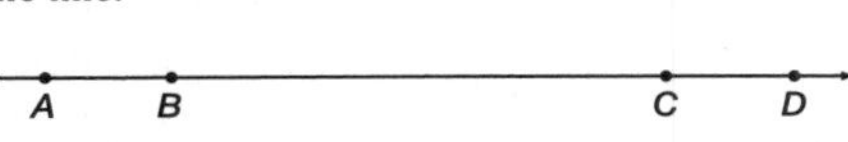

3 Practice

Review with students the labeling of the figures on page 215. Help students work the first exercise. Help students name two or three line segments in Exercise 16. Remind students to always use a straightedge to draw lines or line segments. Then, have students complete the page independently.

4 Assess

Ask students what lines that intersect at right angles are called. (perpendicular lines)

For Mixed Abilities

Common Errors • Intervention

Some students may confuse the geometric symbols for naming a line and a line segment. Draw line RS and line segment RS on the board.

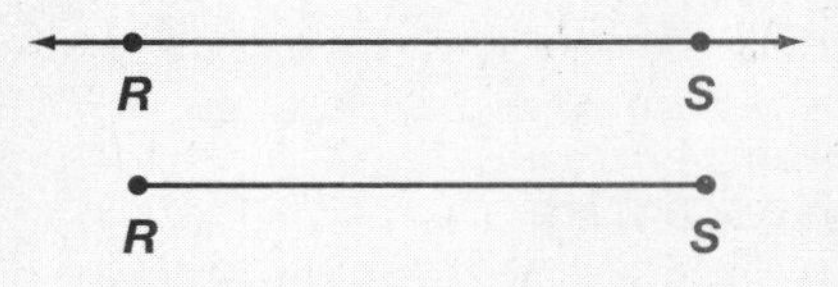

Have students go to the board, discussing and pointing to the points that are on line R and those that are on line segment RS. Discuss how all the points on line segment $\overline{RS}$ are also on line $\overleftrightarrow{RS}$. Also, discuss how line segment $\overline{RS}$ has a beginning and end, whereas line $\overleftrightarrow{RS}$ does not.

Enrichment • Geometry

Have students draw each of the following figures and write an example of an object for each: parallel lines, line segment, point, intersecting lines, line. (Possible answers: a rectangle, a bulletin board.)

More to Explore • Applications

Have students draw a plane figure that, when folded, will make a model of a cube. Explain to students that a plane figure that can be folded to create a solid figure is called a **net**. Ask students to describe real world objects that are cubes. Finally, encourage students to draw nets that will make a model of a pyramid.

ESL/ELL STRATEGIES

When discussing the terms *points*, *lines*, *rays*, and *line segments*, define each term in simple English. If necessary, provide additional explanations with illustrations on the board to clarify other new vocabulary such as *intersect*, *perpendicular*, and *parallel*.

11-2 Angles

pages 217–218

1 Getting Started

Objective
- To identify and draw right, acute, obtuse, and straight angles

Vocabulary
right angle, acute angle, obtuse angle, straight angle

Materials
*flashlight; *straightedge; *rulers

Warm Up • Mental Math
Have students divide the number by 30, then subtract 7.

1. 300 (3)
2. 1,200 (33)
3. 660 (15)
4. 900 (23)
5. 1,800 (53)
6. 33,000 (1,093)
7. 450 (8)
8. 750 (18)

Warm Up • Geometry
Write the following on the board in a column: $\overline{LM}$, $\overleftrightarrow{PQ}$, **line** *TU* **parallel to line** *CD*, **point 2, line segment** *MN*, **line segment** *DE* **intersecting line segment** *RS*. Have students draw and label the identifying qualities of each figure. (See definitions on page 215.)

Name ______________________________

Lesson 11-2

Angles

An **angle** is a figure formed by two rays that intersect at a common endpoint. The common endpoint is called the **vertex** of the angle. We name an angle using one or three of its points.

The angle at the right can be named any one of the following:

angle *CDE* or ∠*CDE*

angle *EDC* or ∠*EDC*

angle *D* or ∠*D*

vertex→ D E C

When you use three points to name an angle, the vertex is always in the middle.

An angle that forms a square corner is called a **right angle**. ∠*XYZ* is a right angle. The □ shows the angle is a right angle. A right angle measures 90°.

X Y Z

An angle that has a measure less than a right angle, or 90°, is an **acute angle**.

An angle that has a measure greater than 90° but less than 180° is an **obtuse angle**.

A **straight angle** measures 180°.

M E G   J C T   B L D

We write the following: ∠*MEG* ∠*JCT* ∠*BLD*

Getting Started

Name and classify the angle as *acute*, *right*, *obtuse*, or *straight*.

1. 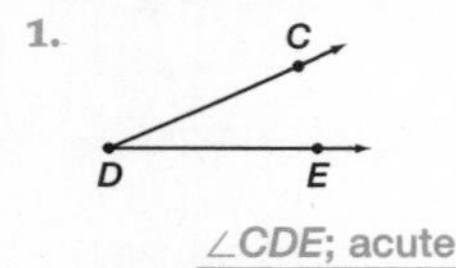

∠CDE; acute

2. G H I
∠GHI; obtuse

3. M N O
∠MNO; right

2 Teach

Introduce the Lesson Have students read with you to learn about an **angle**, a **vertex**, a **right angle**, an **acute angle**, an **obtuse angle**, and a **straight angle**.

- Ask students to name the vertex for the right, acute, obtuse, and straight angles. (Y, E, C, L)

Develop Skills and Concepts Draw and label two rays from a common endpoint and tell students that the two rays form an angle and their common endpoint is called the vertex.

- Explain to students that angles are named using the points on them with the letter that names the vertex in the middle. Tell students that the symbol ∠ can replace the word *angle*. Have students label and name the angle.
- Now, draw a square or rectangle on the board and highlight each corner. Tell students that the angle in each corner is called a right angle.
- Next, place two rulers end to end and hold them up against the board, modeling the three different angles (right, acute, and obtuse). Model the angles after the standard angles: 90°, 45°, and 120°.
- Have students classify the angles. Once they have identified an angle, trace the angle on the board and write the angle classification.
- Have students hold two pencils end to end and create their own angles. Have them trace each angle onto construction paper and classify it. Ask them to share their tracings with the class.

Practice

Name and classify the angle as *acute*, *right*, *obtuse*, or *straight*.

1. 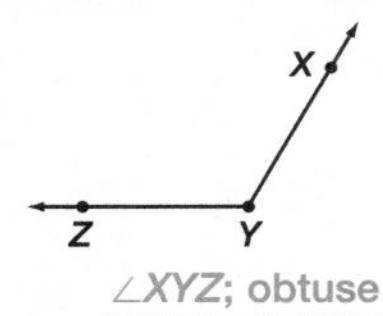

∠XYZ; obtuse

2. 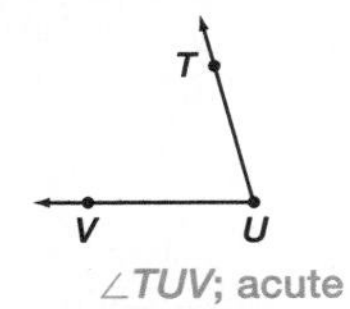

∠TUV; acute

3. 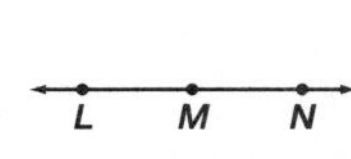

∠LMN; straight

4. 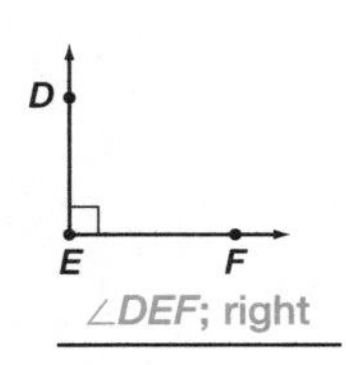

∠DEF; right

5. 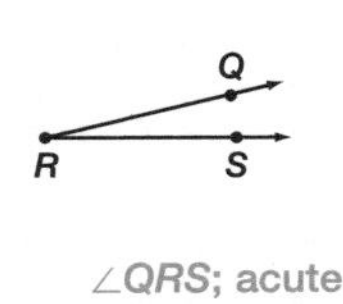

∠QRS; acute

6. 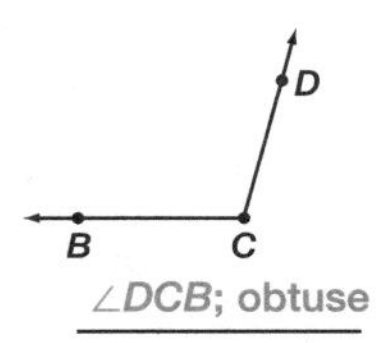

∠DCB; obtuse

Draw and label the angle. Check students' drawings.

7. acute angle *ABC*

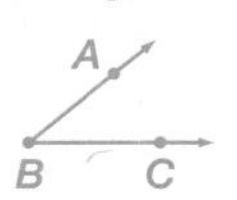

8. obtuse angle *EFG*

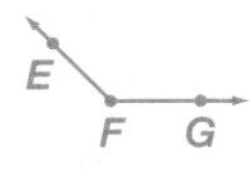

9. straight angle *MNO*

10. right angle *FGH*

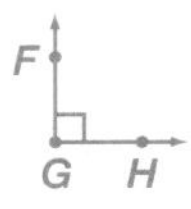

Match the angle to its definition.

11. acute c
12. obtuse d
13. right a
14. straight b

a. measures 90°
b. measures 180°
c. measures less than a right angle
d. measures greater than a right angle and less than 180°

3 Practice

Remind students that angles can be named by using the word *angle* and identifying initials, or by the symbol for angle, ∠, and the identifying initials. Remind students to always use a straightedge to draw lines and line segments. Have students complete the page independently.

4 Assess

Ask students what an angle that measures greater than 90° but less than 180° is called. (an obtuse angle)

For Mixed Abilities

Common Errors • Intervention

Watch for students who do not put the letter for the vertex in the middle when they are naming an angle. Have students work in pairs, drawing two rays with a common endpoint, labeling the endpoint *X*. Discuss how point *X* is the endpoint of both rays and the vertex of the angle. Then, have students put a point, other than the endpoint, on each ray and use these points to name the angle.

Enrichment • Geometry

Have students draw and label an open or closed figure that has six angles, one of which is a right angle. Then, tell them to use a square corner to draw the right angle and the six ways to name each angle in their figure.

More to Explore • Measurement

Ask students to bring in a variety of irregularly shaped containers such as jars, bottles, vases, and so on. Have them estimate and then measure to determine which container has the greatest volume. Then, have them measure the capacity of each container using a graduate marked off in milliliters. Have students arrange the containers by capacity, from least to greatest. Discuss with them how the shape of a container can distort its capacity.

ESL/ELL STRATEGIES

Preteach the words *angle* (the figure formed when two rays come together) and *vertex* (the point at which the two rays come together). Point out these elements in the models on page 217.

11-3 Polygons

pages 219–220

1 Getting Started

Objective

- To identify and draw polygons

Vocabulary

plane figure, polygon, triangle, quadrilateral, pentagon, hexagon, octagon, decagon

Materials

*cube, rulers, compasses, protractors

Warm Up • Mental Math

Tell students to give the total cost for 1 year if the cost is

1. \$600/month (\$7,200)
2. \$20 biweekly (\$520)
3. \$150/quarter year (\$600)
4. \$2,600 semiannually (\$5,200)
5. \$760 biannually (\$380)
6. \$4/week (\$208)
7. \$89/6 months (\$178)
8. \$200 bimonthly (\$1,200)

Warm Up • Geometry

Have students draw and label figures on the board such as two rays with *C* as their vertex, a figure having two square angles, a closed figure having three line segments, and so on.

Name ______________________

Polygons

How can the images on a movie screen help you to understand plane figures?

A **plane figure** is a "flat" shape. It has length and width but no height.

We think of plane figures as being on the surface of things, like the figures drawn on a page or images we see in a movie.

Some plane figures are called **polygons**. Polygons have only straight sides. **Triangles** have three sides; **quadrilaterals** have four sides; **pentagons** have five sides; **hexagons** have six sides; **octagons** have eight sides; and **decagons** have ten sides.

Plane figures have <u>length</u> and <u>width</u>.

Getting Started

Name the kind of polygon. List its line segments and angles.

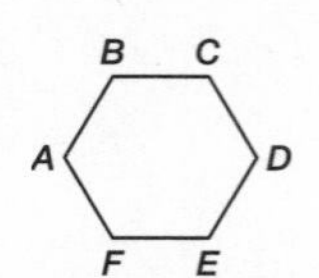

1. kind of polygon
hexagon

2. line segments
$\overline{AB}$ $\overline{DE}$
$\overline{BC}$ $\overline{EF}$
$\overline{CD}$ $\overline{FA}$

3. angles
$\angle A$ $\angle D$
$\angle B$ $\angle E$
$\angle C$ $\angle F$

Draw and label the polygon.

4. triangle *XYZ*

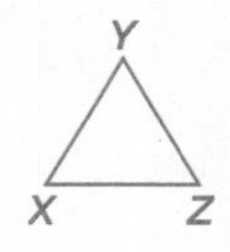

5. quadrilateral *ABCD*

A B
D C

2 Teach

Introduce the Lesson Have a student describe the picture and then read the problem. Ask students what kind of figures will be discussed in this lesson. (plane figures)

- Read about plane figures and polygons with students. Ask what figure has no straight sides. (circle) Have students complete the solution sentence.

Develop Skills and Concepts Ask students if they have ever seen a movie in 3-D. Tell students that the term *3-D* means "three-dimensional" and that length and width are two of the dimensions. Tell students that the images on a 3-D movie also have height or depth and seem touchable.

- Display a cube and tell students that it has three dimensions: length, width, and height. Point to each surface as you tell students that each side has only length and width, so we call it a plane figure.
- Point out some plane figures in the room and tell students that they are called **polygons** because they are plane figures with straight sides.
- Show a circular surface as you tell students that the circle is a plane figure. Ask why a circle is not a polygon. (It has no straight sides.)
- Have students use rulers, compasses, and protractors to experiment with drawing polygons and plane figures.

3 Practice

Remind students that a line segment can be written as $\overline{BC}$ and that they should use a straightedge to draw polygons. Have students complete page 220 independently.

Practice

Name the kind of polygon. List its line segments and angles.

1. 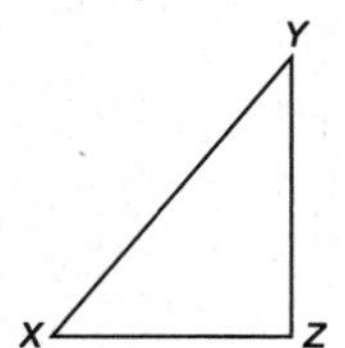

triangle

$\overline{ZY}$	$\angle Z$
$\overline{YX}$	$\angle Y$
$\overline{XZ}$	$\angle X$

2. 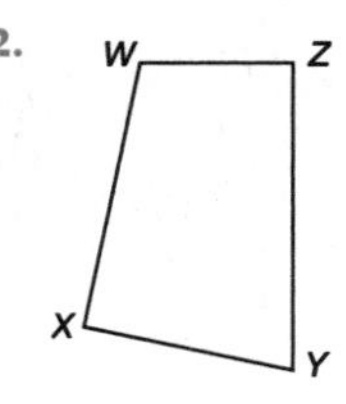

quadrilateral

$\overline{WZ}$	$\angle W$
$\overline{ZY}$	$\angle Z$
$\overline{YX}$	$\angle Y$
$\overline{XW}$	$\angle X$

3. 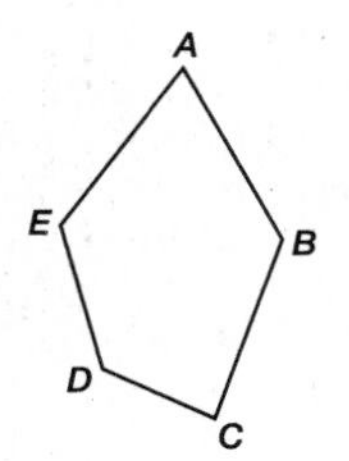

pentagon

$\overline{AB}$	$\angle A$
$\overline{BC}$	$\angle B$
$\overline{CD}$	$\angle C$
$\overline{DE}$	$\angle D$
$\overline{EA}$	$\angle E$

4. 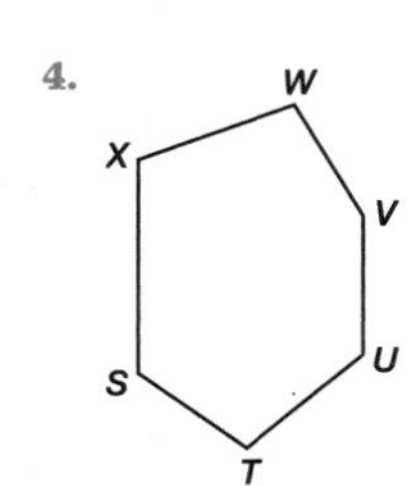

hexagon

$\overline{XW}$	$\angle W$
$\overline{WV}$	$\angle V$
$\overline{VU}$	$\angle U$
$\overline{UT}$	$\angle T$
$\overline{TS}$	$\angle S$
$\overline{SX}$	$\angle X$

Draw and label the polygon.

5. pentagon *ABCDE*

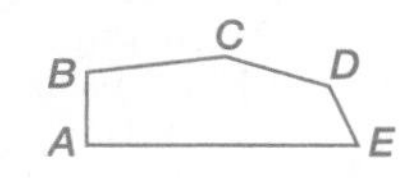

6. quadrilateral *RSTU*

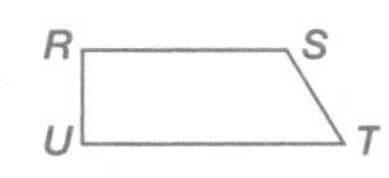

7. octagon *ABCDEFGH*

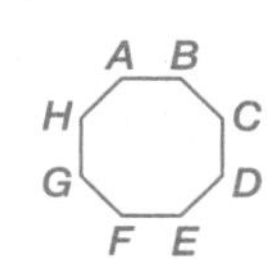

Complete the chart.

	Polygon	Number of Sides	Number of Angles
8.	Quadrilateral	4	4
9.	Pentagon	5	5
10.	Hexagon	6	6
11.	Octagon	8	8
12.	Decagon	10	10

4 Assess

Ask students to define a polygon and give examples of polygons. (A polygon is a figure with only straight sides. Triangles, quadrilaterals, pentagons, hexagons, octagons, and decagons are examples of polygons.)

For Mixed Abilities

Common Errors • Intervention

Some students may have difficulty visualizing plane figures in a three-dimensional world. Have them work in pairs to identify three-dimensional objects in the room and then identify any plane figures on the objects.

Enrichment • Geometry

Tell students to make a list of five plane figures found in each room of their home.

More to Explore • Geometry

Have students draw a triangle, quadrilateral, pentagon, hexagon, and octagon and cut out each polygon.

Have them cut each polygon into two pieces, making sure that both pieces are polygons. Then, have students name the kind of polygon the original figure is and explain why it is classified as such. (Triangles have three sides, quadrilaterals have four sides, pentagons have five sides, hexagons have six sides, and octagons have eight sides.) Ask them what attributes change or do not change when a polygon is cut up. (The size and shape change, but the sides remain straight.)

11-4 Triangles

pages 221–222

1 Getting Started

Objective

- To identify isosceles, equilateral, scalene, right, acute, and obtuse triangles

Vocabulary

triangle, isosceles triangle, equilateral triangle, scalene triangle, right triangle, acute triangle, obtuse triangle

Materials

*geoboards

Warm Up • Mental Math

Have students name the numbers that are 20 more than and 20 less than

1 47 (67, 27)
2 6 tens and 1 one (81, 41)
3. 125 (145, 105)
4. 2,000 (2,020, 1,080)
5. 505 (525, 485)
6. 7 tens and 4 ones (94, 54)

Warm Up • Pencil and Paper

Have students copy and find the difference.

1. 31 − 4 = (27)
2. 25 − 12 = (13)
3. 82 − 7 = (75)
4. 182 − 40 = (142)
5. 220 − 187 = (33)
6. 625 − 292 = (333)

Name ____________________

Triangles

A **triangle** is a polygon with three sides and three angles.

To classify a triangle, we can look at the lengths of its sides and the size of its angles.

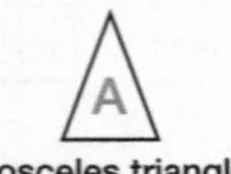

isosceles triangle

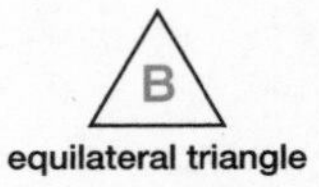

equilateral triangle

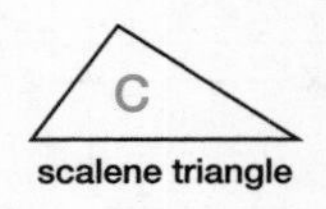

scalene triangle

A triangle with at least two sides or angles that are equal is an **isosceles triangle**. Triangle A is an isosceles triangle.

A triangle that has three sides and angles that are equal is called an **equilateral triangle**. Triangle B is an equilateral triangle.

A triangle that does not have any sides or angles that are equal is called a **scalene triangle**. Triangle C is a scalene triangle.

A triangle can also be classified by its greatest angle measure.

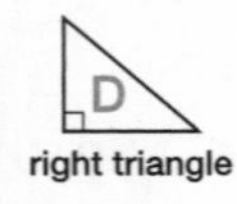

right triangle

acute triangle

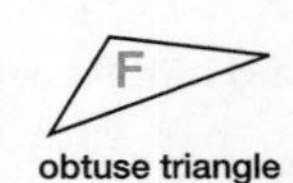

obtuse triangle

A **right triangle** has one right angle.

Triangle D is a right triangle.

An **acute triangle** has three acute angles.

Triangle E is an acute triangle.

An **obtuse triangle** has one obtuse angle.

Triangle F is an obtuse triangle.

Getting Started

Classify the triangle in as many ways as you can.

1. 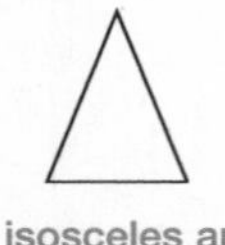
isosceles and acute

2. 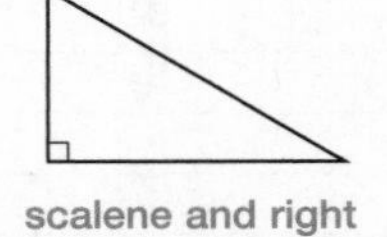
scalene and right

3. isosceles and obtuse

---

2 Teach

Introduce the Lesson Have students follow in their textbook as you read the definition of an isosceles, an equilateral, and a scalene triangle aloud. Draw an illustration of each triangle on the board.

- Read aloud the next section about classifying triangles by their greatest angle measure. Draw a right, an acute, and an obtuse triangle on the board.
- Work with students through each model in their textbook as they fill in the solution sentences. (right, acute, obtuse)

Develop Skills and Concepts Show students how to make an equilateral triangle on a geoboard. Have students make one on their geoboards. Ask volunteers to describe its characteristics. (It has three sides and three angles that are equal.)

- Repeat for each type of triangle in this lesson: isosceles, scalene, right, acute, and obtuse.

3 Practice

Have students complete the exercises on page 222. Remind them that there is more than one way to classify a triangle.

Practice

Identify each triangle in as many ways as you can.

1. 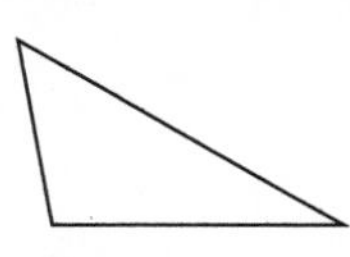
scalene and obtuse

2. equilateral, acute, isosceles

3. scalene and acute

4. 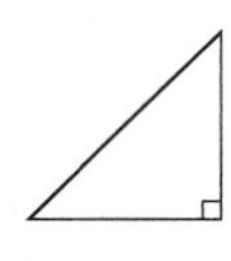
isosceles and right

5. scalene and obtuse

6. 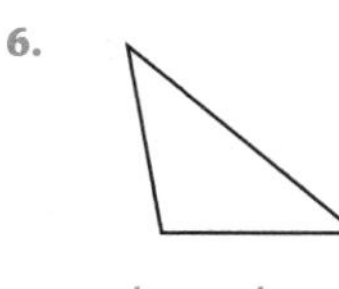
isosceles and obtuse

Draw the triangle. Check students' drawings.

7. isosceles, right triangle

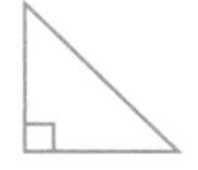

8. scalene, obtuse triangle

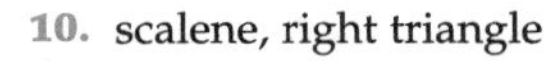

9. equilateral, acute triangle

10. scalene, right triangle

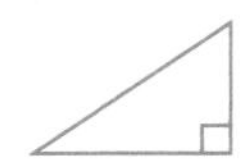

Complete the chart.

	Triangle	Number of Acute Angles	Number of Right Angles	Number of Obtuse Angles
11.	Acute	3	0	0
12.	Right	2	1	0
13.	Obtuse	2	0	1

4 Assess

Draw a triangle with two equal sides and one right angle on the board. Ask students to identify the triangle in as many ways as they can. (isosceles, right)

5 Mixed Review

1. 86 + 5 (91)
2. \$3.25 + \$0.25 (\$3.50)
3. \$25.75 − \$24.23 (\$1.52)
4. 785 − 159 (626)
5. 7 − 3 + 12 (16)
6. 5 × (2 + 7) (45)
7. 3 × (3 + 9) (36)
8. \$40 ÷ 8 (\$5)
9. 78 ÷ 4 (19 R2)
10. 46 ÷ 7 (6 R4)

For Mixed Abilities

Common Errors • Intervention

Some students may have difficulty identifying triangles according to side lengths. Have them use a ruler to measure the sides of a triangle. Remind them to use a right angle as a benchmark when trying to identify a triangle according to its angle.

Enrichment • Geometry

Have students construct three different classifications of triangles out of craft sticks and clay. Tell them that craft sticks can be broken in half to make smaller triangles. Have them identify their triangles in as many ways as they can and ask them to share their findings with the class.

More to Explore • Geometry

Have students work with partners. Give each pair one sheet of construction paper. Have pairs use a ruler to draw an equilateral triangle and ask them to cut it out when they have finished. Remind them to measure the sides of the triangle. Ask, *Can a triangle be both equilateral and isosceles?* (Yes; an equilateral triangle also fits the description of an isosceles triangle.) *Can a triangle be both isosceles and equilateral?* (No; an isosceles triangle does not fit the description of an equilateral triangle; it does not have three equal sides.)

11-5 Quadrilaterals

pages 223–224

1 Getting Started

Objective

- To identify, classify, and draw quadrilaterals

Vocabulary

quadrilateral, parallelogram, rectangle, square, rhombus, trapezoid

Warm Up • Mental Math

Ask students how many of each item will fit.

1. 10-in. plates on 50-in. shelf (5)
2. cars in 2 lots of 32 spaces each (64)
3. quarters in one $5 roll (20)
4. nickels in one $2 roll (40)
5. people in 7 rows of 10 seats (70)
6. 4-in. tiles on a 48-in. counter (12)

Warm Up • Pencil and Paper

Have students halve the number and subtract 8.

1. 52 (18)
2. 120 (52)
3. 534 (259)
4. 300 (142)
5. 36 (10)
6. 900 (442)

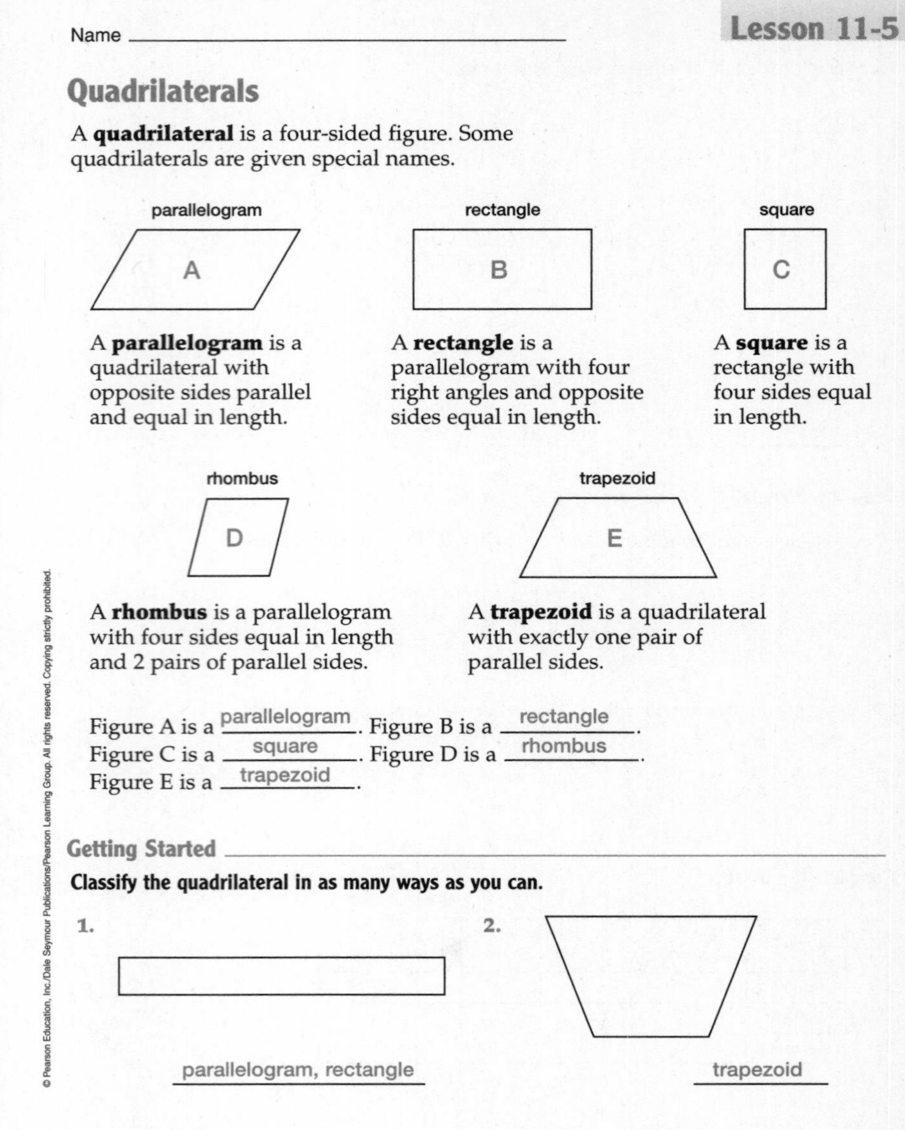

Name ____________________ Lesson 11-5

Quadrilaterals

A **quadrilateral** is a four-sided figure. Some quadrilaterals are given special names.

parallelogram: A — A **parallelogram** is a quadrilateral with opposite sides parallel and equal in length.

rectangle: B — A **rectangle** is a parallelogram with four right angles and opposite sides equal in length.

square: C — A **square** is a rectangle with four sides equal in length.

rhombus: D — A **rhombus** is a parallelogram with four sides equal in length and 2 pairs of parallel sides.

trapezoid: E — A **trapezoid** is a quadrilateral with exactly one pair of parallel sides.

Figure A is a parallelogram. Figure B is a rectangle.
Figure C is a square. Figure D is a rhombus.
Figure E is a trapezoid.

Getting Started

Classify the quadrilateral in as many ways as you can.

1. parallelogram, rectangle
2. trapezoid

2 Teach

Introduce the Lesson Have students read the definition of a quadrilateral. Then, have them look at each illustration and read its accompanying definition.

- Ask, *What are parallel lines?* (two lines that do not intersect) Remind students that a right angle is an angle that forms a square corner and measures 90°.
- Have a volunteer read the solution sentences as students fill them in. (parallelogram, rectangle, square, rhombus, and trapezoid)

Develop Skills and Concepts Explain to students that just as *tri-* in *triangle* means "three," indicating that a triangle has three sides, *quad-* in *quadrilateral* means "four," indicating four sides.

- Have students use graph paper and a ruler to draw their own rectangle, square, rhombus, and parallelogram. Have them cut out and label each shape.

3 Practice

Have students complete the exercises on page 224 independently. Tell students to use a straightedge for Exercises 9 and 10.

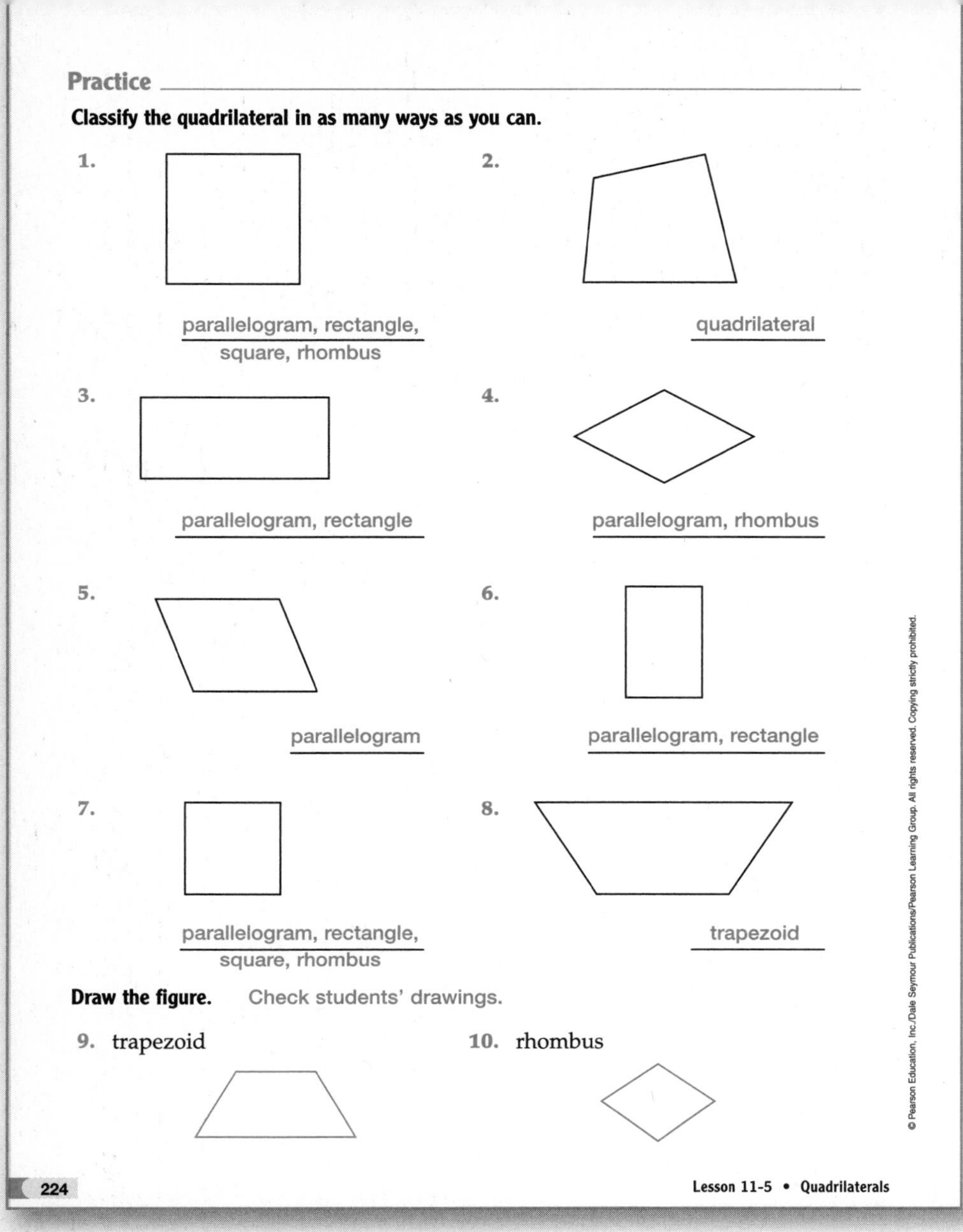

Practice ______

Classify the quadrilateral in as many ways as you can.

1. parallelogram, rectangle, square, rhombus
2. quadrilateral
3. parallelogram, rectangle
4. parallelogram, rhombus
5. parallelogram
6. parallelogram, rectangle
7. parallelogram, rectangle, square, rhombus
8. trapezoid

Draw the figure. Check students' drawings.

9. trapezoid
10. rhombus

4 Assess

Ask students to explain why all squares are rectangles, but all rectangles are not squares. (All squares must have four equal sides, but all rectangles do not have four equal sides.)

5 Mixed Review

1. $5 + (4 + 7)$ (16)
2. $9 + (1 + 0)$ (10)
3. $29 + 59$ (88)
4. $241 - 109$ (132)
5. $956 - 78$ (878)
6. $n \times 8 = 32$ (4)
7. $8 - 3 \times 2 = n$ (2)
8. $38 \div 2$ (19)
9. $21 \div 4$ (5 R1)
10. $52 \div 4$ (13)

For Mixed Abilities

Common Errors • Intervention

Some students may have difficulty understanding the difference between a parallelogram and a rhombus. Remind students that a rhombus has four sides that are equal in length, and a parallelogram has only two opposite sides that are equal in length. Students can use a ruler to measure the sides if they are having difficulty seeing whether the sides are equal in length.

Enrichment • Geometry

Tell students to make quadrilaterals using paper clips and erasers, anchoring the unfolded paper clips to the erasers. Ask them to try and deform the shape. Can they change its shape easily? (yes) Give them more paper clips until they have made their shape rigid. What kinds of shapes make up the quadrilateral when it is rigid? (triangles)

More to Explore • Biography

Grace Chisholm Young was a woman pioneer in the field of mathematics. She was born in England in 1868. Grace was fortunate to be accepted into a girls' boarding school. Her education there prepared her to take university entrance exams. Encouraged by her father, Grace enrolled at Girton College, a women's college.

She applied to graduate school and entered the University at Göttingen, in Germany. She was granted a doctorate, the first official doctorate granted a woman in Germany in any subject area. Besides her research work in mathematics, Grace wrote the *First Book of Geometry*. She felt that to help students learn solid geometric concepts, they should make and handle three-dimensional figures.

11-6 Circles

pages 225–226

1 Getting Started

Objective
- To identify the parts of a circle

Vocabulary
circle, center, chord, diameter, radius

Materials
cardboard, string, tacks

Warm Up • Mental Math
Have students give the product.

1. $29 \times 2 =$ (58)
2. $13 \times 12 =$ (156)
3. $45 \times 7 =$ (315)
4. $21 \times 5 =$ (105)
5. $61 \times 2 =$ (122)
6. $25 \times 5 =$ (125)

Warm Up • Measurement
Have students identify the following:

1. 18 days from now
2. 62 years from today
3. 18 hours from now
4. 3 hours and 20 minutes ago
5. 9 months ago
6. 4 weeks ago

Name ____________________

Circles

A **circle** is a closed figure that is not a polygon. All of the points on the circle are the same distance from the **center** of the circle.

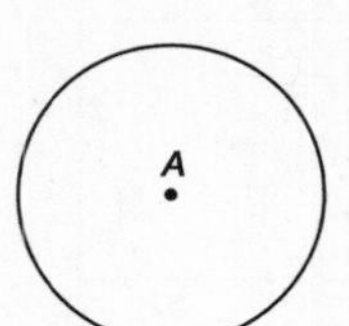

A circle is named by its center.

The circle on the right is named circle *A*.

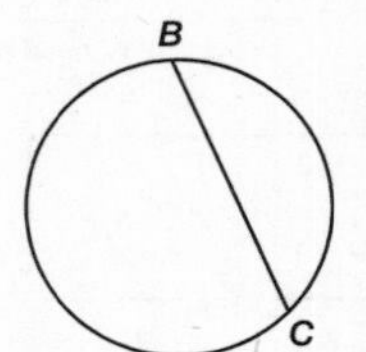

A **chord** is a line segment that connects two points on a circle.

Segment *BC* is a chord.

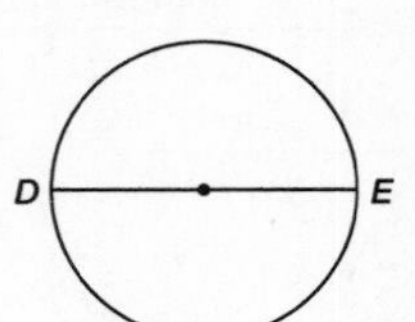

A **diameter** is a chord that passes through the center of the circle.

Segment *DE* is a chord and a diameter.

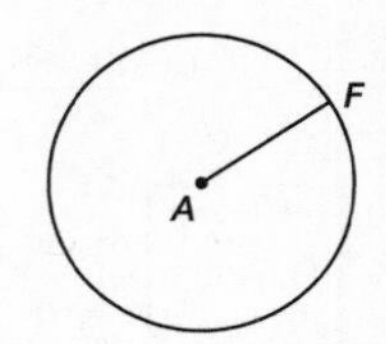

A **radius** is a line segment that connects the center of the circle to any point on the circle.

Segment *AF* is a radius.

Getting Started

Identify and name the part of the circle.

1. 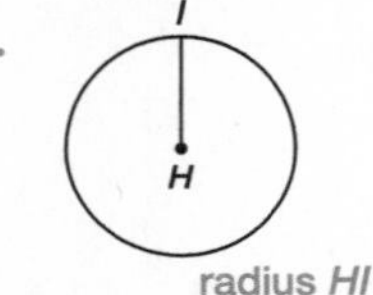

radius *HI*

2. 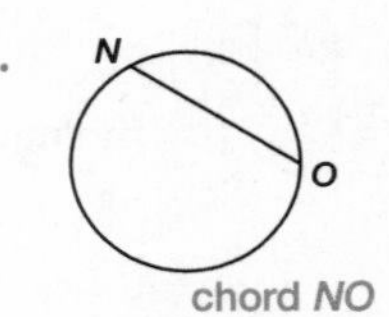

chord *NO*

3. 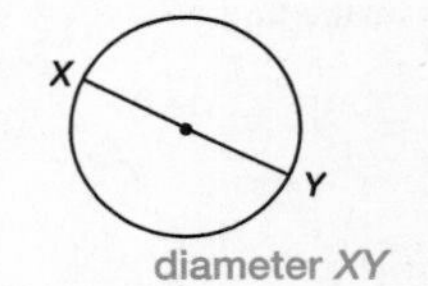

diameter *XY*

2 Teach

Introduce the Lesson Have students read the definition of a circle aloud. As you work through each model with students, reproduce it on the board.

- Show students that the **center** of the circle is a point equidistant from all points along the circle.
- Explain that the **radius** is the distance from the center to a point on the circle. The **diameter** is twice the radius, or the length of a line segment from a point on the circle, through the center, and across to a point on the other side.
- Show students that if they connect two points on the circle with a straight line, they draw a **chord**.
- Have students fill in the solution sentences. (circle *A*, chord, chord, diameter, radius)

Develop Skills and Concepts Have each student use cardboard, string, a tack, and a sheet of paper to draw a circle. Have them put the tack through the paper and cardboard, make a string loop, and put the loop around the tack.

- Explain that if they put a sharp pencil into the loop and swing the pencil around the tack, leaving no slack in the string, they will draw a circle whose radius is the length of the string loop.
- Have them draw and label the center, the radius, the diameter, and a chord.

3 Practice

Have students complete the exercises on page 226 independently.

Practice

Identify and name the part of the circle.

1. 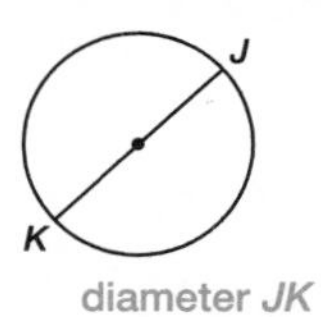

diameter JK

2. 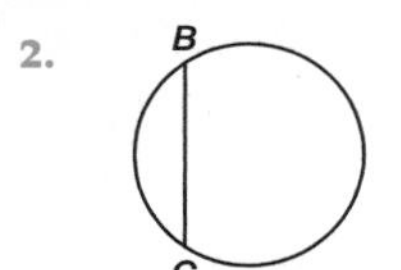

chord BC

3. 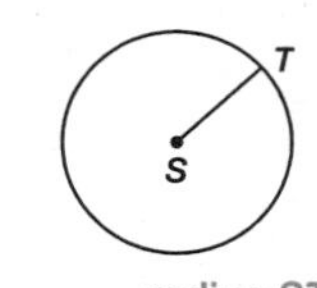

radius ST

4. 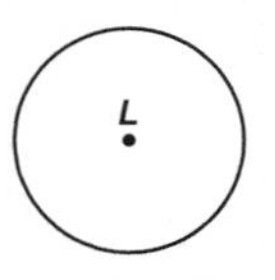

center of circle L

5. 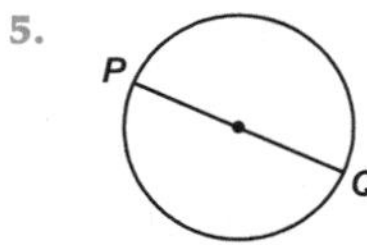

diameter PQ

6. 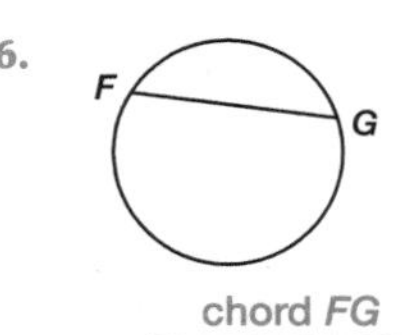

chord FG

Draw and label the figure. Check students' drawings.

7. chord *AB*

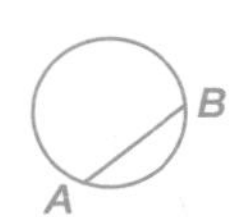

8. radius *DC*

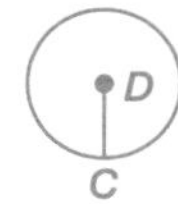

9. diameter *EF*

10. circle with radius *GH* and diameter *IJ*

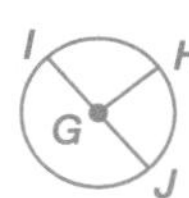

Match the part of the circle to its definition.

11. chord <u>c</u>
12. diameter <u>a</u>
13. radius <u>b</u>

a. a chord that passes through the center of a circle

b. the distance from the center of a circle to any point on the circle

c. a line segment that connects two points on a circle

4 Assess

Draw a circle with a chord on the board. Point to the chord and ask students to identify it. (chord)

5 Mixed Review

1. \$32 + \$15 (\$47)
2. \$85 + \$25 (\$110)
3. \$25 − \$19 (\$6)
4. \$320 − \$159 (\$161)
5. \$12 × 6 (\$72)
6. \$35 × 9 (\$315)
7. \$48 ÷ 12 (\$4)
8. \$15 ÷ 5 (\$3)
9. 49 ÷ 4 (12 R1)
10. 77 ÷ 8 (9 R5)

For Mixed Abilities

Common Errors • Intervention

Some students may confuse radius and diameter. Have them work with partners and a 10-inch circle with the center labeled. Have them use a ruler to draw and measure a radius (5 in.) and a diameter (10 in.). Ask, *How does the length of a radius compare to the length of a diameter?* (It is half as long.) *How does the length of a diameter compare to the length of a radius?* (It is twice as long.)

Enrichment • Measurement

Have students cut circles out of stiff paper. Ask them to mark the center, radius, and diameter. Have them use a ruler to measure the diameter of their circle and ask them to share their findings with the class.

More to Explore • Statistics

Have students make Pascal's triangle and complete nine rows. Have each student graph the numbers appearing in the last line on graph paper.

1
1 1
1 2 1
1 3 3 1
1 4 6 4 1
1 5 10 10 5 1
1 6 15 20 15 6 1
1 7 21 35 35 21 7 1
1 8 28 56 70 56 28 8 1
1 2 3 4 5 6 7 8 9
column

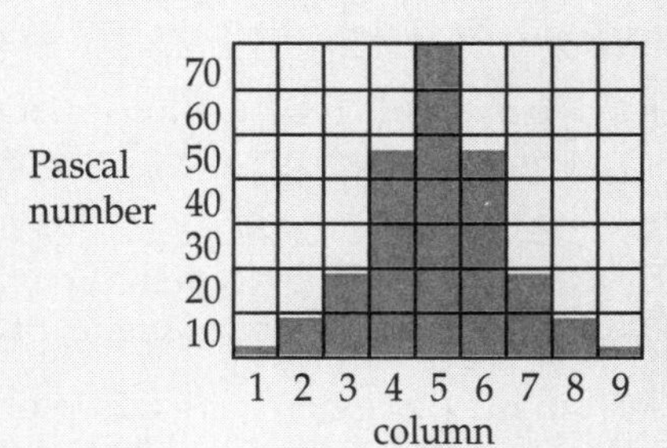

11-7 Solid Figures

pages 227–228

1 Getting Started

Objective

- To identify solid figures

Vocabulary

solid figure, face, vertex, edge

Materials

*wooden cube; *rectangular prism; *pyramid; large cutouts of polygons

Warm Up • Mental Math

Tell students to identify the following:

1. 16 days from now
2. 62 years from today
3. 15 hours from now
4. 2 hours 15 minutes ago
5. 7 months ago
6. hours and minutes yet until 3:45 P.M.
7. 36 years and 1 one month from today
8. 3 weeks ago

Warm Up • Measurement

Show students the wooden cube. Have a student name any length and width for one of its surfaces. Have students find the area and perimeter of that surface. Repeat for the rectangular prism.

Name ____________________

Solid Figures

Many of the objects used in our daily lives are shaped like these solid figures.

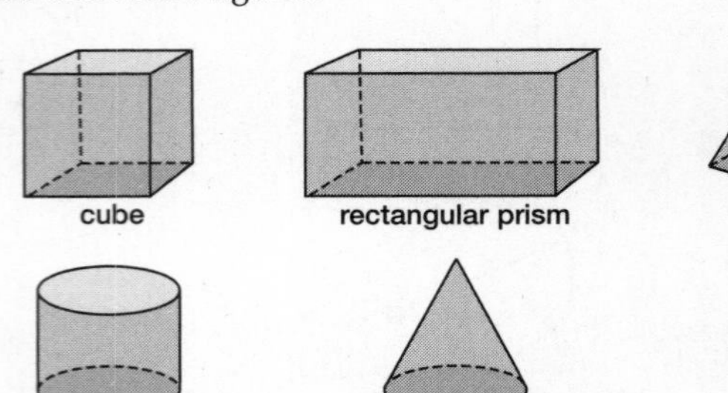

Solid figures have length and width as well as height. Some figures such as cubes, prisms, and pyramids have **faces**, **vertices**, and **edges**.

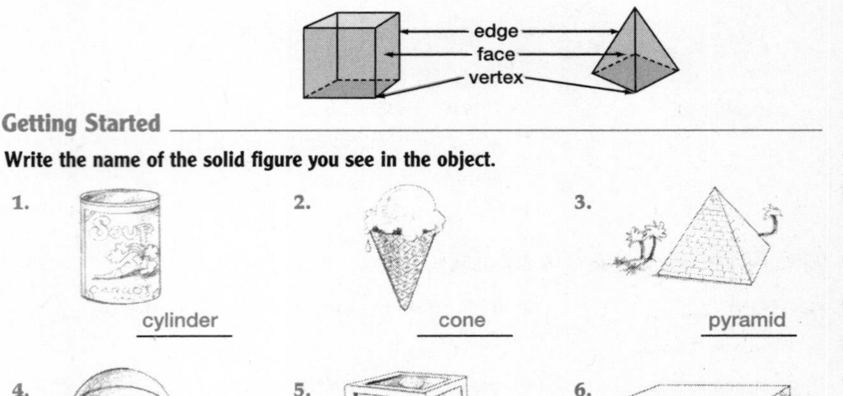

Getting Started

Write the name of the solid figure you see in the object.

1. cylinder
2. cone
3. pyramid
4. sphere
5. cube
6. rectangular prism

2 Teach

Introduce the Lesson Have a student read the opening statement about solid figures. Have students point to each pictured figure as you say its name.

- Have students read the information about solid figures and some properties of cubes, prisms, and pyramids in the model.

Develop Skills and Concepts Remind students that they have been learning about plane, or flat, figures.

- Hold up a wooden cube and point to one surface at a time as you tell students that each surface is a plane figure with length and width. Tell students that if we look at the cube as a whole object, it has length, width, and height. Tell students that we call any figure that has all three dimensions a **solid figure**.
- Now, point to one of the surfaces and tell students that the plane figure, or surface, is called a **face** of the cube. Ask how many faces the cube has. (6)
- Run your finger along the edge of the cube as you tell students that a segment that is the side of a face on a solid is called an **edge**. Have students count the edges on the cube. (12)
- Show a corner on the cube as you tell students that line segments meet at the corners to form a **vertex**. Have students count the number of vertices on the cube. (8)
- Repeat the activity for a rectangular prism and a pyramid.

Practice

Write the name of the solid figure you see in the object.

1. 
cylinder

2. 
cone

3. 
cube

4. 
rectangular prism

5. 
sphere

6. 
pyramid

Complete the table.

7.

Figure	Number of Faces	Number of Edges	Number of Vertices
	6	12	8
	6	12	8
	5	8	5

Now Try This!

These polygons are **symmetrical**. They can be divided into two identical figures by at least one line segment, called an **axis of symmetry**. Draw all axes of symmetry for these polygons. Complete the chart.

	Symmetrical Polygons			
Sides	4	4	6	8
Angles	4	4	6	8
Axes of Symmetry	2	4	4	6

3 Practice

Have students complete the page independently.

Now Try This! Demonstrate the property of symmetry for a triangle. Point out that regular polygons have more than one axis of symmetry. Provide cutouts of the other polygons on a larger scale for students to trace on scratch paper.

4 Assess

Ask students to name three solid figures that have faces, vertices, and edges. (cubes, rectangular prisms, and pyramids)

For Mixed Abilities

Common Errors • Intervention

For students who have difficulty identifying solid figures, have them work with partners to find and name solid figures in the classroom. When appropriate, have them point to a face, vertex, and edge on each model.

Enrichment • Applications

Have students find and identify objects in the classroom that are an example of a rectangular prism and a pyramid. Have students identify the number of edges, faces, and vertices. (rectangular prism: usually, 12 edges, 6 faces, 8 vertices; pyramid: 8 edges, 5 faces, 5 vertices) Then, ask them to identify and count the plane figures that make up each solid figure. (rectangular prism: 2 squares and 4 rectangles; pyramid: 4 triangles and 1 square)

More to Explore • Probability

Explain that while the probability of one event occurring is fairly simple to calculate, the probability of two events occurring is more complicated. Point out that a "tree diagram" can make it easier. Ask students to calculate the probability of drawing a club from a deck of cards and then getting a heads on the flip of a coin. Have them list the possible suits and possible outcomes for the coin flip for each suit.

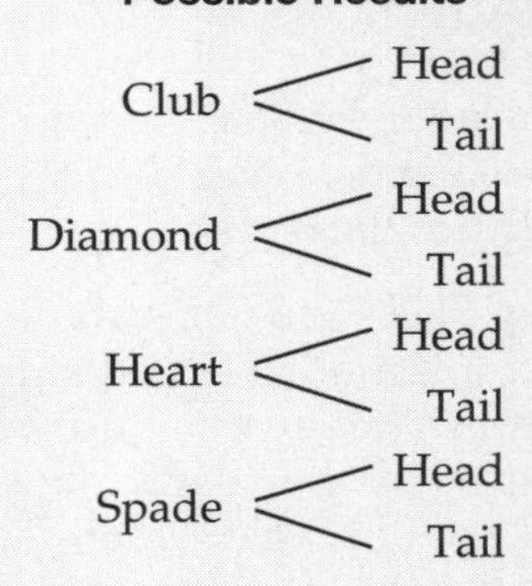

(The probability for clubs and heads is one of eight possible results or $\frac{1}{8}$.)

Ask students why this is called a tree diagram.

11-8 Problem Solving: Act It Out

pages 229–230

1 Getting Started

Objective
- To act out a situation to solve a problem

Materials
56 small objects; 20 pennies; deck of cards; 4 sheets of paper in 4 colors; scissors; 3- and 5-liter containers; 8 different objects; 17 toothpicks; 2 large and 2 small paper clips or similar objects

Warm Up • Mental Math
Ask students how many make the following:

1. boxes of 9 to total 72 (8)
2. seats in each of 6 rows to total 66 (11)
3. 1-in. cookies in 8-by-8-in. pan (64)
4. disks in 4 boxes of 10 each (40)
5. 6 oz cups in 54 oz (9)
6. decades in $1\frac{1}{2}$ centuries (15)
7. $\frac{1}{2}$ inches on a 12-in. ruler (24)

Name ______________________

Problem Solving
Lesson 11-8

Problem Solving: Act It Out

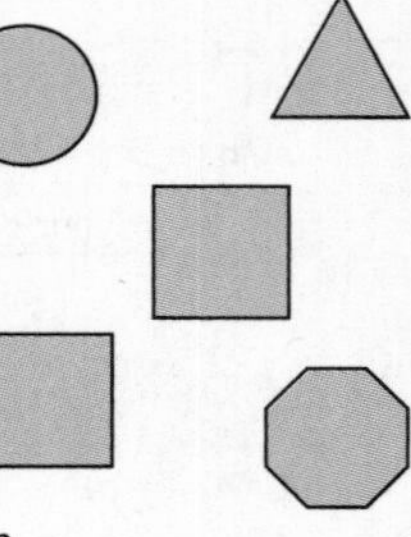

Sunni created a pattern using five figures: a triangle, a rectangle, a square, a circle, and an octagon. The figures are not in that order. The octagon is just to the right of the square. Neither the rectangle nor the triangle is next to the octagon. Neither the triangle nor the rectangle is next to the circle. Neither the circle nor the rectangle is next to the square. Tell the order of the figures in Sunni's pattern.

★SEE
We need to find the order of the five figures in Sunni's pattern.

★PLAN
We can act this problem out by following the clues using the five figures and five empty spaces.

★DO
1. We know the octagon is just to the right of the square.
2. We know the octagon is not next to the rectangle or the triangle, which leaves only the circle.
3. Of the two figures that are left, the rectangle is not next to the square, so we can put the triangle there.
4. The rectangle is the only figure left.

★CHECK
Neither the circle nor the rectangle is next to the square.
Neither the triangle nor the rectangle is next to the circle.
Neither the rectangle nor the triangle is next to the octagon.
The octagon is just to the right of the square.

2 Teach

Introduce the Lesson Remind students about the four-step problem-solving strategy they learned to use for solving problems.

Develop Skills and Concepts Tell students that it is often helpful to act out a situation to find a solution. Tell students that they often use acting to solve problems, such as when they try on shoes to see how they look and feel before buying or when they move furniture in their rooms to see how an arrangement looks. Ask students to give more examples of when they act out a situation.

- Ask students how we can keep track of where we have been, where we are, and where we are going when we act out a situation. (write it down)
- Tell students that keeping a record of our work helps to keep us organized. Tell them that they can write steps in the form of a list, a table, a picture, or a diagram.
- Have students read the problem aloud. Tell students that there is much information given and they need to organize this information into meaningful parts.
- Read the SEE and PLAN step aloud with students. Ask what clue tells us the order of two of the figures. (The octagon is just to the right of the square.) Tell students that this clue gives us a good place to start.
- Have students use manipulatives to act out the problem as they complete the DO step. Tell students that in this problem they may find it more helpful to solve the problem and then record the work.
- Read through the CHECK step with students.

Apply

Act out the problems to help you solve them.

1. Draw the fifth polygon in the pattern.

2. Choose four pieces of colored paper and cut four small circles from each. Arrange the circles in the squares below so that there is a circle in each square and there are no circles of like color in the same row, the same column, or either main diagonal.

B	R	G	Y
Y	G	R	B
R	B	Y	G
G	Y	B	R

B = blue
Y = yellow
R = red
G = green

3. What could be the next shape in the pattern if only solid figures are used and the figure is not one that is already in the pattern?

cylinder

Problem-Solving Strategy: Using the Four-Step Plan

★ **SEE** What do you need to find?
★ **PLAN** What do you need to do?
★ **DO** Follow the plan.
★ **CHECK** Does your answer make sense?

4. A square is not next to a rectangle. A rhombus is not next to a trapezoid. The square is just to the left of the rhombus. Use the clues to draw the pattern.
Answers will vary. One possible response is shown.

5. How many squares can you find in the grid?
Hint: The answer is not 9.
14

6. Draw what the next shape in the pattern could look like.

For Mixed Abilities

More To Explore • Numeration

Have students keep track of examples of Roman numerals they see over the course of a week. Ask students to list the Roman numerals, where they were used, and their Hindu-Arabic equivalents.

Display these lists and have students compare their findings. Ask them to count how many different places Roman numerals were found. Ask students to list places where they think it would be impractical or confusing to use Roman numerals.

3 Apply

Solution Notes

1. Allow students to use rectangular blocks or rectangular cutouts to act out this problem.
2. Help students see that they have 16 circles, 4 of each color. It may be easier for some students to position 4 circles at a time in each row and then position additional circles one at a time until they have met all the qualifications. Students may find solutions other than the following:

R G Y B	B R G Y
B Y G R	Y G R B
G R B Y	R B Y G
Y B R G	G Y B R

3. Help students to list all the solid figures they have learned about in order to solve the problem.

Higher-Order Thinking Skills

4. **Synthesis:** This problem is similar to the one in the model on page 229. Remind students how to use each clue to solve the problem.

5. **Analysis:** Students will need to look at the entire figure to discover the total number of squares. It may be helpful for students to draw over each square using different colored crayons and then count the different colors.

6. **Analysis:** Students should realize that the number of sides of each figure increases by one each time.

Chapter 11 Test

page 231

Items	Objectives
1–2	To identify points, line segments, lines, and rays (see pages 215–216)
3, 7–9	To identify angles: right angles, acute angles, obtuse angles, and straight angles (see pages 217–218)
4–6	To identify intersecting, parallel, and perpendicular lines (see pages 215–216)
10–12	To identify polygons by the number of sides (see pages 219–220)
13–15	To identify quadrilaterals (see pages 223–224)

Name ______________________

Write the name of the figure.

1. A B — line AB or $\overleftrightarrow{AB}$

2. D C — ray CD or $\overrightarrow{CD}$

3. E F G — angle EFG or ∠EFG

Write *intersecting*, *parallel*, or *perpendicular*.

4. C B H K — perpendicular

5. E H G F — intersecting

6. 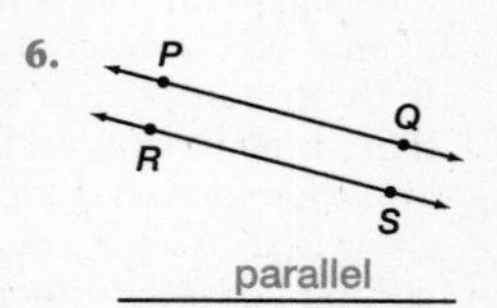

parallel

Name and classify the angle as *acute*, *right*, *obtuse*, or *straight*.

7. E J C — ∠JCE; acute

8. R M B — ∠RMB; straight

9. 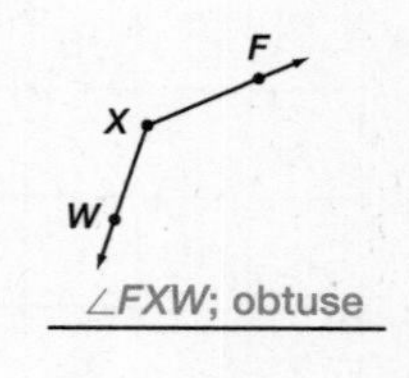

∠FXW; obtuse

Name the kind of polygon.

10. hexagon

11. square

12. 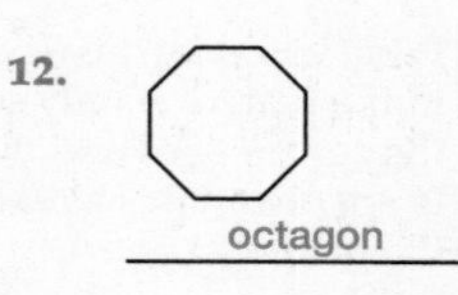
octagon

Name the kind of quadrilateral.

13. parallelogram

14. rhombus

15. trapezoid

Alternate Chapter Test

You may wish to use the Alternate Chapter Test on page 366 of this book for further review and assessment.

CUMULATIVE ASSESSMENT

Circle the letter of the correct answer.

1. Round 4,628 to the nearest thousand.
a. 4,000
(b.) 5,000
c. NG

2. What is the place value of the 9 in 629,206?
a. tens
(b.) thousands
c. hundred thousands
d. NG

3. 4,268 + 3,659
a. 7,817
(b.) 7,927
c. 8,927
d. NG

4. 39,475 + 26,628
(a.) 66,103
b. 56,003
c. 56,093
d. NG

5. 707 − 388
a. 389
b. 419
c. 481
(d.) NG

6. 36,239 − 14,856
a. 22,383
b. 22,623
(c.) 21,383
d. NG

7. Choose the better estimate of volume.
(a.) 1 pint
b. 1 quart

8. $3.25 × 7
a. $21.45
b. $21.75
(c.) $22.75
d. NG

9. 16 × 45
a. 144
b. 490
(c.) 720
d. NG

10. 516 × 25
a. 3,612
b. 12,700
(c.) 12,900
d. NG

11. 27)162
a. 5
(b.) 6
c. 7
d. NG

12. 629 ÷ 34
a. 18
(b.) 18 R17
c. 180 R17
d. NG

13. Name the figure.
a. square
b. rectangle
(c.) parallelogram
d. NG

score

STOP

Cumulative Assessment

page 232

Items	Objectives
1	To round numbers to the nearest hundred (see pages 31–32)
2	To identify place value through hundred thousands (see pages 25–26)
3–4	To add two numbers up to 5 digits (see pages 41–48)
5–6	To subtract up to two 5-digit numbers (see pages 59–62)
7	To determine the appropriate customary unit of capacity (see pages 145–148)
8–10	To multiply up to 3-digit numbers by up to 2-digit numbers including money with and without regrouping (see pages 167–180)
11–12	To divide 3-digit numbers by 2-digit numbers to get up to 2-digit quotients with and without remainders (see pages 196–206)
13	To identify polygons (see pages 219–220)

Alternate Cumulative Assessment

Circle the letter of the correct answer.

1. Round 6,318 to the nearest thousand.
(a) 6,000
b 7,000
c NG

2. What is the place value of the 4 in 429,631?
a ten thousands
(b) hundred thousands
c millions
d NG

3. 6,422 + 3,398
a 9,710
(b) 9,820
c 9,830
d NG

4. 44,378 + 18,644
a 62,942
b 62,932
c 63,042
(d) NG

5. 604 − 276
(a) 328
b 338
c 428
d NG

6. 46,328 − 12,974
a 33,454
b 33,554
(c) 33,354
d NG

7. What would you use to measure gas in a car?
a pints
b quarts
(c) gallons
d NG

8. $4.64 × 8 =
a $32.14
b $36.22
(c) $37.12
d NG

9. 27 × 36 =
a 732
b 243
(c) 972
d NG

10. 623 × 34
a 20,182
b 21,172
c 21,082
(d) NG

11. 32)224
(a) 7
b 7 R1
c 7 R10
d NG

12. 756 ÷ 25
a 3 R6
(b) 30 R6
c 36
d NG

13. What is the name of a six-sided figure?
a pentagon
(b) hexagon
c octagon
d NG

12-1 Similar and Congruent Figures

pages 233–234

1 Getting Started

Objective

- To identify figures that are similar or congruent

Vocabulary

similar, congruent

Materials

*cardboard square; *2 × 3 and 4 × 6 unit cardboard rectangles;* cardboard triangle; *congruent and similar figures; *figures not similar or congruent; transparent grid paper; rulers

Warm Up • Mental Math

Have students give the total cost if gasoline is $1.10/gal.

1. 10 gal ($11)
2. 11 gal ($12.10)
3. 7 gal ($7.70)
4. $10\frac{1}{2}$ gal ($11.55)
5. $2\frac{1}{10}$ gal ($2.31)
6. 20 gal ($22)

Warm Up • Geometry

Dictate the following angles for students to draw on the board and label: $\angle C$, $\angle XYZ$, right angle STU, $\angle R$, and $\angle KLM$. Have students use rulers and trace square corners for the right angle.

Name ________________________________

CHAPTER 12

Using Geometric Figures

Lesson 12-1

Similar and Congruent Figures

Plane figures that have the same shape are called **similar** figures. Plane figures that have the same shape and size are called **congruent** figures. Which triangles are similar to triangle BAC? Which triangle is congruent to triangle BAC?

Grid paper can be used to show that two plane figures are similar.

Triangle ABC is similar to triangles NMO and ZYX.

Tracing paper can be used to show that two plane figures are congruent.

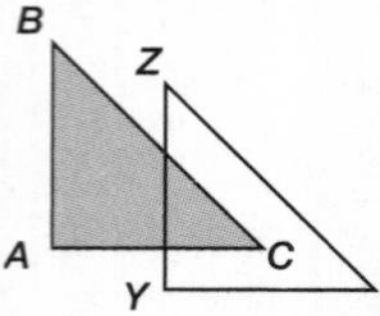

Triangle BAC is congruent to triangle ZYX.

Getting Started

Tell if each pair of figures is similar. Write *yes* or *no*.

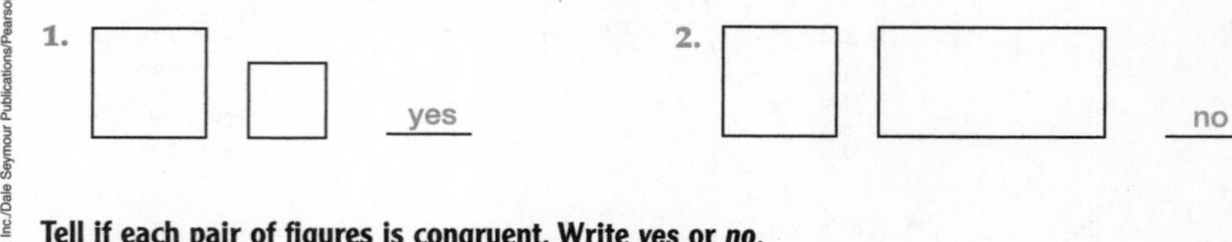

1. yes
2. no

Tell if each pair of figures is congruent. Write *yes* or *no*.

3. no
4. yes

2 Teach

Introduce the Lesson Have a student read the problem aloud and tell what is to be answered. (which triangle is similar to and which is congruent to triangle ABC)

- Ask students what information is known about similar plane figures. (They have the same shape.) Ask what is known about congruent figures. (They have the same shape and size.) Have students read with you as they complete the sentences.

Develop Skills and Concepts Before class, use the cardboard triangle and trace a triangle on the board. Then, rotate the cardboard and trace a second triangle.

- Have students use transparent grid paper to duplicate the figures and cut them out. Tell students to lay one triangle on top of the other and tell if the triangles are the same size and shape. (yes) Tell students that the two triangles are **congruent** because they have the same size and shape.
- Make two tracings of the cardboard square on the board. Tell students that the two squares are also congruent.
- Make tracings of the two rectangles on the board and tell students that the rectangles are called **similar** figures because they have the same shape but not the same size.
- Have students identify congruent and similar plane figures in the classroom. If possible, have students trace the figures on the board to check their identification.

3 Practice

Allow students to use tracing paper to check their work if needed. Have students complete page 234 independently.

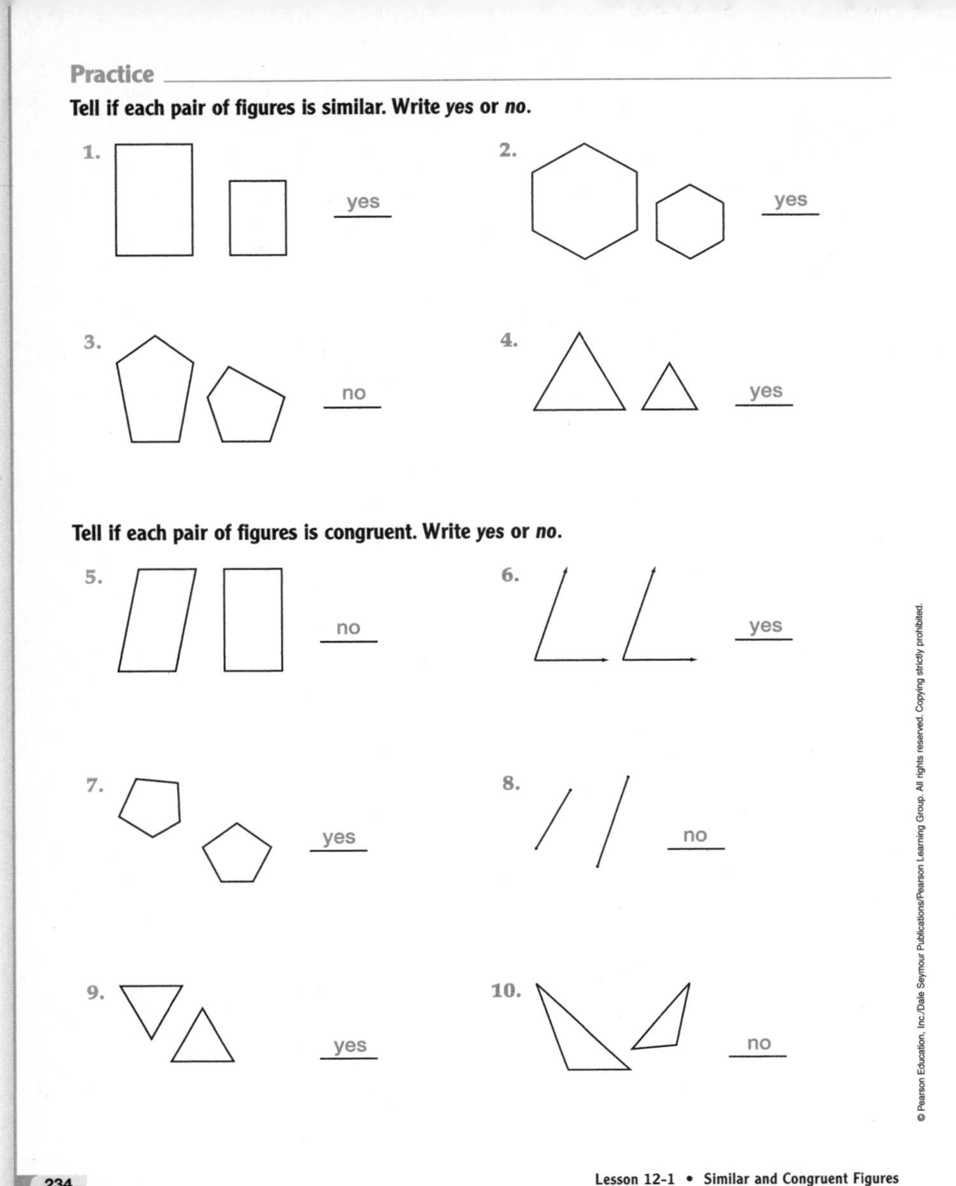
Practice

Tell if each pair of figures is similar. Write *yes* or *no*.

1. yes
2. yes
3. no
4. yes

Tell if each pair of figures is congruent. Write *yes* or *no*.

5. no
6. yes
7. yes
8. no
9. yes
10. no

4 Assess

Ask students to give the definition of similar and congruent figures and draw an example of each. (Similar figures are plane figures that have the same shape. Congruent figures are plane figures that have the same shape and size. Drawings will vary.)

5 Mixed Review

1. \$7.98 ÷ 3 (\$2.66)
2. 600 × 7 (4,200)
3. 2,050 − 923 (1,127)
4. 27 × 436 (11,772)
5. 58,176 + 4,950 (63,126)
6. 738 + 297 + 30 (1,065)
7. 21,753 − 17,247 (4,506)
8. 123 ÷ 6 (20 R3)
9. 608 × 9 (5,472)
10. 57 ÷ 6 (9 R3)

For Mixed Abilities

Common Errors • Intervention

Some students may have difficulty identifying similar and congruent figures. Have them work in pairs with cutout figures, some of which are similar, some congruent, and some with neither relationship. Have them take turns identifying similar and congruent figures.

Enrichment • Geometry

Tell students to use grid paper to draw five plane figures. Then, have them draw and cut out one congruent figure and one similar figure for each of the five plane figures.

More to Explore • Numeration

Have students create their own pictorial number system, for numbers 1 to 10, by using common objects to stand for each number.

Tell students that the object chosen to represent a number must have a reference to that number. For example, sun = one, eyes = two, triangle = three, and so on. Tell them to write a story problem using their number system and give it to a partner to solve.

Extend the activity by having students write their phone numbers and addresses using their systems.

12-2 Symmetry

pages 235–236

1 Getting Started

Objective
- To identify and draw lines of symmetry

Vocabulary
line of symmetry, symmetrical

Materials
rectangular sheets of paper

Warm Up • Mental Math
Have students tell how old each person was in 1994.

1. Katrina, born in 1942 (52)
2. Christine, born in 1976 (18)
3. Ira, born in 1955 (39)
4. Kwame, born in 1967 (27)
5. Marva, born in 1982 (12)
6. Cosmo, born in 1932 (62)

Warm Up • Pencil and Paper
Have students copy the following and find the sum:

1. 76 + 11 = (87)
2. 358 + 12 = (370)
3. 459 + 95 = (554)
4. 385 + 259 = (644)
5. 986 + 853 = (1,839)
6. 1,458 + 187 = (1,645)

Name ______________________ Lesson 12-2

Symmetry

A figure is **symmetrical** if you can fold it so that two parts match exactly. The line where the figure can be folded so that two parts match is called a **line of symmetry**.

The letter *A* is symmetrical.

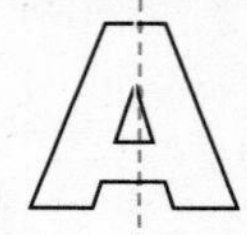

Some figures have more than one line of symmetry.

The letter *H* has two lines of symmetry.

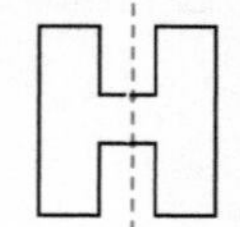

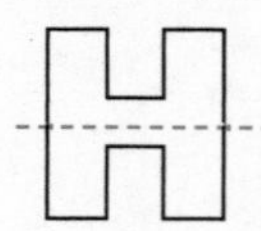

Getting Started

Tell if the dotted line is a line of symmetry. Write *yes* or *no*.

1. no
2. yes

Draw all lines of symmetry.

3. 4. 5.

Draw the lines of symmetry in this figure.
Then, write the number of lines of symmetry this figure has.

6. 2

2 Teach

Introduce the Lesson Ask students to follow in their textbook as you read the introduction. Have them examine each illustration.

- Draw an illustration of the letter *H* on a large sheet of construction paper. Fold it in half vertically. Show that the two halves match. Explain that the fold line is the **line of symmetry**. Ask if it is possible to find another line of symmetry. (yes) Then, fold the paper in half horizontally to show that the other two halves match.

Develop Skills and Concepts Tell students to fold the rectangular sheet of paper in half making sure corners match. Have students unfold their papers and ask, *Do both halves of the paper match exactly?* (yes)

- Explain that when half of a figure can be matched to the other half by folding, the figure is **symmetrical**.

Explain to students that there are many symmetrical letters in the alphabet.

- Give each student three sheets of paper. Have students draw three capital letters that have at least one line of symmetry. (A, B, C, D, E, H, I, K, M, O, T, U, V, W, X, Y) Have them draw a dotted line to mark each line of symmetry. Remind students that they can fold the paper in half to check for symmetry.

3 Practice

Have students complete the exercises on page 236. Remind them that a figure can have more than one line of symmetry.

Practice

Tell if the dotted line is a line of symmetry. Write yes or *no*.

1. 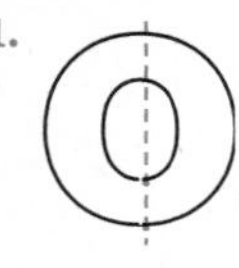
yes

2. 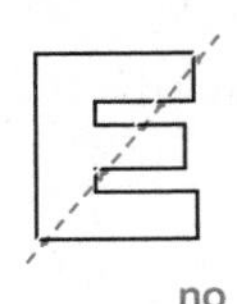
no

3. 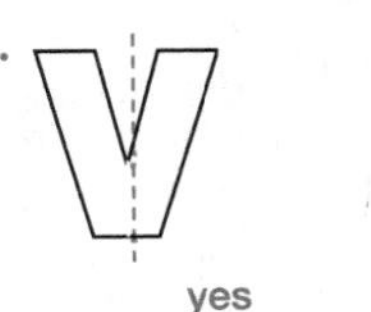
yes

Draw all lines of symmetry.

4. 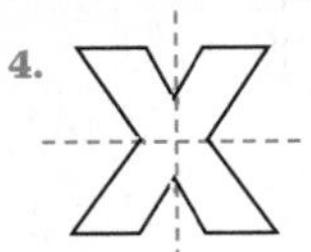

5. 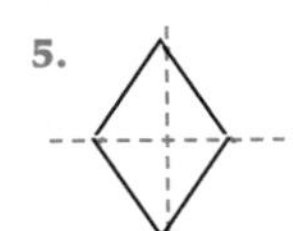

6. 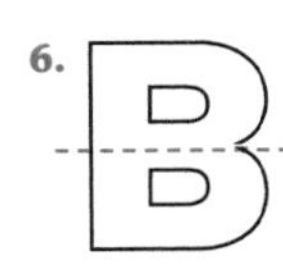

Draw the lines of symmetry in each figure.
Then, write the number of lines of symmetry each figure has.

7. 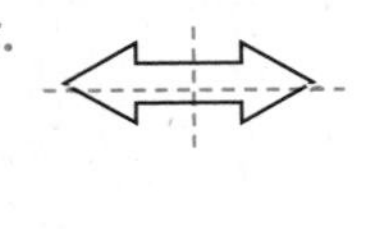
2

8. 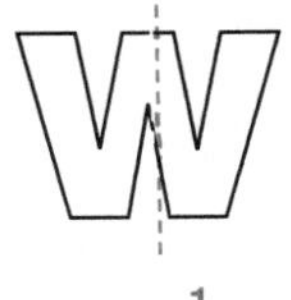
1

9. 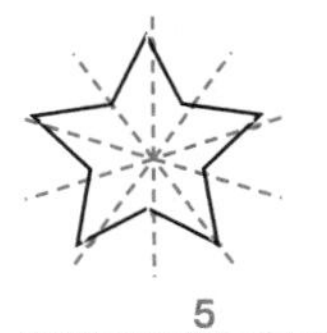
5

Now Try This!

Is the figure symmetrical? If yes, draw all of its lines of symmetry.

1. 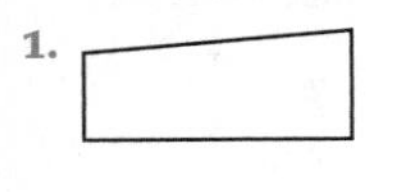
no

2. 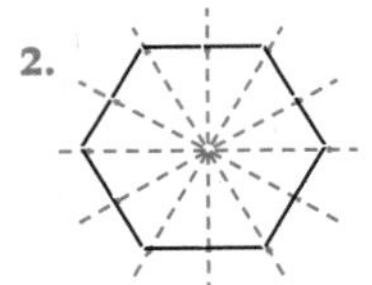
yes

4 Assess

Have students draw a rectangle and an equilateral triangle on construction paper. Have them mark all lines of symmetry with a dotted line. (rectangle, 2; equilateral triangle, 3)

5 Mixed Review

1. 98 ÷ 2 (49)
2. 88 ÷ 11 (8)
3. 64 ÷ 4 (16)
4. 31 ÷ 5 (6 R1)
5. 9 − 3 + 15 (21)
6. 0 × (3 + 7) (0)
7. 2 × (6 + 7) (26)
8. 8 ÷ 1 + 9 (17)
9. 3 + (2 × 9) (21)
10. 6 + (4 × 6) (30)

For Mixed Abilities

Common Errors • Intervention

Some students may have difficulty drawing lines of symmetry. Have them work with partners and a sheet of paper on which are shown all uppercase letters of the alphabet. Have them take turns, one letter at a time, drawing any lines of symmetry. Students should recognize that one way to test symmetry is to cut out the letter and fold it different ways to see halves match.

Enrichment • Application

Have students fold tissue paper and cut out symmetrical patterns with scissors. Suggest that they start with a paper that has two lines of symmetry.

More to Explore • Numeration

Have students devise a list of 15 addition and subtraction problems in which they have left out the operational sign. For example, 62 (+) 53 = 115, 32 (−) 17 = 15, and so on. Then, have students exchange lists with a partner and complete each other's problems. Partners should correct the problems they devised and record the scores.

ESL/ELL STRATEGIES

Write on the board the terms *symmetrical* and *line of symmetry*. As you discuss and define these terms, have students take turns making drawings on the board to illustrate each term.

12-3 Transformations

pages 237–238

1 Getting Started

Objective
- To identify and describe transformations

Vocabulary
transformation, slide, translation, flip, reflection, turn, rotation

Materials
*cardboard trapezoid

Warm Up • Mental Math
Ask students what number tells the following:

1. minutes in 9 hours (540)
2. tens in 9,432 (3)
3. inches in 7 feet (84)
4. sneakers in 6 pairs (12)
5. cups in a pint (2)
6. cookies in 12 dozen (144)

Warm Up • Pencil and Paper
Have students copy the following and find the product:

1. $32 \times 10 =$ (320)
2. $58 \times 17 =$ (986)
3. $37 \times 45 =$ (1,665)
4. $66 \times 10 =$ (660)
5. $49 \times 99 =$ (4,851)
6. $87 \times 78 =$ (6,786)

Name ____________________

Transformations

You can slide, flip, or turn figures and they will remain congruent to each other. Each one of these moves is called a **transformation**.

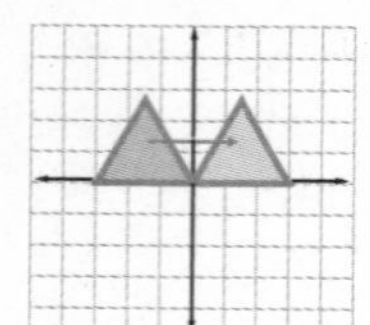

A **slide** moves a figure a given distance in a given direction. Another name for a slide is a **translation**.

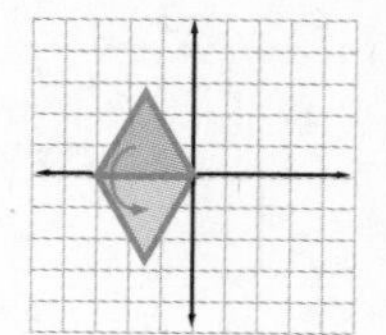

A **flip** moves a figure over a line to get a mirror image. Another name for a flip is a **reflection**.

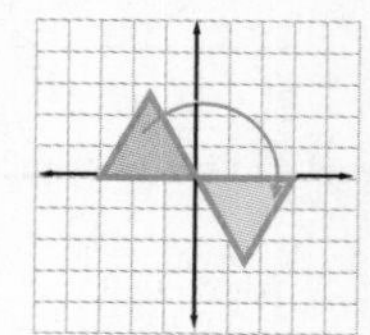

A **turn** rotates a figure clockwise or counterclockwise. Another name for a turn is a **rotation**.

With each move, the triangle did not change its ___shape___ or ___size___.

The original triangle and the resulting triangle are ___congruent___.

Getting Started

Write *translation*, *rotation*, or *reflection* to describe each move.

1. 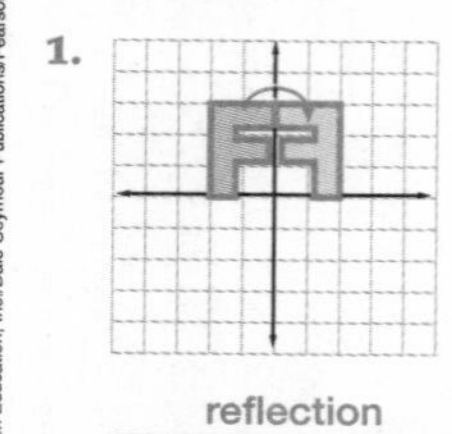

reflection

2. rotation

3. 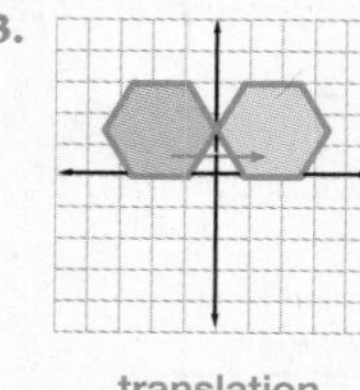

translation, reflection

2 Teach

Introduce the Lesson Discuss some realistic situations in which students moved from one spot to another, for example, slid down a slide, swung on a swing, and so on.

- Ask students to follow in their textbook as you read slowly, stopping to explain or answer questions students might have about the resulting figures for each move.
- Next, have students study each of the transformations, paying special attention to whether the figures are changing. Then, have them complete the solution sentences. (shape, size, congruent)

Develop Skills and Concepts Draw a graph on the board and use the cardboard trapezoid to demonstrate each type of transformation: slide/translation, flip/reflection, and turn/rotation and have students identify the transformations.

- Explain to students that the original figure and the transformed figure in each case are congruent. Remind them that the move does not change the shape or size of the figure.

3 Practice

Have students complete the exercises on page 238. Allow students to work with a partner.

Practice

Write *translation*, *rotation*, or *reflection* to describe each move.

1. 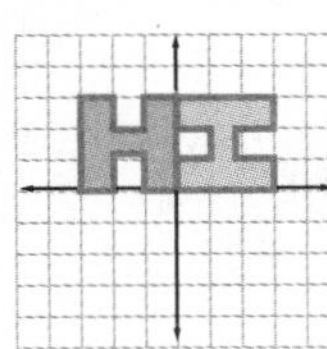
   rotation

2. rotation

3. reflection

4. 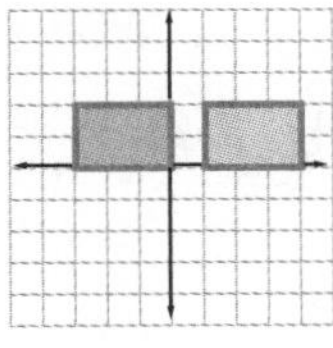
   translation

5. 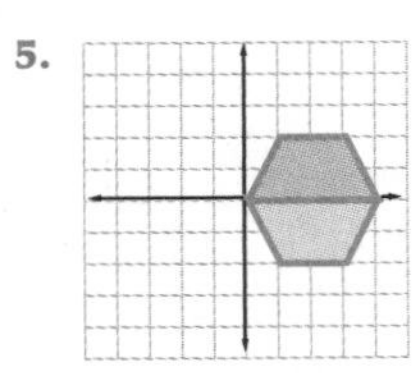
   reflection

6. 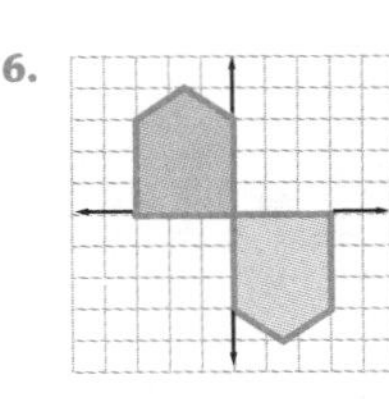
   rotation

7. 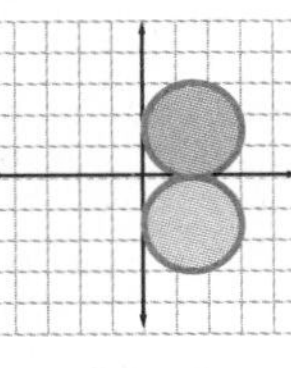
   reflection

8. 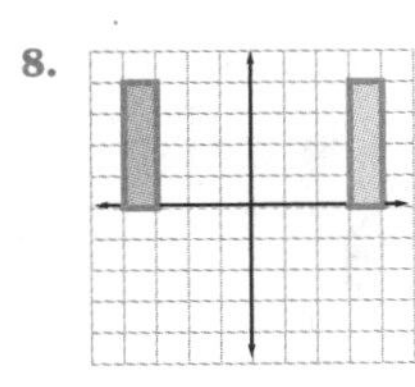
   translation

9. 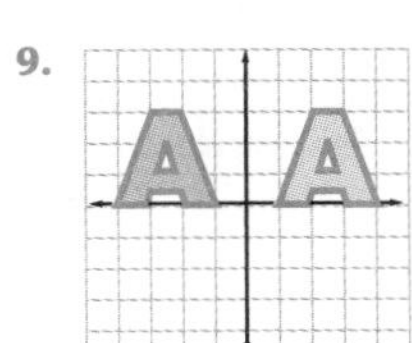

   translation

Now Try This!

1. Name two lowercase letters that can be formed by transforming the letter *b*.
   d, p, or q

2. Name three uppercase letters that look the same after being reflected.
   A, H, I, M, O, T, U, V, W, X, or Y

4 Assess

Draw a graph on the board and use a cutout of a right triangle to demonstrate a slide/translation, a flip/reflection, and a turn/rotation. Have students identify each transformation.

5 Mixed Review

1. $5 - 2 + 22$ (25)
2. $9 - 5 + 15$ (19)
3. $1 \times (3 + 5)$ (8)
4. $5 \times (6 + 3)$ (45)
5. $9 \times (4 + 7)$ (99)
6. $2 \div 1 + 10$ (12)
7. $4 \div 2 + 3$ (5)
8. $6 \div 3 + 52$ (54)
9. $7 + (3 \times 6)$ (25)
10. $11 + (4 \times 9)$ (47)

For Mixed Abilities

Common Errors • Intervention

Some students may have difficulty identifying the difference between transformations. Have them draw and cut out a figure of their choice. Then, ask them to fold a sheet of grid paper in four and mark the center point. Go through each transformation, slide/translation, flip/reflection, and turn/rotation, on the board and have them follow along, moving their figure across the paper as you move your figure across the board. Have them use the terms *slide, flip*, and *turn* instead of *translation, reflection*, and *rotation* until they are more confident with transformations.

Enrichment • Geometry

Have students trace a cardboard triangle cutout on grid paper. Then, have them slide the cutout along the same grid line and trace a second triangle. Ask students to identify the transformation using both terms. (slide/translation)

More to Explore • Geometry

Have students align one side of a trapezoid cutout along a grid line and trace it. Then, have them flip the cutout over the grid line and trace the figure. Remind students that the figure has not changed its shape or size. Have them identify the transformation using both terms. (flip/reflection) Ask them if the original trapezoid is symmetrical with the resulting trapezoid. (yes) Remind them that they can check for symmetry by folding a figure in half.

ESL/ELL STRATEGIES

Provide students with triangular paper cutouts. Define terms *slide, flip, reflection, turn*, and *rotation*. Ask students to demonstrate the meaning of each term by manipulating their triangles. Check their movements for accuracy.

12-4 Perimeter

pages 239–240

1 Getting Started

Objective
- To find the perimeter of a figure

Vocabulary
perimeter

Materials
*cardboard rectangle and square; rulers

Warm Up • Mental Math
Tell students to name the figure or part of a figure that

1. has no height or depth (plane)
2. is 2 lines that cross (intersecting lines)
3. has 2 endpoints (line segment)
4. is a point of an angle (vertex)
5. is formed by 2 intersecting rays (angle)
6. is part of a line (line segment)
7. is a square corner (right angle)
8. has no endpoints (line)

Warm Up • Numeration
Review column addition through 2-digit addends. Have students write and solve the problem 14 + 14 + 14 + 14 using an operation other than simple addition. ($4 \times 14 = 56$) Repeat for 16 + 9 + 9 + 16. [(2×16) + (2×9) = 50] Provide additional problems.

Name ______________________

Perimeter

Dino is fencing in a section of the yard for his dog. How many meters of fencing will Dino need?

We want to know the distance around the dog's section of the yard. This section is shaped like a parallelogram.

The length of the section is __5__ meters.

Its width is __3__ meters.

Because the section is a parallelogram, we know that opposite sides are equal. The distance around a polygon is called its **perimeter**.

To find the perimeter of a polygon, we add the length of all the sides. We add __3__, __5__, __3__, and __5__.

3 + 5 + 3 + 5 = __16__

Dino will need __16__ meters of fencing.

Getting Started

Find the perimeter.

1. 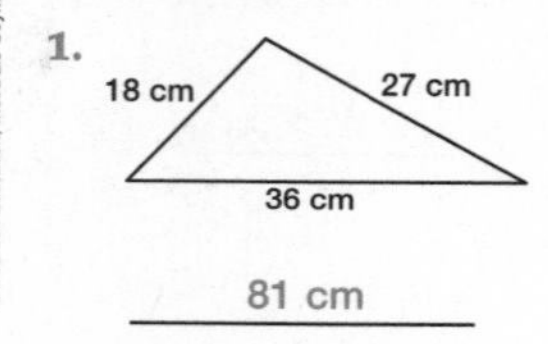

81 cm

2. 16 m

64 m

2 Teach

Introduce the Lesson Ask a student to describe the picture, read the problem aloud, and tell what information is given. (The space to be fenced is 3 meters on one side and 5 meters on another.) Ask what information is needed. (the length of the other two sides)

- Have students read aloud with you as they complete the sentences to solve the problem.

Develop Skills and Concepts Draw a rectangle on the board. Have students identify the figure. (rectangle) Ask students what length the opposite side would be as you write **7 ft** along the length of the rectangle. (7 ft) Ask how we know this. (The figure is a rectangle and opposite sides are congruent.)

- Repeat for a width of **3 ft**. Tell students that they can find the **perimeter**, or the distance around this figure, by adding 7 ft + 7 ft + 3 ft + 3 ft. Have a student find the perimeter. (20 ft)
- Change the dimensions to 13 cm by 28 cm and have students find the perimeter. (82 cm)
- Show students how to find the perimeter by adding two times the length and two times the width.
- Have students find the perimeter of a square when the length of only one side is known. Show students that side + side + side + side is the same as 4 times the measurement of one side.

3 Practice

Have students find the perimeter of each figure. Ask students what information is needed to work Exercise 8. (how many sides a pentagon has; 5) Have students complete the table on page 240 independently.

Practice

Find the perimeter.

1. 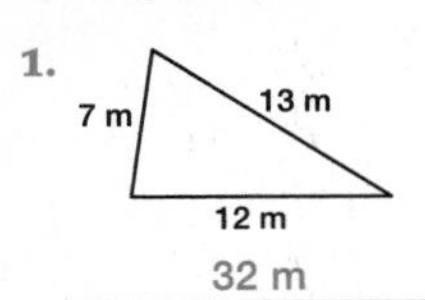

32 m

2. 24 cm
56 cm

160 cm

3. 15 m
50 m
40 m
45 m
24 m

174 m

4. 10 cm
34 cm

88 cm

5. 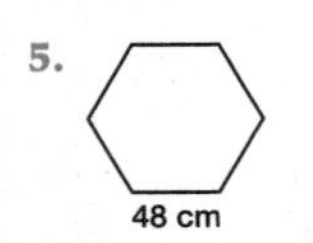

288 cm

6. 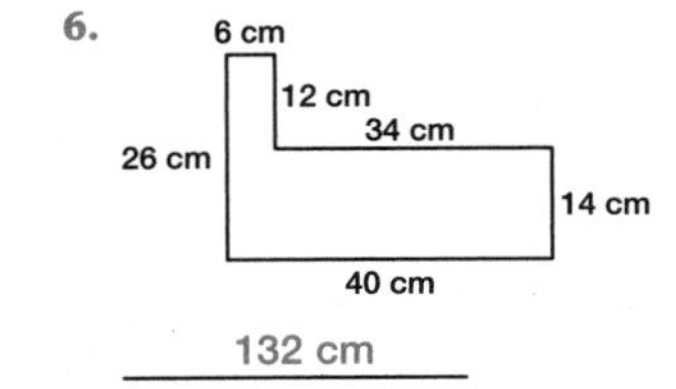

132 cm

Complete the table.

7. 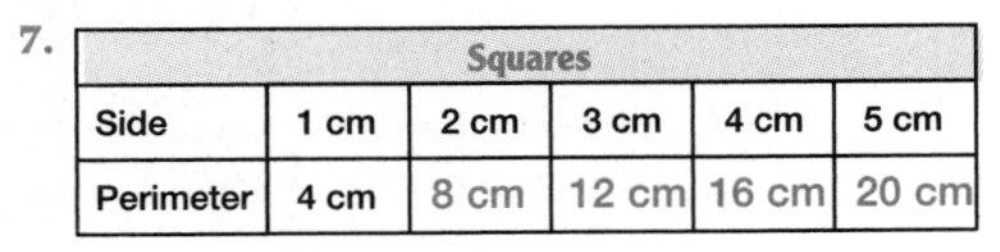

Squares					
Side	1 cm	2 cm	3 cm	4 cm	5 cm
Perimeter	4 cm	8 cm	12 cm	16 cm	20 cm

Problem Solving

Solve each problem.

8. Find the perimeter of a rectangular parking lot that is 28 meters wide and 47 meters long. 150 meters

9. Allison drew a pentagon with all the sides equal in length. One side was 14 centimeters. What was the perimeter of the pentagon? 70 centimeters

4 Assess

Ask students to find the perimeter of a rectangle with a width of 36 yards and a length of 42 yards. (156 yards)

For Mixed Abilities

Common Errors • Intervention

Watch for students who forget to include the measures of all the sides of a figure when finding the perimeter. Have them work with partners and a variety of polygons, using rulers to measure the length of each side. Then, have them write addition sentences and find the perimeter.

Enrichment • Measurement

Have students draw the shape of their room at home. Tell them to be sure to trace a square corner to draw a right angle. Tell them to measure the length of each wall and then find the perimeter of their room.

More to Explore • Logic

Read the following to students for them to solve:

On her way to school, Heather counted 47 trees along the right side of the street. On the way home, she counted 47 trees on the left side of the street. How many trees did Heather count in all? (47 trees; the trees on her right, going to school are on her left when she comes home.)

12-5 Area

pages 241–242

1 Getting Started

Objective
- To find the area of a figure

Vocabulary
area, square units, formula

Materials
various-sized rectangular sheets of grid paper; rulers

Warm Up • Mental Math
Tell students to solve for R.

1. $20 \times R = 100$ (5)
2. $160 \div R = 8$ (20)
3. $8 \times R = 888$ (111)
4. $(3 \times R) + (6 \times R) = 108$ (12)
5. $4 \times R = 240$ (60)
6. $(2 \times R) + (2 \times 10) = 28$ (4)
7. $\frac{1}{2}$ of $R = 720$ (1,440)
8. $R \div 5 = 150$ (750)

Warm Up • Numeration
Draw an array of 3 rows of 6 Xs on the board. Ask students how many rows of Xs there are. (3) Ask how many Xs are in each row. (6) Ask students how to find the total number of Xs. (multiply 3 by 6 or 6 by 3) Repeat for more problems to find the total number when objects are arranged in rows.

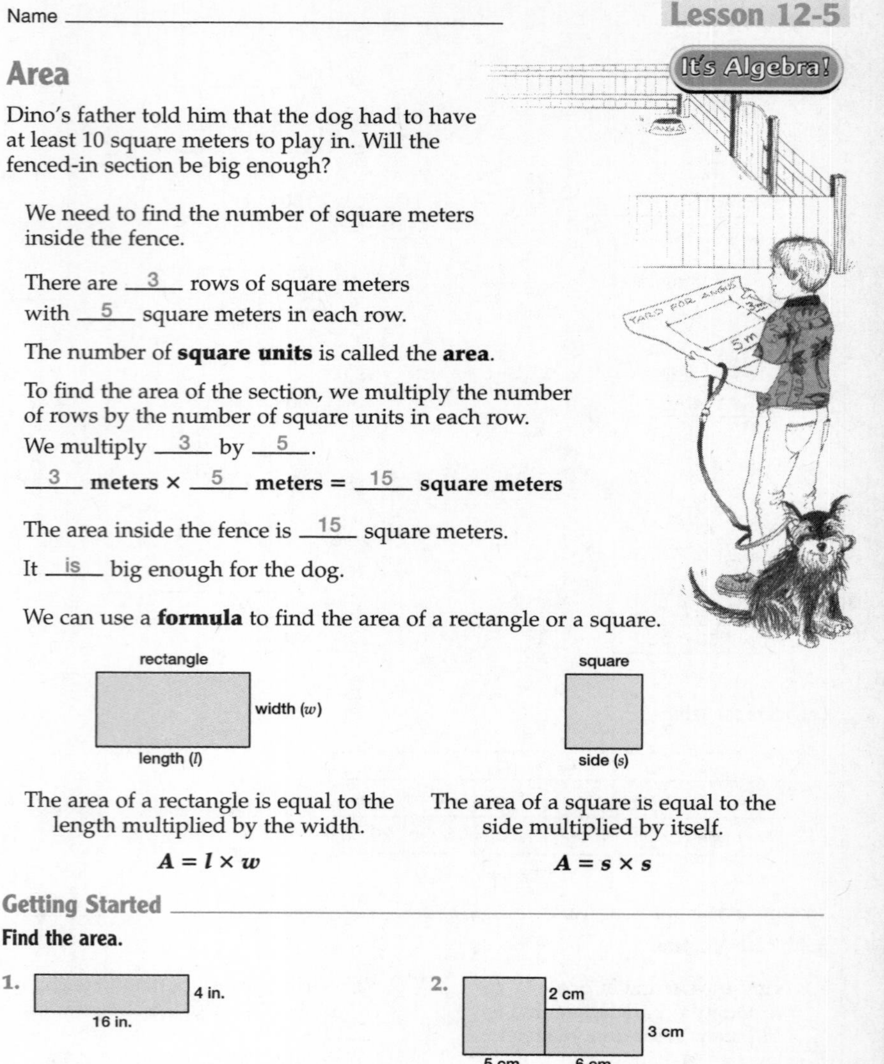

Name ____________________

Lesson 12-5

It's Algebra!

Area

Dino's father told him that the dog had to have at least 10 square meters to play in. Will the fenced-in section be big enough?

We need to find the number of square meters inside the fence.

There are __3__ rows of square meters with __5__ square meters in each row.

The number of **square units** is called the **area**.

To find the area of the section, we multiply the number of rows by the number of square units in each row.
We multiply __3__ by __5__.

__3__ **meters** × __5__ **meters** = __15__ **square meters**

The area inside the fence is __15__ square meters.

It __is__ big enough for the dog.

We can use a **formula** to find the area of a rectangle or a square.

rectangle — width (w), length (l)

square — side (s)

The area of a rectangle is equal to the length multiplied by the width.

$A = l \times w$

The area of a square is equal to the side multiplied by itself.

$A = s \times s$

Getting Started
Find the area.

1. 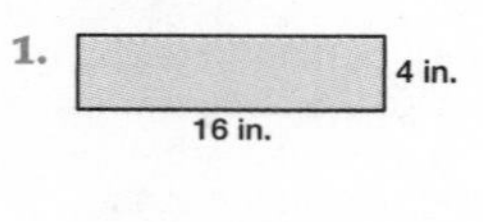
4 in.
16 in.
64 square inches

2. 2 cm, 3 cm, 5 cm, 6 cm
43 square centimeters

2 Teach

Introduce the Lesson Have a student read the problem and tell what needs to be answered. (if the fenced-in section is 10 or more square meters)

- Ask what information is known. (The length is 5 meters and the width is 3 meters.) Ask what is known about the shape of the figure. (It is a rectangle.)
- Have students read the rules in the model aloud and study the figure. Have them fill in the sentences and solve the problem.

Develop Skills and Concepts Show students a sheet of grid paper. Tell them to find the number of units on the whole sheet. Help students see that each unit is a square.

- Tell students that we find the number of square units in all by multiplying the length by the width. Have a student find the number of square units. (Answers will vary.) Write the number on the board as _____ **square units**. Tell students that we call this the **area** and area is always given in square units whether it is square miles, square feet, or square centimeters.
- Have students find the areas of more rectangular sheets of grid paper. Write $A = l \times w$ on the board and tell students that this is the **formula** to find the area of a rectangle.
- Repeat the activity for a square sheet of grid paper and have students discover that the formula for the area of a square is $A = s \times s$.

3 Practice

Write the formulas $A = l \times w$ for a rectangle and $A = s \times s$ for a square on the board. Ask students how they will find the area of just the shaded part in Exercise 4. (find the area of the whole square and subtract the area of the small white square) Tell students that they are to find the area of each rectangle and square in the tables. Have students complete page 242 independently.

Practice

Find the area.

1. 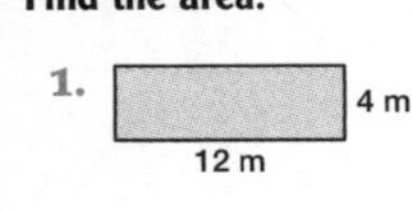

48 sq m

2. 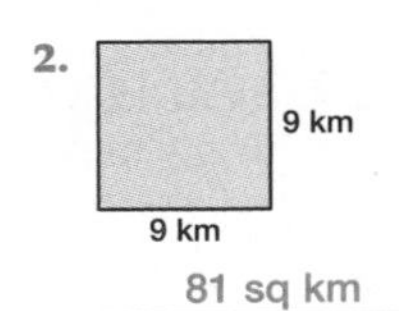

81 sq km

3. 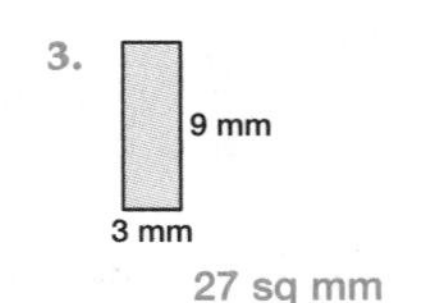

27 sq mm

Find the area of the shaded part of each figure.

4. 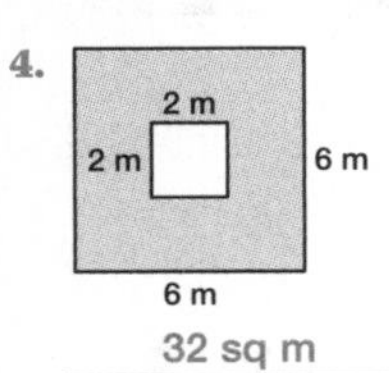

32 sq m

5. 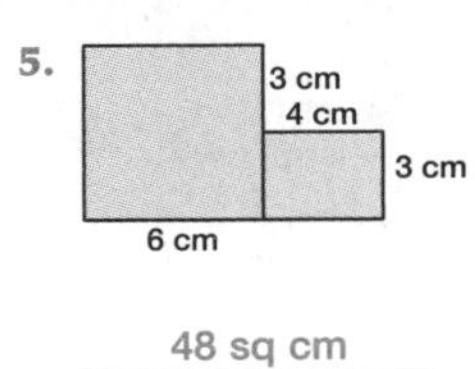

48 sq cm

6. 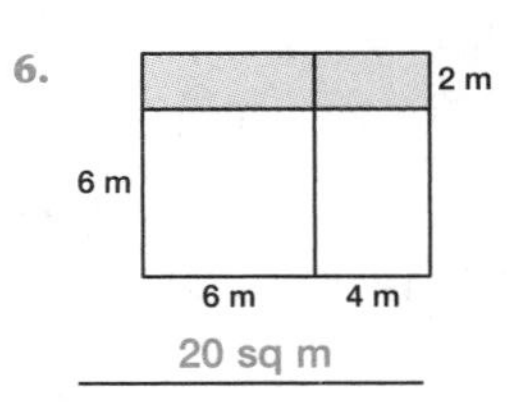

20 sq m

Complete each table.

7.

Rectangles					
Length	6 miles	5 miles	2 miles	5 miles	9 miles
Width	4 miles	4 miles	8 miles	7 miles	2 miles
Area	24 sq mi	20 sq mi	16 sq mi	35 sq mi	18 sq mi

8.

Squares					
Side	2 meters	6 meters	7 meters	10 meters	11 meters
Area	4 sq m	36 sq m	49 sq m	100 sq m	121 sq m

Problem Solving

Solve each problem.

9. Fritz is tiling a floor. The length of the floor is 3 meters. The width is 4 meters. How many square meters of floor does Fritz have to tile?
12 square meters

10. Wayne has a rectangle that is 5 centimeters by 12 centimeters. Rosa has a square that is 8 centimeters on a side. Who has the larger polygon? How much larger is it?
Rosa; 4 square centimeters

4 Assess

Ask students to define area. Then, have them give the formula for finding the area of a rectangle. (the number of square units; $A = l \times w$)

For Mixed Abilities

Common Errors • Intervention

Some students may have difficulty relating multiplication to finding the area of a square or rectangle. Have them work with partners and use grid paper to draw a variety of rectangles and squares that are a whole number of units long and wide. One member of the pair should count the small squares to find the area of each figure and the other should write the multiplication sentence that also gives the area.

Enrichment • Measurement

Tell students to write the formulas for finding the perimeter of a rectangle and a square. Tell them to use these formulas to find the perimeter of a square whose area is 36 square miles and the perimeter of a rectangle whose area is 90 square yards.

More to Explore • Statistics

Give each student a sheet of 1-inch grid paper that is 10 inches on a side. Ask students to look at the pattern you have begun on the board:

1
1 1
1 2 1
1 3 3 1
(1) (4) (6) (4) (1)
(1) (5) (10) (10) (5) (1)

Have students copy the section from the board and continue with the pattern.

This arrangement is called "Pascal's triangle." Point out that each number in the pattern is the sum of the two numbers diagonally above it. Students should also note that each number represents the number of different ways there are to get to that square from the topmost square.

12-6 Volume

pages 243–244

1 Getting Started

Objective
- To find the volume of a rectangular prism

Vocabulary
volume, cubic units

Materials
*rectangular prism; *building cubes

Warm Up • Mental Math
Tell students to find the value of P.

1. $P \times P \times P = 27$ (3)
2. $P \times P \times P \times P = 16$ (2)
3. $P \div P = 1$ (any number)
4. $P \times 30 = 0$ (0)
5. 4 mi $\times P = 320$ sq mi (80 mi)
6. $P \times P \times P = 8$ (2)
7. $8 \times (P + 6) = 64$ (2)
8. 62 ft $\times P = 124$ sq ft (2 ft)

Warm Up • Numeration
Have students copy the following exercises and solve them: $3 \times 6 \times 2$ (36), $8 \times 10 \times 4$ (320), $6 \times 7 \times 9$ (378), $3 \times 3 \times 3$ (27), and $20 \times 20 \times 26$ (10,400).

Name ____________________

Lesson 12-6

It's Algebra!

Volume

One way to measure the volume of a solid figure is in cubic units. How many cubic units are in this prism?

1 cubic unit

We want to find the volume of the prism in **cubic units**.

The prism is __3__ cubic units long, __4__ cubic units wide, and __5__ cubic units high.

To find the **volume**, we multiply the number of cubic units in a row by the number of rows and again by the number of units high.

We multiply __3__ cubic units by __4__ cubic units by __5__ cubic units.

$3 \times 4 \times 5 =$ __60__ cubic units

There are __60__ cubic units in the prism. We can use a formula to find the volume of a prism.

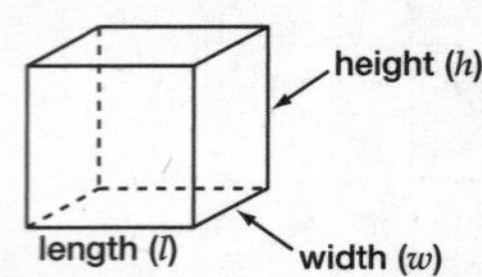

The volume of a prism is equal to the length multiplied by the width multiplied by the height.

$$V = l \times w \times h$$

Getting Started

Find the volume.

1.

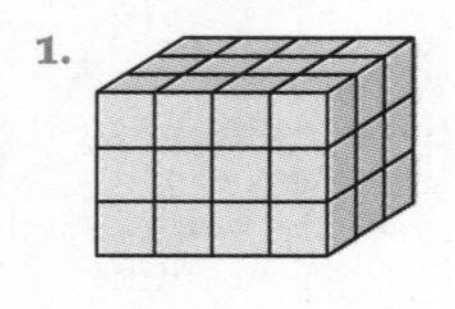

36 cubic units

2.

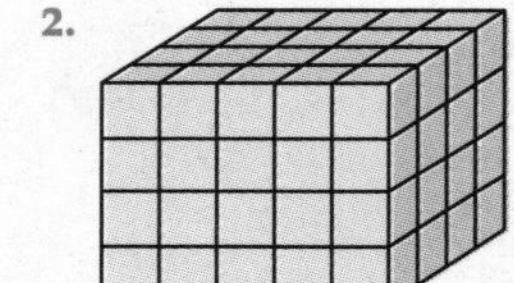

80 cubic units

---

2 Teach

Introduce the Lesson Have a student read the problem aloud and tell what is to be solved. (how many cubic units are in the prism) Have students tell what information is known. (There are 4 rows of 3 units each and there are 5 layers.)

- Have students read and complete the sentences to solve the problem. Tell students to check their solution by finding the number of cubes on each layer and multiplying by 5. Have students read the formula for finding the volume of a prism as you explain it.

Develop Skills and Concepts Show students a rectangular prism. Remind students that a solid figure has length, width, and height. Have students count the solid's faces, vertices, and edges. Tell students that they will now learn how to find the **volume**, or the number of **cubic units** in the whole prism.

- Write the following on the board:
  Length = 7 units
  Width = 6 units
  Height = 4 units
- Tell students that they are to find the number of cubic units in one layer by multiplying the length by the width (42) and then multiplying that number by the number of layers, or the height, of the prism. (168)
- Write $V = l \times w \times h$ on the board as you tell students that the volume of a solid figure is length × width × height. Tell students that volume is reported in cubic units such as cubic yards, cubic meters, and so on.

3 Practice

Remind students to label volumes with cubic units. Have students complete page 244 independently.

Practice

Find the volume.

1. 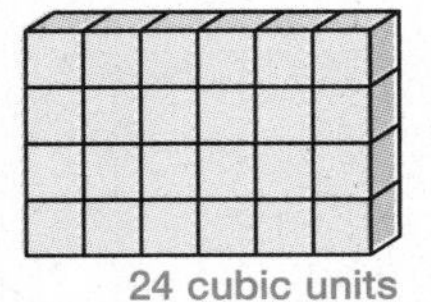
24 cubic units

2. 27 cubic units

3. 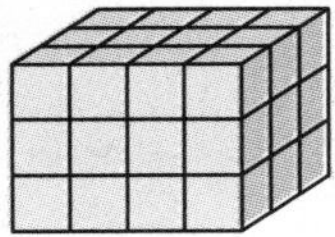
36 cubic units

4. 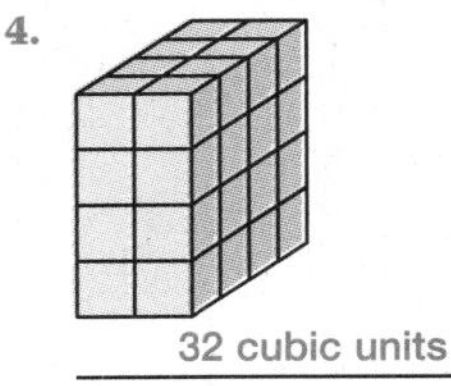
32 cubic units

5. 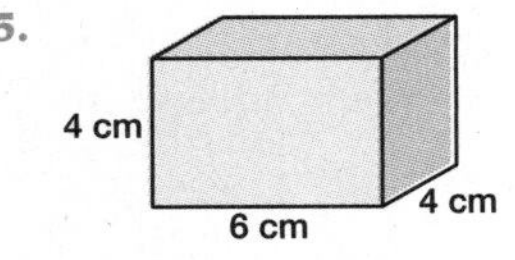

96 cubic centimeters

6. 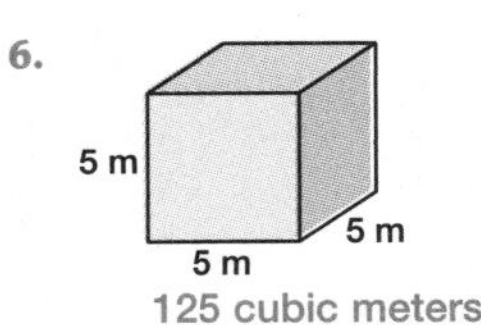

125 cubic meters

7. 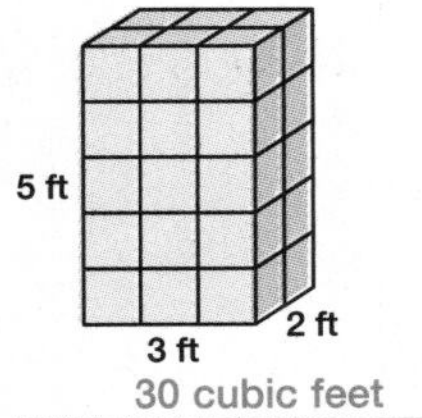

30 cubic feet

8. 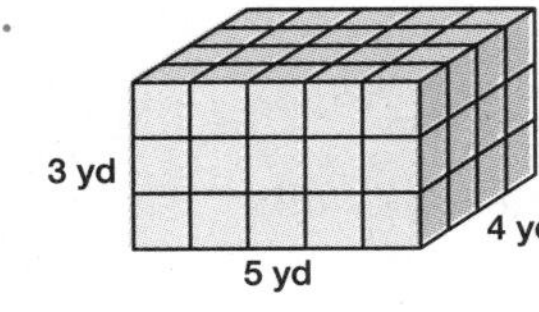

60 cubic yards

Problem Solving

Solve each problem.

9. Mr. Lee has a hot tub. The hot tub is 6 feet long, 6 feet wide, and 3 feet deep. What is the volume of the hot tub?
108 cubic feet

10. Melissa wants to fit 2-inch cubes into a box that is 10 inches long, 8 inches wide, and 4 inches high. How many 2-inch cubes will fit inside the box?
40 cubes

4 Assess

Have students give the formula for the volume of a prism.
(length multiplied by width multiplied by height or $V = l \times w \times h$)

5 Mixed Review

1. 300 × 7 (2,100)
2. 622 ÷ 6 (103 R4)
3. 2,956 × 8 (23,648)
4. 456 ÷ 80 (5 R56)
5. $56.02 − $12.94 ($43.08)
6. $3.42 ÷ 6 ($0.57)
7. 7 × $43.28 ($302.96)
8. 21,256 + 3,825 (25,081)
9. 38 × 24 (912)
10. 18 + 36 + 47 + 23 (124)

For Mixed Abilities

Common Errors • Intervention

Some students may have difficulty understanding the relationship between multiplication and finding the volume of solid figures. Have them work with partners and small cubes to build rectangular prisms of varying sizes. Have them take turns, one counting the number of cubes in a layer and the other multiplying by the number of layers, or the height, to find the volume. Then, have them find the product of the three dimensions and compare it with the number of small cubes in the prism to recognize that they are the same.

Enrichment • Measurement

Tell students to measure three different boxes to the nearest inch or centimeter. Tell them to then find the volume of each box and find the area of the bottom of each box.

More to Explore • Applications

Give students a calorie chart for foods and the activity chart below. Have them record what they eat for one meal and the number of calories in each food. Have them find the total calories for that meal. Then, have students calculate how many hours of their favorite activities they must do to use up all the calories consumed in that one meal.

Activity	Calories Used/Hour
Bicycling	170
Ice skating	300
Jogging	550
Jumping rope	600
Playing baseball	225
Playing football	230
Roller skating	300
Swimming	260
Playing soccer	325

Then, have students calculate the total number of calories consumed for an entire day and the hours of activities needed to burn them up.

12-7 Problem Solving: Restate the Problem

pages 245–246

1 Getting Started

Objective

- To solve problems by restating the problem

Warm Up • Mental Math

Have students name the number.

1. minutes in 6 hours (360)
2. months in a year (12)
3. seconds in 5 minutes (300)
4. freezing Fahrenheit temperature (32°)
5. freezing Celsius temperature (0°)
6. years in 8 decades (80)

Warm Up • Pencil and Paper

Have students copy the following and determine the correct symbol: <, >, or =.

1. $12 \div 4 \bigcirc 66 \div 6$ (<)
2. $18 \div 9 \bigcirc 2 \div 1$ (=)
3. $15 \div 3 \bigcirc 24 \div 8$ (>)
4. $36 \div 4 \bigcirc 54 \div 9$ (>)
5. $40 \div 8 \bigcirc 64 \div 8$ (<)
6. $88 \div 11 \bigcirc 18 \div 9$ (>)

Name ______________________

Problem Solving: Restate the Problem

The perimeter of Lincoln's backyard is 200 feet. The length of the backyard is 20 feet longer than the width. What is the length and the width of Lincoln's backyard?

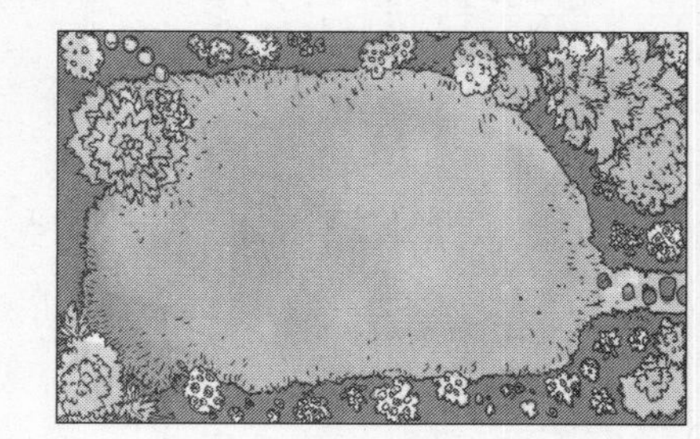

★ SEE

We want to find the length and the width of Lincoln's backyard. We know that the perimeter is 200 feet and that the length is 20 feet longer than the width.

★ PLAN

We want to express this problem in our own way, so that we really understand it.

★ DO

Any way we reword the problem is good, as long as it helps us to understand it better. Because a rectangle has two pairs of congruent sides, we can find the length and the width by dividing the perimeter by 2.

200 feet ÷ 2 = 100 feet

The length plus the width of the backyard is equal to 100 feet. We need to find two numbers that equal 100 with one that is 20 greater than the other.

Numbers That Equal 100	Is One Number Greater Than 20?
50 + 50	No, 50 − 50 = 0
55 + 45	No, 55 − 45 = 10
60 + 40	Yes, 60 − 40 = 20

The length of the backyard is 60 feet and the width is 40 feet.

★ CHECK

60 ft + 60 ft + 40 ft + 40 ft = 200 ft

2 Teach

Introduce the Lesson Remind students about the four-step problem-solving strategy they learned to use for solving problems.

Develop Skills and Concepts Have students read the problem and tell what they are to find. (the length and width of Lincoln's backyard) Read each sentence under the SEE step, asking for volunteers to give the missing information as students fill in the answers.

- Ask another volunteer to read the PLAN sentence.
- Next, help students recall the meaning of congruent (same shape and same size) and that the perimeter is the sum of the length of the sides of the figure.
- Now, read the DO step together. Have students focus on the chart. Tell students that the problem was restated to make it easier to solve. Explain that in the first row, we begin with 50 + 50 and then perform the subtraction to see if there is 20 left over. In the second row, we add 5 to the first addend and subtract 5 from the second addend. Since the difference between these figures is 10, we now add 10 to the first number and subtract 10 from the second number.
- Work through the CHECK step with students.

3 Apply

Solution Notes

1. Ask students to explain what the question is asking. (the length and width of the figure) Have students find a square number that is equal to 36. Remind students that their answer should be in centimeters.
2. Ask students how they can find the depth of the pool. (multiply the length by the width and divide the volume by the product)
3. Make sure students understand that an octagon has eight sides.

Apply

Solve each problem by restating it so you understand it.

1. A square has an area of 36 square centimeters. What is the length and the width of the square?
   length = 6 centimeters; width = 6 centimeters

2. A swimming pool has a volume of 4,000 cubic feet. The pool is 50 feet long and 20 feet wide. The depth of the pool is the same throughout. How deep is the pool?
   4 feet

3. One of the rooms in Tom's house is shaped like an octagon. If the room has a perimeter of 104 feet, how long is each wall?
   13 feet

4. The area of a rectangle is 128 square meters. The length of the rectangle is twice the width. What is the length and the width of the rectangle?
   length = 16 meters; width = 8 meters

Problem-Solving Strategy: Using the Four-Step Plan

- ★ **SEE** What do you need to find?
- ★ **PLAN** What do you need to do?
- ★ **DO** Follow the plan.
- ★ **CHECK** Does your answer make sense?

5. Alexis is putting 1-inch cubes into a box that is 8 inches long and 6 inches wide. If Alexis can fit 144 of the 1-inch cubes into the box, what is the height of the box?
   3 inches

6. The perimeter of a basketball court is 288 feet. The length of the court is 44 feet longer than the width. What is the length and the width of the basketball court?
   94 feet × 50 feet

For Mixed Abilities

More to Explore • Application

Have students calculate the area and perimeter of a rectangle with a length of 7 inches and a width of 5 inches. (The area is 35 sq inches, and the perimeter is 24 inches.) Then, have them double the dimensions and explain what happens to the area and perimeter. (The area quadruples and the perimeter doubles.)

Higher-Order Thinking Skills

4. **Comprehension:** Ask students what clue will help them find the length and width of the rectangle. (The length of the rectangle is twice the width.)
5. **Analysis:** Students must recognize that they need to find the volume of the box by dividing the total number of cubes by the product of the length times the width of the box.
6. **Evaluation:** Divide the perimeter, 288, by 2 to get 144. Then, find two numbers, one number that is 44 larger than the other, that have a sum of 144.

4 Assess

Show students the following figure and ask how many triangles are in it. (5)

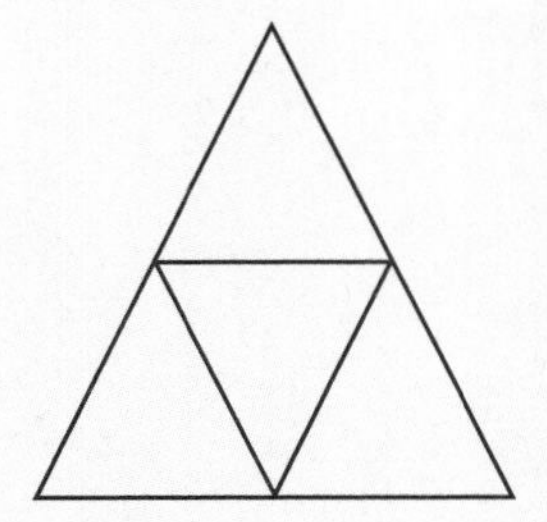

Chapter 12 Test

page 247

Items	Objectives
1–3	To identify figures that are congruent or similar (see pages 233–234)
4–5	To draw lines of symmetry (see pages 235–236)
7–8	To find the perimeter of a polygon (see pages 239–240)
9–10	To find the area of rectangles and squares (see pages 241–242)
11–12	To find the volume of a box (see pages 243–244)

Name ______________________________

Write *congruent* or *similar* to describe each pair of figures.

1. 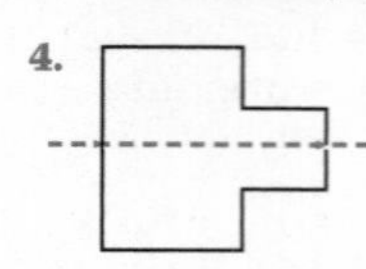
congruent

2. 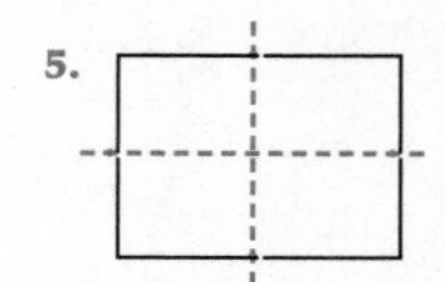
similar

3. 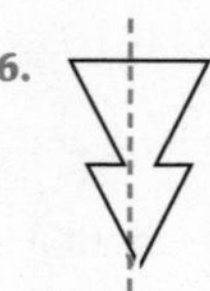
congruent

Draw all lines of symmetry.

4. 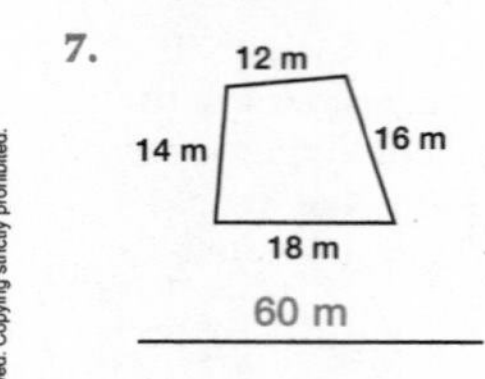

5. 6. 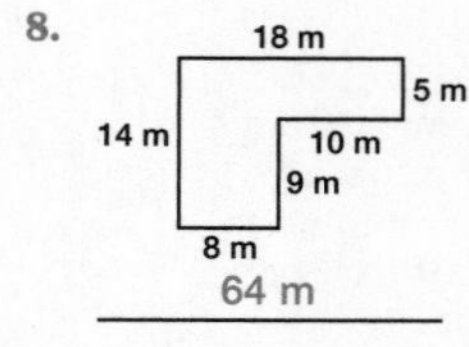

Find the perimeter.

7. 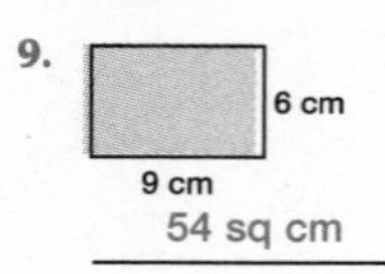

60 m

8. 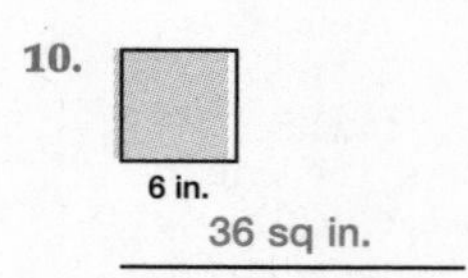

64 m

Find the area.

9. 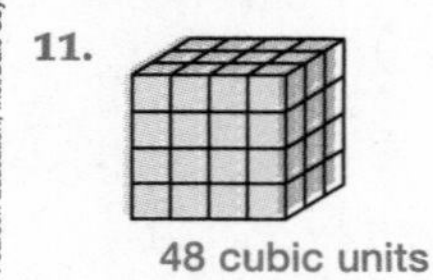

54 sq cm

10. 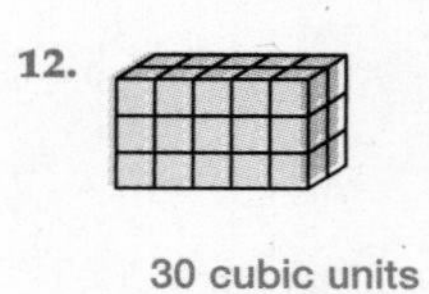

36 sq in.

Find the volume.

11. 48 cubic units

12. 30 cubic units

Alternate Chapter Test

You may wish to use the Alternate Chapter Test on page 368 of this book for further review and assessment.

CUMULATIVE ASSESSMENT

Circle the letter of the correct answer.

1. Round 5,628 to the nearest thousand.
a. 5,000
(b.) 6,000
c. NG

2. What is the place value of the 9 in 409,115?
a. tens
(b.) thousands
c. hundred thousands
d. NG

3. 5,379 + 4,568
a. 10,946
(b.) 9,947
c. 8,927
d. NG

4. 48,556 + 34,739
(a.) 83,295
b. 63,295
c. 73,295
d. NG

5. 908 − 699
a. 389
b. 419
c. 481
(d.) NG

6. 47,128 − 12,945
a. 22,383
b. 24,283
(c.) 34,183
d. NG

7. Choose the better estimate of volume.
a. 1 pint
(b.) quart

8. $4.35 × 6
a. $21.45
b. $24.10
(c.) $26.10
d. NG

9. 15 × 36
a. 440
b. 490
(c.) 540
d. NG

10. 627 × 25
a. 3,612
b. 16,755
(c.) 15,675
d. NG

11. 37)222
a. 5
(b.) 6
c. 7
d. NG

12. 993 ÷ 54
a. 18
(b.) 18 R21
c. 180 R21
d. NG

13. Name the figure.
a. square
b. rectangle
(c.) trapezoid
d. NG

STOP

score

Cumulative Assessment

page 248

Items	Objectives
1	To round numbers to the nearest thousand (see pages 31–32)
2	To identify place value through hundred thousands (see pages 25–26)
3–4	To add two numbers up to 5 digits (see pages 41–42, 47–48)
5–6	To subtract up to 5-digit numbers (see pages 59–62)
7	To determine the appropriate customary unit of capacity (see pages 145–146)
8–10	To multiply up to 3-digit numbers by up to 2-digit numbers including money with and without regrouping (see pages 167–180)
11–12	To divide 3-digit numbers by 2-digit numbers with and without remainders (see pages 197–206)
13	To identify polygons by the number of sides (see pages 219–220)

Alternate Cumulative Assessment

Circle the letter of the correct answer.

1. Round 7,318 to the nearest thousand.
a 6,000
(b) 7,000
c NG

2. What is the place value of the 9 in 929,765?
a ten thousands
(b) hundred thousands
c millions
d NG

3. 5,311 + 4,479
(a) 9,790
b 9,820
c 9,830
d NG

4. 44,378 + 18,664
a 62,942
b 62,932
(c) 63,042
d NG

5. 506 − 288
a 328
b 338
(c) 218
d NG

6. 35,217 − 11,863
a 33,454
(b) 23,354
c 34,354
d NG

7. What would you use to measure water in a pool?
a pints
b quarts
(c) gallons
d NG

8. $4.64 × 8 =
a $52.14
b $56.22
c $51.75
(d) NG

9. 28 × 34 =
a 732
b 243
(c) 952
d NG

10. 534 × 36
a 20,182
(b) 19,224
c 21,082
d NG

11. 33)264
(a) 8
b 7 R1
c 7 R10
d NG

12. 856 ÷ 24
a 3 R6
(b) 35 R16
c 36
d NG

13. What is the name for an eight-sided figure?
a pentagon
b hexagon
(c) octagon
d NG

13-1 Fractional Parts

pages 249–250

1 Getting Started

Objective

- To write a fraction for part of a whole or part of a set

Vocabulary

numerator, denominator

Warm Up • Mental Math

Tell students to find the average.

1. 3, 7, and 8 (6)
2. 40, 50, and 120 (70)
3. 0 and 100 (50)
4. 17 and 7 (12)
5. 20, 30, and 40 (30)
6. 70 and 100 (85)
7. 600 and 900 (750)
8. 1, 2, and 21 (8)

Warm Up • Numeration

Review the division of 3- and 4-digit dividends by 2-digit divisors, by having a student work some problems on the board. Have another student check each problem by multiplying and adding any remainder.

Name ____________________

Fractions

CHAPTER 13

Lesson 13-1

Fractional Parts

Fractions can help you talk about a part of a figure or part of a set of things. What part of this rectangle is green? What part of the set of cars is not green?

The rectangle is divided into __5__ equal parts.

__1__ part is green.

We write $\frac{1}{5}$. ← numerator ← denominator We say **one-fifth.**

__$\frac{1}{5}$__ of the rectangle is green.

There are __3__ cars.

__2__ cars are not green.

We write $\frac{2}{3}$. We say **two-thirds.**

__$\frac{2}{3}$__ of the cars are not green.

Getting Started

Write a fraction to show what part of each figure is green.

1. 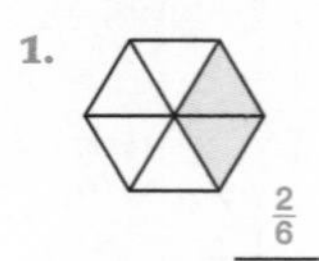 __$\frac{2}{6}$__

2. __$\frac{7}{8}$__

3. __$\frac{3}{8}$__

Write a fraction to show what part of each set is *not* green.

4. 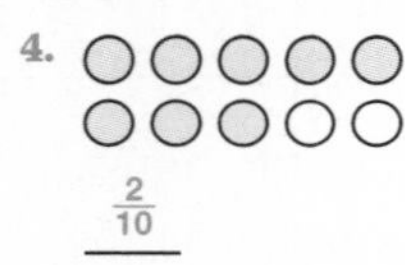 __$\frac{2}{10}$__

5.  __$\frac{3}{5}$__

6. __$\frac{2}{9}$__

2 Teach

Introduce the Lesson Have a student describe the picture and tell what is to be answered. (what part of the rectangle is green and what part of the set of cars is not green)

- Ask what is known about the rectangle. (It has 5 parts that are the same size.) Ask what is known about the set of cars. (There are 3 in all.)
- Have students complete the sentences and answer the questions as they read with you.

Develop Skills and Concepts Have 6 students stand in a group. Tell 1 of the 6 to step away from the group. Ask students how many of the 6 students moved away. (1)

- Write $\frac{1}{6}$ on the board and tell students that we read this fraction as one-sixth. Tell students that the top number is called the **numerator** and it tells what part of the whole group we are talking about. Tell students that the bottom number is called the **denominator** and it tells how many are in the whole group.
- Have another student step away from the group. Ask how many students moved away from the group of 6 as you write $\frac{2}{6}$ on the board. Continue for $\frac{3}{6}$, $\frac{4}{6}$, $\frac{5}{6}$, and $\frac{6}{6}$.
- Draw a rectangle on the board and divide it into 3 equal parts. Have a student shade 1 of the 3 parts and write the fraction to show the shaded part. ($\frac{1}{3}$) Continue for $\frac{2}{3}$ and $\frac{3}{3}$.
- Repeat for other parts of whole sets of objects.

3 Practice

Tell students to read the directions for each section carefully. Have students complete page 250 independently.

Practice

Write a fraction to show what part of each figure is green.

1. 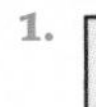 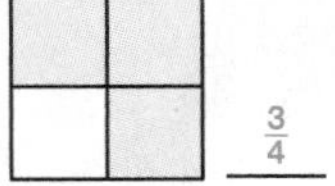 $\frac{3}{4}$

2.  $\frac{4}{6}$

3. 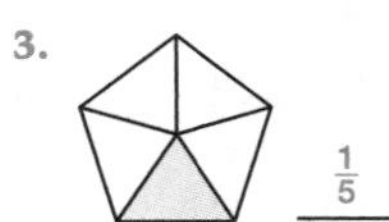 $\frac{1}{5}$

4. 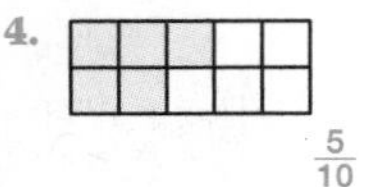 $\frac{5}{10}$

5. 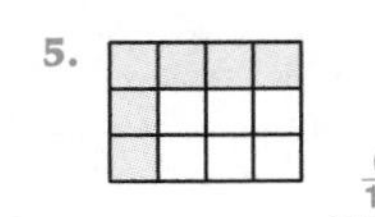 $\frac{6}{12}$

6. 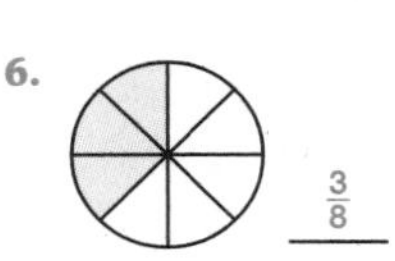 $\frac{3}{8}$

Write a fraction to show what part of each set is *not* green.

7.   $\frac{1}{3}$

8.  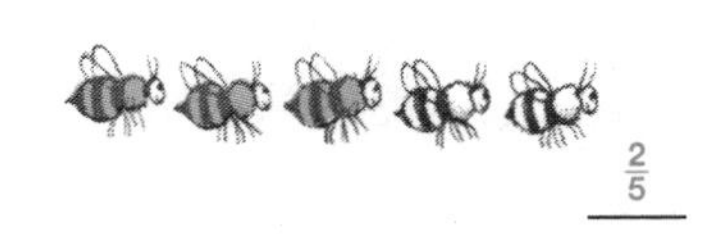 $\frac{2}{5}$

9.  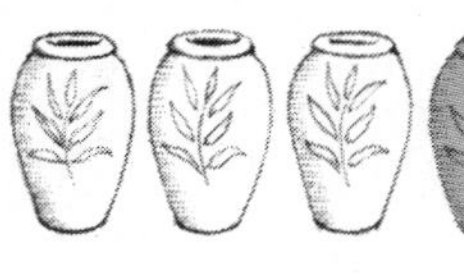 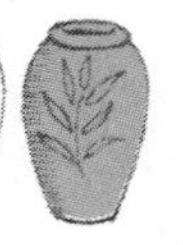 $\frac{3}{4}$

10.   $\frac{3}{8}$

Problem Solving

Solve each problem.

11. What part of the triangle is green?

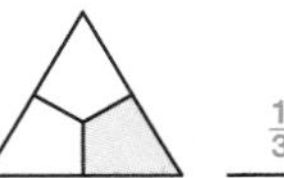 $\frac{1}{3}$

12. What part of the octagon is *not* green?

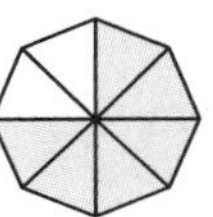 $\frac{2}{8}$

13. What part of the set of coins are pennies?

 $\frac{4}{7}$

14. Write a fraction to show what part of the set of figures are squares.

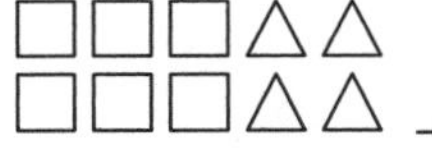 $\frac{6}{10}$

4 Assess

Display 8 quarters and 7 dimes. Ask students what part of the set of coins are quarters. ($\frac{8}{15}$)

5 Mixed Review

1. $8.95 × 6 ($53.70)
2. 30 ÷ 13 (2 R4)
3. 2,075 − 927 (1,148)
4. 65,237 + 30,207 (95,444)
5. 7 + 5 × 7 (42)
6. 320 ÷ 40 (8)
7. 376 + 408 + 27 (811)
8. $56.95 − $28.38 ($28.57)
9. 562 ÷ 3 (187 R1)
10. 52 × 34 (1,768)

For Mixed Abilities

Common Errors • Intervention

Some students may confuse the numerator and denominator when writing a fraction. Have these students work with partners and the following fractions: $\frac{2}{6}$, $\frac{4}{5}$, $\frac{3}{8}$, $\frac{2}{5}$, and $\frac{1}{9}$. Students should take turns identifying the numerator and the denominator of each fraction and drawing a model to illustrate the fraction.

Enrichment • Fractions

Tell students to write and illustrate the fractions that tell the following: the student as a part of his or her family, the student as a part of his or her class, the student as part of his or her class that is girls, the student as part of his or her class that is boys.

More to Explore • Numeration

Put the following example on the board and explain to students that it is possible to multiply with Roman numerals.

XXVII	27
× XVI	× 16
CCCCXXXII	432

Remind students to use the whole value of the Roman numeral when multiplying rather than each digit. Ask students to try and solve these problems. Check each problem with Arabic numerals to make sure the answers are correct.

XIV	14
× III	× 3
(XLII)	(42)
XXV	25
× IV	× 4
(C)	(100)
LV	55
× V	× 5
(CCLXXV)	(275)

Challenge students to write their own multiplication problems using Roman numerals.

13-2 Unit Fractions

pages 251–252

1 Getting Started

Objective
- To find the number of objects in one part of a set

Vocabulary
unit fraction

Materials
counters; division fact cards; graph paper; scissors

Warm Up • Mental Math
Ask students how much more or less than \$10 the following amounts are:

1. \$1.68 (\$8.32 less)
2. 4 × \$2.52 (8¢ more)
3. \$5.75 + \$2 (\$2.25 less)
4. \$10 ÷ \$4 (\$7.50 less)
5. \$8.15 + \$2.50 (65¢ more)
6. \$3.50 × \$2 (\$3 less)
7. \$4.96 + \$7 (\$1.96 more)
8. \$7.50 × 2 (\$5 more)

Warm Up • Numeration
Have students work in pairs with fact cards to review the basic division facts. Have one student give the quotient of a fact with the partner stating the related multiplication fact.

Name ____________________

Lesson 13-2

Unit Fractions

Mark and Nadia are trading seashells for a science project. Nadia gives Mark $\frac{1}{5}$ of her seashells. How many seashells does Mark get from Nadia?

We want to know how many seashells Nadia gives Mark.

Nadia has __15__ seashells.

She gives Mark __$\frac{1}{5}$__ of them.

To find the number of seashells Nadia gave Mark, we need to multiply __15__ by $\frac{1}{5}$.

$\frac{1}{5}$ is called a **unit fraction** because its numerator is one. To multiply by a unit fraction, we simply divide by the denominator.

We divide __15__ by __5__.

15 ÷ 5 = __3__

Nadia gave Mark __3__ seashells.

Getting Started

Find each part.

1. $\frac{1}{3}$ of 9 = __3__
2. $\frac{1}{6}$ of 36 = __6__
3. $\frac{1}{2}$ of 32 = __16__
4. $\frac{1}{4}$ of 48 = __12__
5. $\frac{1}{7}$ of 84 = __12__
6. $\frac{1}{12}$ of 180 = __15__

2 Teach

Introduce the Lesson Have a student describe the picture. Have another student read the problem aloud and tell what is to be solved. (find the number of seashells Nadia gives Mark) Ask what information is known. (Nadia has 15 seashells and gives $\frac{1}{5}$ of them to Mark.)

- Have students complete the sentences as they read with you to solve the problem. Tell them to do the division mentally.
- Have students see if 5 times their solution equals the total number of seashells Nadia had.

Develop Skills and Concepts Have students work in pairs with 12 counters. Have students divide the counters in half so that there are 6 counters in each group. Ask students how many counters are in one group. (6) Tell students that the 12 counters were divided into two equal parts and one of those two parts has 6 in it.

- Tell students that $\frac{1}{6}$ is called a **unit fraction** because its numerator is 1. Tell students that to find $\frac{1}{6}$ of a number, we divide the number by the denominator. Repeat for $\frac{1}{3}$ of 12, $\frac{1}{4}$ of 12, $\frac{1}{6}$ of 12, and $\frac{1}{12}$ of 12.
- Have students put out 18 counters and find $\frac{1}{2}$, $\frac{1}{3}$, $\frac{1}{6}$, $\frac{1}{9}$, and $\frac{1}{18}$ of them.

3 Practice

Remind students to divide by the fraction's denominator to solve the first group of exercises on page 252. Tell students that in some of the word problems, they will use more than one operation. Have students complete the page independently.

Practice

Find each part.

1. $\frac{1}{5}$ of 60 = 12
2. $\frac{1}{8}$ of 64 = 8
3. $\frac{1}{9}$ of 63 = 7
4. $\frac{1}{4}$ of 40 = 10
5. $\frac{1}{3}$ of 36 = 12
6. $\frac{1}{2}$ of 56 = 28
7. $\frac{1}{11}$ of 77 = 7
8. $\frac{1}{4}$ of 96 = 24
9. $\frac{1}{3}$ of 87 = 29
10. $\frac{1}{6}$ of 96 = 16
11. $\frac{1}{12}$ of 72 = 6
12. $\frac{1}{15}$ of 750 = 50

Problem Solving

Solve each problem.

13. Betty bought 36 pencils. She gave $\frac{1}{2}$ to her best friend. How many pencils did Betty give away? 18 pencils

14. Peter's baseball team played 21 games. The team won $\frac{1}{3}$ of the games. How many games did the team win? 7 games

15. There are 32 children in Mr. Chan's class. One day, $\frac{1}{4}$ of the class was absent. How many children were absent? 8 children

16. There are 65 children on the school swimming team. $\frac{1}{5}$ of the team are boys. How many boys are on the team? 13 boys

17. Mr. James planted 56 trees. $\frac{1}{4}$ of the trees were maples. How many trees were maples? 14 trees

18. Mrs. Spencer bought 156 apples. $\frac{1}{6}$ of the apples were spoiled. How many apples were not spoiled? 130 apples

19. Ed lives 924 meters from school. He jogs $\frac{1}{3}$ of the way there and walks the rest. How far does Ed walk? 616 meters

20. A book has 216 pages. Lucia has read $\frac{1}{12}$ of the book. How many pages does Lucia have left to read? 198 pages

4 Assess

Ask students to name five unit fractions. (Answers will vary, but each fraction should have 1 as a numerator.)

For Mixed Abilities

Common Errors • Intervention

Some students may have difficulty finding the number of objects in one part of a set. Have them work with partners to find different parts of 12. Give them strips of paper marked off in 12 equal units. To find $\frac{1}{3}$ of 12, they cut a strip into 3 equal pieces and tell how many units are in 1 piece (4 units). In a similar way, they can find $\frac{1}{2}$, $\frac{1}{4}$, and $\frac{1}{6}$ of 12.

Enrichment • Fractions

Tell students to find out which unit fractions can be multiplied by the number of students in their class to give whole numbers. Explain that $\frac{1}{3}$ of 26 is not a whole number because there is a remainder when 26 is divided by 3.

More to Explore • Geometry

Ask students to collect pictures of bridges and bring them to class. If there is a bridge near the school, suggest that students visit the bridge and draw it as best they can.

Display all the photographs, copies, and drawings. Have students study the display and write a description of the shapes they see repeated in many of the bridges. (As students may already have discovered, a triangle is a very rigid shape. Bridge struts often form patterns of triangles. Arches are a common shape in masonry bridges.)

ESL/ELL STRATEGIES

When discussing fractions, review the meaning and pronunciation of the key terms. Write the terms on the board, capitalizing the stressed syllables: NUMerator, deNOMinator, UNit FRACtion. Have students repeat each term and then explain the meaning in their own words.

13-3 Finding a Fraction of a Number

pages 253–254

1 Getting Started

Objective

- To find a fraction of a number by multiplying and dividing

Vocabulary

non-unit fraction

Materials

$5 and $1 bills

Warm Up • Mental Math

Tell students to answer true or false.

1. A square is a rectangle. (T)
2. Parallel lines do not meet. (T)
3. A line has endpoints. (F)
4. A pyramid has 5 vertices. (T)
5. *Congruent* means "same size and shape." (T)
6. A ray has 1 endpoint. (T)
7. A square has only 2 right angles. (F)
8. The volume of a solid is $l \times w$. (F)

Warm Up • Fractions

Remind students that to multiply by a unit fraction, we divide by the denominator. Have students illustrate and solve the following on the board: $\frac{1}{6}$ of 18 (3), $\frac{1}{4}$ of 16 (4), $\frac{1}{5}$ of 10 (2), $\frac{1}{3}$ of 6 (2), and $\frac{1}{4}$ of 12 (3).

Name ____________________

Finding a Fraction of a Number

Peg looks for sales when she shops for clothes. How much will Peg pay for a blouse?

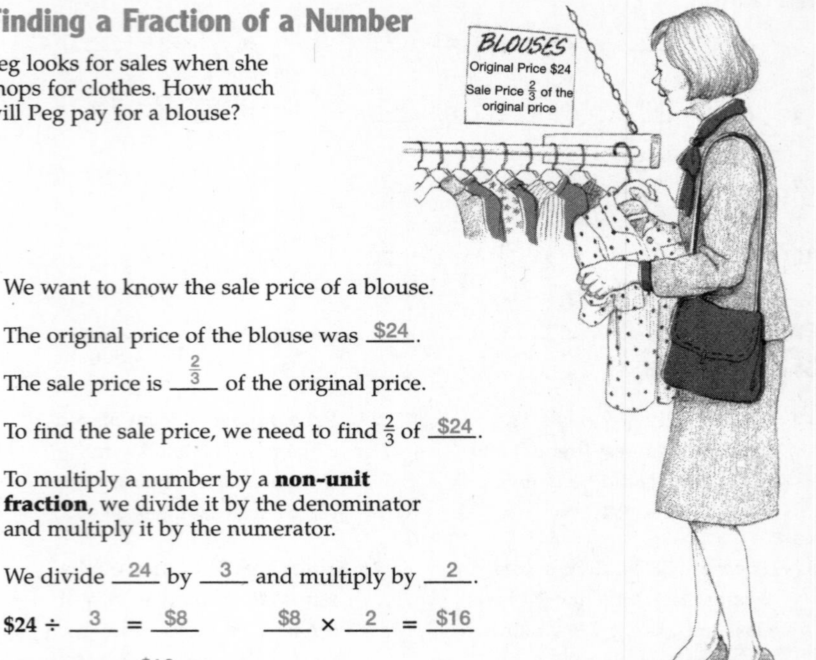

We want to know the sale price of a blouse.

The original price of the blouse was <u>$24</u>.

The sale price is <u>$\frac{2}{3}$</u> of the original price.

To find the sale price, we need to find $\frac{2}{3}$ of <u>$24</u>.

To multiply a number by a **non-unit fraction**, we divide it by the denominator and multiply it by the numerator.

We divide <u>24</u> by <u>3</u> and multiply by <u>2</u>.

$24 ÷ <u>3</u> = <u>$8</u>   <u>$8</u> × <u>2</u> = <u>$16</u>

Peg will pay <u>$16</u> for a blouse.

Getting Started

Find each part.

1. $\frac{3}{4}$ of 16 = <u>12</u>
2. $\frac{5}{8}$ of $72 = <u>$45</u>
3. $\frac{3}{16}$ of 224 = <u>42</u>

Find each sale price.

4. $\frac{5}{6}$ of a price of $36 <u>$30</u>
5. $\frac{3}{5}$ of a price of $365 <u>$219</u>
6. $\frac{2}{3}$ of a price of $672 <u>$448</u>

---

2 Teach

Introduce the Lesson Have a student describe the picture and read the sign. Have a volunteer read the problem and tell what is to be found. (the sale price of the blouse) Ask what information is known. (The original cost was $24 and the sale price is $\frac{2}{3}$ of the original price.)

- Have students read and complete the sentences to solve the problem. Guide them through the division and multiplication steps in the model. Students can check their solution with play money.

Develop Skills and Concepts Write $\frac{1}{3}$ **of 9** on the board and have students illustrate the problem by drawing a set of 9 objects and circling 3 objects of the set. Have students solve the problem. Remind them that $\frac{1}{3}$ is a unit fraction and 9 ÷ 3 = 3.

- Now, write $\frac{2}{3}$ **of 9** and have students illustrate a set of 9, circle 2 of the 3 groups, and tell how many in the set. (6) Tell students that $\frac{2}{3}$ is a **non-unit fraction** because its numerator is not 1 and to find $\frac{2}{3}$ of 9, we divide the 9 by 3 (3) and multiply the quotient by the numerator. (3 × 2 = 6)
- Have students work $\frac{3}{5}$ of 45 (27), $\frac{3}{4}$ of 16 (12), $\frac{7}{8}$ of 56 (49), and $\frac{2}{9}$ of 81 (18).

3 Practice

Have students complete page 254 independently.

Practice

Find each part.

1. $\frac{2}{5}$ of \$25 = \$10
2. $\frac{3}{4}$ of \$36 = \$27
3. $\frac{3}{8}$ of 72 = 27
4. $\frac{5}{6}$ of 54 = 45
5. $\frac{7}{9}$ of 90 = 70
6. $\frac{2}{3}$ of \$30 = \$20
7. $\frac{7}{8}$ of 64 = 56
8. $\frac{3}{4}$ of \$84 = \$63
9. $\frac{5}{11}$ of \$121 = \$55

Find each sale price.

10. $\frac{3}{8}$ of a price of \$56 — \$21
11. $\frac{3}{4}$ of a price of \$28 — \$21
12. $\frac{3}{5}$ of a price of \$25 — \$15
13. $\frac{5}{7}$ of a price of \$49 — \$35
14. $\frac{5}{8}$ of a price of \$80 — \$50
15. $\frac{2}{3}$ of a price of \$33 — \$22
16. $\frac{2}{3}$ of a price of \$57 — \$38
17. $\frac{5}{6}$ of a price of \$126 — \$105
18. $\frac{1}{2}$ of a price of \$96 — \$48

Complete the table.

19.

Sale $\frac{2}{5}$ Off			
Original Price	\$65	\$140	\$585
Sale Price	\$39	\$84	\$351

Problem Solving

Solve each problem.

20. Sweaters are on sale for $\frac{2}{3}$ of the original price. Before the sale, the sweaters were \$42 each. What is the sale price? \$28

21. The original price of a jogging suit was \$80. It is on sale for $\frac{3}{4}$ of the price. How much can be saved if you buy the suit on sale? \$20

4 Assess

Ask students to find the cost of a CD originally priced at \$12.80 and priced now at $\frac{6}{8}$ of the original price. (\$9.60)

5 Mixed Review

1. 206 × 8 (1,648)
2. 535 ÷ 70 (7 R45)
3. 572 − 391 (181)
4. 275 ÷ 67 (4 R7)
5. 26,275 + 3,746 (30,021)
6. \$40.00 − \$25.41 (\$14.59)
7. 19 + 36 + 71 + 43 (169)
8. 897 ÷ 7 (128 R1)
9. 647 × 35 (22,645)
10. 4,302 + 469 (4,771)

For Mixed Abilities

Common Errors • Intervention

Students may forget to multiply by the numerator when finding the fraction of a number. Have students work in pairs with 45 counters. Ask them to find $\frac{1}{5}$ of 45 by arranging the counters in 5 equal groups and counting the number in 1 group (9). Then, ask how they would find $\frac{2}{5}$ (double the number in $\frac{1}{5}$, or 18), how they would find $\frac{3}{5}$ (3 times the number in $\frac{1}{5}$, or 27), and $\frac{4}{5}$ (4 times the number in $\frac{1}{5}$, or 36).

Enrichment • Numeration

Have students cut ads from a newspaper for three items, each with a price of \$20 or more. Tell students to multiply each price by a non-unit fraction whose denominator will go into the price evenly and find the reduced cost.

More to Explore • Geometry

Show students how to make a cube out of straws and pins. Have each student make a cube and a rectangular prism.

cube

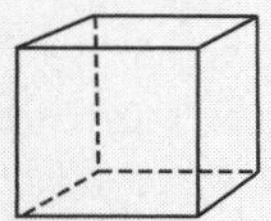

rectangular prism

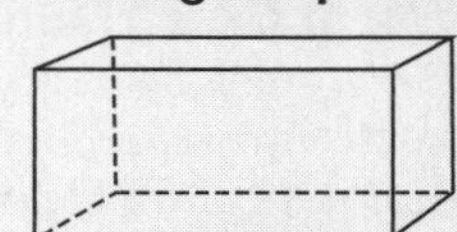

Neither of these figures will be rigid. Ask students to add more straws until they have made each structure rigid. Ask them to generalize what they have to do to make structures rigid. (Add diagonals, dividing each face into triangles.)

13-4 Equivalent Fractions

pages 255–256

1 Getting Started

Objective
- To understand equivalent fractions

Vocabulary
equivalent fractions

Materials
strips of paper; crayons

Warm Up • Mental Math
Ask students how many 4-foot lengths of fence are needed to fence the following:

1. a perimeter of 720 ft (180)
2. a 12-ft square (12)
3. a 16-by-56-ft rectangle (36)
4. a 28-by-24-ft rectangle (26)
5. a perimeter of 364 ft (91)
6. a 32-ft and a 28-ft side (15)
7. a hexagon with each side measuring 24 ft (36)
8. an octagon with each side measuring 8 ft (16)

Warm Up • Fractions
Have a student write any unit fraction on the board and then find that part of a random number you give. Repeat for other unit fractions and then continue the activity for non-unit fractions. Be sure that each number you give is a multiple of the fraction's denominator.

Name ____________________

Equivalent Fractions

Fractions that name the same amount are called **equivalent fractions**. Name two fractions that are equivalent to $\frac{2}{3}$.

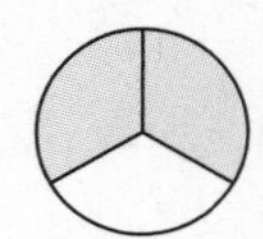
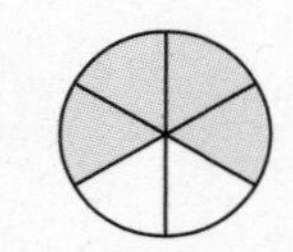
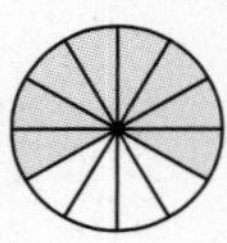

equivalent fractions

We want to find two fractions that are equivalent to $\underline{\frac{2}{3}}$.

Each large rectangle below is the same size. We can shade in the same amount of space in each rectangle to find equivalent **thirds, sixths**, and **twelfths.**

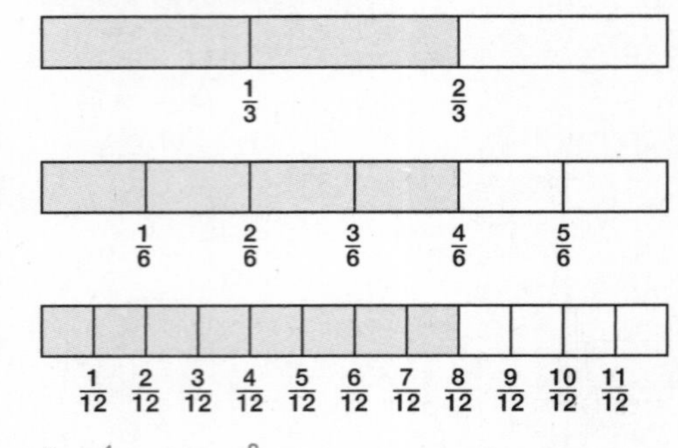

$\frac{2}{3}$, $\underline{\frac{4}{6}}$, and $\underline{\frac{8}{12}}$ are equivalent fractions.

Getting Started

Write the missing numerator.

1. 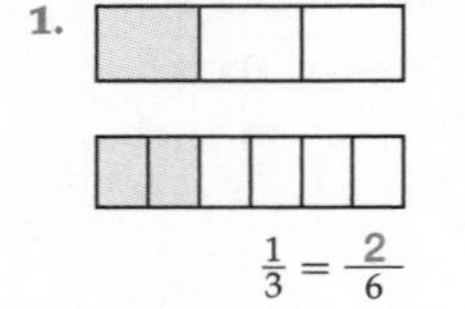
$\frac{1}{3} = \frac{2}{6}$

2. 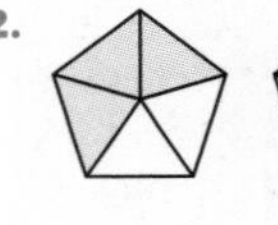 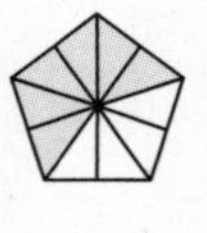
$\frac{3}{5} = \frac{6}{10}$

3. 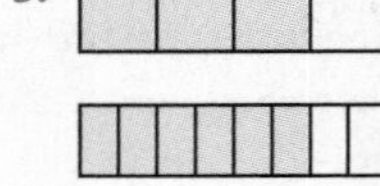
$\frac{3}{4} = \frac{6}{8}$

---

2 Teach

Introduce the Lesson Ask a student to read the introduction aloud and tell what is to be found. (two fractions that name the same amount as $\frac{2}{3}$) Have students tell what part is shaded in each pictured circle. ($\frac{2}{3}$, $\frac{4}{6}$, $\frac{8}{12}$)

- Next, read about the rectangles that follow and have students tell which parts of each rectangle are shaded. Have students complete the solution statement.

Develop Skills and Concepts Have students fold a strip of paper in half and color one of the halves. Tell students to fold the paper in half again and tell how many parts there are in all. (4) Ask students how many parts are colored. (2) Ask what part of the whole is colored. ($\frac{2}{4}$)

- Tell students to fold the strip of paper again and tell how many parts there are in all. (8) Ask what part of the whole is colored. ($\frac{4}{8}$) Write $\frac{1}{2} = \frac{2}{4} = \frac{4}{8}$ on the board and tell students that the fractions $\frac{1}{2}$, $\frac{2}{4}$, and $\frac{4}{8}$ are **equivalent** fractions because each describes the same part of the whole strip of paper.

3 Practice

Remind the students that the denominator tells the number of parts in all and the numerator tells the number of parts we are talking about. Have students complete the page independently.

Practice

Write the missing numerator.

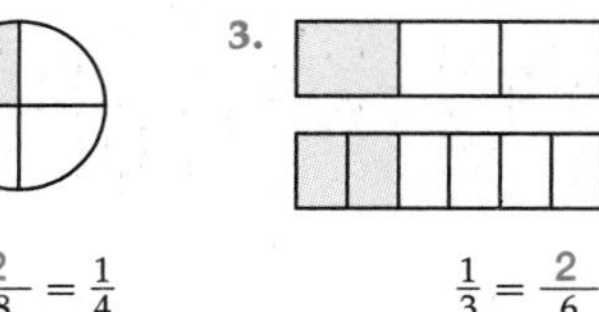

1. $\frac{1}{2} = \frac{2}{4}$

2. $\frac{2}{8} = \frac{1}{4}$

3. $\frac{1}{3} = \frac{2}{6}$

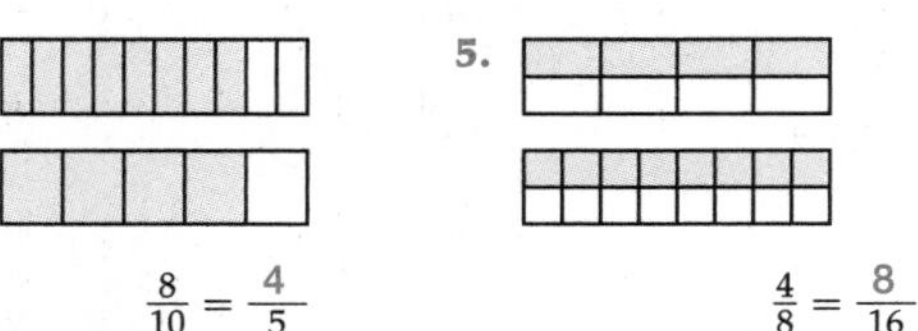

4. $\frac{8}{10} = \frac{4}{5}$

5. $\frac{4}{8} = \frac{8}{16}$

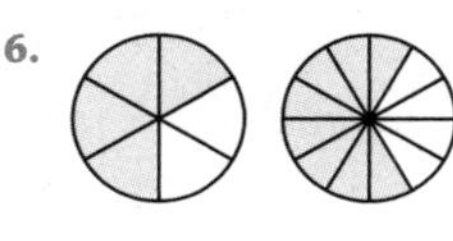

6. $\frac{4}{6} = \frac{8}{12}$

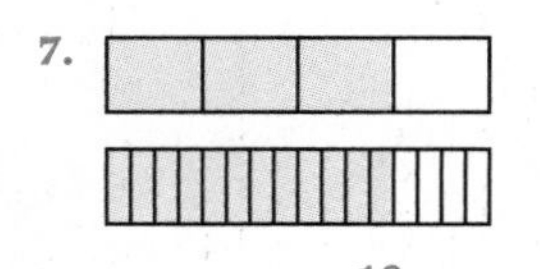

7. $\frac{3}{4} = \frac{12}{16}$

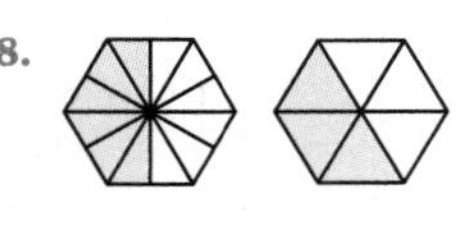

8. $\frac{6}{12} = \frac{3}{6}$

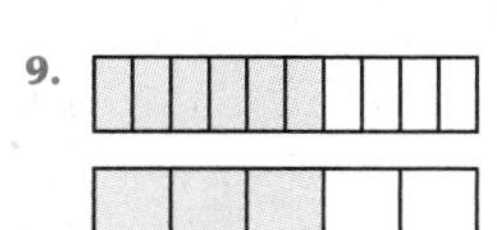

9. $\frac{6}{10} = \frac{3}{5}$

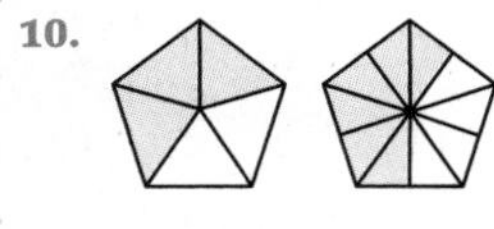

10. $\frac{3}{5} = \frac{6}{10}$

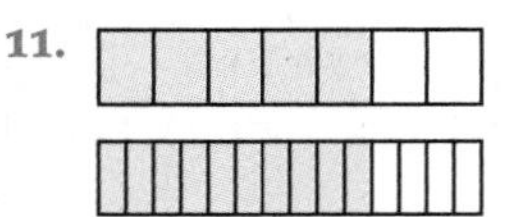

11. $\frac{5}{7} = \frac{10}{14}$

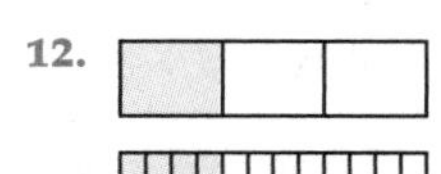

12. $\frac{1}{3} = \frac{4}{12}$

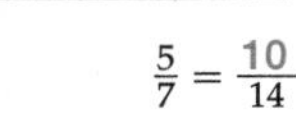

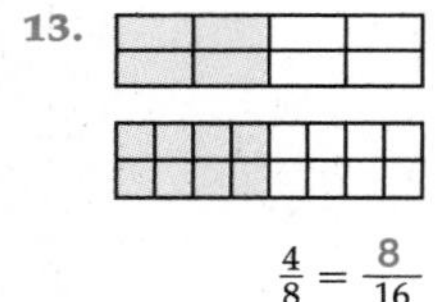

13. $\frac{4}{8} = \frac{8}{16}$

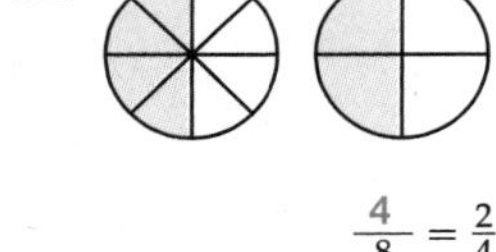

14. $\frac{4}{8} = \frac{2}{4}$

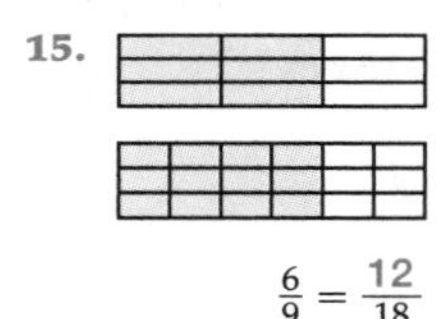

15. $\frac{6}{9} = \frac{12}{18}$

4 Assess

Ask students to tell what equivalent fractions are and give examples. (Equivalent fractions are fractions that name the same amount; examples will vary.)

For Mixed Abilities

Common Errors • Intervention

Students may have difficulty finding a missing numerator when given diagrams for equivalent fractions. Have them work with partners with two rectangles, one separated into 4 equal parts and the other into 16 equal parts. Have them color half of each rectangle. Ask, *What fraction names half of the first rectangle?* ($\frac{2}{4}$) *What fraction names half of the second rectangle?* ($\frac{8}{16}$) Write the following on the board: $\frac{1}{2} = \frac{2}{4} = \frac{8}{16}$.

Enrichment • Fractions

Tell students to find equivalent fractions for the non-shaded portion in each problem on page 256.

More to Explore • Fractions

Have students use the letters of their names as a set. Tell them to write a fraction to answer these questions:

1. What fraction of the letters in your first name are vowels?
2. What fraction of the letters are consonants?
3. What fraction of the letters in your first and last name are vowels?
4. What fraction are consonants?

For the next set of questions, tell students to use the names of all the students in the class.

5. What fraction of the first names in the class begin with the letter J (or any other letter)?
6. What fraction of the first names have four or fewer letters?
7. What fraction of the girls' names have more than five letters?

13-5 Finding Equivalent Fractions

pages 257–258

1 Getting Started

Objective

- To find equivalent fractions by multiplying the numerator and denominator by the same number

Materials

*cardboard rectangle

Warm Up • Mental Math

Ask students who is taller.

1. Joe is 49 in; Uri is 5 ft. (Uri)
2. Beth is 4 ft; Chun is 52 in. (Chun)
3. Zita is 1 m; Amit is 120 cm. (Amit)
4. Mick is $4\frac{1}{2}$ ft; Don is 50 in. (Mick)
5. Mia is $1\frac{1}{2}$ m; Tim is 160 cm. (Tim)
6. Al is 2 yd; Julio is 68 in. (Al)
7. Renee is $3\frac{1}{2}$ ft; Cam is $1\frac{1}{2}$ yd. (Cam)

Warm Up • Number Sense

Write **36** on the board. Ask students how many 2s are in 36 (18) and then how many 3s (12), 4s (9), 6s (6), 9s (4), and 12s (3) are in 36. Repeat for 2s, 3s, 4s, 6s, 8s, 12s, and 24s in 48.

Name ________________________

Finding Equivalent Fractions

Bobbi discovered a shortcut for finding equivalent fractions. Use her shortcut to find the missing numbers.

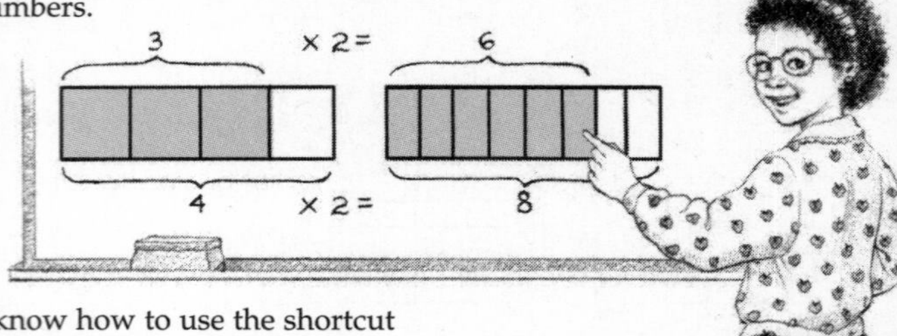

We want to know how to use the shortcut for finding equivalent fractions. We can compare the shaded areas of the rectangles to see what has happened to the numerator and denominator in equivalent fractions.

$$\frac{3}{4} = \frac{6}{8}$$

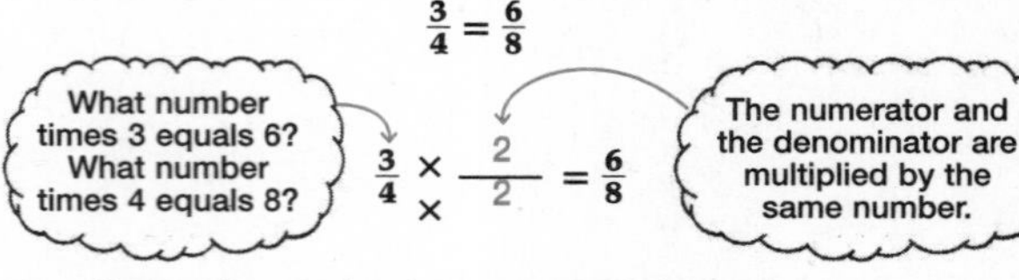

$$\frac{3}{4} \times \frac{2}{2} = \frac{6}{8}$$

Usually, we know the denominator of an equivalent fraction.

What number times 3 equals the denominator 12?

$$\frac{2}{3} = \frac{8}{12}$$

Multiply the numerator by the same number. For this fraction, multiply by 4.

Getting Started

Write the missing numerators.

1. $\frac{1}{4} = \frac{2}{8}$
2. $\frac{1}{3} = \frac{3}{9}$
3. $\frac{1}{6} = \frac{4}{24}$
4. $\frac{2}{3} = \frac{8}{12}$
5. $\frac{3}{4} = \frac{12}{16}$
6. $\frac{4}{5} = \frac{24}{30}$
7. $\frac{1}{2} = \frac{2}{4} = \frac{3}{6} = \frac{4}{8} = \frac{5}{10} = \frac{6}{12} = \frac{7}{14} = \frac{8}{16} = \frac{9}{18}$

2 Teach

Introduce the Lesson Have a student read the problem aloud and tell what is to be done. (use the shortcut to find equivalent fractions) Ask students to tell what is known from the picture. ($\frac{3}{4}$ is equivalent to $\frac{6}{8}$.) Have students read with you as they complete the fractions.

Develop Skills and Concepts Trace the cardboard rectangle on the board to show three congruent rectangles. Shade $\frac{1}{2}$ of the first rectangle and $\frac{2}{4}$ of the second rectangle. Tell students that the same amount of each rectangle is shaded, but there are two times as many parts in the second triangle and therefore two times as many parts are shaded.

- Ask students how much of the first rectangle is shaded. ($\frac{1}{2}$) Ask what part of the second rectangle is shaded. ($\frac{2}{4}$)
- Write $\frac{1}{2} = \frac{2}{4}$ on the board. Tell students that the denominator 4 is two times the denominator 2 in the first fraction, and the numerator 2 is two times the numerator 1.
- Now, shade $\frac{4}{8}$ of the third rectangle and ask students to tell the fraction of the rectangle that is shaded. ($\frac{4}{8}$)
- Write $\frac{1}{2} = \frac{2}{4} = \frac{4}{8}$ on the board. Show students that 8 is 2 × 4 and 4 is 1 × 4.
- Tell students that when we multiply both the numerator and the denominator by the same number, we are multiplying by 1 and multiplying any number by 1 does not change the number's value.
- Have students use the multiplication shortcut to find an equivalent fraction for $\frac{7}{8}$ in sixteenths and for $\frac{6}{7}$ in fourteenths.

3 Practice

Remind students that they will be finding equivalent fractions. Have students complete page 258 independently.

Practice

Write the missing numerators.

1. $\frac{1}{4} = \frac{3}{12}$ 2. $\frac{5}{6} = \frac{15}{18}$ 3. $\frac{3}{5} = \frac{9}{15}$ 4. $\frac{4}{7} = \frac{12}{21}$

5. $\frac{2}{3} = \frac{8}{12}$ 6. $\frac{5}{6} = \frac{20}{24}$ 7. $\frac{5}{8} = \frac{15}{24}$ 8. $\frac{3}{10} = \frac{6}{20}$

9. $\frac{5}{9} = \frac{10}{18}$ 10. $\frac{7}{8} = \frac{56}{64}$ 11. $\frac{3}{9} = \frac{9}{27}$ 12. $\frac{3}{4} = \frac{12}{16}$

13. $\frac{4}{5} = \frac{16}{20}$ 14. $\frac{3}{7} = \frac{12}{28}$ 15. $\frac{5}{8} = \frac{10}{16}$ 16. $\frac{1}{9} = \frac{5}{45}$

17. $\frac{4}{7} = \frac{12}{21}$ 18. $\frac{5}{9} = \frac{30}{54}$ 19. $\frac{3}{4} = \frac{24}{32}$ 20. $\frac{4}{5} = \frac{24}{30}$

21. $\frac{6}{11} = \frac{18}{33}$ 22. $\frac{5}{12} = \frac{30}{72}$ 23. $\frac{3}{16} = \frac{18}{96}$ 24. $\frac{5}{24} = \frac{15}{72}$

25. $\frac{1}{3} = \frac{2}{6} = \frac{3}{9} = \frac{4}{12} = \frac{5}{15} = \frac{6}{18} = \frac{7}{21} = \frac{8}{24} = \frac{9}{27}$

26. $\frac{2}{3} = \frac{4}{6} = \frac{6}{9} = \frac{8}{12} = \frac{10}{15} = \frac{12}{18} = \frac{14}{21} = \frac{16}{24} = \frac{18}{27}$

27. $\frac{1}{4} = \frac{2}{8} = \frac{3}{12} = \frac{4}{16} = \frac{5}{20} = \frac{6}{24} = \frac{7}{28} = \frac{8}{32} = \frac{9}{36}$

28. $\frac{3}{4} = \frac{6}{8} = \frac{9}{12} = \frac{12}{16} = \frac{15}{20} = \frac{18}{24} = \frac{21}{28} = \frac{24}{32} = \frac{27}{36}$

29. $\frac{1}{5} = \frac{2}{10} = \frac{3}{15} = \frac{4}{20} = \frac{5}{25} = \frac{6}{30} = \frac{7}{35} = \frac{8}{40} = \frac{9}{45}$

30. $\frac{1}{6} = \frac{2}{12} = \frac{3}{18} = \frac{4}{24} = \frac{5}{30} = \frac{6}{36} = \frac{7}{42} = \frac{8}{48} = \frac{9}{54}$

4 Assess

Ask students to explain how to use the shortcut for finding equivalent fractions. (multiply the numerator and the demoninator by the same number)

5 Mixed Review

1. 259 ÷ 35 (7 R14)
2. 356 + 4,279 (4,635)
3. 4,000 − 1,356 (2,644)
4. $15.29 + $3.82 + $9.47 ($28.58)
5. 26 × 401 (10,426)
6. 107 ÷ 9 (11 R8)
7. $2.53 × $8 ($20.24)
8. 25,376 + 7,208 (32,584)
9. 78 × 36 (2,808)
10. 14 + 7 × 2 (28)

For Mixed Abilities

Common Errors • Intervention

Some students may not multiply the numerator by the same number used for the denominator. When students are working with a problem such as $\frac{2}{9} = \frac{?}{18}$, ask them to write the factors so that they multiply the numerator of the first fraction by the correct number. For example,

$$\frac{2 \times 2}{9 \times 2} = \frac{4}{18}$$

Enrichment • Fractions

Tell students to write ten equivalent fractions for each of the following: $\frac{4}{10}$, $\frac{7}{12}$, and $\frac{3}{11}$.

More to Explore • Numeration

Have students research the Babylonian number system, 1 to 10. Help students make Babylonian numerals on clay. Have them use a pencil to carve these shapes:

V for ones   ⟨ for tens

Explain that the Babylonians grouped their symbols to make them easier to read. For example,

Have each member of the class write the following numbers on their clay "tablets":

95   44   64   81

This activity could be completed more quickly, if necessary, using paper and pencils or crayons.

13-6 Comparing Fractions

pages 259–260

1 Getting Started

Objective

- To find the greater of two fractions

Warm Up • Mental Math

Have students find the following:

1. $\frac{2}{5}$ of 100 (40)
2. $\frac{3}{10}$ of 20 (6)
3. $\frac{2}{7}$ of 91 (26)
4. $\frac{5}{8}$ of 96 (60)
5. $\frac{6}{9}$ of 108 (72)
6. $\frac{4}{9}$ of 54 (24)
7. $\frac{7}{12}$ of 144 (84)
8. $\frac{1}{100}$ of 1,800 (18)

Warm Up • Fractions

Draw a number line from 0 to 1 on the board. Divide the line into twelfths and write $\frac{3}{12}$ in the correct place. Have students write the additional twelfths as you dictate the fractions in random order. Repeat for a number line divided into forty-eighths.

Name ______________________

Lesson 13-6

Comparing Fractions

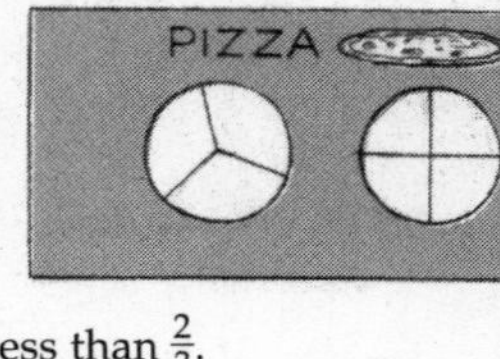

Elaina and Kurt are trying to find out which fraction is greater, $\frac{3}{4}$ or $\frac{2}{3}$. Help them compare the fractions.

We want to know if $\frac{3}{4}$ is greater or less than $\frac{2}{3}$.

We can draw two number lines.

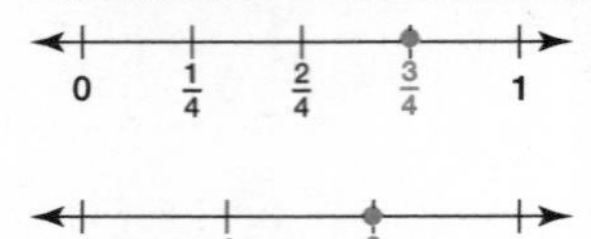

$\frac{3}{4}$ is closer to 1. Therefore, $\frac{3}{4}$ is greater than $\frac{2}{3}$.

We write $\frac{3}{4} > \frac{2}{3}$.

We can also say $\frac{2}{3}$ **is less than** $\frac{3}{4}$. We write $\frac{2}{3} < \frac{3}{4}$.

REMEMBER Numbers to the right on number lines are always greater. Numbers to the left are always less.

We can also find equivalent fractions that have the same denominator. Then, we can compare the numerators.

$\frac{3 \times 3}{4 \times 3} = \frac{9}{12}$ $\quad$ $\frac{2 \times 4}{3 \times 4} = \frac{8}{12}$ $\quad$ $\frac{9}{12} > \frac{8}{12}$

Getting Started

Use the number lines to compare the fractions. Write < or > in the circle.

1. $\frac{1}{2} \; (>) \; \frac{1}{4}$

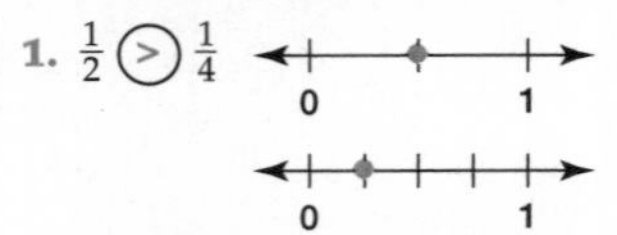

2. $\frac{3}{5} \; (<) \; \frac{2}{3}$

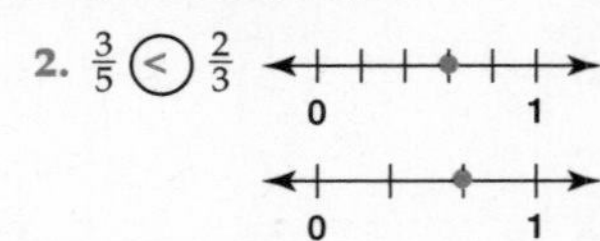

Use equivalent fractions to compare the fractions. Write < or > in the circle.

3. $\frac{3}{4} \; (>) \; \frac{3}{8}$
4. $\frac{3}{8} \; (<) \; \frac{1}{2}$
5. $\frac{2}{3} \; (<) \; \frac{5}{6}$
6. $\frac{5}{6} \; (>) \; \frac{4}{5}$

2 Teach

Introduce the Lesson Have a student read the problem aloud and tell what is to be solved. (find the greater fraction, $\frac{3}{4}$ or $\frac{2}{3}$) Have a student tell about the picture. Have students tell how many slices would be $\frac{3}{4}$ (3) and how many would be $\frac{2}{3}$ (2).

- Have students complete the number lines with you. Ask students what conclusion Elaina and Kurt should come to.
- Read the Remember with students. Then, work through the next section about finding equivalent fractions with the same denominator with students.

Develop Skills and Concepts Remind students that numbers on a number line are always greater to the right and less to the left. Draw two number lines from 0 to 1 on the board. Divide the top number line into fifths and show $\frac{3}{5}$. Divide the bottom number line into tenths and show $\frac{4}{10}$. Tell students that $\frac{3}{5}$ is greater because it is closer to 1.

- Tell students that we can also find out which number is greater by finding a multiple of both 5 and 10 to show $\frac{3}{5}$ and $\frac{4}{10}$ as fractions with the same denominator. Ask students what number is a multiple of 5 and 10. (Answers will vary.) Show students how $\frac{6}{10}$ and $\frac{4}{10}$, $\frac{8}{20}$ and $\frac{12}{20}$, and so on can be used to compare the numbers in order to find the greater number.
- Repeat to compare $\frac{2}{5}$ and $\frac{3}{7}$.

3 Practice

Tell students that in Exercises 1 to 6 they are to put dots on the number lines where the fractions fall and then write > or < in the circle to show how the two fractions compare in size. Have students complete page 260 independently.

Practice

Use the number lines to compare the fractions. Write < or > in the circle.

1. $\frac{1}{3}$ (>) $\frac{1}{5}$

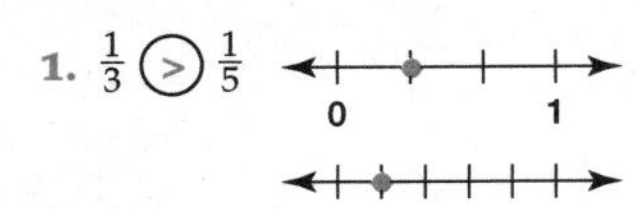

2. $\frac{3}{5}$ (>) $\frac{2}{8}$

3. $\frac{4}{6}$ (<) $\frac{3}{4}$

4. $\frac{1}{2}$ (<) $\frac{6}{10}$

5. $\frac{5}{8}$ (<) $\frac{2}{3}$

6. $\frac{7}{10}$ (>) $\frac{3}{5}$

7. $\frac{7}{8}$ (>) $\frac{8}{10}$

8. $\frac{1}{4}$ (>) $\frac{1}{5}$

Use equivalent fractions to compare the fractions. Write < or > in the circle.

9. $\frac{5}{6}$ (>) $\frac{1}{2}$
10. $\frac{5}{8}$ (>) $\frac{3}{4}$
11. $\frac{2}{3}$ (>) $\frac{3}{12}$
12. $\frac{1}{2}$ (>) $\frac{4}{16}$
13. $\frac{1}{4}$ (>) $\frac{1}{12}$
14. $\frac{1}{4}$ (<) $\frac{3}{8}$
15. $\frac{5}{9}$ (>) $\frac{1}{2}$
16. $\frac{3}{5}$ (<) $\frac{5}{6}$
17. $\frac{5}{12}$ (<) $\frac{1}{2}$
18. $\frac{7}{9}$ (>) $\frac{2}{3}$
19. $\frac{5}{6}$ (>) $\frac{7}{12}$
20. $\frac{3}{4}$ (<) $\frac{4}{5}$
21. $\frac{1}{2}$ (>) $\frac{1}{3}$
22. $\frac{3}{7}$ (>) $\frac{3}{8}$
23. $\frac{2}{3}$ (<) $\frac{3}{4}$
24. $\frac{3}{8}$ (<) $\frac{5}{6}$

Problem Solving

Solve each problem.

25. Cleve ran $\frac{2}{5}$ of a mile. Gary ran $\frac{5}{10}$ of a mile. Who ran farther?

Gary

26. Jennie read for $\frac{3}{4}$ of an hour. Myra read for $\frac{5}{6}$ of an hour. Who read longer?

Myra

4 Assess

Ask students what rule helps them remember how to compare fractions on a number line. (Numbers to the right on number lines are always greater. Numbers to the left are always less.)

For Mixed Abilities

Common Errors • Intervention

Some students may compare just the numerators or just the denominators when they are comparing fractions. Have them work with partners and two fractions such as $\frac{5}{8}$ and $\frac{4}{5}$. Have the students write a set of equivalent fractions for each fraction until they find one from each set with the same denominator. When they compare these, they can see that $\frac{25}{40} < \frac{32}{40}$; therefore $\frac{5}{8} < \frac{4}{5}$.

Enrichment • Fractions

Have students explain why multiplying the numerator and the denominator by the same number does not change the value of a fraction. Then, tell them to apply reasoning to find how many more thousandths are in $\frac{3}{4}$ than in $\frac{3}{5}$.

More to Explore • Logic

Read the following passage to students and tell them to find the logical solution:

A man died and his will decreed his two sons should have a horse race to determine their inheritance. The sons were told to race from their barn to a bridge exactly 3 miles away. The son whose horse reached the bridge last would inherit the entire estate.

After the reading of the will, the sons immediately mounted the two horses and raced toward the bridge at breakneck speed, hoping to be the winner. What would make the sons ride so fast when the winner was to be the one whose horse reached the bridge last? (Each son rode his brother's horse. His own horse would arrive last if he could get his brother's horse to the bridge first.)

13-7 Finding Simplest Terms

pages 261–262

1 Getting Started

Objective

- To reduce a fraction to its lowest terms

Vocabulary

simplified terms

Warm Up • Mental Math

Ask students how many years there are from the following:

1. the present to 2100
2. 1776 to the present
3. 1492 to the present
4. your birth year to 2000
5. the present to 2025
6. 1865 to the present
7. 1945 to the present
8. 1929 to the present

Warm Up • Fractions

Have students write equivalent fractions for $\frac{1}{2}$ and $\frac{1}{3}$ until they find fractions that have the same denominator. ($\frac{3}{6}$, $\frac{2}{6}$) Have students tell if $\frac{1}{2}$ is >, <, or = to $\frac{1}{3}$. (>) Repeat for equivalent fractions for $\frac{1}{9}$ and $\frac{1}{6}$ ($\frac{2}{18}$, $\frac{3}{18}$) and $\frac{1}{3}$ and $\frac{1}{5}$ ($\frac{5}{15}$, $\frac{3}{15}$).

Name ______________________

Finding Simplest Terms

Mr. Granger bought 18 cans of motor oil on sale. What fraction of the case of oil did he buy?

We want to know what part of a case of oil Mr. Granger bought.

Mr. Granger bought <u>18</u> cans of oil.

There are <u>24</u> cans in a case.

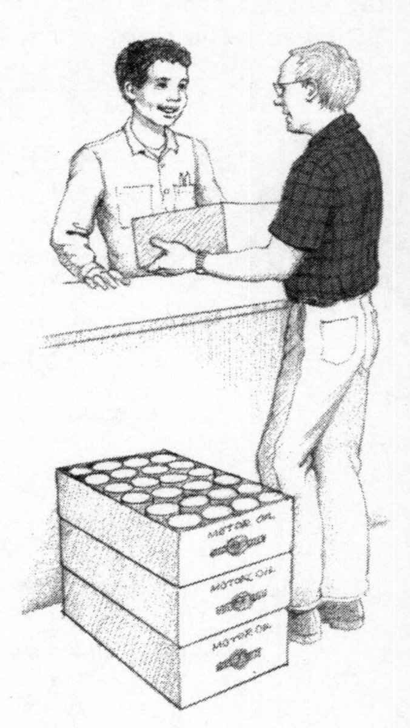

REMEMBER The denominator names the total number of parts, and the numerator names the number of parts you are counting.

$\frac{18}{24} = \frac{\text{the number of cans bought}}{\text{the number of cans in a case}}$

This fraction can be **simplified**. The numerator and the denominator of a fraction are called the **terms** of a fraction. To simplify a fraction, we name it in its lowest terms. We divide the numerator and the denominator by the same non-zero number.

$\frac{18 \div 6}{24 \div 6} = \frac{3}{4}$

A fraction is in its lowest terms when the terms cannot be divided by any common factor other than 1.

Mr. Granger bought <u>$\frac{3}{4}$</u> of a case of oil.

Getting Started

Simplify.

1. $\frac{6}{8} = \frac{3}{4}$
2. $\frac{5}{15} = \frac{1}{3}$
3. $\frac{3}{9} = \frac{1}{3}$
4. $\frac{10}{20} = \frac{1}{2}$
5. $\frac{16}{24} = \frac{2}{3}$
6. $\frac{9}{99} = \frac{1}{11}$
7. $\frac{10}{60} = \frac{1}{6}$
8. $\frac{2}{6} = \frac{1}{3}$
9. $\frac{8}{64} = \frac{1}{8}$
10. $\frac{18}{72} = \frac{1}{4}$

2 Teach

Introduce the Lesson Have a student read the problem and tell what is to be found. (what fraction of the case of oil Mr. Granger bought) Ask what data is known. (Mr. Granger bought 18 cans of oil; a case contains 24 cans.)

- Have students read aloud with you as they complete the sentences to solve the problem. Reinforce the explanation of simplest terms.

Develop Skills and Concepts Remind students that multiplying a fraction by any form of 1 does not change the value of the fraction. Tell students that dividing a fraction by any form of 1 also does not change the value of the fraction.

- Explain to students that the numerator and the denominator of a fraction are called its **terms**. Write $\frac{24}{40}$ on the board and tell students that if there is a number other than 1 that can be divided evenly into both the numerator and the denominator, then the fraction is not **simplified** or in **lowest terms**.
- Ask students if 24 and 40 can both be divided by 2. (yes) Ask students if there is any larger number that can be divided into both 24 and 40. (4, 8) Simplify the fraction by dividing by 2 repeatedly and then show students how much quicker it is to bring a fraction to simplest terms if we find the largest divisor possible.
- Have students find the lowest terms for the fractions $\frac{6}{9}$ ($\frac{2}{3}$), $\frac{16}{20}$ ($\frac{4}{5}$), $\frac{30}{45}$ ($\frac{2}{3}$), and $\frac{17}{51}$ ($\frac{1}{3}$). Then, write various fractions on the board and have students simplify them.

3 Practice

Remind students to divide both terms of the fraction by the same number. Have students complete page 262 independently.

Practice

Simplify.

1. $\frac{10}{15} = \frac{2}{3}$	2. $\frac{6}{9} = \frac{2}{3}$	3. $\frac{4}{12} = \frac{1}{3}$	4. $\frac{5}{10} = \frac{1}{2}$	5. $\frac{6}{18} = \frac{1}{3}$
6. $\frac{4}{20} = \frac{1}{5}$	7. $\frac{4}{24} = \frac{1}{6}$	8. $\frac{4}{16} = \frac{1}{4}$	9. $\frac{8}{12} = \frac{2}{3}$	10. $\frac{9}{18} = \frac{1}{2}$
11. $\frac{6}{12} = \frac{1}{2}$	12. $\frac{14}{16} = \frac{7}{8}$	13. $\frac{15}{25} = \frac{3}{5}$	14. $\frac{6}{10} = \frac{3}{5}$	15. $\frac{3}{12} = \frac{1}{4}$
16. $\frac{16}{20} = \frac{4}{5}$	17. $\frac{9}{12} = \frac{3}{4}$	18. $\frac{24}{48} = \frac{1}{2}$	19. $\frac{8}{16} = \frac{1}{2}$	20. $\frac{8}{32} = \frac{1}{4}$
21. $\frac{10}{12} = \frac{5}{6}$	22. $\frac{16}{24} = \frac{2}{3}$	23. $\frac{4}{8} = \frac{1}{2}$	24. $\frac{27}{36} = \frac{3}{4}$	25. $\frac{12}{16} = \frac{3}{4}$
26. $\frac{16}{48} = \frac{1}{3}$	27. $\frac{14}{21} = \frac{2}{3}$	28. $\frac{20}{25} = \frac{4}{5}$	29. $\frac{40}{100} = \frac{2}{5}$	30. $\frac{26}{52} = \frac{1}{2}$

Now Try This!

To simplify fractions, we find the **greatest common factor** of the numerator and the denominator. The greatest common factor is the greatest common factor of two or more numbers. For example, the fraction $\frac{8}{12}$ can be simplified by finding the **prime factors** of each. Prime factors are **prime numbers** that when multiplied make a product. A prime number is a number greater than zero that has exactly two factors, 1 and the number itself.

8	12
4×2	4×3
$2 \times 2 \times 2$	$2 \times 2 \times 3$

8 and 12 have 2×2 in common, so the greatest common factor is 4.

$$\frac{8}{12} \div \frac{4}{4} = \frac{2}{3}$$

Find the prime factors.

1. 36 $2 \times 2 \times 3 \times 3$
2. 24 $2 \times 2 \times 2 \times 3$
3. 21 3×7

Now Try This! Read the instructions for finding the greatest common factor aloud. Guide students through the model. Have students complete Exercises 1 through 3 independently.

4 Assess

Ask students what the numerator and the denominator of a fraction are called. (terms of a fraction)

For Mixed Abilities

Common Errors • Intervention

Some students may not write the fraction in simplest form because they do not divide by the largest divisor possible. Have students work in pairs with fractions not in lowest terms. For each denominator, ask them to make a list of all the numbers by which the denominator is divisible. Then, have them identify the largest number that is also a divisor of the numerator and use this number to divide both numerator and denominator to write the fraction in simplest terms.

Enrichment • Numeration

Tell students to write three fractions in lowest terms for each of the following denominators: 10, 15, 26, and 40. (Answers will vary.)

More to Explore • Geometry

Give each student a number of plastic straws and straight pins. Tell students to make two figures using straws, fastening them with straight pins.

a tetrahedron and a pyramid

One of these will be rigid (the tetrahedron) and the other will not (the pyramid). Ask students to explain why one figure seems stiff but the other does not. (The square on the bottom of the pyramid is not a rigid figure.) Ask students to find a way to make the pyramid rigid. (Put a strut across the square, dividing it into two rigid triangles.)

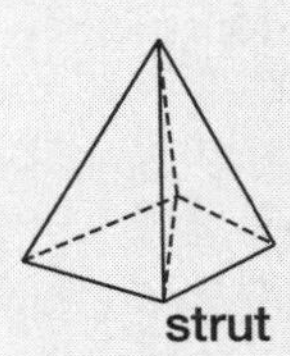

13-8 Mixed Numbers

pages 263–264

1 Getting Started

Objective

- To write a whole or mixed number for a fraction greater than or equal to 1

Vocabulary

mixed number

Materials

*congruent circles in halves, thirds, and so on

Warm Up • Mental Math

Ask students to tell if there is a remainder.

1. 553 ÷ 5 (yes)
2. 78 ÷ 3 (no)
3. 686 ÷ 2 (no)
4. 1,003 ÷ 5 (yes)
5. 703 ÷ 3 (yes)
6. 211 ÷ 2 (yes)
7. 985 ÷ 5 (no)
8. 72,403 ÷ 3 (yes)

Warm Up • Fractions

Have students simplify the following dictated fractions: $\frac{40}{60}$ ($\frac{2}{3}$), $\frac{35}{50}$ ($\frac{7}{10}$), $\frac{12}{18}$ ($\frac{2}{3}$), $\frac{12}{24}$ ($\frac{1}{2}$), $\frac{15}{45}$ ($\frac{1}{3}$), $\frac{18}{72}$ ($\frac{1}{4}$), $\frac{25}{55}$ ($\frac{5}{11}$), $\frac{93}{99}$ ($\frac{31}{33}$), $\frac{54}{72}$ ($\frac{3}{4}$), and $\frac{24}{40}$ ($\frac{3}{5}$).

Name ____________________

Mixed Numbers

Miguel's coach told him to run 9 laps during practice. How many miles did he run?

We want to know how far Miguel ran.

Miguel ran __9__ one-quarter-mile laps.

We can use a number line to help us understand what *quarters* means.

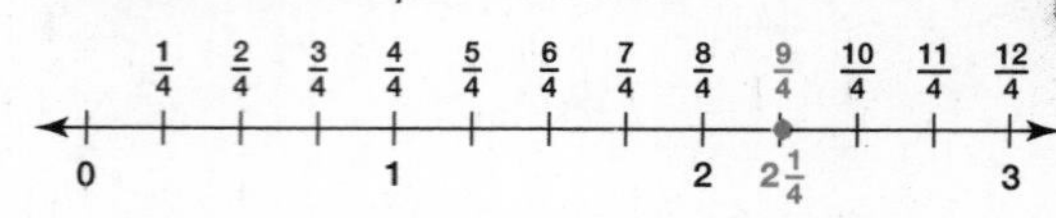

A fraction whose numerator is larger than its denominator can be renamed as a **mixed number**. We divide the numerator by the denominator to rename the fraction as a mixed number. We divide __9__ by __4__.

$\frac{9}{4} = 4\overline{)9}$ → 2 ← whole number; -8; 1 ← number of fourths left

$\frac{9}{4} = 2\frac{1}{4}$ (2 = whole number, $\frac{1}{4}$ = fraction)

$2\frac{1}{4}$ is called a mixed number.

Miguel ran __$2\frac{1}{4}$__ miles.

REMEMBER Simplify fractions when you write a mixed number.

$\frac{16}{6} = 2\frac{4}{6} = 2\frac{2}{3}$ $\frac{15}{3} = 5$

Getting Started

Rename each fraction as a whole or mixed number, and simplify.

1. $\frac{7}{3} = 2\frac{1}{3}$
2. $\frac{5}{2} = 2\frac{1}{2}$
3. $\frac{10}{8} = 1\frac{1}{4}$
4. $\frac{8}{4} = 2$
5. $\frac{16}{10} = 1\frac{3}{5}$

2 Teach

Introduce the Lesson Have a student read the problem aloud and tell what is being asked. (how many miles Miguel ran) Ask what information is known. (Miguel ran 9 laps and each lap is $\frac{1}{4}$ mile.)

- Have students study the number line and read the statements about mixed numbers and simplifying fractions with you. Ask them to complete the solution sentence.

Develop Skills and Concepts Write $\frac{7}{2}$ on the board. Tell students that this fraction is more than the whole number 1. Tell students that we read this fraction as 7 halves and because we know there are 2 halves in 1 whole, we divide the 7 by 2 to see how many wholes are in 7 halves. Show the division to find there are 3 wholes and 1 leftover half. Tell students that $3\frac{1}{2}$ is called a **mixed number**.

- Write $\frac{18}{4}$ on the board. Have a student divide 18 by 4. ($4\frac{2}{4}$) Ask students if the $\frac{2}{4}$ is in simplified terms. (no) Have a student simplify the fraction and then tell the mixed number. ($4\frac{1}{2}$)
- Have students rename, and simplify if necessary, each of the following fractions as whole numbers or mixed numbers: $\frac{14}{4}$ ($3\frac{1}{2}$), $\frac{26}{12}$ ($2\frac{1}{6}$), $\frac{16}{4}$ (4), and $\frac{28}{16}$ ($1\frac{3}{4}$).

3 Practice

Remind students to simplify any fractions that are part of mixed numbers. Have students complete the exercises on page 264 independently.

Practice

Rename each fraction as a whole or mixed number, and simplify.

1. $\frac{5}{4} = 1\frac{1}{4}$	2. $\frac{8}{3} = 2\frac{2}{3}$	3. $\frac{6}{4} = 1\frac{1}{2}$	4. $\frac{7}{2} = 3\frac{1}{2}$	5. $\frac{9}{3} = 3$
6. $\frac{8}{5} = 1\frac{3}{5}$	7. $\frac{12}{8} = 1\frac{1}{2}$	8. $\frac{14}{6} = 2\frac{1}{3}$	9. $\frac{16}{3} = 5\frac{1}{3}$	10. $\frac{18}{4} = 4\frac{1}{2}$
11. $\frac{25}{10} = 2\frac{1}{2}$	12. $\frac{30}{12} = 2\frac{1}{2}$	13. $\frac{30}{9} = 3\frac{1}{3}$	14. $\frac{24}{16} = 1\frac{1}{2}$	15. $\frac{21}{7} = 3$
16. $\frac{44}{8} = 5\frac{1}{2}$	17. $\frac{16}{10} = 1\frac{3}{5}$	18. $\frac{63}{12} = 5\frac{1}{4}$	19. $\frac{40}{6} = 6\frac{2}{3}$	20. $\frac{86}{20} = 4\frac{3}{10}$
21. $\frac{16}{5} = 3\frac{1}{5}$	22. $\frac{14}{10} = 1\frac{2}{5}$	23. $\frac{18}{8} = 2\frac{1}{4}$	24. $\frac{36}{6} = 6$	25. $\frac{42}{9} = 4\frac{2}{3}$
26. $\frac{16}{7} = 2\frac{2}{7}$	27. $\frac{72}{9} = 8$	28. $\frac{9}{5} = 1\frac{4}{5}$	29. $\frac{29}{10} = 2\frac{9}{10}$	30. $\frac{49}{7} = 7$

Now Try This!

It's Algebra!

Fill in the blanks and circles with numbers and signs that correctly complete the pattern.

$9 \times 9 =$ 81	8 + 1 = 9
$9 \times 8 =$ 72	7 + 2 = 9
$9 \times 7 =$ 63	6 (+) 3 = 9
$9 \times 6 =$ 54	5 (+) 4 = 9
$9 \times 5 =$ 45	4 (+) 5 = 9

Write a sentence that describes the pattern.

If we add the digits of the product of 9 and a single-digit number, the sum should be 9.

Now Try This! When the numbers are multiplied by 9, the sum of the digits in the answer is always 9. Have students extend the pattern through 9×2.

4 Assess

Ask students to tell what a mixed number is. (a whole number with a fraction)

For Mixed Abilities

Common Errors • Intervention

When students write a fraction as a mixed number, they may write the whole-number part and forget the fraction part. Have students work in pairs, rewriting a problem such as $\frac{5}{4}$ as $\frac{4}{4} + \frac{1}{4} = 1 + \frac{1}{4}$, or $1\frac{1}{4}$.

Enrichment • Fractions

Tell students to rename the following fractions and simplify if necessary: $\frac{216}{42}$ $(5\frac{1}{7})$, $\frac{378}{189}$ (2), $\frac{500}{200}$ $(2\frac{1}{2})$, $\frac{1000}{300}$ $(3\frac{1}{3})$, and $\frac{896}{7}$ (128).

More to Explore • Biography

Duplicate the following for students:

Emmy Noether was a German mathematician of the twentieth century. She taught mathematics both in Europe and in the United States. Some famous mathematicians she worked with included Weyl, Hasse, Klein, Brauer, Hilbert, and Einstein. She taught at Bryn Mawr and Princeton Universities in the United States, at Moscow University in Russia, and at Göttingen University in Germany.

Using this information about Noether, fill in the missing words below. When you finish, the boxed letters will spell the part of mathematics in which Emmy Noether did her best work.

(B r [a] u e r)	a friend
(W e y [l])	a friend
([G] ö t t i n g e n)	a university in Germany
(E i n s t [e] i n)	a friend
(H i l [b] e r t)	a friend
(P [r] i n c e t o n)	a university in the United States
(B r y n M [a] w r)	a university in the United States

(algebra)

13-9 Comparing Mixed Numbers

pages 265–266

1 Getting Started

Objective
- To compare mixed numbers

Warm Up • Mental Math
Ask if the following questions are true or false:

1. $7 \times 8 = 56$ (T)
2. $10 \times 0 = 10$ (F)
3. 56 can be rounded to 50. (F)
4. $2,246 < 2,217$ (F)
5. $\frac{2}{10} - \frac{2}{10} = 0$ (T)

Warm Up • Numeration
Ask students what number tells the following:

1. $(\frac{1}{4}$ of $20) \times 60$ (300)
2. 7s in 81×7 (81)
3. hours in $2\frac{1}{4}$ days (54)
4. inches in 4 yards (144)
5. zeros in the product of $100 \times 6 \times 10$ (3)
6. $100 \div (\frac{1}{2}$ of $40)$ (5)

Name ______________________

Lesson 13-9

Comparing Mixed Numbers

George and Eric are trying to find out which mixed number is greater, $1\frac{2}{5}$ or $1\frac{1}{3}$.

We want to know which mixed number is greater, $1\frac{2}{5}$ or $1\frac{1}{3}$.

Always compare the whole numbers first. The whole numbers are the same. We can draw two number lines.

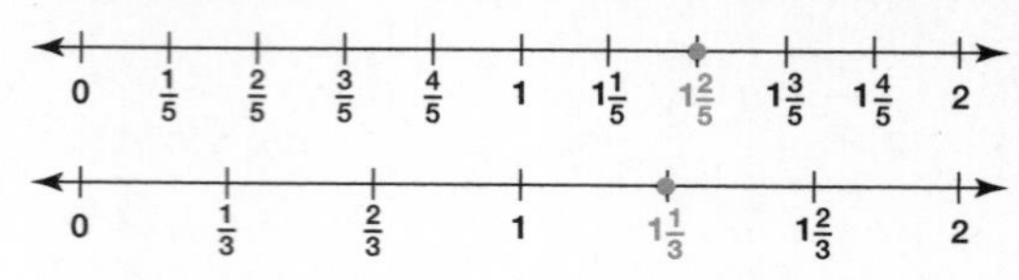

$1\frac{2}{5}$ is closer to 2. Therefore, $1\frac{2}{5}$ is greater than $1\frac{1}{3}$.
We write $\mathbf{1\frac{2}{5} > 1\frac{1}{3}}$.

We can also say $1\frac{1}{3}$ is less than $1\frac{2}{5}$. We write $\mathbf{1\frac{1}{3} < 1\frac{2}{5}}$.

To compare mixed numbers with the same whole number and the same denominator, we can compare the numerators.

Compare $3\frac{3}{8}$ and $3\frac{7}{8}$. $3\frac{3}{8}$ (<) $3\frac{7}{8}$

Getting Started

Use the number lines to compare the mixed numbers. Write < or > in the circle.

1. $4\frac{3}{4}$ (<) $4\frac{4}{5}$

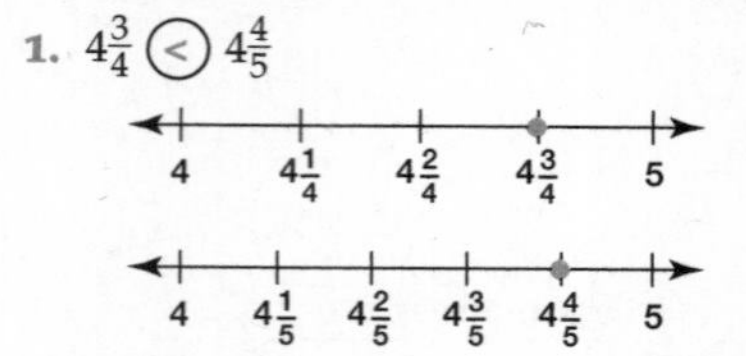

2. $6\frac{1}{6}$ (>) $6\frac{1}{8}$

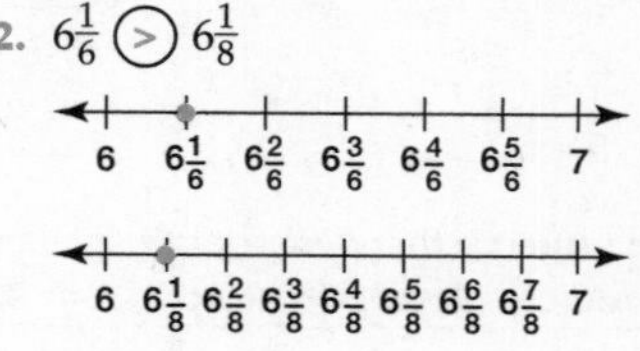

Compare. Write < or > in the circle. Draw a number line to help.

3. $3\frac{1}{4}$ (>) $2\frac{2}{4}$
4. $1\frac{5}{8}$ (>) $1\frac{1}{4}$
5. $4\frac{3}{5}$ (<) $6\frac{1}{8}$
6. $3\frac{2}{3}$ (>) $3\frac{1}{3}$

2 Teach

Introduce the Lesson Have a student read the problem and information sentences above the number lines.

- Now, ask students to study the number lines. Point out that one is divided into fifths and the other into thirds. Have students name the fractions and mixed numbers on each number line.
- Ask, *Which mixed number is closer to 2, $1\frac{2}{5}$ or $1\frac{1}{3}$?* Have a student write the inequality on the board. ($1\frac{2}{5} > 1\frac{1}{3}$) Have students complete the solution sentence. Remind students that when comparing mixed numbers, they must always compare the whole numbers first.
- Explain that when comparing mixed numbers with the same whole number and the same denominator, students can compare numerators. Read and complete the next problem with students.

Develop Skills and Concepts Explain that although number lines can be used to compare mixed numbers, drawings can also be used.

- Have students compare $1\frac{5}{8}$ and $1\frac{3}{4}$. Explain that because the whole numbers are the same, they need to compare the fractions. Draw the following on the board:

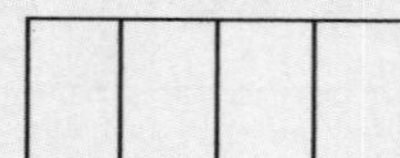

- Ask a volunteer to shade $\frac{5}{8}$ of the first square and another volunteer to shade $\frac{3}{4}$ of the second square. Ask the class to tell which square has the most parts shaded. (the square with $\frac{3}{4}$ shaded)
- Write the following on the board and have a volunteer fill in the answers:

$\frac{5}{8}$ ○ $\frac{3}{4}$ (<) $1\frac{5}{8}$ ○ $1\frac{3}{4}$ (<)

Practice

Use the number lines to compare the mixed numbers. Write > or < in the circle.

1. $7\frac{1}{3}$ ($>$) $7\frac{1}{4}$

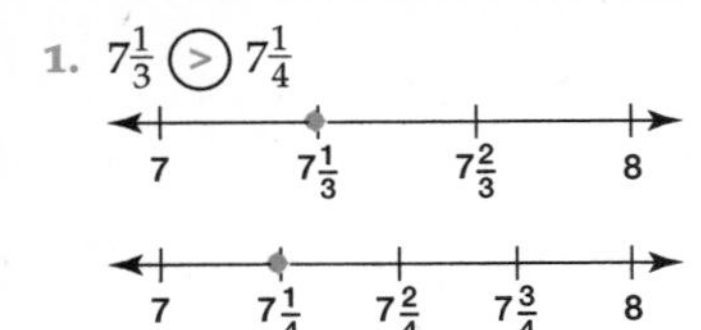

2. $8\frac{5}{8}$ ($>$) $8\frac{3}{10}$

8 $8\frac{1}{8}$ $8\frac{2}{8}$ $8\frac{3}{8}$ $8\frac{4}{8}$ $8\frac{5}{8}$ $8\frac{6}{8}$ $8\frac{7}{8}$ 9

8 $8\frac{1}{10}$ $8\frac{2}{10}$ $8\frac{3}{10}$ $8\frac{4}{10}$ $8\frac{5}{10}$ $8\frac{6}{10}$ $8\frac{7}{10}$ $8\frac{8}{10}$ $8\frac{9}{10}$ 9

3. $1\frac{7}{8}$ ($>$) $1\frac{1}{5}$

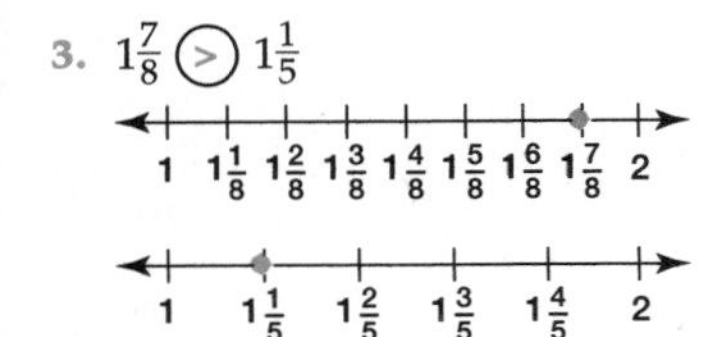

4. $4\frac{1}{2}$ ($<$) $4\frac{3}{5}$

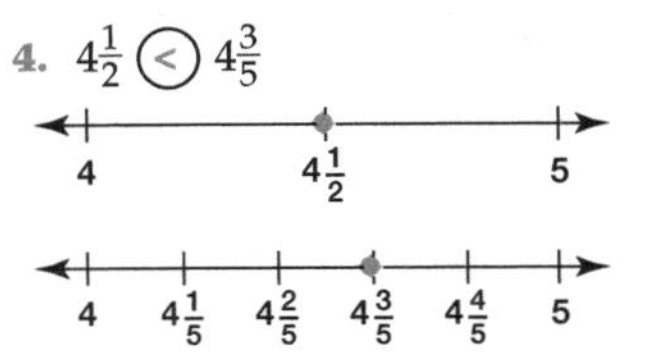

5. $1\frac{1}{4}$ ($>$) $1\frac{1}{6}$

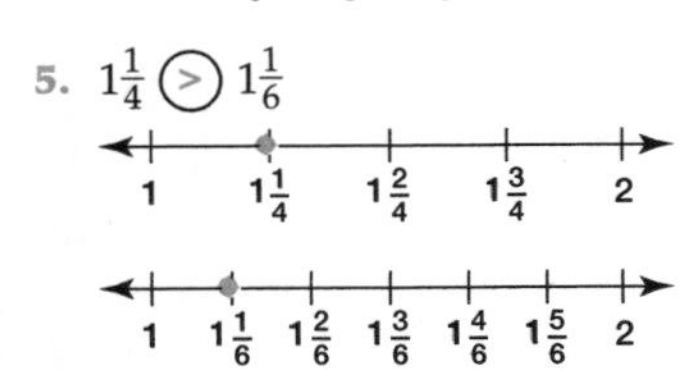

6. $1\frac{3}{4}$ ($<$) $1\frac{5}{6}$

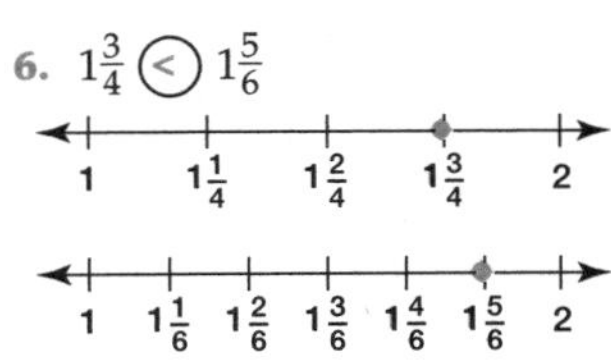

Compare. Write > or < in the circle. Draw a number line to help.

7. $3\frac{2}{3}$ ($>$) $2\frac{3}{4}$
8. $2\frac{2}{5}$ ($>$) $\frac{1}{2}$
9. $1\frac{4}{5}$ ($<$) 2
10. $2\frac{3}{4}$ ($>$) $2\frac{1}{5}$
11. $2\frac{1}{2}$ ($>$) $1\frac{9}{10}$
12. $3\frac{1}{3}$ ($<$) $4\frac{1}{5}$
13. $2\frac{5}{8}$ ($<$) $2\frac{3}{4}$
14. $3\frac{1}{4}$ ($<$) $3\frac{1}{2}$
15. $1\frac{2}{3}$ ($>$) $1\frac{1}{5}$
16. $2\frac{1}{8}$ ($>$) $2\frac{1}{10}$
17. $3\frac{1}{2}$ ($>$) $2\frac{4}{5}$
18. $2\frac{3}{5}$ ($<$) $2\frac{5}{6}$
19. 3 ($>$) $\frac{1}{8}$
20. $2\frac{7}{10}$ ($<$) $3\frac{1}{5}$
21. $2\frac{1}{5}$ ($>$) $2\frac{1}{8}$
22. $3\frac{1}{4}$ ($<$) $3\frac{2}{3}$

Problem Solving

Solve each problem.

23. Tim practiced for $1\frac{1}{2}$ hours. José practiced for $1\frac{3}{4}$ hours. Who practiced longer?
José

24. Frankie's movie lasted $2\frac{1}{4}$ hours. Mom's movie lasted $2\frac{1}{2}$ hours. Whose movie lasted longer?
Mom's

3 Practice

Have students complete the exercises on the page.

4 Assess

Have students name a mixed number that is greater than or less than a given mixed number.

For Mixed Abilities

Common Errors • Intervention

When comparing mixed numbers, some students may compare just the whole numbers. Remind these students that they must also compare the fractions. Have these students practice comparing mixed numbers with partners. Have one student write the mixed numbers one above the other and have the other student first compare the whole numbers and then compare the fractions.

Enrichment • Numeration

Have students devise an arithmetic progression using Roman numerals. Use the following as an example: V, XV, XX, LX, LXV, and so on. (×3, +5) Tell students to exchange progressions with a partner and to have partners provide the next five Roman numerals and identify the common difference that describes the progression. Have students repeat the activity.

13-10 Fractions in Measurement

pages 267–268

1 Getting Started

Objective
- To write mixed numbers for lengths measured to the nearest quarter inch

Materials
*large paper ruler marked in $\frac{1}{4}$ inches; rulers marked in $\frac{1}{4}$ inches

Warm Up • Mental Math
Have students give an equivalent fraction.

1. $\frac{1}{2}$ ($\frac{2}{4}$ and so on)
2. $\frac{1}{3}$ ($\frac{2}{6}$ and so on)
3. $\frac{1}{4}$ ($\frac{2}{8}$ and so on)
4. $\frac{1}{5}$ ($\frac{2}{10}$ and so on)
5. $\frac{2}{3}$ ($\frac{4}{6}$ and so on)
6. $\frac{3}{4}$ ($\frac{9}{12}$ and so on)
7. $\frac{5}{8}$ ($\frac{10}{16}$ and so on)
8. $\frac{1}{10}$ ($\frac{10}{100}$ and so on)

Warm Up • Fractions
Have students write a whole or simplified mixed number for the following dictated fractions: $\frac{9}{4}$ ($2\frac{1}{4}$), $\frac{17}{4}$ ($4\frac{1}{4}$), $\frac{22}{4}$ ($5\frac{1}{2}$), $\frac{36}{4}$ (9), and $\frac{42}{4}$ ($10\frac{1}{2}$).

Name ______________________ **Lesson 13-10**

Fractions in Measurement

Find the length of the pen to the nearest quarter inch.

We use a ruler marked in quarter inches.

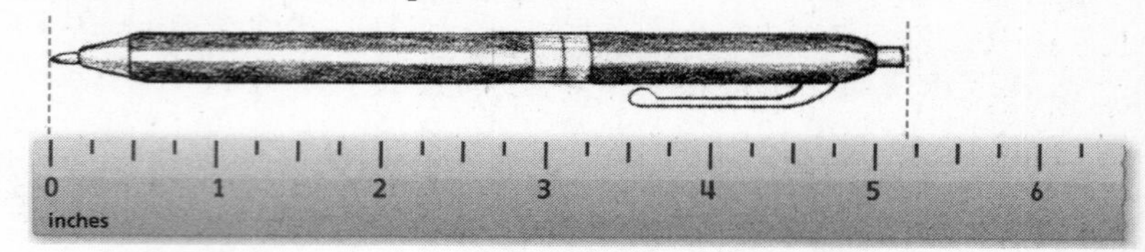

The length of the pen is between __5__ and __6__ inches. It is closer to __5__ inches.
To the nearest $\frac{1}{4}$ inch, the length of the pen is __$5\frac{1}{4}$__ inches.

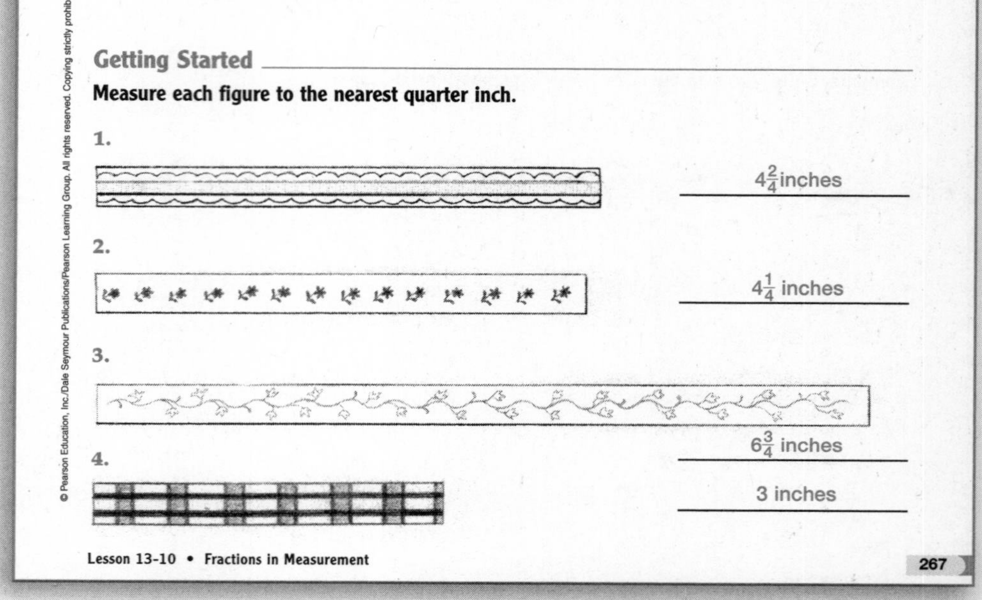

Getting Started

Measure each figure to the nearest quarter inch.

1. $4\frac{2}{4}$ inches
2. $4\frac{1}{4}$ inches
3. $6\frac{3}{4}$ inches
4. 3 inches

2 Teach

Introduce the Lesson Have a student read the problem and tell what is to be done. (measure the pen to the nearest $\frac{1}{4}$ inch) Have students complete the sentences as they read aloud with you to solve the problem.

Develop Skills and Concepts Post the large paper ruler on the board. Ask students how many quarter-inch spaces are between each two numbers on the ruler. (4) Tell students that each of those spaces measures a quarter inch.

- Write $\frac{4}{4} = 1$ on the board to remind students that 4 fourths equals 1 whole. Tell students that 4 quarter inches equal 1 whole inch. Have students find various points such as $2\frac{1}{2}$, $3\frac{3}{4}$, and $5\frac{1}{4}$ inches on the large ruler.
- Use a pointer to locate a point on the ruler that is approximately $5\frac{5}{16}$ inches and have students tell if the point is closer to $5\frac{1}{4}$ inches or $5\frac{1}{2}$ inches. ($5\frac{1}{4}$)
- Repeat for other measurements to be found to the nearest quarter inch. Include some points closer to the whole inch for students to name the measurement closest to a whole number and zero fourths.

3 Practice

Remind students to record each measurement to the nearest quarter inch. Have students complete page 268 independently.

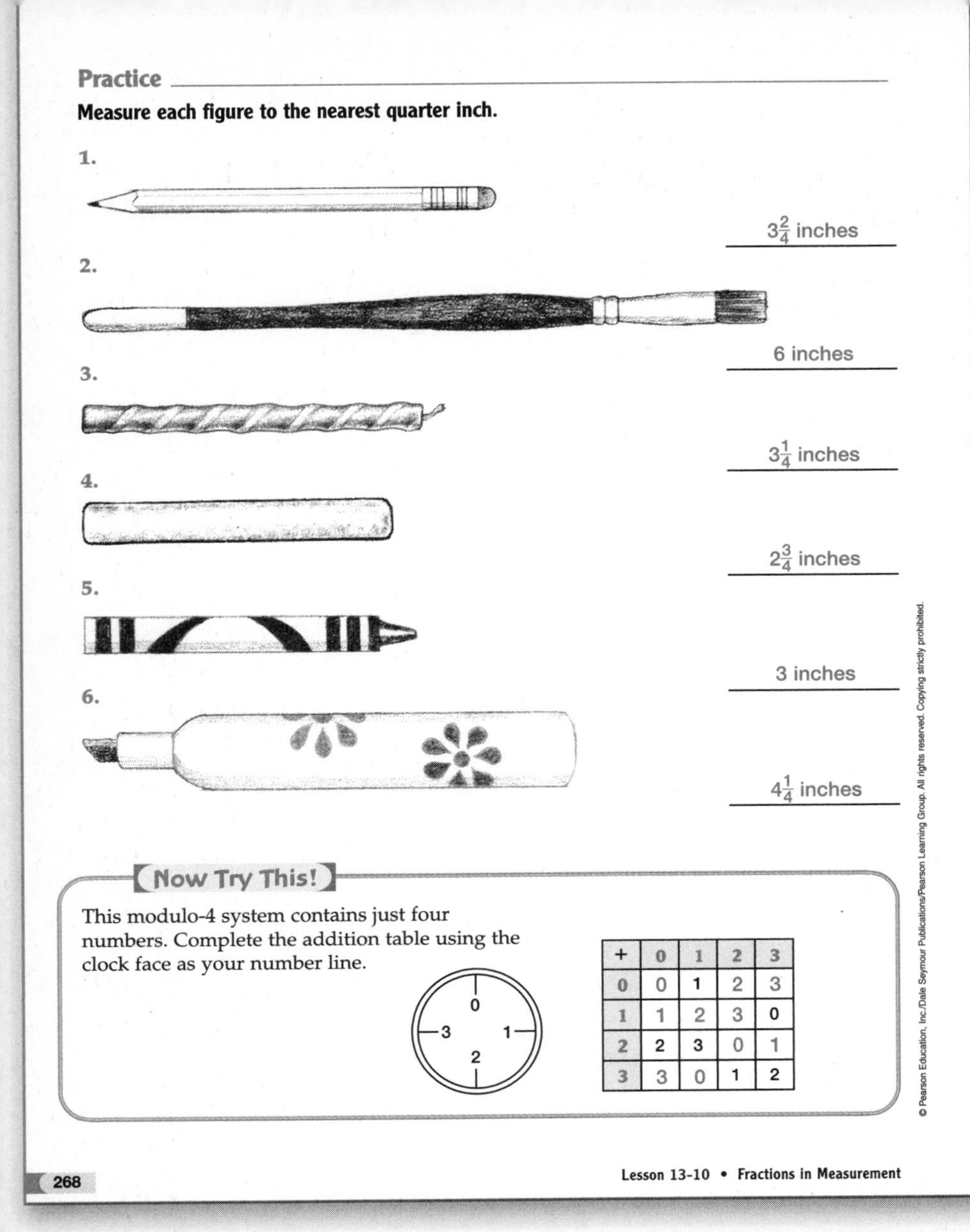

Practice

Measure each figure to the nearest quarter inch.

1. $3\frac{2}{4}$ inches

2. 6 inches

3. $3\frac{1}{4}$ inches

4. $2\frac{3}{4}$ inches

5. 3 inches

6. $4\frac{1}{4}$ inches

Now Try This!

This modulo-4 system contains just four numbers. Complete the addition table using the clock face as your number line.

+	0	1	2	3
0	0	1	2	3
1	1	2	3	0
2	2	3	0	1
3	3	0	1	2

Now Try This! Define a modulus as a closed system of numbers. To demonstrate the use of the clock face as a closed number line, perform addition by starting at the first addend and moving clockwise as many places as the second addend. Students can construct a modulo-5 clock face and complete its addition table.

4 Assess

Have students measure the width of their textbook to the nearest quarter inch. ($8\frac{1}{2}$ inches)

For Mixed Abilities

Common Errors • Intervention

Students may have difficulty deciding to which quarter inch a measure is closer. Have them work with partners to practice drawing line segments given directions such as the following: Draw a line segment that is between $4\frac{3}{4}$ and 5 inches but closer to $4\frac{3}{4}$ inches.

Enrichment • Measurement

Tell students to draw a map, and place and label points that are the following distances from each other: $\frac{9}{4}$ inches, $1\frac{3}{4}$ inches, $5\frac{1}{4}$ inches, and $\frac{12}{4}$ inches. (Because distances are to the quarter inch, it is not necessary to say "to the nearest quarter inch.")

More to Explore • Statistics

Present the class with a list of anonymous scores for a test, perhaps this week's spelling test. List the scores in random order.

Have them group the scores in five-point categories: 0–5, 6–10, 11–15, 16–20, 21–25, 26–30, 31–35, 36–40, and so on. Now, have them tally the number of scores within each five-point range.

Explain that the number of scores in each category is called the **frequency**.

Discuss with students and have them list on the board when this kind of frequency tally might be made. (monitoring traffic on a roadway, tracking season totals for a basketball team, and so on)

ESL/ELL STRATEGIES

Explain that when talking about fractions, the words *quarter* and *fourth* are both correct. We can say "three-fourths" or "three quarters." However, when counting things (using ordinal numbers), we always use *fourth*; for example, "Wednesday is the *fourth* day of the week."

13-11 Ratios

pages 269–270

1 Getting Started

Objective

- To use a ratio to compare two quantities

Vocabulary

ratio

Materials

counters in 2 colors

Warm Up • Mental Math

Have students compare $\frac{7}{8}$ and $\frac{3}{4}$ and tell which fraction is greater. Repeat for more comparisons of simple fractions.

Warm Up • Algebra

Have students solve for n.

1. $\frac{5}{8} = \frac{n}{16}$ (10)
2. $\frac{2}{3} = \frac{n}{18}$ (12)
3. $\frac{n}{5} = \frac{4}{10}$ (2)
4. $\frac{1}{2} = \frac{n}{100}$ (50)
5. $\frac{4}{n} = \frac{8}{20}$ (10)
6. $\frac{7}{6} = 1 + \frac{n}{6}$ (1)
7. $2 + \frac{5}{6} = \frac{17}{n}$ (6)
8. $\frac{1}{n} = \frac{4}{8}$ (2)

Name ______________________

Ratios

A **ratio** is a comparison of two quantities. The ratio of black pins to all pins on the bulletin board is 3 to 9. This can also be written as the fraction $\frac{3}{9}$. The ratio of black pins to white pins is 3 to 4, or $\frac{3}{4}$. What is the ratio of green pins to black pins?

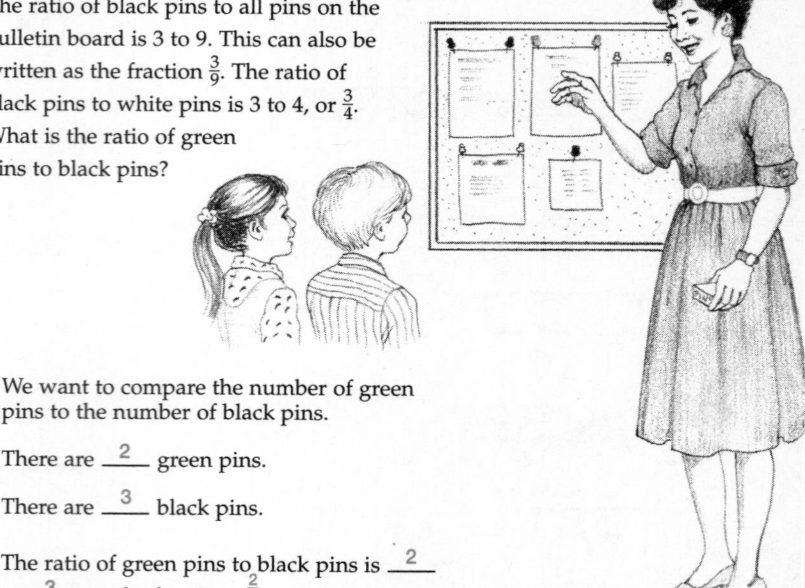

We want to compare the number of green pins to the number of black pins.

There are <u>2</u> green pins.

There are <u>3</u> black pins.

The ratio of green pins to black pins is <u>2</u> to <u>3</u>, or the fraction <u>$\frac{2}{3}$</u>.

Getting Started

Write each comparison as a fraction.

1. white pins to black pins $\frac{4}{3}$
2. white pins to all pins $\frac{4}{9}$
3. green pins to black pins $\frac{2}{3}$
4. black pins to green pins $\frac{3}{2}$

Write each comparison as a ratio.

5. green pins to all pins <u>2 to 9</u>
6. green pins to white pins <u>2 to 4</u>

2 Teach

Introduce the Lesson Have a student read the problem aloud and tell what is to be found. (the ratio of green pins to black pins) Ask what information is given. (There are 3 black pins and 2 green pins.)

- Ask students to tell what is known about a ratio. (It is a comparison of two quantities and can be written as a fraction.) Have students read with you as they complete the sentences.

Develop Skills and Concepts Draw the following on the board:

XXX RR
XX R

- Tell students that we want to compare the number of Xs to the number of Rs. Write the following on the board:

X R
5 to 3 or $\frac{5}{3}$

Tell students that the ratio of 5 to 3 or $\frac{5}{3}$ compares the Xs to the Rs. Now, tell students to compare the number of Rs to Xs. Explain that we give the number of Rs first and then the number of Xs. Write the following on the board:

R X
3 to 5 or $\frac{3}{5}$

- Next, have students compare the number of Rs to all the letters. (3 to 8 or $\frac{3}{8}$) Repeat for the number of Xs to all letters. (5 to 8 or $\frac{5}{8}$) Add more Xs and/or Rs and have students write ratios to compare the new numbers.

3 Practice

Have students complete page 270 independently.

Practice

Write each comparison as a fraction.

1. white marbles to all marbles $\frac{6}{12}$
2. white marbles to green marbles $\frac{6}{4}$
3. black marbles to green marbles $\frac{2}{4}$
4. green marbles to white marbles $\frac{4}{6}$
5. black marbles to all marbles $\frac{2}{12}$
6. green marbles to black marbles $\frac{4}{2}$
7. white marbles to black marbles $\frac{6}{2}$
8. green marbles to all marbles $\frac{4}{12}$

Write each comparison as a ratio.

9. soccer balls to baseball mitts 3 to 2
10. baseball mitts to baseballs 2 to 4
11. baseballs to all sports equipment 4 to 9
12. baseballs to soccer balls 4 to 3
13. soccer balls to all sports equipment 3 to 9
14. soccer balls to baseballs 3 to 4
15. soccer balls and baseball mitts to all sports equipment 5 to 9
16. baseball mitts to soccer balls and baseballs 2 to 7

4 Assess

Draw 5 circles, 3 squares, and 4 triangles on the board. Ask students for the ratio of circles to all the figures, circles to squares, and circles to triangles. ($\frac{5}{12}$, $\frac{5}{3}$, $\frac{5}{4}$)

5 Mixed Review

1. 2,056 ÷ 81 (25 R31)
2. 320 × 42 (13,440)
3. $\frac{3}{5}$ of 25 (15)
4. 6,021 − 4,317 (1,704)
5. $\frac{4}{5} = \frac{?}{35}$ (28)
6. 3,658 + 23 + 176 (3,857)
7. 253 ÷ 5 (50R3)
8. $\frac{?}{4} = \frac{12}{16}$ (3)
9. 8 × 700 (5,600)
10. 3,156 + 6,080 (9,236)

For Mixed Abilities

Common Errors • Intervention

When students write ratios, they may not write the numbers in the correct order. Have them work in pairs with counters. Have one student lay out some counters, using two colors, and have the other student write a ratio that compares one color to the other and then vice versa. Have them discuss why the order of the numbers is important.

Enrichment • Ratio

Tell students to write a ratio to compare each of the following: males to females in their family, females to males in their family, males to the total number of members in their family, females to the total number of members in their family, and males and females in their classroom. Remind them to remember to include their teacher in the count.

More to Explore • Geometry

Give each student a piece of corrugated cardboard, a sheet of paper, two pushpins, and a small loop of string. Have students put the sheet of paper on top of the cardboard and put the two pushpins through both paper and cardboard. Tell them to put the loop of string around the pushpins. Explain that by putting a pencil into the loop and slowly moving the pencil around the pushpins, they will draw an ellipse. Some will say it is an oval.

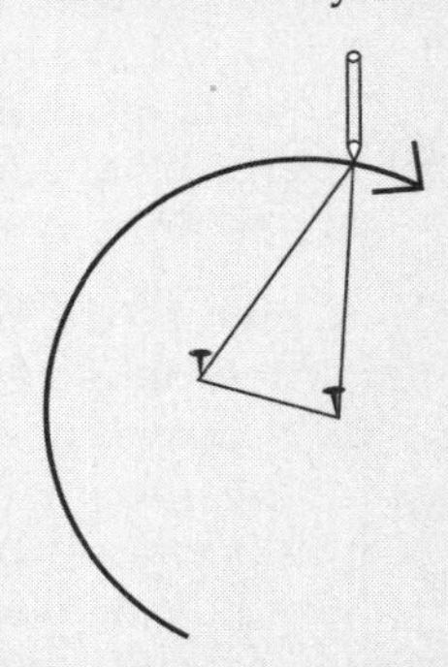

Explain that an ellipse is defined by two "foci," or fixed points, which are the pushpins in this activity.

13-12 Problem Solving: Solve a Simpler Problem

pages 271–272

1 Getting Started

Objective

- To solve a problem by solving a simpler or related problem first

Warm Up • Mental Math

Have students identify each stage of problem solving.

1. Carry out the plan. (Do)
2. Look for careless errors. (Check)
3. Ask if the answer is reasonable. (Check)
4. Decide on the operations to be used. (Plan)
5. Find a solution. (Do)
6. Decide what is to be solved. (See)
7. Perform the last stage. (Check)
8. State facts that are known. (See)

Name ______________________________

Problem Solving: Solve a Simpler but Related Problem

Carlos has $1.20. Half of the coins are nickels and half are dimes. What is the value of his nickels?

★**SEE**

We want to find the value of Carlos's nickels.

We know that the value of all his coins is $1.20.

Half of the coins are nickels and half are dimes.

★**PLAN**

We want to express this problem as simply as we can so that we can understand it. Because Carlos has only nickels and dimes, we should start with a simpler problem. We should make a table of our results to see if we can find a pattern that would help us solve the problem.

★**DO**

Number of Nickels	Number of Dimes	Value of Nickels	Value of Dimes	Total Value of the Coins
3	3	$0.15	$0.30	$0.45
6	6	$0.30	$0.60	$0.90
7	7	$0.35	$0.70	$1.05
8	8	$0.40	$0.80	$1.20

Carlos has 8 nickels and 8 dimes.

★**CHECK**

$$5¢ \times 8 = 40¢ \qquad 10¢ \times 8 = 80¢ \qquad 80¢ + 40¢ = \$1.20$$

2 Teach

Introduce the Lesson Remind students that problems can be solved by the four-step problem-solving strategy.

Develop Skills and Concepts Ask students to think about the last time they became confused when someone told them something. Tell students that in order to understand what we hear, we often reword it in our own words and try to simplify it. Ask students if they have ever read something that was confusing. Tell students that again, we find ourselves simplifying what we have read. We often need to reread something several times before we can simplify it.

- Have a student read the problem aloud and tell what is to be found. (the value of Carlos' nickels) Tell students that a first reading of this problem leads us to assume that because the number of nickels and dimes is the same, then their values would also be the same. So, we would take half of $1.20 for a 60¢ value of the nickels. Tell students that we must keep in mind, though, that a nickel's value is $\frac{1}{2}$ that of a dime, so we cannot just take half of $1.20.
- Have students read aloud with you through the SEE, PLAN, and DO steps and complete the table to solve the problem. Have students complete the CHECK stage to verify their solution.

3 Apply

Solution Notes

Have students simplify each problem before beginning to solve it.

1. Careful reading will help students see that the number of pinchworms is unnecessary information. The important information is that pinchworms double their number every minute. If the jar is full in 40 minutes, then it is half-full in 1 minute less than 40 minutes.
2. Students should see that they should draw a map to locate the three cities in relation to each other.
3. At first glance, the students are likely to respond that Kyle spent $8. Upon closer reading, they should see that the ball cost $2.50 because it was half as much as the bat. So Kyle spent $7.50.
4. Help students restate the cost of 13 stamps as 26¢.

Apply

Solve each problem.

1. Suppose there are 20 pinchworms in the bottom of a jar. Pinchworms multiply so fast that they double their number every minute. If it takes 40 minutes for the pinchworms to fill the jar, how long will it take them to fill half the jar?
   39 minutes

2. It is 34 miles from Bay City to Hamilton. It is 28 miles from Hamilton to Glenville. Glenville is on the road between Bay City and Hamilton. How far is it from Bay City to Glenville?
   6 miles

3. Kyle bought a bat for $5.00. Then, he bought a ball costing half as much as the bat. How much did Kyle spend?
   $7.50

4. If you can buy 13 stamps for a cent and a quarter, what is the cost of one stamp?
   $0.02

5. If 5 boys can paint 5 garages in 5 days, how many boys would it take to paint 25 garages in 25 days?
   5 boys

Problem-Solving Strategy: Using the Four-Step Plan

- ★ **SEE** What do you need to find?
- ★ **PLAN** What do you need to do?
- ★ **DO** Follow the plan.
- ★ **CHECK** Does your answer make sense?

6. I went to the store and bought several items that were the same price. I bought as many items as the number of total cents in the cost of each item. My bill was $1.44. How many items did I buy?
   12 items

7. Read Exercise 2 again. Write a problem about three towns where the distance between the nearest two towns is 5 miles and the distance between the two towns farthest apart is 15 miles.
   Answers will vary.

8. Explain how you would find the perimeter of a rectangle if the length were given in feet and the width in yards.
   See Solution Notes.

For Mixed Abilities

More to Explore • Measurement

Tell students that a painter must be able to estimate the amount of paint needed to complete a job. The label on a can of paint tells them how much surface the paint will cover. For example, the paint might cover 36 square meters or 106 square feet.

Have students bring in paint can labels or ask a paint company to send labels from various products to the classroom. Have students take measurements of their classroom or hallway and calculate how much of a specific paint it would take to paint that area. Next, have them figure out how many of a specific can size they will need, reminding them that they may end up with some paint left over. Then, ask what other jobs require this type of estimation.

5. Help students restate the problem to see that the 5 boys paint 1 garage each day and therefore the 5 boys can paint 25 garages in 25 days.
6. Students need to restate the problem to realize that the cost of each item equals the number of items bought. Thus they are looking for a number that, multiplied by itself, is 144.

Higher-Order Thinking Skills

7. **Synthesis:** Answers will vary. A sample answer is "The distance between A and B is 5 miles. Peter Rabbit runs from A to C, a distance of 15 miles. How far is it from B to C?"
8. **Synthesis:** Either change the length to a number of yards or the width to a number of feet.

4 Assess

Give students the following problem:

Kyle bought a bat for $5.00 and a ball that costs $3.00 more than the bat. How much did he spend? ($13.00)

Chapter 13 Test

page 273

Items	Objectives
1–2	To write a fraction for part of a whole or part of a set (see pages 249–250)
3	To find the number of objects in one part of a set (see pages 251–252)
4–5	To find a fraction of a number by multiplying and dividing (see pages 253–254)
6–9	To find equivalent fractions by multiplying (see pages 257–258)
10–13	To find the greater of two fractions (see pages 259–260)
14–17	To reduce a fraction to lowest terms (see pages 261–262)
18–21	To write a whole or mixed number for a fraction greater than or equal to 1 (see pages 263–264)
22–23	To write mixed numbers for lengths measured to the nearest quarter inch (see pages 267–268)
24–25	To use ratios to compare two quantities (see pages 269–270

Alternate Chapter Test

You may wish to use the Alternate Chapter Test on page 370 of this book for further review and assessment.

Name ____________________

CHAPTER 13 TEST

Write the fraction.

1. What part of the square is green? 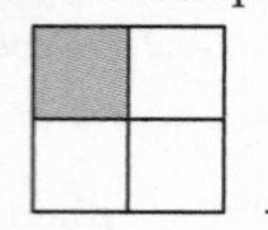 $\frac{1}{4}$

2. What part of the coins are *not* pennies?    $\frac{2}{5}$

Find the part.

3. $\frac{1}{5}$ of 25 = 5

4. $\frac{3}{4}$ of 24 = 18

5. $\frac{7}{8}$ of a price of $48 — $42

Write the missing numerator.

6. $\frac{3}{5} = \frac{6}{10}$ 7. $\frac{3}{8} = \frac{6}{16}$ 8. $\frac{2}{3} = \frac{6}{9}$ 9. $\frac{5}{6} = \frac{20}{24}$

Write < or > in the circle.

10. $\frac{2}{3} > \frac{5}{9}$ 11. $\frac{1}{2} > \frac{3}{8}$ 12. $\frac{7}{12} < \frac{3}{4}$ 13. $\frac{2}{3} > \frac{1}{4}$

Simplify.

14. $\frac{6}{10} = \frac{3}{5}$ 15. $\frac{4}{8} = \frac{1}{2}$ 16. $\frac{9}{12} = \frac{3}{4}$ 17. $\frac{14}{16} = \frac{7}{8}$

Write each fraction as a mixed number.

18. $\frac{7}{2} = 3\frac{1}{2}$ 19. $\frac{8}{3} = 2\frac{2}{3}$ 20. $\frac{16}{5} = 3\frac{1}{5}$ 21. $\frac{15}{12} = 1\frac{1}{4}$

Measure each figure to the nearest quarter inch.

22. 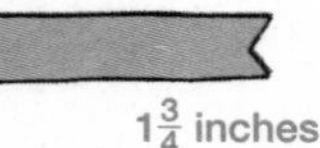 $1\frac{3}{4}$ inches

23. 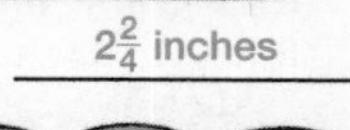 $2\frac{2}{4}$ inches

Write the comparison as a fraction.

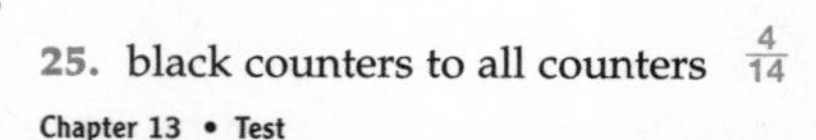

24. green counters to black and gray counters $\frac{5}{9}$

25. black counters to all counters $\frac{4}{14}$

CUMULATIVE ASSESSMENT

Circle the letter of the correct answer.

1. Round 925 to the nearest hundred.
a. 800
(b.) 900
c. NG

2. What is the place value of the 6 in 473,160?
(a.) tens
b. hundreds
c. thousand
d. NG

3. 3,416 + 7,396
a. 10,702
b. 10,712
(c.) 10,812
d. NG

4. 23,089 + 29,716
a. 42,805
b. 52,705
c. 52,795
(d.) NG

5. 1,836 − 929
(a.) 907
b. 913
c. 917
d. NG

6. 28,016 − 19,951
a. 11,945
b. 17,065
(c.) 8,065
d. NG

7. Choose the better estimate of weight.
(a.) 28 grams
b. 28 kilograms

8. $4.26 × 8
a. $31.08
b. $32.68
(c.) $34.08
d. NG

9. 424 × 28
a. 5,040
b. 5,172
c. 11,772
(d.) NG

10. 144 ÷ 9
a. 1 R5
b. 1 R6
c. 1 R7
(d.) NG

11. 890 ÷ 26
a. 3 R11
b. 34
(c.) 34 R6
d. NG

12. Find the perimeter.

6 cm
9 cm

a. 12 cm
b. 18 cm
(c.) 30 cm
d. NG

13. Find the area.
a. 14 units
(b.) 20 sq units
c. 24 sq units
d. NG

STOP

score

Cumulative Assessment

page 274

Items	Objectives
1	To round numbers to the nearest hundred (see pages 31–32)
2	To identify place value through hundred thousands (see pages 25–26)
3–4	To add two numbers up to 5 digits (see pages 41–48)
5–6	To subtract two numbers up to 5 digits (see pages 55–65)
7	To determine the appropriate metric unit of weight (see pages 155–156)
8–9	To multiply 3-digit numbers by up to 2-digit numbers including money (see pages 167–180)
10–11	To divide 3-digit numbers by up to 2-digit numbers to get a quotient with and without remainders (see pages 197–202)
12	To find the perimeter of a rectangle (see pages 239–240)
13	To find the area of a rectangle (see pages 241–242)

Alternate Cumulative Assessment

Circle the letter of the correct answer.

1. Round 693 to the nearest hundred.
(a) 600
b 700
c NG

2. What is the place value of the 9 in 629,873?
a hundreds
(b) thousands
c ten thousands
d NG

3. 2,327 + 6,486
a 8,713
b 8,803
(c) 8,813
d NG

4. 25,217 + 38,887
a 63,104
b 64,004
c 64,094
(d) NG

5. 3,742 − 1,816
(a) 1,926
b 1,936
c 2,126
d NG

6. 46,025 − 28,864
(a) 17,161
b 17,261
c 12,126
d NG

7. What unit would you use to measure a sack of flour?
a gram
(b) kilogram

8. $6.47 × 7 =
a $42.29
b $44.89
(c) $45.29
d NG

9. 643 × 26 =
a 15,618
b 16,618
(c) 16,718
d NG

10. 8)136
(a) 17
b 17 R2
c 28 R8
d NG

11. 649 ÷ 23
a 28
(b) 28 R5
c 28 R8
d NG

12. Find the perimeter of a rectangle that is 5 cm by 8 cm.
a 20 cm
b 40 cm
c 91 cm
(d) NG

13. Find the area of a 4-cm square.
a 8 sq cm
(b) 16 sq cm
c 32 sq cm
d NG

CHAPTER 14

14-1 Adding Fractions With Like Denominators

pages 275–276

1 Getting Started

Objective

- To add fractions having like denominators

Materials

fraction circles

Warm Up • Mental Math

Have students tell the larger fraction.

1. $\frac{1}{2}$ or $\frac{1}{3}$ $(\frac{1}{2})$
2. $\frac{3}{5}$ or $\frac{4}{10}$ $(\frac{3}{5})$
3. $\frac{5}{10}$ or $\frac{3}{5}$ $(\frac{3}{5})$
4. $\frac{4}{15}$ or $\frac{6}{30}$ $(\frac{4}{15})$
5. $\frac{1}{4}$ or $\frac{2}{12}$ $(\frac{1}{4})$
6. $\frac{2}{3}$ or $\frac{4}{5}$ $(\frac{4}{5})$
7. $\frac{1}{2}$ or $\frac{17}{30}$ $(\frac{17}{30})$
8. $\frac{2}{7}$ or $\frac{1}{14}$ $(\frac{2}{7})$

Warm Up • Fractions

Have students identify fractions as you show 2 of 8 parts of a circle, 3 of 5 parts, and so on. Then, dictate fractions and have students show the correct number of parts of the appropriate fraction circle.

Name ____________________

Adding and Subtracting Fractions

CHAPTER 14

Lesson 14-1

Adding Fractions With Like Denominators

Ellen spent the afternoon at the zoo. She walked from the elephant's cage to the monkey's house and on to the lion's den. How far did Ellen walk?

We want to find how many miles Ellen walked.

She walked $\underline{\frac{2}{10}}$ of a mile from the elephant's cage to the monkey's house and $\underline{\frac{3}{10}}$ of a mile more to the lion's den.

To find the total distance Ellen walked, we add $\underline{\frac{2}{10}}$ and $\underline{\frac{3}{10}}$.

We can use a number line.

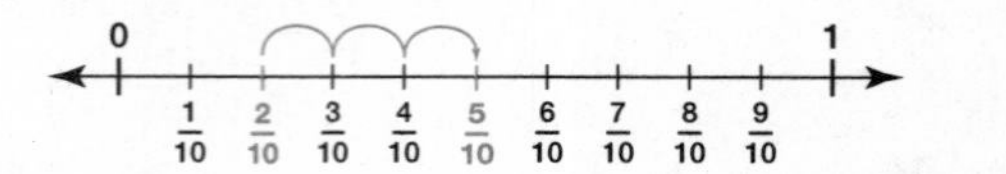

$\frac{5}{10} = \frac{1}{2}$

We can also add the numerators of fractions with like denominators and put the sum over the denominator. We simplify the answer if necessary.

$\frac{2}{10} + \frac{3}{10} = \frac{2+3}{10} = \frac{5}{10} = \frac{1}{2}$

$$\begin{array}{r} \frac{2}{10} \\ + \frac{3}{10} \\ \hline \frac{5}{10} = \frac{1}{2} \end{array}$$

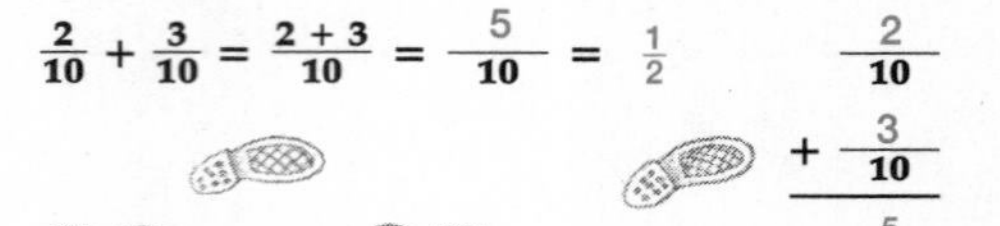

Ellen walked a total of $\underline{\frac{1}{2}}$ of a mile to reach the lion's den.

Getting Started

Add. Simplify if necessary.

1. $\frac{5}{12} + \frac{6}{12} = \frac{11}{12}$
2. $\frac{3}{8} + \frac{3}{8} = \frac{3}{4}$
3. $\begin{array}{r} \frac{3}{5} \\ + \frac{1}{5} \\ \hline \frac{4}{5} \end{array}$
4. $\begin{array}{r} \frac{5}{16} \\ + \frac{7}{16} \\ \hline \frac{3}{4} \end{array}$

2 Teach

Introduce the Lesson Have students describe the picture. Ask a student to read the problem and tell what is to be answered. (how far Ellen walked at the zoo) Ask students what information is known. (It is $\frac{2}{10}$ of a mile from the elephant's cage to the monkey's house and $\frac{3}{10}$ of a mile from the monkey's house to the lion's den.)

- Have students read with you as they complete the sentences and study the number line.
- Have students read the information below the number line aloud. Guide them through the fraction addition steps in the model. Have students use the fraction circles to check their work and complete the solution sentence.

Develop Skills and Concepts Write $\frac{2}{5} + \frac{1}{5}$ on the board. Ask students to name the denominator in each addend. (5 or fifths)

- Draw a number line from 0 to 1 on the board and write $\frac{0}{5}$ through $\frac{5}{5}$ along it. Have a student locate $\frac{2}{5}$ on the number line and then move $\frac{1}{5}$ more to the right and name the stopping place. $(\frac{3}{5})$ Tell students that when fractions have the same denominator, we can add the numerators and write that sum over the denominator.
- Write $\frac{2}{5} + \frac{1}{5} = \frac{2+1}{5} = \frac{3}{5}$ on the board. Write $\frac{2}{5} + \frac{1}{5} = \frac{3}{5}$ vertically on the board.
- Repeat the procedure for $\frac{4}{7} + \frac{2}{7}$, $\frac{3}{9} + \frac{3}{9}$, and so on. Have students simplify their answers if necessary.

3 Practice

Remind students to simplify their sums if necessary. Have students complete page 276 independently.

Practice

Add. Simplify if necessary.

1. $\frac{3}{8} + \frac{2}{8} = \frac{5}{8}$
2. $\frac{5}{7} + \frac{1}{7} = \frac{6}{7}$
3. $\frac{3}{9} + \frac{3}{9} = \frac{2}{3}$
4. $\frac{7}{12} + \frac{4}{12} = \frac{11}{12}$
5. $\frac{3}{10} + \frac{5}{10} = \frac{4}{5}$
6. $\frac{2}{6} + \frac{1}{6} = \frac{1}{2}$
7. $\frac{5}{12} + \frac{3}{12} = \frac{2}{3}$
8. $\frac{3}{16} + \frac{1}{16} = \frac{1}{4}$
9. $\frac{1}{6} + \frac{3}{6} = \frac{2}{3}$
10. $\frac{5}{9} + \frac{2}{9} = \frac{7}{9}$
11. $\frac{2}{5} + \frac{2}{5} = \frac{4}{5}$
12. $\frac{5}{11} + \frac{3}{11} = \frac{8}{11}$
13. $\frac{1}{4} + \frac{2}{4} = \frac{3}{4}$
14. $\frac{5}{8} + \frac{1}{8} = \frac{3}{4}$
15. $\frac{5}{10} + \frac{1}{10} = \frac{3}{5}$
16. $\frac{4}{12} + \frac{5}{12} = \frac{3}{4}$
17. $\frac{1}{6} + \frac{3}{6} = \frac{2}{3}$
18. $\frac{3}{16} + \frac{5}{16} = \frac{1}{2}$
19. $\frac{7}{12} + \frac{3}{12} = \frac{5}{6}$
20. $\frac{4}{8} + \frac{1}{8} = \frac{5}{8}$
21. $\frac{7}{9} + \frac{1}{9} = \frac{8}{9}$
22. $\frac{5}{12} + \frac{3}{12} = \frac{2}{3}$
23. $\frac{1}{3} + \frac{1}{3} = \frac{2}{3}$
24. $\frac{3}{7} + \frac{2}{7} = \frac{5}{7}$
25. $\frac{4}{10} + \frac{4}{10} = \frac{4}{5}$
26. $\frac{5}{8} + \frac{2}{8} = \frac{7}{8}$
27. $\frac{3}{16} + \frac{9}{16} = \frac{3}{4}$
28. $\frac{1}{5} + \frac{3}{5} = \frac{4}{5}$

Problem Solving

Solve each problem.

29. Wally lives $\frac{2}{5}$ of a mile from school. It is another $\frac{1}{5}$ of a mile from school to the store. How far is it from Wally's house to the store?
$\frac{3}{5}$ of a mile

30. Rochelle painted $\frac{3}{8}$ of a mural. Martha painted $\frac{1}{8}$ of it. How much of the mural did the girls paint?
$\frac{1}{2}$ of the mural

4 Assess

Ask students how to add fractions with the same denominator. (Add the numerators, write that sum over the denominator, and simplify the answer if necessary.)

For Mixed Abilities

Common Errors • Intervention

When adding fractions, students may add both the numerators and the denominators to find the sum.

Incorrect	Correct
$\frac{2}{5} + \frac{1}{5} = \frac{3}{10}$	$\frac{2}{5} + \frac{1}{5} = \frac{3}{5}$

Have students work in pairs to model addition problems. They should use a circle to model each fraction and then count to find the number of fractional parts in the sum.

Enrichment • Fractions

Tell students to write five different problems of two fraction addends so that each problem has a sum of $\frac{11}{12}$.

More to Explore • Probability

Have students compare a probability that they calculate with a probability they discover through experiment.

Remove all the face cards from a deck of cards and ask students the probability of drawing a 5 from the deck. (4 out of 40, or $\frac{1}{10}$) Give each pair of students one of these decks. Ask them to check the theoretical probability by conducting an experiment. Have them shuffle the deck, draw a card, record the number, replace it, shuffle and draw again until they have drawn 100 cards. Tell students to express the results of their experiment as a fraction: the number of 5s drawn over 100. Have them compare this with the theoretical probability, $\frac{1}{10}$ or $\frac{10}{100}$.

ESL/ELL STRATEGIES

When introducing common denominators, ask students what the word *common* means. They may suggest that it means "ordinary" or "usual." Explain that in mathematical terms, a common denominator is one that is exactly the same in two or more different fractions.

14-2 Subtracting Fractions With Like Denominators

pages 277–278

1 Getting Started

Objective

- To subtract fractions having like denominators

Materials

fraction circles

Warm Up • Mental Math

Ask students to tell how many.

1. $\frac{2}{5}$ of 15 (6)
2. $\frac{1}{5} + \frac{3}{5}$ ($\frac{4}{5}$)
3. $60 \times \frac{1}{2}$ of 2 (60)
4. $16 \times (6 \div 3)$ (32)
5. $\frac{5}{8} + \frac{5}{8}$ ($1\frac{1}{4}$)
6. months in $\frac{1}{3}$ of 1 year (4)
7. lowest terms for $\frac{12}{16}$ ($\frac{3}{4}$)
8. lowest terms for $\frac{35}{90}$ ($\frac{7}{18}$)

Warm Up • Fractions

Review addition of fractions with like denominators by having students work the following problems on the board: $\frac{4}{15} + \frac{10}{15}$ ($\frac{14}{15}$), $\frac{3}{12} + \frac{6}{12}$ ($\frac{3}{4}$), $\frac{9}{13} + \frac{1}{13}$ ($\frac{10}{13}$), $\frac{4}{9} + \frac{2}{9}$ ($\frac{2}{3}$), and $\frac{7}{16} + \frac{5}{16}$ ($\frac{3}{4}$).

Name ______________________

Lesson 14-2

Subtracting Fractions With Like Denominators

Barry, Cal, and Craig challenged each other to an 8-minute run around the school track. How much farther did Barry run than Cal?

We want to know how much farther Barry ran than Cal.

Barry ran $\frac{7}{8}$ of a mile. Cal ran $\frac{5}{8}$ of a mile in the same period of time.

To find the difference in the distance Barry and Cal ran, we subtract $\frac{5}{8}$ from $\frac{7}{8}$.

We can use a number line.

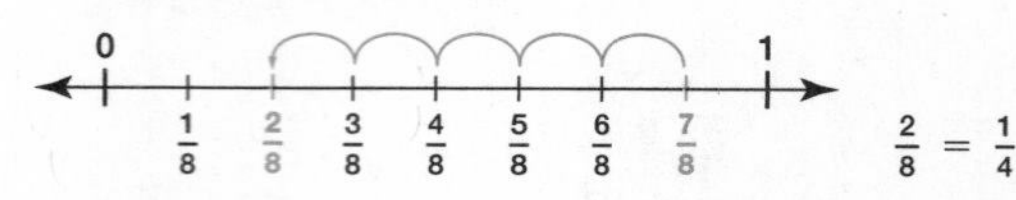

$\frac{2}{8} = \frac{1}{4}$

We can also subtract the numerators of fractions with like denominators and write the difference over the denominator. We simplify the answer if necessary.

$$\frac{7}{8} - \frac{5}{8} = \frac{7-5}{8} = \frac{2}{8} = \frac{1}{4}$$

$$\begin{array}{r} \frac{7}{8} \\ -\ \frac{5}{8} \\ \hline \frac{2}{8} = \frac{1}{4} \end{array}$$

Barry ran $\frac{1}{4}$ of a mile farther than Cal.

Getting Started

Subtract. Simplify if necessary.

1. $\frac{4}{5} - \frac{1}{5} = \frac{3}{5}$
2. $\frac{9}{10} - \frac{3}{10} = \frac{3}{5}$
3. $\frac{7}{8} - \frac{7}{8} = 0$
4. $\frac{8}{12} - \frac{5}{12} = \frac{1}{4}$

2 Teach

Introduce the Lesson Ask a student to read the problem and tell what is to be solved. (how much farther Barry ran than Cal) Ask students what information is given. (In 8 minutes, Barry ran $\frac{7}{8}$ of a mile, Cal ran $\frac{5}{8}$ of a mile, and Craig ran $\frac{6}{8}$ of a mile.) Ask what, if any, of this information is unnecessary to solve the problem. (8 minutes and Craig ran $\frac{6}{8}$ of a mile)

- Have students read with you as they complete the sentences. Have them study the number line and the rule, and then do the subtraction in the model. Have students use fraction circles to check their work.

Develop Skills and Concepts Draw a number line from 0 to 1 on the board and write $\frac{0}{9}$ through $\frac{9}{9}$ along it. Write $\frac{8}{9} - \frac{3}{9}$ on the board and ask students to name the denominator in each fraction. (9 or ninths)

- Have a student locate $\frac{8}{9}$ on the number line, go backward $\frac{3}{9}$, and name the stopping place. ($\frac{5}{9}$) Tell students that when fractions have the same denominator, we can subtract the numerator and write that difference over the denominator. Write $\frac{8}{9} - \frac{3}{9} = \frac{8-3}{9} = \frac{5}{9}$ on the board. Write $\frac{8-3}{9}$ vertically on the board.
- Repeat the procedure for $\frac{7}{12} - \frac{1}{12}$, $\frac{9}{10} - \frac{4}{10}$, and so on. Have students simplify their answers if necessary.

3 Practice

Remind students to simplify their answers if necessary. Have students complete page 278 independently.

Practice

Subtract. Simplify if necessary.

1. $\frac{4}{5} - \frac{1}{5} = \frac{3}{5}$
2. $\frac{5}{6} - \frac{2}{6} = \frac{1}{2}$
3. $\frac{3}{11} - \frac{2}{11} = \frac{1}{11}$
4. $\frac{12}{16} - \frac{8}{16} = \frac{1}{4}$
5. $\frac{5}{8} - \frac{2}{8} = \frac{3}{8}$
6. $\frac{7}{12} - \frac{4}{12} = \frac{1}{4}$
7. $\frac{6}{7} - \frac{3}{7} = \frac{3}{7}$
8. $\frac{5}{6} - \frac{1}{6} = \frac{2}{3}$
9. $\frac{7}{9} - \frac{4}{9} = \frac{1}{3}$
10. $\frac{4}{5} - \frac{3}{5} = \frac{1}{5}$
11. $\frac{6}{8} - \frac{1}{8} = \frac{5}{8}$
12. $\frac{11}{12} - \frac{5}{12} = \frac{1}{2}$
13. $\frac{13}{16} - \frac{5}{16} = \frac{1}{2}$
14. $\frac{9}{10} - \frac{1}{10} = \frac{4}{5}$
15. $\frac{5}{7} - \frac{2}{7} = \frac{3}{7}$
16. $\frac{7}{9} - \frac{3}{9} = \frac{4}{9}$
17. $\frac{5}{8} - \frac{1}{8} = \frac{1}{2}$
18. $\frac{7}{11} - \frac{3}{11} = \frac{4}{11}$
19. $\frac{11}{16} - \frac{9}{16} = \frac{1}{8}$
20. $\frac{7}{8} - \frac{3}{8} = \frac{1}{2}$
21. $\frac{7}{10} - \frac{2}{10} = \frac{1}{2}$
22. $\frac{12}{15} - \frac{9}{15} = \frac{1}{5}$
23. $\frac{11}{12} - \frac{2}{12} = \frac{3}{4}$
24. $\frac{2}{3} - \frac{1}{3} = \frac{1}{3}$
25. $\frac{7}{16} - \frac{3}{16} = \frac{1}{4}$
26. $\frac{8}{15} - \frac{3}{15} = \frac{1}{3}$
27. $\frac{5}{10} - \frac{2}{10} = \frac{3}{10}$
28. $\frac{9}{16} - \frac{7}{16} = \frac{1}{8}$

Problem Solving

Solve each problem.

29. Cindy has $\frac{5}{8}$ of a liter of milk. She drinks $\frac{3}{8}$ of a liter with her lunch. How much milk does Cindy have left?
$\frac{1}{4}$ of a liter

30. Clay did $\frac{9}{10}$ of his homework correctly. David did $\frac{7}{10}$ of his homework correctly. How much more homework did Clay do correctly than David did?
$\frac{1}{5}$ more

4 Assess

Ask students how to subtract fractions with the same denominators. (subtract the numerators, write the difference over the denominator, and simplify the answer if necessary)

5 Mixed Review

1. $\frac{3}{8} = \frac{?}{24}$ (9)
2. 368 × 29 (10,672)
3. Simplify. $\frac{16}{20}$ ($\frac{4}{5}$)
4. $6.93 × 4 ($27.72)
5. 508 − 39 (469)
6. 25,176 + 18,927 (44,103)
7. 723 × 47 (33,981)
8. Simplify. $\frac{24}{9}$ ($2\frac{2}{3}$)
9. 31,205 − 16,571 (14,634)
10. 638 ÷ 27 (23 R17)

For Mixed Abilities

Common Errors • Intervention

When subtracting fractions, some students may use the procedure for adding instead. Have students work in pairs to model problems. For a problem such as $\frac{5}{8} - \frac{2}{8}$, they should divide a rectangle into 8 parts and color 5 of them. Then, they can cross out 2 of the parts to show subtraction and count to find that $\frac{3}{8}$ is left.

Enrichment • Fractions

Have students write five different fraction subtraction problems so that each problem has a difference of $\frac{2}{13}$.

More to Explore • Numeration

Explain that computers use a special system of numbers and letters because they understand only two signals: electric current on and electric current off. Ask students to look up the binary number system in the library. Give each student a place-value chart and ask students to fill in the place values for binary numbers place value through 512.

512	256	128	64	32	16	8	4	2	1

Explain that in each place, combinations of only two digits, 0 and 1, are possible. Ask them to write the binary numbers for these numbers:

7 (111) 3 (11) 20 (10100)

36 (100100) 14 (1110) 73 (1001001)

14-3 Adding and Subtracting Fractions

pages 279–280

1 Getting Started

Objective

- To add or subtract fractions and simplify the answer

Warm Up • Mental Math

Have students complete the comparison: 6 is to 2 as 12 is to 4 as

1. 18 is to (6)
2. 120 is to (40)
3. 72 is to (24)
4. 3 is to (1)
5. 24 is to (8)
6. 30 is to (10)
7. 99 is to (33)
8. 63 is to (21)

Warm Up • Numeration

Have students write a whole or mixed number for each of the following dictated fractions: $\frac{16}{4}$ (4), $\frac{10}{3}$ $(3\frac{1}{3})$, $\frac{16}{5}$ $(3\frac{1}{5})$, $\frac{22}{10}$ $(2\frac{1}{5})$, and $\frac{17}{13}$ $(1\frac{4}{13})$.

Name ____________________

Lesson 14-3

Adding and Subtracting Fractions, Simplifying

Leah is going skiing on Tuesday if fresh snow falls in the mountains. She checks with the Weather Bureau on Monday before she packs her gear. How much new snow fell over the weekend?

We want to know how many inches of new snow fell on Saturday and Sunday.

It snowed $\frac{9}{10}$ of an inch on Saturday, and $\frac{5}{10}$ of an inch on Sunday.

To find the total amount of snowfall for both days, we add $\frac{9}{10}$ and $\frac{5}{10}$.

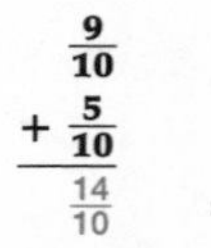

$$\begin{array}{r} \frac{9}{10} \\ + \frac{5}{10} \\ \hline \frac{14}{10} \end{array} = 1\frac{4}{10} = 1\frac{2}{5}$$

Simplifying fractions includes changing fractions to whole or mixed numbers if necessary.

It snowed $1\frac{2}{5}$ inches over the weekend.

Getting Started

Add or subtract. Simplify.

1. $\frac{3}{5} + \frac{4}{5} = 1\frac{2}{5}$
2. $\frac{7}{8} - \frac{1}{8} = \frac{3}{4}$
3. $\frac{3}{4} + \frac{3}{4} = 1\frac{1}{2}$
4. $\frac{19}{12} - \frac{13}{12} = \frac{1}{2}$
5. $\frac{5}{6} + \frac{4}{6} = 1\frac{1}{2}$
6. $\frac{9}{16} - \frac{1}{16} = \frac{1}{2}$
7. $\frac{1}{2} - \frac{1}{2} = 0$
8. $\frac{17}{10} - \frac{3}{10} = 1\frac{2}{5}$
9. $\frac{3}{10} + \frac{7}{10} = 1$

2 Teach

Introduce the Lesson Have a student read the problem aloud and tell what is to be found. (the amount of new snow that fell over the weekend) Have students tell what information is given in the problem or picture. ($\frac{9}{10}$ of an inch of snow fell on Saturday and $\frac{5}{10}$ of an inch fell on Sunday.)

- Have students complete the sentences. Guide them through the simplification steps in the model to solve the problem. Have students draw a number line in tenths to check their answer.

Develop Skills and Concepts Tell students that fraction answers must always be given in the simplest form possible. Explain that this may mean simplifying a fraction to its lowest terms or changing a fraction to a whole or mixed number.

- Write $\frac{14}{5}$ on the board and ask students how this fraction would be simplified. (change to mixed number) Have a student change the fraction to $2\frac{4}{5}$. Ask if $2\frac{4}{5}$ can be further simplified. (no)
- Have students talk through the work to solve $\frac{4}{5} + \frac{3}{5}$. ($\frac{7}{5} = 1\frac{2}{5}$) Repeat for $\frac{19}{10} - \frac{8}{10}$, $\frac{14}{2} + \frac{1}{2}$, $\frac{13}{8} - \frac{5}{8}$, and so on.

3 Practice

Remind students to write their answers in the simplest form. Have students complete page 280 independently.

Practice

Add or subtract. Simplify.

1. $\frac{5}{8} + \frac{6}{8} = 1\frac{3}{8}$
2. $\frac{11}{9} - \frac{1}{9} = 1\frac{1}{9}$
3. $\frac{11}{12} - \frac{3}{12} = \frac{2}{3}$
4. $\frac{7}{5} + \frac{3}{5} = 2$
5. $\frac{19}{8} - \frac{9}{8} = 1\frac{1}{4}$
6. $\frac{3}{4} + \frac{6}{4} = 2\frac{1}{4}$
7. $\frac{5}{6} + \frac{5}{6} = 1\frac{2}{3}$
8. $\frac{11}{15} - \frac{6}{15} = \frac{1}{3}$
9. $\frac{9}{10} + \frac{6}{10} = 1\frac{1}{2}$
10. $\frac{4}{7} + \frac{6}{7} = 1\frac{3}{7}$
11. $\frac{18}{9} - \frac{12}{9} = \frac{2}{3}$
12. $\frac{2}{3} + \frac{2}{3} = 1\frac{1}{3}$
13. $\frac{18}{10} - \frac{2}{10} = 1\frac{3}{5}$
14. $\frac{14}{15} + \frac{4}{15} = 1\frac{1}{5}$
15. $\frac{10}{12} + \frac{8}{12} = 1\frac{1}{2}$
16. $\frac{7}{10} + \frac{9}{10} = 1\frac{3}{5}$
17. $\frac{9}{16} - \frac{5}{16} = \frac{1}{4}$
18. $\frac{9}{5} + \frac{3}{5} = 2\frac{2}{5}$
19. $\frac{7}{12} + \frac{8}{12} = 1\frac{1}{4}$
20. $\frac{17}{12} - \frac{3}{12} = 1\frac{1}{6}$
21. $\frac{7}{8} - \frac{1}{8} = \frac{3}{4}$
22. $\frac{9}{12} + \frac{7}{12} = 1\frac{1}{3}$
23. $\frac{10}{6} + \frac{4}{6} = 2\frac{1}{3}$
24. $\frac{19}{10} + \frac{11}{10} = 3$
25. $\frac{17}{12} - \frac{13}{12} = \frac{1}{3}$
26. $\frac{13}{4} + \frac{13}{4} = 6\frac{1}{2}$
27. $\frac{15}{16} + \frac{11}{16} = 1\frac{5}{8}$
28. $\frac{13}{6} - \frac{3}{6} = 1\frac{2}{3}$

Problem Solving

Solve each problem.

29. Bonnie drank $\frac{5}{8}$ of a quart of water before the basketball game and $\frac{7}{8}$ of a quart after. How much water did Bonnie drink?
$1\frac{1}{2}$ quarts

30. A jacket button is $\frac{9}{8}$ inches wide. The buttonhole is $\frac{13}{8}$ inches wide. How much wider is the buttonhole?
$\frac{1}{2}$ of an inch

4 Assess

Write $\frac{20}{15} - \frac{5}{15}$ on the board. Ask students to find and simplify the difference if necessary. ($\frac{15}{15} = 1$)

For Mixed Abilities

Common Errors • Intervention

When adding and subtracting fractions, students may perform the opposite operation because they did not read the problem carefully. As they work each problem in a set of mixed exercises, have them first circle the symbol for the operation to make them more aware of what they are to do.

Enrichment • Numeration

Tell students to make a table to show mixed or whole numbers that result when 460 is divided into halves, thirds, fourths, fifths, sixths, sevenths, eighths, ninths, and tenths.

More to Explore • Geometry

Give each student a full set of tangrams consisting of 2 small, 1 medium, 2 large triangles, 1 square, and 1 parallelogram. Tell students to make the following figures using all 7 pieces:

1. 1 large square
2. 2 smaller, but equal, squares
3. a triangle
4. a rectangle
5. a pentagon

Ask them to trace the outline of each figure on paper when it is complete. Encourage students to move on to another part of the assignment if they have trouble with any one figure.

14-4 Adding Fractions With Unlike Denominators

pages 281–282

1 Getting Started

Objective

- To add two fractions when one denominator is a multiple of the other denominator

Warm Up • Mental Math

Tell students to name a mixed number for the following:

1. 2 h 15 min ($2\frac{1}{4}$)
2. 3 h 12 min ($3\frac{1}{5}$)
3. 11 h 5 min ($11\frac{1}{12}$)
4. 4 h 2 min ($4\frac{1}{30}$)
5. 10 h 20 min ($10\frac{1}{3}$)
6. 70 min ($1\frac{1}{6}$)
7. 375 min ($6\frac{1}{4}$)
8. 200 min ($3\frac{1}{3}$)

Warm Up • Fractions

Have students complete the following problems on the board to review equivalent fractions:

$\frac{1}{2} = \frac{(2)}{4} = \frac{(3)}{6} = \frac{(4)}{8} = \frac{5}{(10)}$

$\frac{1}{3} = \frac{2}{(6)} = \frac{(3)}{9} = \frac{4}{(12)}$

$\frac{1}{4} = \frac{(2)}{8} = \frac{3}{(12)} = \frac{(4)}{16} = \frac{5}{(20)}$

Name ______________________

Lesson 14-4

Adding Fractions With Unlike Denominators

Following the veterinarian's advice, Tony is keeping track of his new kitten's weight gain. How many pounds did his kitten gain during the first 2 weeks?

We want to find the kitten's weight gain for the first 2 weeks.

It gained $\frac{1}{2}$ of a pound the first week and $\frac{3}{4}$ of a pound the second week.

To find how many pounds the kitten gained in 2 weeks, we add $\frac{1}{2}$ and $\frac{3}{4}$.

Find out if the fractions have like denominators.	Find equivalent fractions with like denominators.	Add the fractions. Simplify.

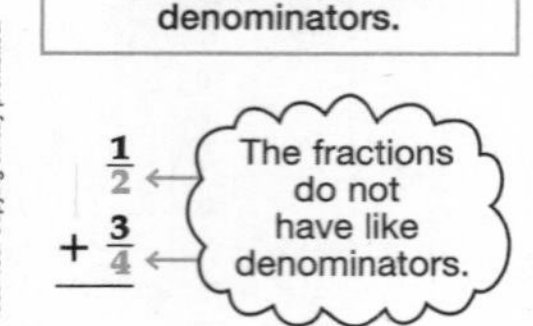

$\frac{1}{2} + \frac{3}{4}$

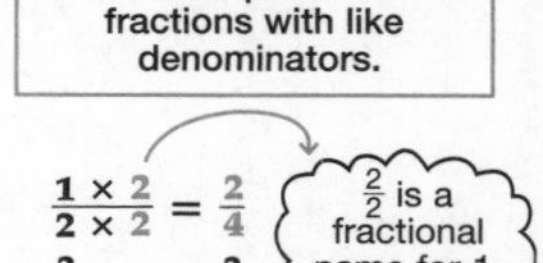

$\frac{1 \times 2}{2 \times 2} = \frac{2}{4}$

$+\frac{3}{4} = \frac{3}{4}$

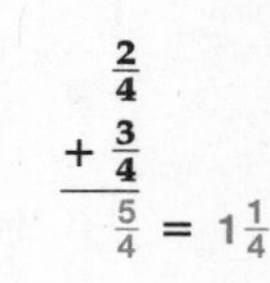

$\frac{2}{4} + \frac{3}{4} = \frac{5}{4} = 1\frac{1}{4}$

Tony's kitten gained $1\frac{1}{4}$ pounds after 2 weeks.

Getting Started

Add. Simplify if necessary.

1. $\frac{3}{5} + \frac{3}{10}$ = $\frac{9}{10}$
2. $\frac{2}{3} + \frac{5}{6}$ = $1\frac{1}{2}$
3. $\frac{3}{8} + \frac{1}{4}$ = $\frac{5}{8}$
4. $\frac{7}{12} + \frac{2}{3}$ = $1\frac{1}{4}$

Copy and add.

5. $\frac{1}{2} + \frac{3}{4}$ $1\frac{1}{4}$
6. $\frac{2}{5} + \frac{3}{20}$ $\frac{11}{20}$
7. $\frac{1}{6} + \frac{3}{12}$ $\frac{5}{12}$

2 Teach

Introduce the Lesson Have students describe the picture and tell what is to be found. (the weight gain of the kitten after 2 weeks) Ask what information will be used from the chart. ($\frac{1}{2}$ lb gain in Week 1 and $\frac{3}{4}$ lb gain in Week 2)

- Have students complete the sentences. Guide them through the problems in the model to find the kitten's weight gain in the first 2 weeks.
- Have students use a number line marked in halves and fourths to check their answer.

Develop Skills and Concepts Tell students that when we add two fractions that have different denominators, we see if the larger denominator is a multiple of the other.

- Write $\frac{2}{3} + \frac{3}{6}$ on the board and ask students if 6 is a multiple of 3. (yes) Tell students that we then find an equivalent fraction for $\frac{2}{3}$ that has 6 as its denominator. ($\frac{4}{6}$)
- Write $\frac{4}{6} + \frac{3}{6}$ on the board and have students add and simplify the answer. ($1\frac{1}{6}$)
- Repeat the procedure to solve $\frac{6}{12} + \frac{4}{6}$ ($1\frac{1}{6}$), $\frac{5}{8} + \frac{13}{16}$ ($1\frac{7}{16}$), $\frac{7}{9} + \frac{2}{3}$ ($1\frac{4}{9}$), and $\frac{5}{20} + \frac{9}{10}$ ($1\frac{3}{20}$).

3 Practice

Remind students to check their work to be sure each answer is in simplest form. Have students complete page 282 independently.

Practice

Add. Simplify if necessary.

1. $\frac{1}{3} + \frac{1}{6}$ $\frac{1}{2}$
2. $\frac{4}{2} + \frac{3}{8}$ $2\frac{3}{8}$
3. $\frac{7}{12} + \frac{3}{4}$ $1\frac{1}{3}$
4. $\frac{2}{3} + \frac{5}{9}$ $1\frac{2}{9}$
5. $\frac{7}{8} + \frac{3}{4}$ $1\frac{5}{8}$
6. $\frac{5}{12} + \frac{5}{6}$ $1\frac{1}{4}$
7. $\frac{3}{5} + \frac{7}{10}$ $1\frac{3}{10}$
8. $\frac{3}{14} + \frac{2}{7}$ $\frac{1}{2}$
9. $\frac{7}{15} + \frac{3}{5}$ $1\frac{1}{15}$
10. $\frac{9}{10} + \frac{1}{2}$ $1\frac{2}{5}$
11. $\frac{12}{5} + \frac{7}{20}$ $2\frac{3}{4}$
12. $\frac{9}{16} + \frac{8}{4}$ $2\frac{9}{16}$

Copy and add.

13. $\frac{3}{4} + \frac{7}{2}$ $4\frac{1}{4}$
14. $\frac{2}{3} + \frac{7}{9}$ $1\frac{4}{9}$
15. $\frac{5}{8} + \frac{1}{4}$ $\frac{7}{8}$
16. $\frac{7}{10} + \frac{3}{5}$ $1\frac{3}{10}$
17. $\frac{11}{12} + \frac{3}{4}$ $1\frac{2}{3}$
18. $\frac{4}{3} + \frac{5}{12}$ $1\frac{3}{4}$
19. $\frac{2}{3} + \frac{9}{6}$ $2\frac{1}{6}$
20. $\frac{7}{15} + \frac{4}{5}$ $1\frac{4}{15}$
21. $\frac{7}{9} + \frac{7}{18}$ $1\frac{1}{6}$
22. $\frac{9}{16} + \frac{7}{8}$ $1\frac{7}{16}$
23. $\frac{9}{20} + \frac{9}{10}$ $1\frac{7}{20}$
24. $\frac{4}{9} + \frac{2}{3}$ $1\frac{1}{9}$
25. $\frac{9}{2} + \frac{3}{16}$ $4\frac{11}{16}$
26. $\frac{8}{10} + \frac{4}{5}$ $1\frac{3}{5}$
27. $\frac{13}{12} + \frac{1}{6}$ $1\frac{1}{4}$
28. $\frac{5}{7} + \frac{11}{21}$ $1\frac{5}{21}$

Problem Solving

Solve each problem.

29. Jackie's dog lost $\frac{2}{3}$ of a pound in 1 week. The next week, her dog lost $\frac{5}{6}$ of a pound. How much weight did Jackie's dog lose altogether? $1\frac{1}{2}$ pounds

30. Lanny used $\frac{1}{2}$ of a roll of red paper and $\frac{5}{8}$ of a roll of green paper to wrap gifts. How much paper did Lanny use? $1\frac{1}{8}$ rolls

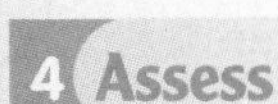

4 Assess

Ask students to solve $\frac{3}{4} + \frac{6}{8}$. ($1\frac{1}{2}$)

5 Mixed Review

1. 27×205 (5,535)
2. $\frac{21}{24} = \frac{?}{8}$ (7)
3. $\frac{7}{8} - \frac{5}{8}$ ($\frac{1}{4}$)
4. $986 \div 42$ (23 R20)
5. Simplify. $\frac{18}{27}$ ($\frac{2}{3}$)
6. $365 + 201 + 879$ (1,445)
7. $\frac{1}{4} + \frac{3}{4}$ (1)
8. \$420.00 − \$253.21 (\$166.79)
9. Simplify. $\frac{13}{6}$ ($2\frac{1}{6}$)
10. 8 × \$14.72 (\$117.76)

For Mixed Abilities

Common Errors • Intervention

When adding fractions, some students may find a like denominator but forget to change the numerators. Have them make diagrams to show the equivalent fractions and then perform the operations correctly.

Enrichment • Fractions

Tell students to write and solve an addition problem of fractions whose like denominator is 45. Have them repeat for a like denominator of 72.

More to Explore • Measurement

Have students put their books into bookbags and weigh them using a bathroom scale. After all the bookbags have been weighed, ask students to find the total weight of the bookbags.

Now, ask students to name animals that weigh about the same as a single book and about the same as the total weight of the books. Have them write the name of these animals on the board.

Have students do research to find how many of their estimated weights corresponded to the actual weight of the animals they listed.

14-5 Estimating Fraction Sums

pages 283–284

1 Getting Started

Objective
- To estimate sums of fractions

Material
fraction strips

Warm Up • Mental Math
Have students work various single-digit column addition problems such as $2 + 6 + 3 + 0 + 4 =$ _____.

Warm Up • Calculators
Review the operations of addition, subtraction, multiplication, and division on the calculator by having students work the following codes:

1. $26 \times 3 \div 2 =$ (39)
2. $(47 + 3) \div 2 =$ (25)
3. $784 \times 16 =$ (12,544)
4. $45 \div 3 \times 20 =$ (300)
5. $17.64 × 50 = ($882)

Name ______________________

Estimating Fraction Sums

Janice has $\frac{2}{7}$ of a yard of fabric and Sam has $\frac{4}{10}$ of a yard of fabric. Do they together have more or less than a yard of fabric?

We can use a number line to help estimate the sum of fractions with unlike denominators.

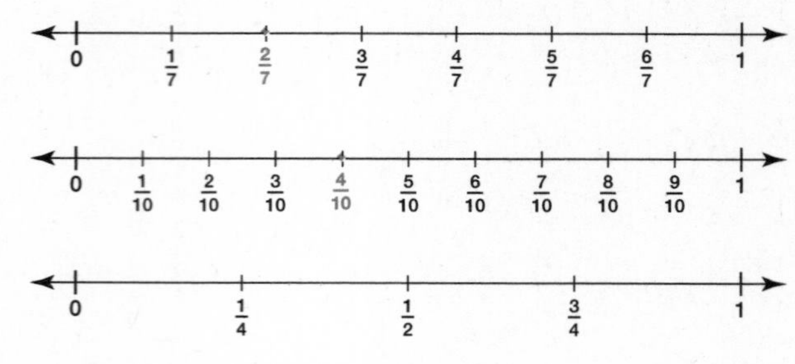

$\frac{2}{7}$ is less than $\frac{1}{2}$ and $\frac{1}{2}$ is less than 1.

$\frac{4}{10}$ is less than $\frac{1}{2}$ and $\frac{1}{2}$ is less than 1.

Both fractions are less than $\frac{1}{2}$ yard. Together they do not equal 1 yard.

Janice and Sam have less than a yard of fabric.

Getting Started

Compare. Write >, <, or = in the circle.

1. $\frac{9}{10}$ (>) $\frac{1}{2}$
2. $\frac{5}{12}$ (<) $\frac{1}{2}$
3. $\frac{3}{9}$ (<) $\frac{1}{2}$
4. $\frac{4}{8}$ (=) $\frac{1}{2}$

Use the number lines above to estimate whether each sum is > or < 1. Show your work.

5. $\frac{5}{7} + \frac{3}{7}$ (>) 1
6. $\frac{1}{2} + \frac{2}{10}$ (<) 1
7. $\frac{5}{10} + \frac{1}{4}$ (<) 1

2 Teach

Introduce the Lesson Read the problem aloud and ask a student to tell what the problem is asking. (whether Janice and Sam together have more or less than a yard of fabric) Explain that in order to round $\frac{2}{7}$ and $\frac{4}{10}$, students need to compare each fraction to $\frac{1}{2}$.

- If the fraction is $\frac{1}{2}$ or greater, round it to the next whole number. If the fraction is less than $\frac{1}{2}$, round it down to the previous whole number.
- Draw the three number lines in the text on the board. Have a student circle $\frac{2}{7}$ on the number line marked in sevenths. Ask students if $\frac{2}{7}$ is greater than or less than $\frac{1}{2}$. (less) Have another student circle $\frac{4}{10}$ on the number line marked in tenths. Ask students if $\frac{4}{10}$ is greater than or less than $\frac{1}{2}$. (less) Help students see that putting $\frac{2}{7}$ and $\frac{4}{10}$ together will still not equal 1 whole.
- Have a student read the section under the number lines and the solution sentence.

Develop Skills and Concepts Give students three fraction strips. One representing 1 whole, another representing eighths, and a third representing fifths. Have students align the strips one under the other on their desks in the following order: 1 whole strip on top, the eighths strip in the middle, and the fifths strip on the bottom.

- Write $\frac{2}{5} + \frac{3}{8}$ ○ **1** in vertical form on the board.
- Next, have a student mark off two-fifths on the strip that is divided into fifths and three-eighths on the strip that is divided into eighths. Have students decide if $\frac{2}{5}$ is greater than or less than $\frac{1}{2}$. (<) Then, have students decide if $\frac{3}{8}$ is greater than or less than $\frac{1}{2}$. (<) Ask a volunteer to come to the board and fill in the circle. (<)

Practice

Compare. Write >, <, or = in the circle.

1. $\frac{9}{12}$ (>) $\frac{1}{2}$
2. $\frac{2}{6}$ (<) $\frac{1}{2}$
3. $\frac{1}{9}$ (<) $\frac{1}{2}$
4. $\frac{2}{4}$ (=) $\frac{1}{2}$
5. $\frac{3}{4}$ (>) $\frac{1}{2}$
6. $\frac{1}{3}$ (<) $\frac{1}{2}$
7. $\frac{2}{5}$ (<) $\frac{1}{2}$
8. $\frac{6}{9}$ (>) $\frac{1}{2}$
9. $\frac{6}{8}$ (>) $\frac{1}{2}$
10. $\frac{4}{5}$ (>) $\frac{1}{2}$
11. $\frac{5}{10}$ (=) $\frac{1}{2}$
12. $\frac{1}{4}$ (<) $\frac{1}{2}$

Use the number lines to estimate whether each sum is > or < 1. Show your work.

13. $\frac{5}{5} + \frac{3}{5}$ (>) 1
14. $\frac{1}{4} + \frac{2}{5}$ (<) 1
15. $\frac{3}{4} + \frac{1}{8}$ (<) 1
16. $\frac{4}{8} + \frac{1}{2}$ (=) 1
17. $\frac{4}{5} + \frac{3}{4}$ (>) 1
18. $\frac{2}{8} + \frac{2}{5}$ (<) 1

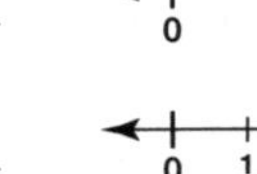

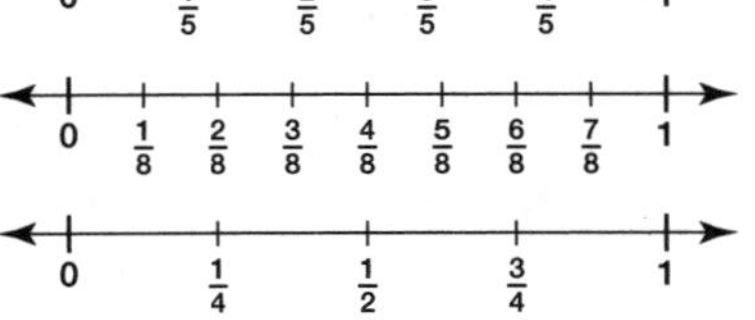

0, $\frac{1}{5}$, $\frac{2}{5}$, $\frac{3}{5}$, $\frac{4}{5}$, 1

0, $\frac{1}{8}$, $\frac{2}{8}$, $\frac{3}{8}$, $\frac{4}{8}$, $\frac{5}{8}$, $\frac{6}{8}$, $\frac{7}{8}$, 1

0, $\frac{1}{4}$, $\frac{1}{2}$, $\frac{3}{4}$, 1

Now Try This!

Steve figured out a shortcut to help Mr. Smalley find the sale prices quickly. He figured that if a customer was going to save $\frac{3}{5}$ of the original price, he or she would pay $\frac{2}{5}$ of it. Steve used this calculator code to find the sales price of a stereo.

750 [×] 2 [÷] 5 [=] [300]

Complete the table to show that Steve's shortcut works every time.

Original Price	Fraction Off	Sale Price
$870	$\frac{2}{3}$	$290
$850	$\frac{1}{4}$	$637.50
$648	$\frac{5}{9}$	$288
$1,026	$\frac{5}{6}$	$171
$460	$\frac{3}{10}$	$322

3 Practice

Have students complete page 284 independently.

Now Try This! Encourage students to think of the original price as the "one whole thing or 100%." Remind them that 1 can be expressed as a fraction, where the numerator and denominator are the same. Thus, the original price of $870 in the first exercise is 100%, or 1 or $\frac{3}{3}$. If we do not pay $\frac{2}{3}$ of the $870, the $\frac{1}{3}$ that remains is the cost.

4 Assess

Ask students to write instructions explaining how to estimate fraction sums. (Instructions will vary but should reflect lesson concepts.)

For Mixed Abilities

Common Errors • Intervention

Watch for students who have difficulty rounding fractions that are close to $\frac{1}{2}$, such as $\frac{3}{8}$ or $\frac{5}{8}$. Have students work in groups. Have each group write the following fractions on a slip of paper: $\frac{3}{5}$, $\frac{3}{7}$, $\frac{4}{7}$, $\frac{3}{8}$, $\frac{5}{8}$, $\frac{4}{9}$, $\frac{5}{9}$, $\frac{5}{12}$, and $\frac{7}{12}$. Provide each group with fraction strips. Have a student pick a fraction and use the fraction strip to decide if the fraction is less than or greater than $\frac{1}{2}$. Have one student in the group record the results of the investigation.

Enrichment • Numeration

Have students write their birth heights and present heights in inches. Then, tell them to find the number of inches they have grown since birth in feet and inches.

More to Explore • Statistics

Ask students to compose a Pascal's triangle. Have them complete nine rows. Have each student graph the numbers appearing in the last line on grid paper.

1
1 1
1 2 1
1 3 3 1
1 4 6 4 1
1 5 10 10 5 1
1 7 21 35 35 21 7 1
1 8 28 56 70 56 28 8 1
1 2 3 4 5 6 7 8 9

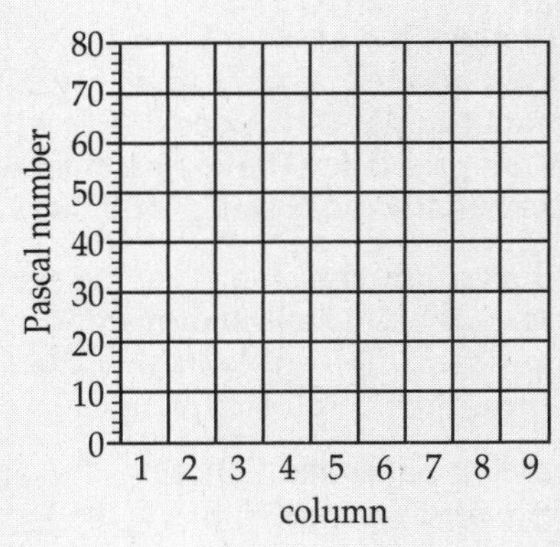

14-6 Subtract Fractions With Unlike Denominators

pages 285–286

1 Getting Started

Objective
- To subtract two fractions when one denominator is a multiple of the other denominator

Materials
*fraction circles, fraction bars

Warm Up • Mental Math
Ask what fraction names the following part of a year:

1. 1 month $(\frac{1}{12})$
2. 3 months $(\frac{1}{4})$
3. 5 days $(\frac{5}{365}$ or $\frac{1}{73})$
4. 10 months $(\frac{5}{6})$
5. 9 months $(\frac{3}{4})$

Ask what mixed number names the following months in years:

6. 18 months $(1\frac{1}{2})$
7. 25 months $(2\frac{1}{12})$
8. 40 months $(3\frac{1}{3})$

Warm Up • Fractions
Have students solve the following problems to review subtraction of fractions: $\frac{2}{3} - \frac{1}{3}$ $(\frac{1}{3})$, $\frac{5}{4} - \frac{3}{4}$ $(\frac{1}{2})$, $\frac{9}{5} - \frac{4}{5}$ (1), $\frac{11}{2} - \frac{7}{2}$ (2), and $\frac{20}{12} - \frac{2}{12}$ $(1\frac{1}{2})$.

Name ____________________

Subtracting Fractions With Unlike Denominators

Dina is making cookies to give to the new neighbors. The recipe calls for $\frac{7}{8}$ of a cup of peanut butter. She has measured out $\frac{1}{2}$ of a cup. How much more peanut butter does Dina need?

We want to find how much more peanut butter Dina needs to measure.

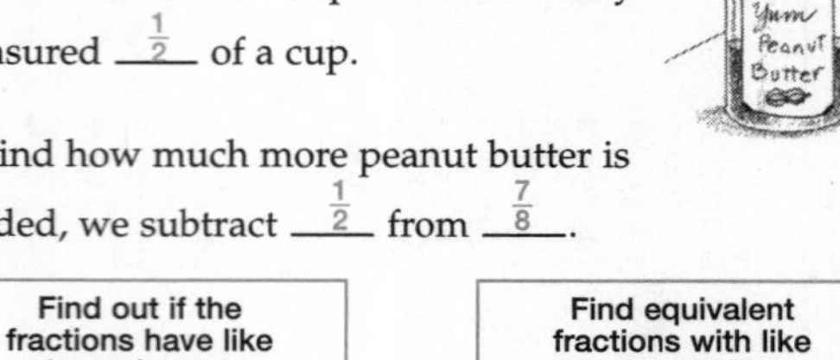

She needs $\underline{\frac{7}{8}}$ of a cup and has already measured $\underline{\frac{1}{2}}$ of a cup.

To find how much more peanut butter is needed, we subtract $\underline{\frac{1}{2}}$ from $\underline{\frac{7}{8}}$.

Find out if the fractions have like denominators.	Find equivalent fractions with like denominators.	Subtract the fractions.

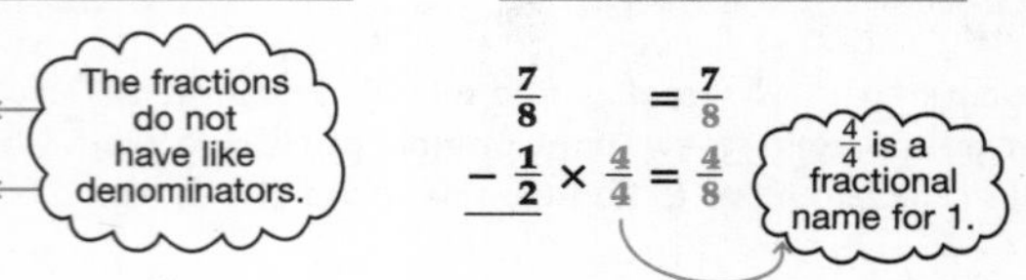

$\frac{7}{8}$ $-\frac{1}{2}$ — The fractions do not have like denominators.

$\frac{7}{8} = \frac{7}{8}$; $-\frac{1}{2} \times \frac{4}{4} = \frac{4}{8}$ — $\frac{4}{4}$ is a fractional name for 1.

$\frac{7}{8} - \frac{4}{8} = \frac{3}{8}$

Dina needs $\underline{\frac{3}{8}}$ of a cup more peanut butter.

Getting Started

Subtract.

1. $\frac{5}{6} - \frac{1}{3}$ $\frac{1}{2}$
2. $\frac{7}{12} - \frac{1}{2}$ $\frac{1}{12}$
3. $\frac{15}{16} - \frac{3}{4}$ $\frac{3}{16}$
4. $\frac{7}{9} - \frac{2}{3}$ $\frac{1}{9}$

Copy and subtract.

5. $\frac{5}{8} - \frac{1}{2}$ $\frac{1}{8}$
6. $\frac{11}{12} - \frac{2}{3}$ $\frac{1}{4}$
7. $\frac{17}{15} - \frac{4}{5}$ $\frac{1}{3}$
8. $\frac{6}{3} - \frac{7}{12}$ $1\frac{5}{12}$

2 Teach

Introduce the Lesson Have a student read the problem aloud and tell what is to be found. (how much more peanut butter Dina needs) Ask what information is given in the problem. (The recipe calls for $\frac{7}{8}$ of a cup and Dina has already measured $\frac{1}{2}$ of a cup.)

- Have students complete the sentences. Guide them through the subtraction steps in the model to solve the problem. Tell students that their solution plus $\frac{1}{2}$ of a cup must equal $\frac{7}{8}$ of a cup.

Develop Skills and Concepts Tell students that if fractions have unlike denominators, we look to see if the larger denominator is a multiple of the other.

- Write $\frac{7}{9} - \frac{1}{3}$ on the board. Ask if 9 is a multiple of 3. (yes) Have a student find the number of ninths in $\frac{1}{3}$. $(\frac{3}{9})$ Remind students that when the denominators are the same, we merely subtract the numerators. Have a student subtract $\frac{3}{9}$ from $\frac{7}{9}$. $(\frac{4}{9})$
- Repeat for more problems where one fraction's denominator is a multiple of the other denominator.

3 Practice

Remind students to find a like denominator, then find an equivalent fraction, and finally to subtract. Tell students to simplify answers if necessary. Have students complete page 286 independently.

Practice

Subtract.

1. $\frac{3}{4} - \frac{1}{2}$ $\frac{1}{4}$
2. $\frac{5}{8} - \frac{1}{4}$ $\frac{3}{8}$
3. $\frac{7}{2} - \frac{9}{10}$ $2\frac{3}{5}$
4. $\frac{7}{9} - \frac{2}{3}$ $\frac{1}{9}$
5. $\frac{9}{6} - \frac{7}{12}$ $\frac{11}{12}$
6. $\frac{8}{5} - \frac{5}{10}$ $1\frac{1}{10}$
7. $\frac{9}{12} - \frac{2}{3}$ $\frac{1}{12}$
8. $\frac{11}{4} - \frac{7}{8}$ $1\frac{7}{8}$
9. $\frac{15}{6} - \frac{7}{12}$ $1\frac{11}{12}$
10. $\frac{17}{10} - \frac{4}{5}$ $\frac{9}{10}$
11. $\frac{13}{5} - \frac{9}{15}$ 2
12. $\frac{7}{6} - \frac{2}{3}$ $\frac{1}{2}$

Copy and subtract.

13. $\frac{7}{9} - \frac{1}{3}$ $\frac{4}{9}$
14. $\frac{8}{3} - \frac{5}{6}$ $1\frac{5}{6}$
15. $\frac{7}{5} - \frac{8}{10}$ $\frac{3}{5}$
16. $\frac{15}{16} - \frac{7}{8}$ $\frac{1}{16}$
17. $\frac{11}{12} - \frac{1}{4}$ $\frac{2}{3}$
18. $\frac{9}{4} - \frac{3}{8}$ $1\frac{7}{8}$
19. $\frac{7}{3} - \frac{5}{9}$ $1\frac{7}{9}$
20. $\frac{9}{2} - \frac{7}{4}$ $2\frac{3}{4}$
21. $\frac{8}{5} - \frac{13}{10}$ $\frac{3}{10}$
22. $\frac{13}{16} - \frac{1}{4}$ $\frac{9}{16}$
23. $\frac{16}{12} - \frac{5}{4}$ $\frac{1}{12}$
24. $\frac{11}{6} - \frac{4}{3}$ $\frac{1}{2}$
25. $\frac{24}{15} - \frac{7}{5}$ $\frac{1}{5}$
26. $\frac{16}{3} - \frac{26}{9}$ $2\frac{4}{9}$
27. $\frac{7}{4} - \frac{7}{8}$ $\frac{7}{8}$
28. $\frac{25}{12} - \frac{5}{4}$ $\frac{5}{6}$

Problem Solving

Solve each problem.

29. A recipe for custard calls for $\frac{3}{4}$ of a cup of milk. Joel has only $\frac{1}{2}$ of a cup of milk. How much milk does Joel need to borrow? $\frac{1}{4}$ of a cup

30. Naomi lives $\frac{3}{5}$ of a mile from Deven. She has walked $\frac{5}{10}$ of a mile so far. How much farther must she walk to get to Deven's? $\frac{1}{10}$ of a mile

4 Assess

Write the following on the board for students to solve: $\frac{9}{16} - \frac{1}{2}$. ($\frac{1}{16}$)

For Mixed Abilities

Common Errors • Intervention

When subtracting fractions, students may subtract the numerators and the denominators. Have them work with partners to model the problems using diagrams or fraction bars to show the equivalent fractions with common denominators.

Enrichment • Fractions

Tell students to write and solve a subtraction problem of two fractions in which the minuend is $\frac{3}{4}$ and the common denominator is 24. Then, have them repeat for a minuend of $\frac{7}{9}$ and a common denominator of 36.

More to Explore • Logic

Have students line up ten coins and assign each coin a number, 1 to 10, for identification purposes. Tell them that the object of the game is to form five stacks of two coins each.

To do this, they must always move one coin over two coins, either stacked or single. While this goal may be accomplished a number of ways, ask students to limit themselves to five moves. (Possible answers: Place number 7 on 10, 4 on 8, 6 on 2, 9 on 5, and 1 on 3. Place number 4 on 1, 7 on 3, 5 on 9, 2 on 6, and 10 on 8.)

14-7 Finding Common Multiples

pages 287–288

1 Getting Started

Objective

- To find a common multiple to add or subtract fractions

Vocabulary

least common multiple

Warm Up • Mental Math

Dictate the following:

1. 78×2 (156)
2. $6{,}000 - 28$ (5,972)
3. $800 \div 40$ (20)
4. $(10 \times 60) \div 30$ (20)
5. $(54 \div 6) + (72 \div 8)$ (18)
6. $(\frac{1}{12}$ of $72) \times 6$ (36)
7. $(48 \times 2) \div 6$ (16)
8. $\frac{2}{3}$ of 24 (16)

Warm Up • Multiples

Have a student write the first six multiples of numbers 1 through 10 on the board as students count by 2s, 3s, 4s, 5s, 6s, 7s, 8s, and 9s. Have students find the least common multiple of 2, 3, 4, 5, and so on.

Name ______________________

Lesson 14-7

Finding Common Multiples

Marty spends her allowance on school supplies, food, and entertainment. What part of her allowance does she spend on school supplies and entertainment?

Marty's Allowance

$\frac{2}{3}$ school supplies

food $\frac{1}{12}$

$\frac{1}{4}$ entertainment

We want to know what fractional part of her money Marty spends on school supplies and entertainment.

Marty spends $\underline{\frac{2}{3}}$ of her allowance on school supplies and $\underline{\frac{1}{4}}$ of her allowance on entertainment.

To find the total portion spent on these two items, we add $\underline{\frac{2}{3}}$ and $\underline{\frac{1}{4}}$.

We need to find the smallest possible number that is a multiple of both 3 and 4. This is called the **least common multiple**.

Multiples of 3: 3, 6, 9, **12**, 15, . . .

Multiples of 4: 4, 8, **12**, 16, . . .

The least common multiple of 3 and 4 is $\underline{12}$.
Use the least common multiple as the common denominator.
Find the equivalent fractions and add.

$$\frac{2 \times 4}{3 \times 4} = \frac{8}{12}$$

$$+\frac{1 \times 3}{4 \times 3} = \frac{3}{12}$$

$$\frac{11}{12}$$

Marty spends $\underline{\frac{11}{12}}$ of her allowance on school supplies and entertainment.

Getting Started

Add or subtract.

1. $\frac{3}{5} + \frac{1}{3}$ $\frac{14}{15}$
2. $\frac{5}{6} - \frac{1}{4}$ $\frac{7}{12}$
3. $\frac{2}{3} + \frac{1}{2}$ $1\frac{1}{6}$

Copy and add or subtract.

4. $\frac{21}{3} - \frac{11}{9}$ $5\frac{7}{9}$
5. $\frac{1}{7} + \frac{1}{3}$ $\frac{10}{21}$
6. $\frac{3}{4} + \frac{5}{6}$ $1\frac{7}{12}$
7. $\frac{8}{6} - \frac{2}{5}$ $\frac{14}{15}$

---

2 Teach

Introduce the Lesson Have a student read the problem and tell what is to be solved. (the part of Marty's allowance spent on school supplies and entertainment) Ask what information is known. (Marty spends $\frac{2}{3}$ on school supplies, $\frac{1}{12}$ on food, and $\frac{1}{4}$ on entertainment.) Ask what information is not needed to solve the problem. ($\frac{1}{12}$ on food)

- Have students read with you as they complete the sentences and read the list of multiples in the model. Tell students that their solution plus $\frac{1}{12}$ must equal 1 whole. Have them complete the addition steps and solution sentence.

Develop Skills and Concepts Write $\frac{1}{2} + \frac{1}{5}$ on the board. Ask students if one denominator is a multiple of the other. (no) Tell students that they must find a denominator that is a multiple of both 2 and 5.

- Have students write the multiples of 2 and 5 through 30 on the board. Ask a student to circle the common multiples (10, 20, 30) and underline the least common multiple (10). Tell students that we call 10 the **least common denominator** because it is the lowest common multiple of 2 and 5. Write $\frac{1}{2} = \frac{(5)}{10}$ and $\frac{1}{5} = \frac{(2)}{10}$ on the board. Have a student find equivalent fractions and then add. ($\frac{7}{10}$)
- Repeat for $\frac{7}{9} + \frac{2}{6}$ ($1\frac{1}{9}$), $\frac{2}{3} - \frac{1}{4}$ ($\frac{5}{12}$), $\frac{20}{4} - \frac{2}{5}$ ($4\frac{3}{5}$), $\frac{4}{7} + \frac{5}{3}$ ($2\frac{5}{21}$), and $\frac{11}{12} - \frac{4}{6}$ ($\frac{1}{4}$).

3 Practice

Remind students to find the least common multiple to create equivalent fractions and then add or subtract and simplify if necessary. Have students complete page 288 independently.

Practice

Add or subtract.

1. $\frac{1}{5} + \frac{1}{3}$ $\frac{8}{15}$
2. $\frac{1}{4} + \frac{5}{6}$ $1\frac{1}{12}$
3. $\frac{2}{3} - \frac{1}{2}$ $\frac{1}{6}$
4. $\frac{3}{8} - \frac{1}{6}$ $\frac{5}{24}$
5. $\frac{3}{4} + \frac{4}{5}$ $1\frac{11}{20}$
6. $\frac{7}{3} - \frac{3}{4}$ $1\frac{7}{12}$
7. $\frac{3}{4} + \frac{5}{8}$ $1\frac{3}{8}$
8. $\frac{11}{12} - \frac{2}{3}$ $\frac{1}{4}$
9. $\frac{7}{8} - \frac{3}{16}$ $\frac{11}{16}$
10. $\frac{3}{2} + \frac{2}{3}$ $2\frac{1}{6}$
11. $\frac{7}{10} + \frac{4}{5}$ $1\frac{1}{2}$
12. $\frac{18}{5} - \frac{9}{4}$ $1\frac{7}{20}$

Copy and add or subtract.

13. $\frac{1}{2} + \frac{1}{9}$ $\frac{11}{18}$
14. $\frac{7}{4} - \frac{5}{3}$ $\frac{1}{12}$
15. $\frac{5}{6} - \frac{4}{9}$ $\frac{7}{18}$
16. $\frac{7}{8} + \frac{5}{3}$ $2\frac{13}{24}$
17. $\frac{2}{3} - \frac{1}{5}$ $\frac{7}{15}$
18. $\frac{11}{12} - \frac{5}{6}$ $\frac{1}{12}$
19. $\frac{13}{9} - \frac{25}{36}$ $\frac{3}{4}$
20. $\frac{5}{9} + \frac{7}{6}$ $1\frac{13}{18}$
21. $\frac{1}{2} - \frac{2}{5}$ $\frac{1}{10}$
22. $\frac{3}{5} + \frac{1}{4}$ $\frac{17}{20}$
23. $\frac{11}{6} - \frac{8}{9}$ $\frac{17}{18}$
24. $\frac{9}{8} + \frac{4}{3}$ $2\frac{11}{24}$
25. $\frac{7}{8} + \frac{5}{6}$ $1\frac{17}{24}$
26. $\frac{7}{10} + \frac{4}{5}$ $1\frac{1}{2}$
27. $\frac{5}{8} + \frac{1}{12}$ $\frac{17}{24}$
28. $\frac{9}{4} - \frac{6}{7}$ $1\frac{11}{28}$

Problem Solving

Use the chart on the right to solve each problem.

29. How much rain fell on Monday and Wednesday? $1\frac{7}{20}$ inches
30. How much more rain fell on Monday than on Tuesday? $\frac{1}{12}$ of an inch

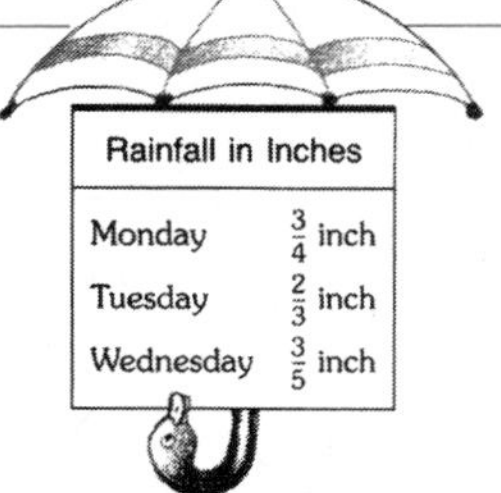

Rainfall in Inches	
Monday	$\frac{3}{4}$ inch
Tuesday	$\frac{2}{3}$ inch
Wednesday	$\frac{3}{5}$ inch

4 Assess

Write the following on the board and ask students to solve it : $\frac{35}{3} - \frac{21}{9}$. $(9\frac{1}{3})$

For Mixed Abilities

Common Errors • Intervention

Some students may have difficulty subtracting when one of the two denominators is not the common denominator. Have them work with partners to list multiples for both numbers in pairs such as 3 and 7 or 6 and 9 until they find a common multiple, 21 and 18, respectively. Then, have them use the data to work problems such as the following:

$\frac{5}{7} - \frac{1}{3}$ $(\frac{8}{21})$ and $\frac{7}{9} - \frac{1}{6}$ $(\frac{11}{18})$

Enrichment • Graphing

Tell students to draw a circle graph to show how they spend the rest of a day if they are in school $\frac{1}{4}$ of their day and asleep $\frac{1}{3}$ of their day. Have them write and solve an addition problem that checks their graph to be sure it totals one whole day.

More to Explore • Estimation

This activity will help students sharpen their estimation and measurement skills. Collect a variety of objects that are measured by varying units, such as those suggested in the following questions. Tell students to estimate the answer as you hold up each item. (Answers will vary.)

1. How much does this full box of salt weigh?
2. How many pages are in this volume of the encyclopedia? How many pages are in the entire set?
3. How many inches long is this baseball bat?
4. What is the circumference of this baseball in centimeters?
5. How many of these crayons, put end to end, would equal 2 feet?

Have volunteers measure each object to find the exact answer and write these measurements on the board. Have students compare their estimates to the measurements.

14-8 Adding Mixed Numbers

pages 289–290

1 Getting Started

Objective
- To add two mixed numbers

Warm Up • Mental Math
Tell students to name two fractions that equal the following:

1. 1 ($\frac{1}{1}$, $\frac{2}{2}$, and so on)
2. 2 ($\frac{2}{1}$, $\frac{4}{2}$, and so on)
3. 13 ($\frac{13}{1}$, $\frac{26}{2}$, and so on)
4. 4 ($\frac{4}{1}$, $\frac{8}{2}$, and so on)
5. 15 ($\frac{15}{1}$, $\frac{30}{2}$, and so on)
6. 9 ($\frac{9}{1}$, $\frac{18}{2}$, and so on)
7. 11 ($\frac{11}{1}$, $\frac{22}{2}$, and so on)
8. 8 ($\frac{8}{1}$, $\frac{16}{2}$, and so on)

Warm Up • Fractions
Review addition of fractions by having students solve the following problems at the board: $\frac{2}{3} + \frac{1}{2}$ ($1\frac{1}{6}$), $\frac{7}{8} + \frac{3}{4}$ ($1\frac{5}{8}$), $\frac{5}{6} + \frac{7}{8}$ ($1\frac{17}{24}$), and $\frac{11}{12} + \frac{5}{8}$ ($1\frac{13}{24}$).

Name ______________________________

Lesson 14-8

Adding Mixed Numbers

Betsy answers the telephone in the afternoons at the library. How many hours does Betsy's timecard show she has worked this week?

Betsy Kline	
Day	Hours
Monday	$3\frac{3}{4}$
Tuesday	$2\frac{1}{2}$
Wednesday	
Thursday	

We want to know Betsy's part-time hours.

Betsy worked $\underline{3\frac{3}{4}}$ hours on Monday and $\underline{2\frac{1}{2}}$ hours on Tuesday.

To find her total hours, we add $\underline{3\frac{3}{4}}$ and $\underline{2\frac{1}{2}}$.

Add the fractions. Add the whole numbers.

$$3\frac{3}{4} = 3\frac{3}{4}$$
$$+\ 2\frac{1}{2} = 2\frac{2}{4}$$
$$5\frac{5}{4}$$

Simplify the answer by regrouping.

$$5\frac{5}{4} = 5 + \frac{4}{4} + \frac{1}{4}$$
$$5 + 1 + \frac{1}{4} = 6\frac{1}{4}$$

REMEMBER A fraction with the same numerator and denominator is equal to 1.

Betsy has worked $\underline{6\frac{1}{4}}$ hours this week.

Getting Started

Add. Simplify if necessary.

1. $3\frac{1}{2} + 7\frac{7}{8}$ = $11\frac{3}{8}$
2. $5\frac{2}{3} + 2\frac{1}{6}$ = $7\frac{5}{6}$
3. $9\frac{1}{2} + 5\frac{2}{3}$ = $15\frac{1}{6}$
4. $16\frac{1}{6} + 25\frac{5}{9}$ = $41\frac{13}{18}$

Copy and add.

5. $10\frac{2}{3} + 4\frac{1}{6}$ $14\frac{5}{6}$
6. $5\frac{1}{9} + 5\frac{3}{18}$ $10\frac{5}{18}$
7. $8\frac{2}{5} + 9\frac{8}{10}$ $18\frac{1}{5}$
8. $11\frac{1}{4} + 2\frac{1}{3}$ $13\frac{7}{12}$

2 Teach

Introduce the Lesson Have a student read the problem aloud and tell what is to be found. (the number of hours Betsy worked this week) Ask students what information is known. (She worked $3\frac{3}{4}$ hours on Monday and $2\frac{1}{2}$ hours on Tuesday.)

- Have students complete the sentences and the addition steps in the model with you. Have students use a number line to check their addition.

Develop Skills and Concepts Write **4 + 6** vertically on the board and have students give the sum. (10)

- Write $\frac{2}{3} + \frac{3}{4}$ vertically on the board and have a student work the problem. ($1\frac{5}{12}$)
- Write $4\frac{2}{3} + 6\frac{3}{4}$ vertically on the board and tell students that we add mixed numbers just like we add whole numbers and fractions. Tell students that we add the fractions first because their sum may be 1 whole or more. Ask students the sum of the fractions in twelfths. ($\frac{17}{12}$) Tell students that we know that $\frac{17}{12}$ is $\frac{12}{12}$ plus $\frac{5}{12}$. Remind students that $\frac{12}{12}$ equals 1. Tell students to record the $\frac{5}{12}$ and regroup the $\frac{12}{12}$ for 1 whole, which is added to 4 + 6 for 11. Record the 11 and ask students to read the mixed number. ($11\frac{5}{12}$)
- Repeat the procedure for $6\frac{2}{7} + 2\frac{3}{7}$ ($8\frac{5}{7}$), $5\frac{3}{8} + 4\frac{7}{8}$ ($10\frac{1}{4}$), and $2\frac{1}{2} + \frac{5}{8}$ ($3\frac{1}{8}$).

3 Practice

Remind students to add the fractions first and regroup any fractions that equal a whole number. Have students work the exercises on page 290 independently.

Practice

Add. Simplify if necessary.

1. $5\frac{1}{6} + 6\frac{1}{2}$ — $11\frac{2}{3}$
2. $4\frac{2}{3} + 7\frac{5}{8}$ — $12\frac{7}{24}$
3. $6\frac{3}{4} + 5\frac{1}{3}$ — $12\frac{1}{12}$
4. $9\frac{1}{2} + 7\frac{1}{8}$ — $16\frac{5}{8}$
5. $16\frac{3}{5} + 9\frac{3}{4}$ — $26\frac{7}{20}$
6. $6\frac{7}{8} + 15\frac{5}{6}$ — $22\frac{17}{24}$
7. $1\frac{2}{3} + 6\frac{5}{8}$ — $8\frac{7}{24}$
8. $6\frac{3}{5} + 7\frac{3}{4}$ — $14\frac{7}{20}$

Copy and add.

9. $5\frac{3}{8} + 2\frac{1}{2}$ $7\frac{7}{8}$
10. $9\frac{7}{16} + 3\frac{7}{8}$ $13\frac{5}{16}$
11. $4\frac{2}{3} + 16\frac{3}{4}$ $21\frac{5}{12}$
12. $7\frac{3}{5} + 15\frac{1}{4}$ $22\frac{17}{20}$
13. $6\frac{7}{10} + 16\frac{3}{4}$ $23\frac{9}{20}$
14. $14\frac{7}{10} + 8\frac{5}{6}$ $23\frac{8}{15}$
15. $9\frac{7}{8} + 7\frac{11}{12}$ $17\frac{19}{24}$
16. $7\frac{2}{3} + 9\frac{5}{9}$ $17\frac{2}{9}$
17. $6\frac{5}{6} + 18\frac{3}{7}$ $25\frac{11}{42}$
18. $2\frac{4}{5} + 6\frac{3}{8}$ $9\frac{7}{40}$
19. $5\frac{7}{8} + 7\frac{5}{7}$ $13\frac{33}{56}$
20. $3\frac{2}{3} + 6\frac{3}{5}$ $10\frac{4}{15}$

Now Try This!

Use the diagram to answer each statement. Write *true* or *false*.

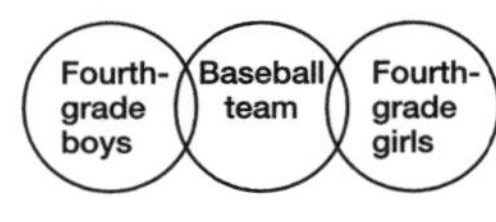

1. All of the fourth-grade boys play on the baseball team. false
2. Only fourth graders play on the baseball team. false
3. There are boys and girls on the team. true
4. Some fourth graders do not play on the team. true

Now Try This! Discuss Venn diagrams. Have students write one more true statement and one more false statement for the diagram after they have completed the exercises.

4 Assess

Ask students what a fraction with the same numerator and denominator is equal to. (1)

For Mixed Abilities

Common Errors • Intervention

Some students may add the fractions but forget to add the whole numbers. Have students work in pairs to solve the problem $\frac{6}{5} + \frac{7}{5}$, writing the answer in simplified form. ($2\frac{3}{5}$) Next, ask them to rewrite $\frac{6}{5}$ and $\frac{7}{5}$ as mixed numbers and add ($1\frac{1}{5} + 1\frac{2}{5} = 2\frac{3}{5}$) and compare their answers to see that they are the same. Have them repeat the procedure for $\frac{9}{4} + \frac{3}{8}$. ($2\frac{5}{8}$)

Enrichment • Numeration

Tell students to write a problem of two mixed numbers whose sum is $25\frac{4}{5}$ if one of the addends is $4\frac{1}{20}$.

More to Explore • Logic

Dictate the following situation for students to solve:

One day Jane White, Amy Black, and Beth Brown were walking to school. "Look at our skirts," said one girl. "One is white, one is black, and one is brown, just like our names."

"Yes,"said the girl with the black skirt, "but none of us is wearing the skirt color to match our own name."

"I noticed that too," said Jane White.

Can you identify what color skirt each girl is wearing? (Jane White, brown skirt; Amy Black, white skirt; Beth Brown, black skirt. Jane White isn't wearing white because that would match her name and she isn't wearing black because she was talking to the girl wearing black. So, her skirt has to be brown. Amy Black can't be wearing her name color and she can't be wearing brown because Jane's skirt is brown. So, Amy must be wearing white. That leaves the black skirt for Beth Brown.)

14-9 Adding Three Mixed Numbers

pages 291–292

1 Getting Started

Objective

- To add three mixed numbers

Warm Up • Mental Math

Have students name two common multiples for the following:

1. 2 and 3 (6, 12, and so on)
2. 2 and 5 (10, 20, and so on)
3. 2 and 4 (4, 8, and so on)
4. 2 and 6 (6, 12, and so on)
5. 2 and 7 (14, 28, and so on)
6. 3 and 4 (12, 24, and so on)
7. 3 and 5 (15, 30, and so on)
8. 3 and 6 (6, 12, and so on)

Warm Up • Numeration

Write the following on the board: $2\frac{2}{3} + 4\frac{5}{6}$ ($7\frac{1}{2}$), $9\frac{3}{5} + 11\frac{1}{3}$ ($20\frac{14}{15}$), and $24\frac{5}{6} + 43\frac{7}{8}$ ($68\frac{17}{24}$). Have a student talk through each problem as it is worked. Continue for similar problems.

Name ______________________

Adding Three Mixed Numbers

Marianne built a model trading fort as part of her project on fur traders in Colonial America. Find the perimeter of her triangular fort.

We want to know the perimeter of Marianne's fort.

The sides of the triangle measure $4\frac{3}{4}$ inches, $3\frac{1}{2}$ inches, and $6\frac{3}{5}$ inches.

To find the perimeter, we add $4\frac{3}{4}$, $3\frac{1}{2}$,

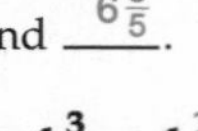

and $6\frac{3}{5}$.

$$\begin{aligned} 4\frac{3}{4} &= 4\frac{15}{20} \\ 3\frac{1}{2} &= 3\frac{10}{20} \\ +\ 6\frac{3}{5} &= 6\frac{12}{20} \\ \hline 13\frac{37}{20} &= 14\frac{17}{20} \end{aligned}$$

The perimeter of Marianne's fort is $14\frac{17}{20}$ inches.

Getting Started

Add.

1. $2\frac{1}{2} + 3\frac{5}{8} + 2\frac{1}{4} = 8\frac{3}{8}$
2. $3\frac{1}{3} + 6\frac{5}{9} + 4\frac{5}{6} = 14\frac{13}{18}$
3. $9\frac{1}{2} + 3\frac{4}{5} + 7\frac{5}{6} = 21\frac{2}{15}$
4. $7\frac{5}{12} + 9\frac{3}{8} + 4\frac{5}{6} = 21\frac{5}{8}$

Copy and add.

5. $10\frac{3}{4} + 2\frac{1}{2} + 7\frac{1}{4}$ $20\frac{1}{2}$
6. $5\frac{1}{3} + 11\frac{5}{6} + \frac{1}{12}$ $17\frac{1}{4}$
7. $11\frac{1}{4} + 3\frac{2}{3} + 2\frac{1}{2}$ $17\frac{5}{12}$

2 Teach

Introduce the Lesson Have a student read the problem aloud and tell what is to be found. (the perimeter of the fort) Ask what information is given. (The triangle has sides of $4\frac{3}{4}$, $3\frac{1}{2}$, and $6\frac{3}{5}$ inches.) Ask students how to find the perimeter of a triangle. (side + side + side)

- Have students complete the sentences and addition steps in the model with you. Have students use a number line to check their solution.

Develop Skills and Concepts Write **2 + 3 + 5** vertically on the board. Have a student solve the problem. (10) Tell students that just as we can add several whole numbers, we can add several fractions.

- Write $\frac{1}{2} + \frac{1}{3} + \frac{1}{4}$ vertically on the board and tell students that we must find the least common multiple for these fractions. Have a student list the multiples of 2, 3, and 4 on the board to find the least common multiple. (12)
- Have a student find the number of twelfths in $\frac{1}{2}$, $\frac{1}{3}$, and $\frac{1}{4}$ and write the equivalent fractions. ($\frac{6}{12}$, $\frac{4}{12}$, $\frac{3}{12}$) Have students tell the sum in simplest form. ($1\frac{1}{12}$) Remind students that the answer is a mixed number because it is more than 1.
- Have students solve more addition problems of three simple fractions or mixed numbers with unlike denominators.

3 Practice

Remind students to find a common multiple for all three fractions. Have students complete the exercises on page 292 independently.

Practice

Add. Simplify if necessary.

1. $\begin{array}{r} 5\frac{1}{5} \\ 7\frac{3}{10} \\ +\ 4\frac{9}{20} \\ \hline 16\frac{19}{20} \end{array}$
2. $\begin{array}{r} 6\frac{2}{3} \\ 5\frac{1}{6} \\ +\ 7\frac{5}{9} \\ \hline 19\frac{7}{18} \end{array}$
3. $\begin{array}{r} 3\frac{2}{5} \\ 4\frac{1}{2} \\ +\ 9\frac{5}{10} \\ \hline 17\frac{2}{5} \end{array}$
4. $\begin{array}{r} 2\frac{5}{8} \\ 3\frac{5}{12} \\ +\ 4\frac{1}{4} \\ \hline 10\frac{7}{24} \end{array}$
5. $\begin{array}{r} 7\frac{5}{9} \\ 6\frac{1}{3} \\ +\ 4\frac{5}{6} \\ \hline 18\frac{13}{18} \end{array}$
6. $\begin{array}{r} 7\frac{3}{4} \\ 2\frac{1}{3} \\ +\ 11\frac{5}{6} \\ \hline 21\frac{11}{12} \end{array}$
7. $\begin{array}{r} 8\frac{5}{8} \\ 7\frac{1}{2} \\ +\ 9\frac{2}{3} \\ \hline 25\frac{19}{24} \end{array}$
8. $\begin{array}{r} 16\frac{3}{4} \\ 2\frac{1}{2} \\ +\ 5\frac{2}{3} \\ \hline 24\frac{11}{12} \end{array}$

Copy and add.

9. $5\frac{1}{8} + 6\frac{2}{3} + 9\frac{1}{2}$ $21\frac{7}{24}$
10. $7\frac{3}{4} + 5\frac{3}{8} + 7\frac{1}{2}$ $20\frac{5}{8}$
11. $9\frac{2}{3} + 4\frac{3}{5} + 6\frac{7}{10}$ $20\frac{29}{30}$
12. $4\frac{1}{3} + 8\frac{5}{6} + 12\frac{5}{9}$ $25\frac{13}{18}$
13. $16\frac{1}{2} + 5\frac{2}{3} + 8\frac{1}{4}$ $30\frac{5}{12}$
14. $7\frac{5}{6} + 9\frac{1}{2} + 8\frac{2}{3}$ 26

Now Try This!

Unscramble each word, writing one letter in each box. To find the mystery word, unscramble the circled letters.

ATIRO	(R)	A	T	I	O						
DERECU	R	E	D	U	(C)	E					
MYSIPFLI	S	(I)	M	P	L	I	(F)	Y			
MERUNATOR	N	U	M	E	R	(A)	T	O	R		
EVTELQUAIN	E	Q	U	I	V	A	L	E	(N)	(T)	
WLETOS MREST	L	(O)	W	E	S	T	T	E	R	M	S

Mystery Word: F R A C T I O N

Now Try This! Tell students that they are looking for words from their vocabulary list for fractions.

4 Assess

Ask students how adding three mixed numbers is different from adding three whole numbers. (If the denominators of mixed numbers are different, we find the least common multiple and write equivalent fractions. Then, we add the fractions, regrouping if necessary. Finally, we add the whole numbers.)

For Mixed Abilities

Common Errors • Intervention

Students may have difficulty keeping their work organized because there are many different steps. Have them work with partners, solving problems for which the first student finds the common denominator and the equivalent fractions and the partner adds and writes the sum in simplest form. Then, have them trade roles.

$4\frac{5}{6} + 2\frac{1}{8} + 1\frac{3}{4}$ $(8\frac{17}{24})$

$3\frac{2}{3} + 12\frac{5}{6} + 26\frac{2}{9}$ $(42\frac{13}{18})$

$14\frac{5}{7} + 2\frac{1}{14} + 1\frac{1}{2}$ $(18\frac{2}{7})$

$2\frac{6}{11} + 1\frac{5}{22} + 15\frac{1}{2}$ $(19\frac{3}{11})$

Enrichment • Applications

Tell students to use the stock market pages of a newspaper to find the total cost of one share of each of three different stocks on a particular day.

More to Explore • Geometry

Give students geoboards, rubber bands, and grid paper. Tell them to make as many figures as possible, each with an area of 8 square units (or square inches, depending on your geoboards). Remind them to be careful when counting fractional squares. Have them draw each figure they make on the grid paper and calculate its perimeter.

ESL/ELL STRATEGIES

As you encounter them, review the meaning of the key terms in each lesson. In this lesson, you might focus on *mixed numbers*, *perimeter*, and *triangular*. Put a numerical or visual example of each key term on the board and use the terms as you explain and discuss the example.

14-10 Problem Solving: Write an Open Sentence

pages 293–294

1 Getting Started

Objective

- To write an open sentence to solve a problem

Warm Up • Mental Math

Have students name a unit fraction whose value is less than the following:

1. $\frac{1}{10}$ ($\frac{1}{11}$, $\frac{1}{12}$, and so on)
2. $\frac{1}{2}$ ($\frac{1}{3}$, $\frac{1}{4}$, and so on)
3. $\frac{1}{6}$ ($\frac{1}{5}$, $\frac{1}{4}$, and so on)
4. $\frac{1}{3}$ ($\frac{1}{4}$, $\frac{1}{5}$, and so on)
5. $\frac{1}{4}$ ($\frac{1}{5}$, $\frac{1}{6}$, and so on)

2 Teach

Introduce the Lesson Remind students of the four-step problem-solving strategy they learned to use for solving problems.

Develop Skills and Concepts Tell students that 2 plus another number equals 30. Tell students that we want to know the other number. Write the following on the board: **2 + some number = 30.**

- Tell students that it would be easier if we could use a symbol to talk about the unknown number. For example, write **2 + *y* = 30** on the board. Tell students that this is called an open sentence because there is an unknown quantity. Ask students what number plus 2 will equal 30. (28)
- Now, tell students that 20 plus 2 times another number equals 30. Explain that we write an open sentence using a letter for the unknown number. Write **20 + 2 × *s* = 30** on the board. Tell students that we can simplify 2 × *s* by writing it 2*s* as you write **20 + 2*s* = 30** on the board.
- Then, tell students that if 2*s* + 20 = 30 then 2*s* would equal 30 − 20 or 10 as you write **2*s* = 10** on the board. Ask what number times 2 equals 10. (5) Write ***s* = 5** on the board. Tell students that we must check the work as you write **20 + 2 × 5 = 30** on the board. Remind students that we work multiplication first as you write **20 + 10 = 30** on the board.

Name ____________________

Problem Solving: Write an Open Sentence

It's Algebra!

A mouse walks around a rectangular-shaped piece of cheese whose length is twice its width. If the mouse walks a total of 12 inches, what are the dimensions of the piece of cheese?

★ SEE

We want to find the length and the width of the piece of cheese.

The perimeter of the cheese is __12__ inches.

The length of the cheese is double that of its width.

If the width were 1 inch, the length would be __2__ inches.

★ PLAN

We make a sketch of the rectangle of cheese with notation. If we label the width with the letter *w*, then the length is 2 × *w* This is written: ***2w***.

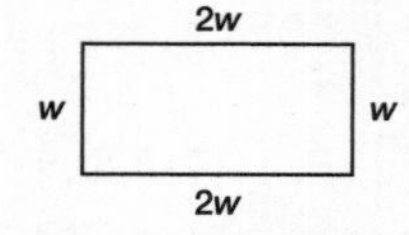

The mouse walks $2w + w + 2w + w$.

★ DO

The **open sentence** that represents the distance the mouse walks is:

$$2w + w + 2w + w = 12$$
$$6w = 12 \quad \text{(We add all the } w\text{s. } 2 + 1 + 2 + 1 = 6)$$
$$w = 2 \quad \text{(We reason if } 6 \times w = 12, \text{ the value of } w \text{ must be 2 because } 6 \times 2 = 12.)$$

If $w = 2$, then $2w$ = __4__.

The width of the cheese is __2__ inches and the length is __4__ inches.

★ CHECK

We can check by adding the lengths and widths together.

4 + 2 + 4 + 2 = __12__

- Have a student read the problem in the textbook aloud and tell what is to be found. (the dimensions of the piece of cheese) Ask what information is known. (The cheese is a rectangular shape, its length is 2 times its width, and total distance around it is 12 inches.)
- Remind students that 2 times an unknown number can be written as 2*w*. Have students read with you and complete the SEE, PLAN, and DO sections to solve the problem. Have them CHECK their solution.

3 Apply

Solution Notes

1. Let *P* = perimeter.
   Let *s* = length of side.
   $P = 4s$
   $P = 4 \times 12$ inches
   $P = 48$ inches

Apply

Write an open sentence to solve each problem.

1. What is the perimeter of a square checkerboard that is 12 inches on a side?
   See Solution Notes; 48 inches
2. Find the distance around a square pasture that is 45 meters on each side.
   See Solution Notes; 180 meters
3. There are 12 months in a year. Write an open sentence to find how many months are in 10 years. How many months are in 100 years?
   See Solution Notes; 120 months; 1,200 months
4. A basketball game is played during which a team scored 55 points. Twenty baskets were made and each is worth two points. A free throw is worth one point. Write an open sentence to help find how many free throws were made.
   See Solution Notes; 15 free throws
5. The sum of two numbers is 90. The first number is twice the second number. What are the two numbers?
   See Solution Notes; 60, 30
6. Jason has saved 3 nickels, 8 dimes, and an unknown number of quarters. The total amount he has saved is $2.20. Write an open sentence that will help determine the number of quarters Jason has.
   See Solution Notes; 5 quarters

Problem-Solving Strategy: Using the Four-Step Plan

- ★ **SEE** What do you need to find?
- ★ **PLAN** What do you need to do?
- ★ **DO** Follow the plan.
- ★ **CHECK** Does your answer make sense?

7. Read Exercise 4 again and the open sentence that you wrote. What if the team scored 60 points? How would you change your open sentence to fit this new situation?
   See Solution Notes. Answers will vary.
8. Read Exercise 5 again. Rewrite the exercise so that the second number is twice the first number. Then, explain how this changes the answer.
   See Solution Notes.
9. Write an addition sentence to show two like fractions that have a sum with a different denominator.
   See Solution Notes. Answers will vary.
10. Explain how you could write many different number sentences where 1 is the sum of two like fractions.
    See Solution Notes. Answers will vary.

For Mixed Abilities

More to Explore • Logic

Dictate the following and have students answer the questions:

Matthew shares his bedroom with his younger brother, who gets up after Matthew goes to school. This presents a problem for Matthew because he has to pick out his school clothes in the dark. One of his drawers is filled with an even number of unpaired black socks and white socks.

What is the smallest number of socks he must pull from the drawer to make sure he will have a matched pair? How many socks must he pull out to make sure he has two pairs of matching socks? If Matthew is going camping and needs 9 pairs of socks, what is the least number he must pull out?

Ask students to identify the pattern they see. (3, 5, 19; if Matthew needs x number of pairs of socks then double x plus 1 is the number of socks he needs.)

2. Let P = perimeter of pasture.
   Let s = length of side.
   $P = 4s$
   $P = 4 \times 45$
   $P = 180$ meters
3. Let m = number of months.
   Let y = number of years.
   $m = 12y$
   $m = 12 \times 10$
   $m = 120$ months
   $m = 12 \times 100$
   $m = 1{,}200$ months
4. Let f = number of free throws.
   Let b = number of baskets.
   $2b + f = 55$
   $2 \times 20 + f = 55$
   $40 + f = 55$
   $f = 15$ free throws
5. Let a = second number.
   $2a + a = 90$
   $3a = 90$
   $a = 30$
6. Let $n = 5$.
   Let $d = 10$.
   Let $q = 25$.
   $3n + 8d + ?q = 220$
   $(3 \times 5) + (8 \times 10) + (?25) = 220$
   $15 + 80 + (?25) = 220$
   $95 + (?25) = 220$
   $?25 = 125$
   $? = 5$

Higher-Order Thinking Skills

7. **Analysis:** Any correct answer will, in some way, indicate changing 55 in the open sentence to 60.
8. **Synthesis:** This really does not change the answer; the two numbers are still 30 and 60.
9. **Synthesis:** The answer will be any two fractions with like denominators with a sum that can be simplified, such as $\frac{1}{8} + \frac{3}{8} = \frac{1}{2}$.
10. **Analysis:** Show fractions where the sum of the numerators equals the denominator, for example, $\frac{1}{4} + \frac{3}{4} = 1$.

Chapter 14 Test

page 295

Items	Objectives
1, 3, 6–8	To add fractions with like denominators (see pages 275–276)
2, 4, 5	To subtract fractions with like denominators (see pages 277–278)
10, 14	To subtract two fractions when one denominator is a multiple of the other denominator (see pages 285–286)
9, 11–13, 15–16, 21–22	To add or subtract two fractions with unlike denominators (see pages 287–288)
17–20, 23–24	To add two mixed numbers (see pages 289–290)

Name ____________________

CHAPTER 14 TEST

Add or subtract. Simplify.

1. $\frac{3}{5} + \frac{1}{5}$ — $\frac{4}{5}$
2. $\frac{7}{8} - \frac{4}{8}$ — $\frac{3}{8}$
3. $\frac{2}{7} + \frac{4}{7}$ — $\frac{6}{7}$
4. $\frac{5}{9} - \frac{4}{9}$ — $\frac{1}{9}$
5. $\frac{11}{12} - \frac{5}{12}$ — $\frac{1}{2}$
6. $\frac{5}{10} + \frac{3}{10}$ — $\frac{4}{5}$
7. $\frac{1}{8} + \frac{5}{8}$ — $\frac{3}{4}$
8. $\frac{5}{6} + \frac{1}{6}$ — 1
9. $\frac{2}{3} + \frac{3}{4}$ — $1\frac{5}{12}$
10. $\frac{7}{8} - \frac{1}{4}$ — $\frac{5}{8}$
11. $\frac{3}{5} + \frac{1}{3}$ — $\frac{14}{15}$
12. $\frac{7}{4} + \frac{5}{6}$ — $2\frac{7}{12}$
13. $\frac{9}{8} - \frac{5}{6}$ — $\frac{7}{24}$
14. $\frac{11}{10} - \frac{4}{5}$ — $\frac{3}{10}$
15. $\frac{9}{6} + \frac{7}{9}$ — $2\frac{5}{18}$
16. $\frac{15}{12} - \frac{9}{8}$ — $\frac{1}{8}$
17. $3\frac{5}{8} + 2\frac{3}{4}$ — $6\frac{3}{8}$
18. $7\frac{2}{3} + 8\frac{1}{4}$ — $15\frac{11}{12}$
19. $6\frac{3}{4} + 6\frac{1}{6}$ — $12\frac{11}{12}$
20. $3\frac{2}{3} + 9\frac{1}{8}$ — $12\frac{19}{24}$
21. $\frac{1}{2} + \frac{3}{7}$ — $\frac{13}{14}$
22. $\frac{2}{3} - \frac{2}{7}$ — $\frac{8}{21}$
23. $4\frac{3}{5} + 2\frac{1}{2}$ — $7\frac{1}{10}$
24. $7\frac{3}{10} + 5\frac{5}{6}$ — $13\frac{1}{6}$

Alternate Chapter Test

You may wish to use the Alternate Chapter Test on page 372 of this book for further review and assessment.

Circle the letter of the correct answer.

1. Round 6,296 to the nearest thousand.
   (a.) 6,000
   b. 7,000
   c. NG

2. What is the place value of the 7 in 736,092?
   a. hundreds
   b. thousands
   (c.) hundred thousands
   d. NG

3. $5,629 + 24,751$
   a. 29,380
   b. 30,370
   (c.) 30,380
   d. NG

4. $8,096 - 1,998$
   a. 7,008
   b. 7,098
   c. 7,902
   (d.) NG

5. Choose the better estimate of weight.
   (a.) 1 gram
   b. 1 kilogram

6. $\$3.26 \times 6$
   a. $18.26
   b. $18.56
   (c.) $19.56
   d. NG

7. 67×53
   a. 536
   b. 3,451
   (c.) 3,551
   d. NG

8. $8\overline{)265}$
   a. 3 R1
   b. 30 R1
   (c.) 33 R1
   d. NG

9. $760 \div 42$
   a. 1 R4
   (b.) 18 R4
   c. 180 R4
   d. NG

10. Find the perimeter.
    12 in. 12 in. 15 in.
    a. 24 in.
    b. 27 in.
    (c.) 39 in.
    d. NG

11. Find the area.
    (a.) 24 sq units
    b. 48 sq units
    c. 48 units
    d. NG

12. $\frac{2}{3}$ of $48 = n$
    $n = ?$
    a. 8
    b. 16
    (c.) 32
    d. NG

score

STOP

Cumulative Assessment

page 296

Items	Objectives
1	To round numbers (see pages 31–32)
2	To identify place value (see pages 25–26)
3	To add two numbers up to 5 digits (see pages 47–48)
4	To subtract two 4-digit numbers (see pages 61–62)
5	To determine the appropriate metric unit of weight (see pages 155–156)
6–7	To multiply two 2-digit numbers including money (see pages 167–168, 177–178)
8–9	To divide 3-digit numbers by up to 2-digit numbers (see pages 201–202)
10	To find the perimeter of a parallelogram (see pages 239-240)
11	To find the area of a rectangle (see pages 241-242)
13	To find the fraction of a number (see pages 253–254)

Alternate Cumulative Assessment

Circle the letter of the correct answer.

1. Round 4,500 to the nearest thousand.
   a 4,000
   (b) 5,000
   c NG

2. What is the place value of the 2 in 326,985?
   a hundreds
   b thousands
   (c) ten thousands
   d NG

3. $4,765 + 34,527$
   a 38,292
   b 38,282
   (c) 39,292
   d NG

4. $9,084 - 2,276$
   (a) 6,808
   b 6,818
   c 7,818
   d NG

5. What unit would you use to weigh a pencil?
   (a) ounces
   b pounds
   c tons
   d NG

6. $\$7.35 \times 7$
   a $49.45
   b $51.15
   c $51.55
   (d) NG

7. $32 \times 74 =$
   (a) 2,368
   b 2,378
   c 2,468
   d NG

8. $9\overline{)380}$
   a 4 R2
   b 40 R2
   (c) 42 R2
   d NG

9. $496 \div 32 =$
   (a) 15 R16
   b 105 R16
   c 150 R16
   d NG

10. Find the perimeter of a triangle that is 9 cm by 9 cm by 10 cm.
    a 81 cm
    (b) 28 cm
    c 36 cm
    d NG

11. Find the area of a 6-m square.
    a 13 sq m
    b 26 sq m
    (c) 36 sq m
    d NG

12. $\frac{3}{4}$ of 84
    a 21
    b 28
    (c) 63
    d NG

15-1 Tenths

pages 297–298

1 Getting Started

Objective
- To understand and write tenths

Vocabulary
decimal, tenths

Warm Up • Mental Math
Dictate the following and ask students to solve:
1. $\frac{2}{9}$ of 81 (18)
2. 4,261 + 9 tens − 6 ones (4,345)
3. (280 ÷ 14) × 2 (40)
4. $\frac{1}{2} + \frac{1}{6}$ ($\frac{2}{3}$)
5. $\frac{5}{7} - \frac{1}{14}$ ($\frac{9}{14}$)
6. $2\frac{1}{2} + 6\frac{1}{3}$ ($8\frac{5}{6}$)
7. LCM of 4, 3, and 6 (12)
8. LCM of 9, 3, and 6 (18)

Warm Up • Fractions
Dictate the following fractions for students to write on the board: $3\frac{5}{10}$, $4\frac{6}{10}$, $13\frac{9}{10}$, $27\frac{4}{10}$, and $9\frac{3}{10}$. Now, have students dictate fractions for other students to write on the board.

Name ______________________

Decimals

CHAPTER 15

Lesson 15-1

Tenths

What decimal represents the number of panels Rita has painted so far?

We want to know how many panels Rita painted so far.

Each panel has 10 equal parts.

Rita has painted 2 complete panels and 6 parts of another panel.

To write the number of panels painted as a decimal, we rename the mixed number $2\frac{6}{10}$.

We write $2\frac{6}{10}$ as the decimal **2.6**.

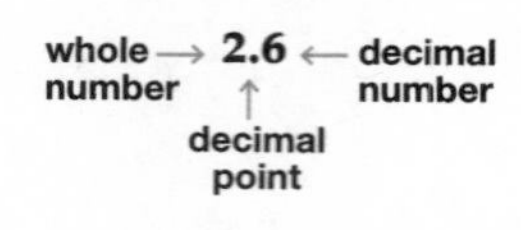

ones	tenths
2 .	6

decimal point

We say **two and six tenths**.

Rita has painted 2.6 panels so far.

Together the whole number and the decimal number are called a decimal.

2.6, 0.4, and 4.0 are examples of decimals.

Getting Started

Write the decimal for the green part.

1. 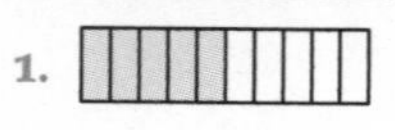

0.5

2. 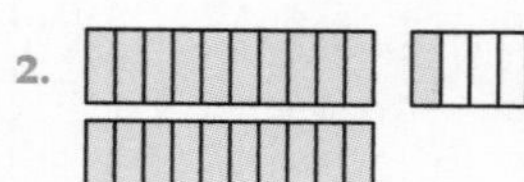 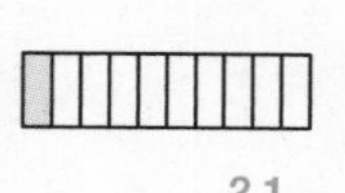

2.1

2 Teach

Introduce the Lesson Have a student read the problem and tell what is to be found. (the decimal that represents the number of panels Rita painted so far) Ask what information is given in the picture. (Two panels have 10 parts painted and 1 panel has 6 of 10 parts painted.) Have students complete the information sentences.

- Read the rules and definitions in the model with them. Read and discuss the sections about decimals that name parts less than 1 and writing whole numbers as decimal numbers.

Develop Skills and Concepts Write $1\frac{2}{10}$ on the board. Have a student read the mixed number. Tell students that any fraction that has 10 as the denominator can be written as a **decimal**.

- Write **1.2** on the board and tell students that $1\frac{2}{10}$ can be written as 1, a decimal point, and then the numerator of the fraction.
- Write $4\frac{5}{10}$ and **4.5** on the board. Tell students that the decimal point separates the whole number from the fraction. Explain that we say the word *and* for the decimal point when reading a decimal.
- Write $\frac{1}{10}$ and **0.1** on the board. Tell students that a decimal number that is less than 1 whole is written with a zero as a placeholder in the whole-number place.
- Write **4.0** on the board. Tell students that a whole number can be written as a decimal with zero as a placeholder in the tenths place.
- Write the following on the board: **three and seven tenths**, **3.7**, and $3\frac{7}{10}$. Dictate decimals for students to write the number words, the decimal number, and the fraction.

3 Practice

Read the directions for each exercise section with students and have them complete page 298 independently.

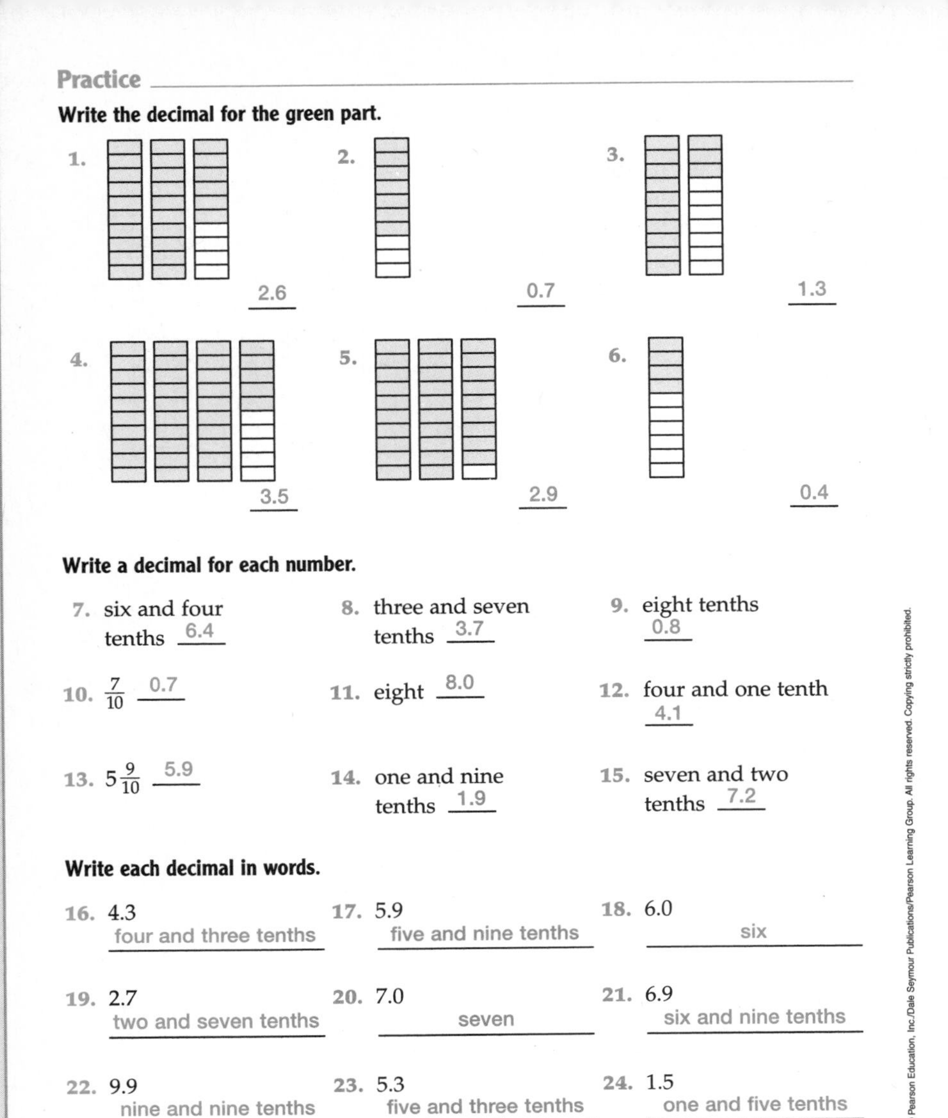

Practice

Write the decimal for the green part.

1. 2.6
2. 0.7
3. 1.3
4. 3.5
5. 2.9
6. 0.4

Write a decimal for each number.

7. six and four tenths 6.4
8. three and seven tenths 3.7
9. eight tenths 0.8
10. $\frac{7}{10}$ 0.7
11. eight 8.0
12. four and one tenth 4.1
13. $5\frac{9}{10}$ 5.9
14. one and nine tenths 1.9
15. seven and two tenths 7.2

Write each decimal in words.

16. 4.3 four and three tenths
17. 5.9 five and nine tenths
18. 6.0 six
19. 2.7 two and seven tenths
20. 7.0 seven
21. 6.9 six and nine tenths
22. 9.9 nine and nine tenths
23. 5.3 five and three tenths
24. 1.5 one and five tenths

4 Assess

Ask students to write $116\frac{9}{10}$ as a decimal. (116.9)

For Mixed Abilities

Common Errors • Intervention

Some students may forget that the decimal point separates the whole number from tenths and write six and seven tenths as 0.67. Have these students work with partners, taking turns writing the decimal for each of the following on a place-value chart:

six tenths	(0.6)
four and six tenths	(4.6)
three tenths	(0.3)
five and three tenths	(5.3)
nine tenths	(0.9)
one and nine tenths	(1.9)

After writing each decimal number, partners should discuss whether the notation is correct and explain.

Enrichment • Decimals

Tell students to draw a picture to show the following decimals: 0.4, 2.9, 4.3, and 6.0.

More to Explore • Probability

One way of expressing the probability of two events occurring is to create an ordered pair, symbols that express the different ways in which some event could take place. Give students the example of throwing a 6 on one number cube and then drawing a heart from a deck of cards.

Have them use the symbols 1, 2, 3, 4, 5, and 6 as the first number in the ordered pair, representing all the possible outcomes of the throw of one number cube. Ask them to use the letters *H*, *S*, *D*, and *C* (heart, spade, diamond, and club) for the four ways of drawing one suit from a pack containing four suits. Ask one student to write the ordered pair describing a throw of 3 and a draw of a diamond. (3, D) Have students list all the ordered pairs for throwing a number cube and drawing a card. [(1, H), (1, S), (1, D), (1, C), (2, H), (2, S), (2, D), (2, C), (3, H), (3, S), (3, D), (3, C), (4, H), (4, S), (4, D), (4, C), (5, H), (5, S), (5, D), (5, C), (6, H), (6, S), (6, D), (6, C)]

15-2 Hundredths

pages 299–300

1 Getting Started

Objective

- To understand and write hundredths

Vocabulary

hundredths

Warm Up • Mental Math

Ask students if all the factors of 8 are 1, 2, 4, and 8, what are all the factors of the following:

1. 6 (1, 2, 3, 6)
2. 3 (1, 3)
3. 4 (1, 2, 4)
4. 5 (1, 5)
5. 7 (1, 7)
6. 9 (1, 3, 9)
7. 10 (1, 2, 5, 10)
8. 12 (1, 2, 3, 4, 6, 12)

Warm Up • Numeration

Review decimals in tenths by having students write each of the following in words and as a decimal: $2\frac{5}{10}$, $\frac{9}{10}$, $14\frac{1}{10}$, $5\frac{3}{10}$, $6\frac{2}{10}$, $7\frac{8}{10}$, and $261\frac{4}{10}$.

Name ______________________

Hundredths

Dick is making a mosaic tile picture to enter in the art contest. What decimal represents the part of the picture that Dick has tiled so far?

We want to know what decimal represents the part of the picture that is tiled so far.

The mosaic will be made up of 100 tiles.

Dick has used 46 of the tiles so far.

To find the decimal that represents the part of the picture that is tiled, we rename the fraction $\frac{46}{100}$.

We write $\frac{46}{100}$ as the decimal **0.46**.

whole number → **0.46** ← decimal number
↑ decimal point

ones	tenths	hundredths
0 .	4	6

↑ decimal point

We say **forty-six hundredths**.

Dick has finished 0.46 of the mosaic tile picture so far.

Getting Started

Write the decimal for each green part.

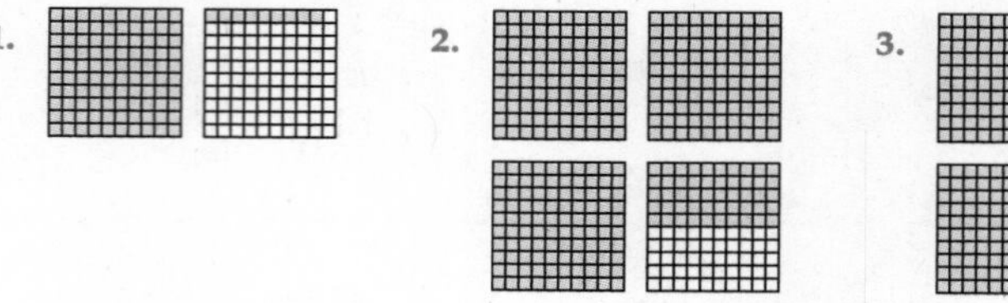

1. 1.09
2. 3.50
3. 2.85

Write a decimal for each number.

4. six and twelve hundredths 6.12
5. $3\frac{36}{100}$ 3.36

Write the decimal in words.

6. 0.08 eight hundredths

2 Teach

Introduce the Lesson Have a student read the problem aloud and tell what is to be found. (the decimal that represents how much of the picture is tiled) Ask students what information is given in the picture. (There are 10 rows of 10 squares or 100 squares in all. Forty-six squares are tiled.)

- Have students read with you as they complete the sentences. Help them put the number correctly in a place-value chart and complete the solution sentence.

Develop Skills and Concepts Write $\frac{42}{100}$ on the board. Tell students that just as we can write $\frac{4}{10}$ as 0.4, we can also write any fraction that has a denominator of 100 as a decimal.

- Write **0.42** on the board and tell students that we read this decimal just like the fraction of $\frac{42}{100}$. Write **forty-two hundredths** on the board and have a student read it.

Write the following on the board:

Ones	.	Tenths	Hundredths
0	.	4	2
(1)	.	(3)	(9)
(1)	.	(5)	(0)
(10)	.	(0)	(1)

- Dictate the following decimals for students to write as a fraction and in number words: 1.39, 1.50, and 10.01.

3 Practice

Remind students that a zero must be placed in the tenths place if there are 9 or fewer hundredths. Have students complete page 300 independently.

Practice

Write the decimal for each green part.

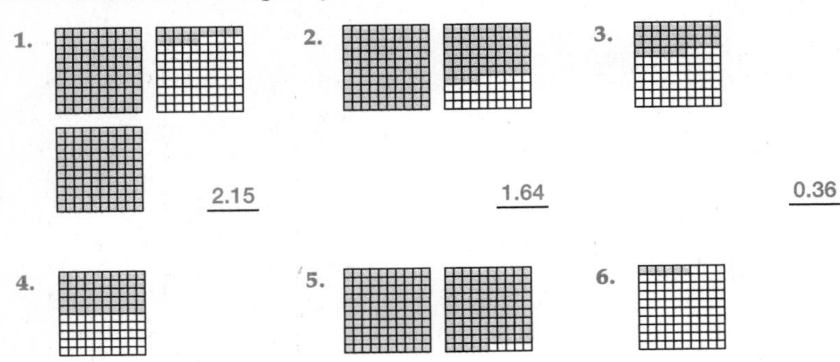

1. 2.15
2. 1.64
3. 0.36
4. 0.50
5. 1.95
6. 0.06

Write a decimal for each number.

7. four and twenty-five hundredths 4.25
8. six and nine hundredths 6.09
9. $14\frac{12}{100}$ 14.12
10. five hundredths 0.05
11. seven and thirty-seven hundredths 7.37
12. seventy-eight hundredths 0.78

Write each decimal in words.

13. 3.29 three and twenty-nine hundredths
14. 6.01 six and one hundredths
15. 3.09 three and nine hundredths
16. 7.50 seven and fifty hundredths
17. 0.16 sixteen hundredths
18. 0.03 three hundredths

4 Assess

Ask students to write $244\frac{5}{100}$ as a decimal. (244.05)

For Mixed Abilities

Common Errors • Intervention

Some students may omit zero as a placeholder and write a number like four and six hundredths as 4.6. Have them work in pairs to take turns and write the decimal numbers for the following on a place-value chart:

seven tenths	(0.7)
seven hundredths	(0.07)
four and seven tenths	(4.7)
four and seven hundredths	(4.07)
four tenths	(0.4)
four hundredths	(0.04)
nine and four tenths	(9.4)
nine and four hundredths	(9.04)

Then, partners should discuss whether the notation is correct and explain.

Enrichment • Numeration

Draw a picture to explain why \$1.35 is the same as 1.35 dollars and $1\frac{35}{100}$ dollars.

More to Explore • Measurement

Ask students to bring in a variety of irregularly shaped containers, such as jars, bottles, vases, and so on. Have them estimate and then measure to determine which container has the greatest volume. Next, have them measure the capacity of each container using a graduate marked off in milliliters. Have students arrange the containers by capacity, from least to greatest. Discuss with them how the shape of a container can distort its capacity.

ESL/ELL STRATEGIES

When introducing simple decimals, write some examples on the board and have ESL students practice reading them aloud. For example, 2.6 is read "two and six tenths."

15-3 Thousandths

pages 301–302

1 Getting Started

Objective
- To understand and write thousandths

Vocabulary
thousandths

Warm Up • Mental Math
Tell students to find the average.

1. $2\frac{1}{2}$ and $3\frac{1}{2}$ (3)
2. $4\frac{1}{4}$, $7\frac{1}{4}$, and $\frac{2}{4}$ (4)
3. 6, 9, and 3 (6)
4. $4\frac{2}{3}$ and $15\frac{1}{3}$ (10)
5. 200, 300, and 850 (450)
6. $\frac{7}{12}$ and $1\frac{5}{12}$ (1)
7. $3\frac{1}{4}$ and $\frac{3}{4}$ (2)
8. 68 and 100 (84)

Warm Up • Decimals
Review decimals in tenths and hundredths by having students write the following decimals on a place-value chart showing ones, tenths, and hundredths: 1.4, 2.66, 0.38, 8.2, 5.7, and 0.16.

Name ______________________ Lesson 15-3

Thousandths

Mr. Tecumseh buys supplies for the school cafeteria. He is purchasing a quantity of meat. How much meat is he buying?

We want to understand the reading on the scale.

The scale reads 16.258 pounds.

We can use a place-value chart to help understand the weight of the meat.

tens	ones	tenths	hundredths	thousandths
1	6 .	2	5	8

decimal point

We say **sixteen and two hundred fifty-eight thousandths.**

Mr. Tecumseh bought 16.258 pounds of meat.

Getting Started

Write the place value of each green digit.

1. 76.259 hundredths
2. 116.326 tenths
3. 4.405 ones

Write the place value of the 4 in each number.

4. 390.421 tenths
5. 26.045 hundredths
6. 5.324 thousandths

Write the decimal for each number.

7. two and sixteenth thousandths 2.016
8. seven thousandths 0.007

Write each decimal in words.

9. 29.752 twenty-nine and seven hundred fifty-two thousandths
10. 5.308 five and three hundred eight thousandths

---

2 Teach

Introduce the Lesson Have a student read the problem and tell what is to be found. (the amount of meat being bought)

- Have students complete the sentences. Read the place-value chart together. Then, have them read the number words and complete the solution sentence.

Develop Skills and Concepts Remind students that we write $\frac{48}{100}$ as **0.48** and $\frac{4}{10}$ as **0.4** as you write the fractions and decimals on the board.

- Write $\frac{238}{1000}$ on the board and tell students we write thousandths as decimals also. Write **0.238** on the board and tell students that we read this decimal just like the fraction of $\frac{238}{1000}$. Write **two hundred thirty-eight thousandths** on the board and have a student read it.
- Now, write **0.860** on the board and tell students that the 8 is in the tenths place, the 6 is in the hundredths place, and the zero is in the thousandths place. Have a student read the number.
- Write **6.016** on the board and have students tell the place value of each number. (6 ones, 0 tenths, 1 hundredth, 6 thousandths)
- Repeat for the following decimal numbers: 12.001, 8.206, 46.200, 0.103, and 4.555. Help students write and read each decimal number in number words.

3 Practice

Remind students to be careful about spelling. Have them complete page 302 independently

Practice

Write the place value of each green digit.

1. 13.26 tenths
2. 67.154 thousandths
3. 7.041 tenths
4. 216.107 hundredths
5. 2.478 hundredths
6. 23.840 thousandths

Write the place value of the 6 in each number.

7. 21.684 tenths
8. 16.273 ones
9. 143.86 hundredths
10. 4.156 thousandths
11. 29.465 hundredths
12. 3.006 thousandths

Write the decimal for each number.

13. five and fifty-seven thousandths 5.057
14. twelve and one hundred thirty-nine thousandths 12.139
15. two hundred and four hundred eight thousandths 200.408
16. eighty-one and seven thousandths 81.007
17. fifty-three and seventy-five thousandths 53.075

Write each decimal in words.

18. 16.25 sixteen and twenty-five hundredths
19. 3.246 three and two hundred forty-six thousandths
20. 7.915 seven and nine hundred fifteen thousandths
21. 12.021 twelve and twenty-one thousandths
22. 14.6 fourteen and six tenths
23. 29.003 twenty-nine and three thousandths
24. 114.218 one hundred fourteen and two hundred eighteen thousandths
25. 16.255 sixteen and two hundred fifty-five thousandths
26. 39.608 thirty-nine and six hundred eight thousandths

4 Assess

Ask students the place value of the 2 in 99.642. (thousandths)

5 Mixed Review

1. $\frac{2}{3} + \frac{5}{8}$ ($1\frac{7}{24}$)
2. 957 × 24 (22,968)
3. Simplify $\frac{28}{36}$. ($\frac{7}{9}$)
4. 1,300 ÷ 48 (27 R4)
5. $\frac{7}{9} - \frac{1}{6}$ ($\frac{11}{18}$)
6. 295 ÷ 7 (42 R1)
7. $\frac{7}{8}$ of 40 (35)
8. $3\frac{3}{5} + 2\frac{3}{4}$ ($6\frac{7}{20}$)
9. $250.00 − $176.32 ($73.68)
10. 65,176 + 84 + 2,174 (67,434)

For Mixed Abilities

Common Errors • Intervention

Some students may have trouble writing decimal numbers correctly. Have them work in pairs to take turns writing on a place-value chart the following numbers as you dictate them:

- one hundred eighty-two and one thousandth (182.001)
- forty-five and five hundred one thousandth (45.501)
- three hundred seventy-eight thousandths (0.378)
- ninety-seven and eighty thousandths (97.080)
- five and seven hundred sixty-seven thousandths (5.767)

After a student writes each decimal number, the partners should discuss whether the decimal notation is correct.

Enrichment • Measurement

Have students make a chart that shows why 1 milliliter is 0.001 of a meter and 1 decimeter is 0.1 of a meter.

More to Explore • Logic

Read or duplicate the following passage for students to solve:

A girl saw two boys in the park who looked exactly alike. She asked, "Are you boys twins?"

"No, we are not twins," one boy replied.

"That's right," said the other. "We have the same parents, we were born on the same day of the same year, but we're not twins."

How can this be so? (The boys are two members of a set of triplets.)

ESL/ELL STRATEGIES

Provide students with intensive practice in reading complex decimal numbers out loud. Point out that we use the word *and* after the whole number—thirty-six and two hundred fifty-eight thousandths. Also, practice pronouncing the *ths* sound at the end of the words *tenths*, *hundredths*, and *thousandths*.

15-4 Comparing and Ordering Decimals

pages 303–304

1 Getting Started

Objective

- To compare and order decimal numbers through hundredths

Warm Up • Mental Math

Have students name the amount of money.

1. 1.02 ($1.02)
2. 15.10 ($15.10)
3. 0.76 (76¢)
4. 4.10 ($4.10)
5. 19.36 ($19.36)
6. 7.70 ($7.70)
7. 3.90 ($3.90)
8. 0.15 (15¢)

Warm Up • Place Value

Review place value by writing numbers such as **405.06** on the board. Ask students to name the value of each digit. Now, write **405 ○ 406** on the board and have a student write >, <, or = in the circle to show the relationship of the numbers. (<) Repeat for more whole number comparisons.

Name ______________________

Comparing and Ordering Decimals

Three friends competed in the Special Olympics. One of their best events was the 100-meter dash. In what order did they finish the race?

100-meter dash
Bert 42 13.6
Dennis 12 13.65
Daryle 8 13.57

We want to know in what order the boys finished the race.

To find their order, we can arrange 13.6, 13.65, and 13.57 on a number line.

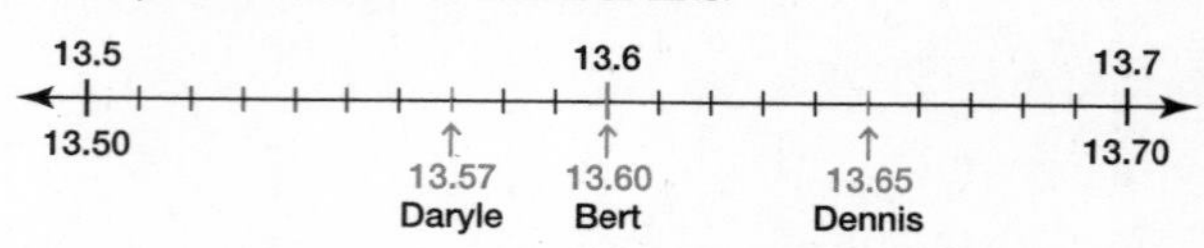

Zeros to the far right of a decimal number do not change its value. So, **13.60 = 13.6**.

We can also compare these times one place value at a time.

13.60 13.65 13.57	13.60 13.65 13.57	13.60 13.65 13.57	13.60 13.65	13.65
Same number of tens	Same number of ones	13.57 has fewer tenths; 13.57 is the least.	13.60 has fewer hundredths. 13.60 < 13.65	13.65 is the greatest.

Daryle finished the race first. Bert finished second, and Dennis finished third.

Getting Started

Write >, <, or = in each circle.

1. 9.35 (>) 9.27
2. 3.09 (<) 3.90
3. 7.5 (=) 7.50
4. 0.3 (>) 0.2
5. 4.24 (<) 4.28
6. 8.6 (<) 8.66

Write the numbers in order from least to greatest.

7. 2.3, 1.67, 1.95 — 1.67, 1.95, 2.3
8. 7.1, 6.9, 7 — 6.9, 7, 7.1
9. 3.5, 3.49, 3.2 — 3.2, 3.49, 3.5

2 Teach

Introduce the Lesson Have a student read the problem aloud and tell what is to be found. (who finished first, second, and third in the race) Talk with students about the Special Olympics for disabled persons. Ask what information is given about the boys' times in the 100-meter dash. (Bert ran in 13.6 seconds, Dennis ran in 13.65 seconds, and Daryle ran in 13.57 seconds.)

- Have students complete the sentences. Guide them through the comparisons in the model. Have students write each decimal as a mixed number to check their solution.

Develop Skills and Concepts Write **2.6** and **2.60** on the board and tell students that the values of these numbers are the same since zeros to the far right of a decimal number do not change its value.

- Write **8.1** and **8.12** on the board and tell students that we want to know which number is larger. Tell students that we must write 8.1 in hundredths in order to compare its value to 8.12. Have a student write 8.1 in hundredths. (8.10) Ask which is larger. (8.12) Ask why. (Both numbers have 8 ones and 1 tenth, but 8.12 has 2 hundredths, whereas 8.10 has no hundredths.)
- Write **8.1 ○ 8.12** on the board and have a student write the symbol to compare the numbers. (<)
- Repeat for comparing 6.72 and 6.6, and 3.02 and 3.1. Now, have students compare 1.26, 1.1, and 1.2.

3 Practice

Remind students that to compare decimals, the decimal numbers must all be in the same form, tenths or hundredths. Have students complete the exercises on page 304 independently.

Practice

Write >, <, or = in each circle.

1. 5.64 (<) 5.78
2. 3.21 (<) 3.30
3. 5.71 (>) 5.17
4. 9.03 (<) 9.30
5. 9.50 (=) 9.5
6. 2.39 (>) 2.35
7. 8.6 (<) 8.68
8. 9.23 (<) 9.32
9. 0.6 (=) 0.60
10. 4.75 (>) 4.6
11. 8.25 (<) 8.3
12. 0.16 (<) 0.17

Write the numbers in order from least to greatest.

13. 4.6, 3.5, 3.9 — 3.5, 3.9, 4.6
14. 6.26, 6.38, 6.16 — 6.16, 6.26, 6.38
15. 8.15, 8.1, 8 — 8, 8.1, 8.15
16. 7.1, 7.15, 7.09 — 7.09, 7.1, 7.15
17. 4.36, 4.16, 5.03 — 4.16, 4.36, 5.03
18. 9.2, 9.14, 9.27 — 9.14, 9.2, 9.27
19. 18.21, 18.12, 18.22 — 18.12, 18.21, 18.22
20. 37.08, 38.7, 37.8 — 37.08, 37.8, 38.7

Now Try This!

Complete the magic square and then make one of your own.

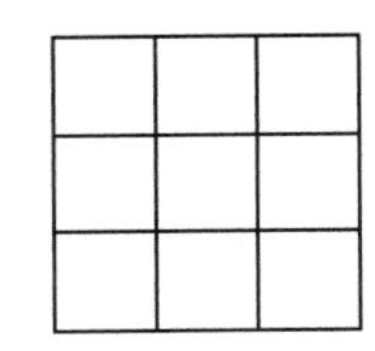

16	2	12
6	10	14
8	18	4

Answers will vary.

The sum of each row, column, and diagonal in the first square is 30.

Now Try This! Tell students to name the numbers in the squares that are on the opposite sides of the center square. (6, 14; 2, 18) The sums of these two pairs are the same. (20) Tell students to use this model to complete a square with 5 in the center and a sum of 15.

4 Assess

Have students put the following numbers in order from least to greatest: 10, 12.8, and 12.08. (10, 12.08, 12.8)

For Mixed Abilities

Common Errors • Intervention

Some students may have difficulty comparing numbers when they are shown side-by-side. Have them rewrite the numbers, one above the other, aligning the decimal points. Then, they can compare the numbers one place at a time, from left to right, until they reach the place where the digits differ, at which point they know that one number is greater than the other. Remind students that they can annex zeros for missing digits to the right of the decimal point, as when comparing 0.5 and 0.58 (0.50 < 0.58).

Enrichment • Decimals

Have students write ten decimal numbers in tenths or hundredths. Working in pairs, have one partner place the numbers in order from least to greatest. Have the other partner check the work. Have them change roles and repeat the activity.

More to Explore • Logic

Tell students that memory tricks, called mnemonics, can help them remember important numbers such as their telephone number. Share these examples:

741776 might be remembered as July (7th month) 4, 1776.

Now, have students develop ways to remember the following sequence of numbers. Remember that the answers given are only one mnemonic technique. Have students share their different solutions.

1. 725-7336 (7 − 2 = 5 + 7 = 12 × 3 = 36)
2. 164366 (The square root of 16 equals 4. The square root of 36 is 6.)
3. 75353105 (7 × 5 = 35 × 3 = 105)
4. 392411 (3 × 3 = 9, 2 × 2 = 4, 1 × 1 = 1)
5. student's phone number
6. student's ZIP Code

15-5 Rounding Decimals

pages 305–306

1 Getting Started

Objective

- To round decimals to the nearest tenth, hundredth, or whole number

Warm Up • Mental Math

Ask students how many there are.

1. boxes of 6 to total 36 (6)
2. seats in each of 9 rows to total 99 (11)
3. 1-in. cookies in two 8 × 8-in. pan (128)
4. CDs in 8 boxes of 10 each (80)
5. 6-oz cups in 72 oz (12)
6. decades in $5\frac{1}{2}$ centuries (55)
7. quarter inches on a 12-in. ruler (48)

Warm Up • Paper and Pencil

Ask students to round the number to the nearest ten, hundred, and thousand.

1. 508 (510; 500; 1,000)
2. 6,172 (6,170; 6,200; 6,000)
3. 798 (800; 800; 1,000)
4. 823 (820; 800; 1,000)
5. 2,003 (2,000; 2,000; 2,000)

Name ______________________

Rounding Decimals

Measurements are never exact. They are only approximate numbers. Rounding decimal numbers, such as weight measurements, makes them easier to add or subtract.

Round the weight of each purchase to the nearest tenth.

We want to round the decimal numbers 15.74 and 15.854 to the nearest tenth.

To round a decimal, look at the digit to the right of the place you want to round to.

If the digit to the right is 0, 1, 2, 3, or 4, the digit we are rounding to stays the same and all the digits to the right are dropped.

15.74

The digit to the right of the tenths place is 4. So, the digit we are rounding stays the same.

If the digit to the right is 5, 6, 7, 8, or 9, add 1 to the digit you are rounding and drop all digits to the right.

15.854

The digit to the right of the tenths place is 5. So, we add 1 to the digit we are rounding.

Rounded to the nearest tenth, Ling bought 15.7 pounds of potatoes and Sandy bought 15.9 pounds of squash.

Getting Started

Round each decimal to the nearest tenth, hundredth, and whole number.

1. 5.838
   tenth 5.8
   hundredth 5.84
   whole number 6
2. 3.651
   tenth 3.7
   hundredth 3.65
   whole number 4
3. 8.147
   tenth 8.1
   hundredth 8.15
   whole number 8

2 Teach

Introduce the Lesson Read the problem aloud and ask a student to read the information sentences. Have a volunteer tell what students are being asked. (Round 15.74 lb and 15.854 lb to the nearest tenth.)

- Work through the model with students, stopping to answer any questions.
- Read the solution sentence aloud as students complete it. (15.7; 15.9)

Develop Skills and Concepts Tell students that decimals can be rounded to any place value: tenths, hundredths, thousandths, and so on. Ask, *What place must we look at to round a number to the nearest hundredth?* (thousandths)

- Explain that if the number to the right of the hundredths is less than 5, they will drop all the digits to the right to round to the nearest hundredth. If the number to the right is 5 or more, they will increase the digit in the hundredths position by one and drop all digits to the right.
- Have a student round 42.634 to the nearest hundreth. (42.63) Ask another student to round 45.827 to the nearest hundreth. (45.83)
- Put the following chart on the board and have students complete it:

Number	Nearest Whole Number	Nearest Tenth	Nearest Hundredth
7.225	(7)	(7.2)	(7.23)
10.989	(11)	(11)	(10.99)
29.651	(30)	(29.7)	(29.65)

Practice

Round each decimal to the nearest tenth.

1. 4.382 4.4
2. 17.32 17.3
3. 36.29 36.3
4. 0.367 0.4
5. 8.88 8.9
6. 0.98 1.0
7. 9.725 9.7
8. 82.42 82.4
9. 105.48 105.5

Round each decimal to the nearest hundredth.

10. 2.648 2.65
11. 3.928 3.93
12. 14.083 14.08
13. 31.115 31.12
14. 0.082 0.08
15. 312.253 312.25
16. 6.481 6.48
17. 4.289 4.29
18. 78.190 78.19
19. 74.998 75.0
20. 5.042 5.04
21. 22.319 22.32

Round each decimal to the nearest whole number.

22. 16.25 16
23. 0.254 0
24. 18.4 18
25. 0.863 1
26. 135.43 135
27. 59.92 60

Problem Solving

Solve each problem.

28. Patrice says that 5.69 rounds to 6. Miguel says that it rounds to 5.7. Who is correct: Patrice, Miguel, both, or neither?

    both

29. Kitty says her race time rounded to the nearest second was 11 seconds. Her actual time was given in hundredths. What is the slowest time Kitty could have run?

    11.49 seconds

3 Practice

Have students complete the exercises on the page independently.

4 Assess

Ask students which place they need to look at in order to round to the indicated place.

1. nearest tenth (hundredth)
2. nearest hundredth (thousandth)
3. nearest whole number (tenth)

For Mixed Abilities

Common Errors • Intervention

Some students may round decimals incorrectly because they forget to drop the digits to the right of the place to which they are rounding. First, have them draw an arrow over the digit in the place to which they are rounding. Once they have decided whether to keep that digit the same or make it one greater, have them drop all of the digits to the right of the digit under the arrow.

Enrichment • Numeration

Have students devise an arithmetic progression using Roman numerals. Use the following as an example: V, XV, XX, LX, LXV, and so on. (× 3, + 5) Tell them to exchange progressions with a partner and then have the partner provide the next five Roman numerals and identify the common difference that describes the progression. Have students repeat the activity.

More to Explore • Probability

Have students draw a tree diagram for a three-stage event. Ask them to illustrate the combinations of possible outcomes when a coin is tossed three times. If necessary, help them by asking what the possibilities are for the first toss. (heads, tails) Have them start the tree diagram with those labels. Ask what the possibilities are for the second toss (heads, tails) and for the third toss (heads, tails). Ask them to complete a tree diagram.

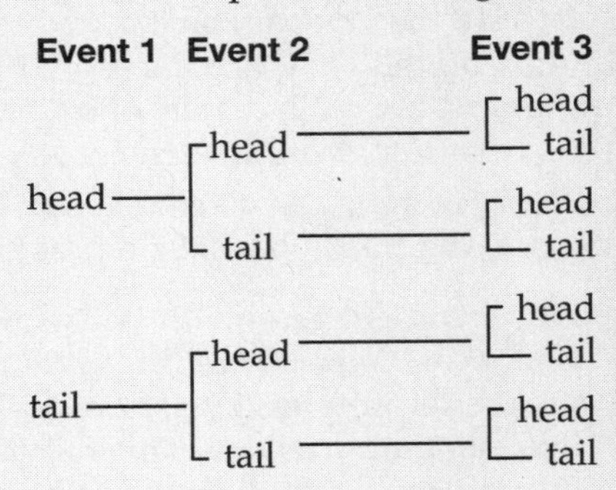

Ask how many possible outcomes there are for this three-flip experiment. (8)

15-6 Adding Decimals

pages 307–308

1 Getting Started

Objective

- To add decimals through thousandths

Warm Up • Mental Math

Have students name the number.

1. $\frac{1}{2}$ less than 1 ($\frac{1}{2}$)
2. $\frac{2}{6}$ of 12 (4)
3. $\frac{7}{8}$ less than $\frac{15}{8}$ (1)
4. $\frac{1}{15}$ more than $\frac{1}{30}$ ($\frac{1}{10}$)
5. $\frac{1}{4} + \frac{1}{3}$ ($\frac{7}{12}$)
6. $\frac{1}{2} + \frac{1}{4} + \frac{1}{4}$ (1)
7. $\frac{2}{5}$ of 400 (160)
8. $3\frac{1}{10}$ more than $6\frac{1}{10}$ ($9\frac{1}{5}$)

Warm Up • Addition

Review addition of whole numbers and money amounts by having students write and solve the following dictated exercises at the board: 38 + 27 (65), 235 + 579 (814), 45 + 83 + 72 (200), \$56.12 + \$25.49 (\$81.61), and \$18.02 + \$46.58 + \$7.89 (\$72.49).

Name ______________________

Adding Decimals

Iron is an important mineral in each person's diet. It is recommended that young people, 11 to 14 years old, have at least 18 milligrams of iron daily. How much iron is found in a meal of sirloin steak and spinach only?

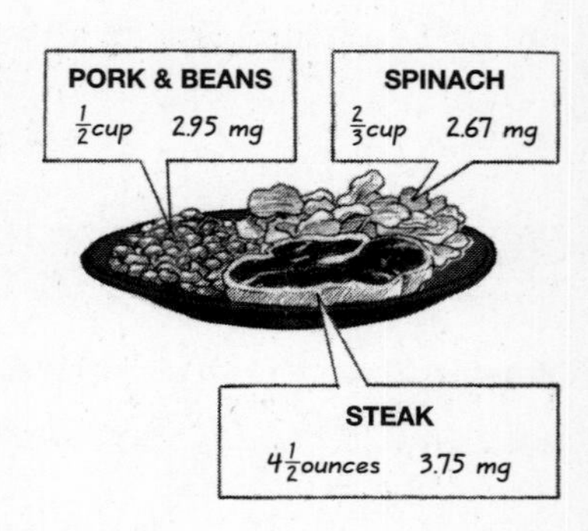

We want to find the amount of iron in a meal of steak and spinach only.

The steak has 3.75 milligrams of iron.

The spinach has 2.67 milligrams of iron.

To find the total milligrams of iron, we add 3.75 and 2.67.

Line up the decimal points and place values. Add the hundredths. Regroup if needed.	Add the tenths. Regroup if needed. Write the decimal point.	Add the ones.
3.75 + 2.67 = 2 (regroup 1)	3.75 + 2.67 = .42 (regroup 1 1)	3.75 + 2.67 = 6.42

The steak and the spinach contain 6.42 milligrams of iron.

REMEMBER When adding decimals, it is important to align the decimal points and place values.

7.4 + 8.62 should be written the following way:

$$\begin{array}{r} 7.4 \\ +\ 8.62 \\ \hline 16.02 \end{array}$$

Getting Started

Add.

1. 4.9 + 2.6 = 7.5
2. 5.83 + 2.16 = 7.99
3. 17.59 + 8.6 = 26.19
4. 37.25 + 18.77 = 56.02

Copy and add.

5. 39.2 + 18.5 (57.7)
6. 47.3 + 21.19 (68.49)
7. 13.6 + 92.5 + 53.8 (159.9)

2 Teach

Introduce the Lesson Have a student read the problem aloud and tell what is to be solved. (the number of milligrams of iron in a meal of sirloin steak and spinach) Ask what information is given. ($4\frac{1}{2}$ oz of steak has 3.75 mg of iron, $\frac{2}{3}$ c of spinach has 2.67 mg of iron, $\frac{1}{2}$ cup of pork and beans has 2.95 mg of iron.)

- Have students read with you as they complete the sentences. Guide them through the addition steps in the model, emphasizing correct placement of the decimal point. Tell students to write the decimals as mixed numbers and then add to check their work.

Develop Skills and Concepts Write **2.8 + 3.67** on the board and tell students that to add decimals, we write the numbers vertically so that the decimal points are aligned and tenths are under tenths and so on. Remind students that we add money that way also.

- Now, write **2.8 + 3.67** vertically on the board and tell students that there are only 7 hundredths, so 7 is recorded in the hundredths place. Tell students that 8 tenths plus 6 tenths equals 14 tenths, or 1 whole and 4 tenths, so 4 is recorded in the tenths place. Tell students that the decimal point is placed in line under those in the addends, and the whole numbers plus the 1 that was regrouped are added. Have students read the answer. (6.47)
- Have students work the following exercises for more practice: 5.67 + 7.2 (12.87), 9.65 + 24.58 (34.23), 6.35 + 1.261 + 4.376 (11.987), and 1.203 + 2.02 + 4.5 (7.723).

3 Practice

Remind students to regroup when necessary. Have students complete page 308 independently.

Practice

Add.

1. 5.3 + 2.4 = 7.7	2. 9.2 + 3.5 = 12.7	3. 7.72 + 6.13 = 13.85	4. 3.64 + 8.3 = 11.94
5. 7.8 + 3.9 = 11.7	6. 4.86 + 2.71 = 7.57	7. 39.5 + 8.64 = 48.14	8. 17.763 + 26.389 = 44.152
9. 7.8, 13.2, + 15.7 = 36.7	10. 26.4, 19.7, + 38.6 = 84.7	11. 9.87, 11.586, + 27.197 = 48.653	12. 52.174, 48.8, + 19.78 = 120.754

Copy and add.

13. 39.7 + 18.58
58.28

14. 72.69 + 28.36
101.05

15. 78.15 + 87.85
166

16. 38.09 + 27.284
65.374

17. 46.9 + 52.48
99.38

18. 37.75 + 39.28
77.03

19. 58.164 + 9.283
67.447

20. 15.96 + 83.48
99.44

21. 10.36 + 15.4 + 12.75
38.51

22. 39.75 + 48.16 + 58.03
145.94

23. 87.51 + 93.78 + 48.62
229.91

Problem Solving

Solve each problem.

24. Lee swam the first lap of the butterfly-stroke race in 25.48 seconds. She swam the second lap in 27.59 seconds. What was Lee's total time?
53.07 seconds

25. A container holds 3.25 liters of liquid. Another container holds 4.65 liters. How much liquid can both containers hold when they are filled?
7.9 liters

4 Assess

Ask students what they must remember to do with the decimal points and place values when adding decimals.
(make sure they are aligned)

For Mixed Abilities

Common Errors • Intervention

When students are adding decimal numbers, they may add the decimal parts to the right of the decimal point and the whole numbers to the left of the decimal point separately.

Incorrect	Correct
7.8	7.8
+ 4.5	+ 4.5
11.13	12.3

Have students work the exercises on a place-value chart, renaming from right to left, just as they do when adding whole numbers.

Enrichment • Numeration

Tell students to find the average of 3.26, 5.686, and 3.054. (4)

More to Explore • Statistics

Bring in an example of a public opinion poll reported in the newspaper. Read the results of the poll and explain that there are companies, like the Harris Polls, that do nothing but ask people their opinions. When they have sampled enough people, they report on the results of the poll and this tells us what people think about some issue.

Ask the class to work in groups to write an opinion poll about a topic important in their school. Ask them to write a number of questions related to the topic they select. Give them time to conduct their survey at lunch or before school. Have groups report on the results of their survey. They should include in their report the frequency of each answer to the questions they asked.

15-7 Subtracting Decimals

pages 309–310

1 Getting Started

Objective

- To subtract decimals through thousandths

Warm Up • Mental Math

Ask students if 80 minutes is $1\frac{1}{3}$ hours, how many hours or what part of an hour is the following:

1. 4 minutes ($\frac{1}{15}$)
2. 30 minutes ($\frac{1}{2}$)
3. 90 minutes ($1\frac{1}{2}$)
4. 190 minutes ($3\frac{1}{6}$)
5. 40 minutes ($\frac{2}{3}$)
6. 15 minutes ($\frac{1}{4}$)
7. 45 minutes ($\frac{3}{4}$)
8. 127 minutes ($2\frac{7}{60}$)

Warm Up • Subtraction

Review subtraction of whole numbers and money amounts by having students solve the following exercises at the board: 68 − 35 (33), 75 − 48 (27), $36.71 − $17.95 ($18.76), 508 − 289 (219), and $6.25 − $1.97 ($4.28).

Name ______________________

Subtracting Decimals

A barometer is used to help forecast the weather. As the barometer rises, the weather becomes clear and dry. How much did the barometer change from noon to 9:00 P.M.?

Barometer Measurements	
Noon	76.28 cm
3:00 P.M.	75.38 cm
6:00 P.M.	74.75 cm
9:00 P.M.	74.57 cm

We want to know the change in the barometer between noon and 9 P.M.

At noon, the barometer read __76.28__ centimeters.

At 9:00 P.M., the barometer read __74.57__ centimeters.

To find the total change, we subtract __74.57__ from __76.28__.

Line up the decimal points and place values. Subtract the hundredths. Regroup if needed.	Subtract the tenths. Regroup if needed. Write the decimal point.	Subtract the ones.
76.28 − 74.57 = 1	76.28 − 74.57 = .71	76.28 − 74.57 = 1.71

The barometer dropped __1.71__ centimeters.

REMEMBER Align the decimal points and place values before you subtract.

Getting Started

Subtract.

1. 8.9 − 3.6 = 5.3
2. 14.25 − 11.21 = 3.04
3. 89.145 − 17.964 = 71.181
4. 57.04 − 29.17 = 27.87

Copy and subtract.

5. 82.7 − 39.9 = 42.8
6. 485.18 − 210.16 = 275.02
7. 280.36 − 198.75 = 81.61

---

2 Teach

Introduce the Lesson Have a student read the problem and tell what is to be found. (the barometer change from noon to 9:00 P.M.) Ask what information is needed from the picture. (The noon reading was 76.28 cm and the 9:00 reading was 74.57 cm.)

- Have students complete the sentences as they read aloud with you. Guide them through the subtraction steps in the model. Tell students that their answer plus the 9:00 reading should equal the noon reading.

Develop Skills and Concepts Tell students that the decimal points must be aligned when subtracting decimals just as when adding decimals.

- Write **12.078 − 4.269** vertically on the board and talk through the subtraction, regrouping, and decimal point placement as you work the problem. Have a student read the answer. (7.809)
- Dictate more exercises for students to write and solve at the board. Have students read their answers.

3 Practice

Remind students to align the decimal point in the difference with those in the minuend and subtrahend. Have students complete page 310 independently.

Practice

Subtract.

1. 4.7 − 3.2 = 1.5	2. 8.9 − 2.5 = 6.4	3. 7.84 − 6.31 = 1.53	4. 9.79 − 5.16 = 4.63
5. 41.56 − 14.81 = 26.75	6. 59.27 − 28.93 = 30.34	7. 38.28 − 26.57 = 11.71	8. 67.35 − 48.72 = 18.63
9. 74.117 − 29.585 = 44.532	10. 87.68 − 52.99 = 34.69	11. 38.046 − 19.787 = 18.259	12. 51.168 − 38.274 = 12.894

Copy and subtract.

13. 8.36 − 4.51 3.85	14. 12.97 − 7.83 5.14	15. 19.21 − 11.75 7.46	16. 27.36 − 18.58 8.78
17. 73.09 − 5.78 67.31	18. 35.59 − 14.96 20.63	19. 68.25 − 19.68 48.57	20. 40.65 − 29.76 10.89
21. 92.17 − 37.82 54.35	22. 80.034 − 27.389 52.645	23. 73.17 − 29.98 43.19	24. 53.706 − 19.851 33.855

Problem Solving

Solve each problem.

25. Jim's long jump measured 4.26 meters. Dave's was 5.03 meters long. How much longer was Dave's jump?
0.77 meters

26. A paint can held 5.36 liters of paint. The handyman poured out 4.75 liters. How much paint was left in the can?
0.61 liters

27. A pair of slacks costs $29.79. A shirt costs $7.95. How much will one pair of slacks and two shirts cost?
$45.69

28. How much change will be left if a suit costing $89.50 is paid for with $100 bill?
$10.50

4 Assess

Have students solve 165.834 − 77.945. (87.889)

For Mixed Abilities

Common Errors • Intervention

When copying decimals to subtract, some students may not align the decimal point correctly.

Incorrect	Correct
82.46 − 5.21 30.36	82.46 − 5.21 77.25

Have students work their exercises on a place-value chart, making sure to align the decimal points correctly.

Enrichment • Applications

Tell students to consult an almanac to find the improvement in Roger Hornsby's 1925 batting average over his 1923 batting average.

More to Explore • Statistics

Have students bring in an example of an opinion poll reported in a newspaper or a news magazine.

Have the class work together to design a short questionnaire that asks the same question as one of the national polls they have seen.

Divide the class into groups and ask them to sample different portions of the school population. Explain that professional pollsters call the questionnaire an instrument. When all the results are in, tally the responses to the questions on the board.

Have students show the results of their survey and those results of the national survey on a bar graph and compare them.

15-8 Adding and Subtracting Decimals

pages 311–312

1 Getting Started

Objective

- To add and subtract decimals through thousandths

Warm Up • Mental Math

Dictate the following:

1. $\frac{2}{10} + \frac{4}{10} + \frac{6}{10}$ $(1\frac{1}{5})$
2. $\frac{7}{15} + \frac{8}{15}$ (1)
3. $\frac{9}{2} + \frac{3}{2}$ (6)
4. 3.2 + 6.6 (9.8)
5. 2.10 + 3.36 (5.46)
6. $1\frac{4}{5} + \frac{2}{10}$ (2)
7. $3\frac{2}{10} + 15\frac{1}{10} + \frac{9}{10}$ $(19\frac{1}{5})$
8. $\frac{12}{11} - \frac{1}{11}$ (1)

Warm Up • Numeration

Review addition and subtraction of money by having students work the following exercises at the board:

1. \$48.01 − \$19.73 (\$28.28)
2. \$32.56 + \$15.95 (\$48.51)
3. \$23.50 − \$19.70 (\$3.80)
4. \$54.90 + \$26.19 (\$81.09)
5. \$46.00 − \$19.26 (\$26.74)

Name ______________________

Adding and Subtracting Decimals

Two animals known for their speed are the quarter horse and the greyhound. Both were timed over a distance of a quarter mile. How much faster is the quarter horse?

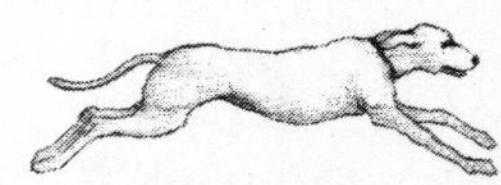

Quarter-Mile Speeds	
Quarter horse	47.5 mph
Greyhound	39.35 mph

We want to know how much faster the quarter horse ran the quarter mile.

The quarter horse ran <u>47.5</u> miles per hour.

The greyhound ran <u>39.35</u> miles per hour.

To compare the two speeds, we subtract <u>39.35</u> from <u>47.5</u>.

REMEMBER Writing a zero to the far right of a decimal number does not change its value.

$$\begin{array}{r} 47.5 \\ -\ 39.35 \end{array} \longrightarrow \begin{array}{r} 47.50 \\ -\ 39.35 \\ \hline 8.15 \end{array}$$

The quarter horse ran <u>8.15</u> miles per hour faster than the greyhound.

Getting Started

Add or subtract.

1. $\begin{array}{r} 9.7 \\ +\ 11.39 \\ \hline 21.09 \end{array}$
2. $\begin{array}{r} 27.56 \\ -\ 13.9 \\ \hline 13.66 \end{array}$
3. $\begin{array}{r} 54.275 \\ +\ 13.3 \\ \hline 67.575 \end{array}$
4. $\begin{array}{r} 96 \\ -\ 12.85 \\ \hline 83.15 \end{array}$

Copy and add or subtract.

5. 14.68 − 7.32 7.36
6. 9.63 + 2.7 + 3 15.33
7. (12 − 7.5) + 6.2 10.7

2 Teach

Introduce the Lesson Have a student read the problem aloud and tell what is to be solved. (how much faster the quarter horse is than the greyhound) Ask what information is given in the picture and the problem. (The quarter horse ran the quarter mile at a speed of 47.5 miles per hour and the greyhound ran at 39.35 miles per hour.)

- Have students read with you as they complete the sentences. Guide them through the placement of the zero in the model subtraction step to solve the problem.

Develop Skills and Concepts Remind students that adding a zero to the far right of the decimal point does not change its value.

- Write **67.1** on the board and ask a student to write the number in hundredths. (67.10) Have a student subtract 17.26 from 67.10 and tell the difference. (49.84)
- Have a student write **80.2 − 17.261** on the board and tell what must be done before the subtraction can be done. (add zeros to write 80.2 in thousandths as 80.200) Have a student talk through the subtraction and tell the difference. (62.939)
- Have a student write **40.21 + 36.9** on the board. Remind students that 36.9 has no hundredths so the total hundredths is 1. Have a student complete the exercise. (77.11)
- Have students work the following exercises at the board for additional practice: 25.4 − 7.781 (17.619), 29 − 4.6 (24.4), (15 − 2.9) + 6.19 (18.29), and (9.24 + 6.596) − 8 (7.836).

3 Practice

Remind students to add zeros to the right of the minuend if the subtrahend is extended to more places. Have students complete page 312 independently.

Practice

Add or subtract.

1. 5.6 + 7.15 = 12.75
2. 6.23 + 5.9 = 12.13
3. 9.27 − 3.1 = 6.17
4. 8.5 − 6.15 = 2.35
5. 7.24 + 11.6 = 18.84
6. 5.28 − 1.965 = 3.315
7. 8 + 3.5 = 11.5
8. 9.27 − 6.9 = 2.37
9. 17.21 − 9 = 8.21
10. 25.3 + 48.68 = 73.98
11. 76.467 + 80.984 = 157.451
12. 14 − 9.52 = 4.48

Copy and add or subtract.

13. 39 + 6.58 45.58
14. 27.39 − 19.8 7.59
15. 4.26 + 8.2 + 9.5 21.96
16. 17.9 − 11.764 6.136
17. 15.2 + 36.48 + 9 60.68
18. (28.39 + 14.6) − 9.8 33.19
19. (48 − 16.3) + 8.25 39.95
20. (47.8 − 15.27) − 7.3 25.23
21. (43 + 18.12) − 6.9 54.22

Problem Solving

Solve each problem.

22. How much more does the sugar packet weigh than the salt? 4.85 grams

23. Find the total snowfall. 8.08 cm

Snowfall	
Monday	3.6 cm
Tuesday	2.48 cm
Wednesday	2 cm

24. About how much longer is the knife than the fork? about 1 cm

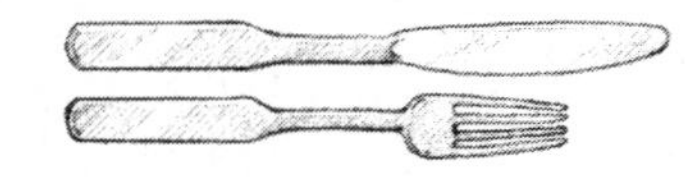

25. Find the perimeter. 49.9 cm

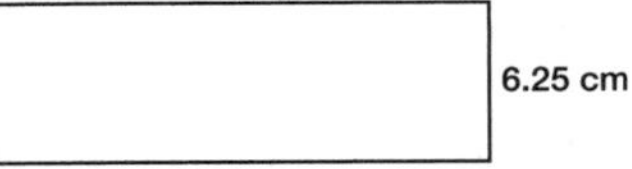

4 Assess

What happens to the value of a decimal number when zero is written to the far right? (The value of the number does not change.)

For Mixed Abilities

Common Errors • Intervention

Some students may not align decimal numbers correctly in an addition or subtraction exercise. Have them align the decimal points and then annex zeros so that there are the same number of places to the right of the decimal point in both of the numbers in the exercise. For example,

$$\begin{array}{r} 8.5 \\ -\ 6.15 \\ \hline \end{array} \qquad \begin{array}{r} 8.50 \\ -\ 6.15 \\ \hline 2.35 \end{array}$$

Enrichment • Applications

Tell students to consult an almanac to find out how much less yearly precipitation Mobile, Alabama, has than Hawaii's Mount Waialeale. They should discover that Mt. Waialeale is the rainiest place in the world, receiving 460 inches per year.

More to Explore • Measurement

Tell students that a painter must be able to estimate the amount of paint needed to complete a job. The label on a paint can tells them how much surface it will cover. For example, it might cover 36 square meters or 106 square feet.

Have students bring in paint can labels or ask a paint company to send labels from various products. Have students take measurements of their classroom or hallway and calculate how much of a specific paint it would take to paint that area. Then, have them figure how many of a specific can size they will need, reminding them they may end up with some paint left over. Then, ask what other jobs would require this type of estimation.

15-9 Multiplying a Whole Number and a Decimal

pages 313–314

1 Getting Started

Objective

- To multiply a whole number by a decimal through hundredths

Warm Up • Mental Math

Have students name the fraction to compare the following:

1. 4 boys to 9 girls ($\frac{4}{9}$)
2. 9 boys to 4 girls ($\frac{9}{4}$)
3. 6 moms to 8 dads ($\frac{6}{8}$)
4. 22 students to 1 teacher ($\frac{22}{1}$)
5. 2 teachers to 39 students ($\frac{2}{39}$)
6. 5 houses to 9 cars ($\frac{5}{9}$)
7. boys to girls in your class
8. girls to boys in your class

Warm Up • Multiplication

Review multiplication through 3-digit numbers by 2-digit numbers.

Name ______________________

Multiplying a Whole Number and a Decimal

Joe and Melissa are using a square table to display their fresh vegetables at the town centennial. They want to hang bunting along 3 sides of the table. How many meters of bunting do they need?

We want to know the length of 3 sides of a square table.

One side of the table measures 3.14 meters.

To find the length of 3 sides, we multiply the length of one side by 3.

We multiply 3.14 by 3.

Multiply the same as whole numbers from right to left.

```
   1
  3.14 ← 2 decimal places
×    3 ← 0 decimal places
  9.42 ← 2 decimal places
```

The product has the same number of decimal places as the sum of the deicmal places in the factors.

Joe and Melissa need 9.42 meters of bunting.

Getting Started

Place the decimal point in each product.

1. 8.12 × 5 = 40.60
2. 12.9 × 3 = 38.7
3. 79 × 2.5 = 197.5
4. 7.08 × 56 = 396.48

Copy and multiply.

5. 9.43 × 7 — 66.01
6. 14.3 × 12 — 171.6
7. 17 × 0.39 — 6.63
8. 2.06 × 37 — 76.22

2 Teach

Introduce the Lesson Have a student read the problem aloud and tell what is to be found. (how many meters of bunting are needed for 3 sides of the table) Ask what information is known. (One side of the table is 3.14 meters.)

- Have students read with you and complete the sentences. Guide them through the multiplication step in the model to solve the problem. Tell students to add to check their solution.

Develop Skills and Concepts Write **2.64** on the board and ask students to tell how many decimal places there are. (two) Repeat the question for 16.1 (one), 29.26 (two), 19 (zero), and 95.06 (two).

- Write **16.24 × 6** vertically on the board and have students do the multiplication. (97.44) Ask students how many decimal places are in the factors. (two)
- Tell students that when multiplying decimals, we count the number of decimal places in the factors and make sure the same number of decimal places are in the product. Have students read the product.
- Repeat the procedure for 62 × 2.4 (148.8), 7.6 × 8 (60.8), 40.02 × 58 (2321.16), 761 × 0.12 (91.32), and 1.26 × 18 (22.68).

3 Practice

Remind students that the number of decimal places in the answer must be the same as the number in the products. Have students complete page 314 independently.

Practice

Place the decimal point in each product.

1. 7.29 × 8 = 58.32	2. 14.6 × 6 = 87.6	3. 85 × 4.1 = 348.5	4. 7.48 × 21 = 157.08
5. 15.2 × 7 = 106.4	6. 3.25 × 9 = 29.25	7. 9.6 × 9 = 86.4	8. 7.05 × 8 = 56.40

Multiply.

9. 8.8 × 6 52.8	10. 15 × 2.9 43.5	11. 281 × 5.3 1489.3	12. 8.65 × 9 77.85
13. 3.75 × 17 63.75	14. 6.85 × 60 411	15. 325 × 0.28 91	16. 5.9 × 36 212.4

Copy and multiply.

17. 96 × 3.5 336	18. 8.7 × 27 234.9	19. 3.25 × 9 29.25	20. 12 × 7.36 88.32
21. 18 × 0.29 5.22	22. 4.8 × 37 177.6	23. 0.73 × 9 6.57	24. 85 × 7.83 665.55
25. 5.9 × 372 2194.8	26. 28.9 × 6 173.4	27. 1.75 × 8 14	28. 85 × 0.16 13.6

Problem Solving

Solve each problem.

29. A dictionary weighs 2.15 pounds. How much does a stack of 4 dictionaries weigh?
8.6 pounds

30. Find the area of a rectangle that has a length of 6.75 inches and a width of 9 inches.
60.75 square inches

4 Assess

Ask students how many decimal places the product of the following exercise will have: 60.333 × 5. (three decimal places)

5 Mixed Review

1. 2,153 × 72 (155,016)
2. $\frac{3}{8} + \frac{1}{6}$ ($\frac{13}{24}$)
3. (46 − 13.8) + 7.23 (39.43)
4. Simplify $\frac{21}{49}$. ($\frac{3}{7}$)
5. 457 ÷ 16 (28 R9)
6. 16 + 19.354 (35.354)
7. $\frac{7}{9} - \frac{1}{3}$ ($\frac{4}{9}$)
8. 1,953 + 15,828 (17,781)
9. 11.235 − 6.47 (4.765)
10. $90.00 − $65.22 ($24.78)

For Mixed Abilities

Common Errors • Intervention

Some students may count the decimal places from the left in the numeral for the product instead of from the right to place the decimal point. Have them count the number of decimal places in the factors before they multiply and write this number in a circle beside the place for the product with an arrow to show the direction in which to count.

```
 7.29
×   8
      ←②
```

Enrichment • Numeration

Tell students to match each exercise to its product.

.04 × 62	2,480
.062 × 4	2,480
620 × 4	0.248
6.2 × 40	2.48
62 × 40	248
0.4 × 62	2,480
6.2 × 400	24.8

More to Explore • Applications

Have students research the names of the cities where the last five summer Olympics were held. Have them find the distance each city is from their hometown.

Now, divide students into groups and have them complete the following activities using the data:

1. Order the cities from closest to farthest away.
2. Find the total distance traveled by someone from the students' hometown attending all five Olympics.
3. Choose one of the cities and prepare a numeration report about it that might include the following data: population, annual rainfall, denomination of currency, total attendance at Olympic events, and so on.

15-10 Multiplying Decimals

pages 315–316

1 Getting Started

Objective

- To multiply two decimal numbers through hundredths

Warm Up • Mental Math

Have students find the equivalent fraction.

1. $\frac{1}{2}$ in tenths $(\frac{5}{10})$
2. $\frac{2}{3}$ in twelfths $(\frac{8}{12})$
3. $\frac{9}{12}$ in fourths $(\frac{3}{4})$
4. $\frac{2}{7}$ in fourteenths $(\frac{4}{14})$
5. $\frac{4}{6}$ in thirds $(\frac{2}{3})$
6. $\frac{75}{100}$ in fourths $(\frac{3}{4})$
7. $\frac{1}{10}$ in hundredths $(\frac{10}{100})$
8. $\frac{1}{10}$ in thousandths $(\frac{100}{1000})$

Warm Up • Numeration

Review multiplication of a whole number and a decimal by having students solve the following exercises at the board: 6.4 × 7 (44.8), 25.7 × 15 (385.5), 68 × 1.7 (115.6), and 2.09 × 53 (110.77). Now, dictate the same digits but change the decimal point to show that the digits of the product remains the same but the decimal point may change.

Name ______________________

Multiplying Decimals

A barrel of maple syrup holds 1.5 times as much syrup as a pail. How much syrup does a barrel hold?

PAIL 4.5 L

BARREL ? L

We want to know how much maple syrup a barrel holds.

A pail holds 4.5 liters of syrup.

A barrel holds 1.5 times as much syrup.

To find how much a barrel can hold, we multiply 4.5 by 1.5.

Multiply the same as whole numbers from right to left.	The product has the same number of decimal places as the sum of the decimal places in the factors.
4.5 × 1.5 225 45 675	4.5 ⟵ 1 place × 1.5 ⟵ 1 place 225 45 6.75 ⟵ 2 places

A barrel holds 6.75 liters of syrup.

Getting Started

Multiply.

1. 3.7 × 2.4 = 8.88
2. 8.9 × 5 = 44.5
3. 3.8 × 0.7 = 2.66
4. 0.9 × 0.6 = 0.54

Copy and multiply.

5. 18 × 3.5 = 63
6. 0.24 × 3 = 0.72
7. 12.6 × 1.9 = 23.94
8. 8.5 × 6.8 = 57.8

2 Teach

Introduce the the Lesson Have a student read the problem aloud and tell what is to be found. (the amount of syrup one barrel holds) Ask what information is given in the problem and the picture. (A pail holds 4.5 L and a barrel holds 1.5 times that amount.)

- Have students read with you as they complete the sentences. Guide them through the multiplication steps in the model to solve the problem.

Develop Skills and Concepts Remind students that in multiplying decimals, we find the product of the factors and count the decimal places in the factors because the product must have the same number of decimal places as there are in the exercise.

- Write **2.6 × 2.1** on the board and ask students how many decimal places will be in the product. (two)
- Repeat for 9 × 0.62 (two), 605 × 2.6 (one), 26.9 × 8.2 (two), 4.6 × 20 (one), and 6.4 × 29.9 (two). Now, have students work the exercises and place the decimal points in their products. (48.98; 1,573; 220.58; 92; 191.36)

3 Practice

Remind students that the product must have the same number of decimal places as the total places in the exercise. Have students complete page 316 independently.

Practice

Multiply.

1. 4.6 × 2.7 = 12.42	2. 3.26 × 8 = 26.08	3. 12.7 × 6 = 76.2	4. 14.3 × 2.5 = 35.75
5. 7.09 × 5 = 35.45	6. 16 × 8.5 = 136	7. 4.9 × 5.7 = 27.93	8. 0.92 × 7 = 6.44
9. 0.8 × 0.6 = 0.48	10. 36.2 × 1.4 = 50.68	11. 49 × 7.3 = 357.7	12. 46.8 × 4.3 = 201.24
13. 96.4 × 3.3 = 318.12	14. 0.75 × 7.6 = 5.7	15. 4.48 × 9 = 40.32	16. 36.9 × 0.8 = 29.52

Copy and multiply.

17. 6.3 × 4.8 30.24	18. 16 × 0.15 2.4	19. 1.6 × 9 14.4	20. 37 × 8.4 310.8
21. 0.7 × 0.7 0.49	22. 8.3 × 7.6 63.08	23. 27 × 4.9 132.3	24. 12 × 0.9 10.8
25. 64 × 6.8 435.2	26. 0.67 × 23 15.41	27. 1.59 × 3 4.77	28. 62.5 × 7.6 475

Problem Solving

Solve each problem.

29. A can of nails weighs 4.8 pounds. How much do 2.5 cans weigh?
12 pounds

30. Find the area of a rectangle that is 5.2 centimeters long and 7.6 centimeters wide.
39.52 square centimeters

Now Try This!

If a year's supply of pencils for the world, laid end to end, circled Earth 5 times, how many times would an $8\frac{3}{4}$ year's supply circle Earth? Write your answer as a decimal.
43.75 times

Now Try This! Guide students in converting $8\frac{3}{4}$ to 8.75.

4 Assess

Ask students to solve 8.16 × 9. (73.44)

For Mixed Abilities

Common Errors • Intervention

Some students may confuse the form for addition with the form for multiplication and align the decimal points in the factors and product.

Incorrect	Correct
3.7 × 2.4 88.8	3.7 × 2.4 8.88

Correct students by having them indicate in writing before they multiply how many decimal places will be in the product.

Enrichment • Numeration

Ask students to find which two factors have a product of 270.06.

4200	64.3	4.2
6.43	643	6430
42	420	0.42

(4.2 × 64.3, 42 × 6.43, 0.42 × 643)

More to Explore • Probability

Give each group of two students a deck of cards with all the face cards removed. There will be 40 cards in each deck. Ask them to shuffle and draw a card, replacing the card and shuffling after each draw until they get a 7. Have them record the number of cards drawn before they get a 7 and then shuffle and draw again. Ask them to repeat this process 20 times. Have students record their data in a logical way.

Show the groups how to find the average number of cards drawn by adding all the cards drawn and dividing by 20. List each group's average on the board. Ask students to look at the averages and predict the number of cards drawn before they get a 7 another time.

15-11 Problem Solving: Use Logical Reasoning

pages 317–318

1 Getting Started

Objective
- To solve problems using logical reasoning

Warm Up • Mental Math
Have students simplify the following fractions:

1. $\frac{2}{6}$ $(\frac{1}{3})$
2. $\frac{4}{8}$ $(\frac{1}{2})$
3. $\frac{4}{10}$ $(\frac{2}{5})$
4. $\frac{3}{9}$ $(\frac{1}{3})$
5. $\frac{6}{9}$ $(\frac{2}{3})$
6. $\frac{5}{10}$ $(\frac{1}{2})$
7. $\frac{4}{24}$ $(\frac{1}{6})$
8. $\frac{8}{20}$ $(\frac{2}{5})$

Warm Up • Numeration
Tell students to write an equation using 3 numbers and any operation to arrive at the following numbers:

1. 42 (50 − 10 + 2 and so on)
2. 100 (50 + 25 + 25 and so on)
3. 1 ($\frac{1}{3} + \frac{1}{3} + \frac{1}{3}$ and so on)
4. 4.1 (10 − 3.9 − 2 and so on)
5. 8 (7.4 + 2.6 − 2 and so on)
6. 75 (50 + 100 − 75 and so on)
7. $5.00 ($20 − $10 − $5 and so on)
8. 3,000 (1,500 + 1,000 + 500 and so on)

Name ______________________________

Problem Solving: Use Logical Reasoning

At a fishing tournament, Amos, Paul, Belinda, and Rosa each caught a fish. Rosa caught the largest fish. Belinda's fish weighed 5.5 kilograms. Amos's fish weighed less than Paul's. What was the weight of each person's fish?

Fishing Tournament Entries
15.8 kilograms
9.7 kilograms
5.5 kilograms
10.1 kilograms

★SEE

We want to know the weight of each person's fish.

We know that Rosa caught the largest fish.

Belinda's fish weighed 5.5 kg. Amos's fish weighed less than Paul's.

★PLAN

We can use logical reasoning to help us find the answer. We need to make a table and use the clues to fill it in.

★DO

Fishing Tournament Entries				
	5.5 kg	9.7 kg	10.01 kg	15.8 kg
Rosa	No	No	No	Yes
Belinda	Yes	No	No	No
Amos	No	Yes	No	No
Paul	No	No	Yes	No

★CHECK

Is Rosa's fish the largest? yes

Does Belinda's fish weigh 5.5 kilograms? yes

Is 9.7 less than 10.1? yes

Does Amos's fish weigh less than Paul's? yes

2 Teach

Introduce the Lesson Remind students of the four-step method for solving problems.

Develop Skills and Concepts Have a student read the problem aloud. Read the SEE step together and have students fill in the missing information. Explain that the clues will help students decide if Amos's fish weighs less than Paul's fish.

- Read the PLAN step aloud. Tell students that they will read each clue one at a time and use the information in the clue to write *yes* or *no* under each weight in the table.
- Work through the DO step one clue at a time with students. Ask students why *no* is written under 5.5 kg, 9.7 kg, and 10.01 kg for Rosa. (Rosa caught the largest fish, and these fish weigh less than 15.8 kg.) Ask students why *yes* is written under 9.7 kg for Amos. (Amos's fish weighed less than Paul's fish, and Paul's fish weighed 10.01 kg.)

3 Apply

Solution Notes

1. Some students may make a table to solve this problem. After solving the problem, talk about the number of each item and the relationship the numbers have with the solution. (4 × 3 × 2 = 24)
2. Have students label the corners of the cube and use the corner letters to denote possible trips: ADHG, ADCG, ABFG, ABCG, AEFG, AEHG.
3. Students may use a more abstract notation to solve this problem. Discuss the pattern that is evident in the solution. (6 + 5 + 4 + 3 + 2 + 1 = 21)

Apply

Use logical reasoning to solve each problem.

1. Jodi's favorite clothes include four sweatshirts, three pairs of jeans, and two pairs of tennis shoes. How many days in a row can she wear a different outfit using her favorite clothes?
   24 days

2. Arthur Ant wants to crawl from A to G on a cube. How many different 3-sided trips can he make? He must stay on an edge of the cube at all times.
   6 trips

3. The Chess Club is having a tournament. Each of the 7 members will play every other member one time. The champion will be the member with the most wins. How many games will be played?
   21 games

4. At Washington School, for every 7 students who walked to school, 25 rode a school bus. At Lincoln School, the ratio was 9 to 30. At which school is the ratio of walkers to riders greater? Prove your answer.
   at Lincoln School
   See Solution Notes.

Problem-Solving Strategy: Using the Four-Step Plan

- ★ **SEE** What do you need to find?
- ★ **PLAN** What do you need to do?
- ★ **DO** Follow the plan.
- ★ **CHECK** Does your answer make sense?

5. Fran and Betsy gathered blackberries and filled an 8-quart pail. They want to divide the berries equally. Besides the 8-quart pail, they have a 5-quart pail and a 3-quart pail. Describe how they can divide the berries equally.
   See Solution Notes.

6. Tamara has a 3-ounce and a 7-ounce container. Her recipe calls for exactly 5 ounces of water. Using only these two containers, show how Tamara can measure exactly 5 ounces of water.
   See Solution Notes.

7. Tom and Juan were playing a game making up special numbers. To make up the first number, Tom said 3 and Juan said 5. To make up another number, Tom said 6 and Juan said 10. Then, Tom said 9 and Juan said 15. What kind of numbers are they making up?
   See Solution Notes.

For Mixed Abilities

More to Explore • Creative Drill

Have each student bring an old calendar to class. Have students complete any or all of the following activities using their calendars:

1. Draw a ring around the multiples of 7. What pattern do you see? (The multiples of 7 are in a row.) Will this pattern always be the same? (yes)
2. Complete the following column addition: the sum of all Tuesdays; the sum of all Fridays; the sum of the last full week of the month.
3. Box in a 3 x 3 square on the calendar and find the average of the numbers in the box. Repeat for another 3 x 3 box. What pattern do you see? (The average is the number in the center square.) Test the pattern in another 3 x 3 box.

Challenge students to come up with their own calendar activities.

4. Proofs will vary but might include answers such as the fraction $\frac{7}{25}$ is greater than the fraction $\frac{9}{30}$, or the quotient of $25 \div 7$ is greater than the quotient of $30 \div 9$.

Higher-Order Thinking Skills

5. **Analysis:** The goal is to get exactly 4 quarts into two of the larger containers by moving the berries from container to container in the following way. Begin by filling the 5-quart container. There will be 3 quarts left over in the 8-quart container. Then, from the 5-quart container fill the 3-quart container, and so on. This is shown in the table below.

8-quart container	5-quart container	3-quart container
3	5	0
3	2	3
6	2	0
6	2	2
1	5	2
1	4	3
4	4	0

6. **Synthesis:** Students will recognize that this problem is similar to problem 5. One solution is 7, 0; 4, 3; 4, 0; 1, 3; 1, 0; 0, 1; 7, 1; and 5, 3.
7. **Analysis:** They are making up equivalent fractions; Tom is naming numerators, and Juan is naming denominators.

4 Assess

Ask students how a table can help them solve a problem like the one on the first page of this lesson. (It can keep track of all the clues by writing yes and no in the different spaces. When only a few spaces are left, it is easy to solve the problem.)

Chapter 15 Test

page 319

Items	Objectives
1–8	To identify the place value of digits in a decimal through thousandths (see pages 297–302)
9–16	To compare and order decimals through hundredths (see pages 303–304)
17–24	To add and subtract decimals through thousandths (see pages 307–312)
25, 30, 32	To multiply a whole number and a decimal (see pages 313–314)
26–29 31	To multiply two decimals (see pages 315–316)

Name ____________________

CHAPTER 15 TEST

Write the place value of the 4 in each number.

1. 84.21 ones
2. 156.43 tenths
3. 20.148 hundredths
4. 47.296 tens

Write the place value of the 7 in each number.

5. 639.74 tenths
6. 387.28 ones
7. 200.07 hundredths
8. 583.957 thousandths

Write >, <, or = in each circle.

9. 3.46 (<) 3.64
10. 8.4 (=) 8.40
11. 7.3 (<) 7.39
12. 0.231 (>) 0.22
13. 7.05 (<) 7.5
14. 0.13 (>) 0.031
15. 65.1 (>) 65.09
16. 8.7 (=) 8.70

Add.

17. $5.9 + 3.6 = 9.5$
18. $27.25 + 48.397 = 75.647$
19. $96.5 + 48.76 = 145.26$
20. $53.47 + 16.7 = 70.17$

Subtract.

21. $8.9 - 2.5 = 6.4$
22. $37.038 - 15.969 = 21.069$
23. $38.7 - 19.58 = 19.12$
24. $89.25 - 27.5 = 61.75$

Multiply.

25. $3.2 \times 8 = 25.6$
26. $12.6 \times 0.4 = 5.04$
27. $13.7 \times 0.6 = 8.22$
28. $8.9 \times 4.7 = 41.83$
29. $24.8 \times 7.2 = 178.56$
30. $1.6 \times 7 = 11.2$
31. $5.1 \times 0.06 = 0.306$
32. $0.09 \times 7 = 0.63$

Alternate Chapter Test

You may wish to use the Alternate Chapter Test on page 374 of this book for further review and assessment.

Circle the letter of the correct answer.

1. Round 4,500 to the nearest thousand.
   a. 4,000
   (b.) 5,000
   c. NG

2. 13,296 + 8,758
   a. 21,954
   (b.) 22,054
   c. 22,064
   d. NG

3. 9,241 − 8,658
   (a.) 583
   b. 1,417
   c. 1,583
   d. NG

4. Choose the better estimate of length.
   (a.) 2 cm
   b. 2 m

5. 409 × 6
   (a.) 2,454
   b. 2,456
   c. 24,054
   d. NG

6. $16\overline{)328}$
   a. 2 R8
   (b.) 20 R8
   c. 28
   d. NG

7. Find the perimeter. (square, 16 ft)
   a. 16 sq ft
   b. 64 sq ft
   c. 48 ft
   (d.) NG

8. Find the area. (rectangle, 6 cm by 9 cm)
   a. 15 cm
   b. 15 sq cm
   c. 30 sq cm
   (d.) NG

9. $\frac{3}{4}$ of 16 = ?
   a. 4
   b. 8
   (c.) 12
   d. NG

10. Simplify $\frac{16}{24}$.
    a. $\frac{1}{2}$
    (b.) $\frac{2}{3}$
    c. $\frac{3}{4}$
    d. NG

11. $\frac{2}{3} + \frac{3}{5}$
    a. $\frac{5}{8}$
    b. $1\frac{4}{5}$
    (c.) $1\frac{4}{15}$
    d. NG

12. $\frac{5}{8} - \frac{1}{8}$
    (a.) $\frac{1}{2}$
    b. $\frac{3}{4}$
    c. 4
    d. NG

STOP

score

Cumulative Assessment

page 320

Items	Objectives
1	To round numbers (see pages 31–32)
2	To add two numbers up to 5 digits (see pages 41–48)
3	To subtract two 4-digit numbers (see pages 61–62)
4	To determine the appropriate metric unit of length (see pages 149–152)
5	To multiply up to 3-digit numbers with regrouping (see pages 169–170, 177–178)
6	To divide numbers with remainders (see pages 193–194)
7–8	To find the perimeter, find the area (see pages 241–242)
9	To find a fraction of a number (see pages 253–254)
10	To reduce fractions to the lowest terms (see pages 261–262)
11–12	To add and subtract fractions (see pages 281–282, 285–286)

Alternate Cumulative Assessment

Circle the letter of the correct answer.

1. Round 6,421 to the nearest thousand.
   a 6,400
   b 6,500
   (c) NG

2. 12,284 + 9,869
   a 21,153
   b 22,053
   c 22,143
   (d) NG

3. 6,130 − 4,796
   (a) 1,334
   b 1,344
   c 1,434
   d NG

4. What unit of length would you use to measure your finger?
   (a) cm
   b dm
   c m
   d NG

5. 306 × 8 =
   (a) 2,448
   b 2,488
   c 2,528
   d NG

6. $14\overline{)347}$
   a 36
   (b) 24 R11
   c 23 R1
   d NG

7. Find the perimeter of an equilateral triangle with 6-cm sides.
   a 10 cm
   b 13 cm
   (c) 18 cm
   d NG

8. Find the area of a rectangle that is 5 ft by 12 ft.
   a 34 sq ft
   b 50 sq ft
   (c) 60 sq ft
   d NG

9. $\frac{2}{3}$ of 18 =
   a 6
   b 9
   (c) 12
   d NG

10. Simplify $\frac{24}{36}$.
    a $\frac{1}{2}$
    b $\frac{1}{3}$
    c $\frac{3}{4}$
    (d) NG

11. $\frac{3}{8} + \frac{2}{6} =$
    (a) $\frac{17}{24}$
    b $\frac{3}{4}$
    c $\frac{19}{24}$
    d NG

12. $\frac{6}{14} - \frac{4}{14} =$
    a $\frac{1}{14}$
    (b) $\frac{1}{7}$
    c $\frac{3}{14}$
    d NG

16-1 Tallies and Bar Graphs

pages 321–322

1 Getting Started

Objective

- To make and interpret tally charts and double bar graphs

Vocabulary

tally, double bar graph

Materials

*large grid paper; graph paper

Warm Up • Mental Math

Have students complete the comparison: 14 is to 2 as

1. 63 is to (9)
2. 21 is to (3)
3. 84 is to (12)
4. 560 is to (80)
5. 4,900 is to (700)
6. 91 is to (13)
7. 7 is to (1)
8. 42 is to (6)

Warm Up • Numeration

Have students work in pairs. Ask a student to write two numbers for another student to compare. Have the second student write >, <, or = to show the comparison. Have students change roles for more practice. Encourage students to use fractions and decimals as well as whole numbers.

Name ____________________

CHAPTER 16

Graphing and Probability

Lesson 16-1

Tallies and Bar Graphs

Mr. Ryan is keeping a **tally** of the different types of books fourth and fifth graders check out of the library. Use the tally chart to complete the double bar graph. A **double bar graph** uses two colored bars to compare two groups of data. Which type of book is most popular for each grade?

Library Books Checked Out

Type of Book	Fourth Graders	Fifth Graders
Mystery	卌 卌 |||	卌 卌 卌
Biography	卌 |||	卌 卌	
Sports	卌 卌		卌 |||
Romance	卌 卌 卌		卌 卌 |||

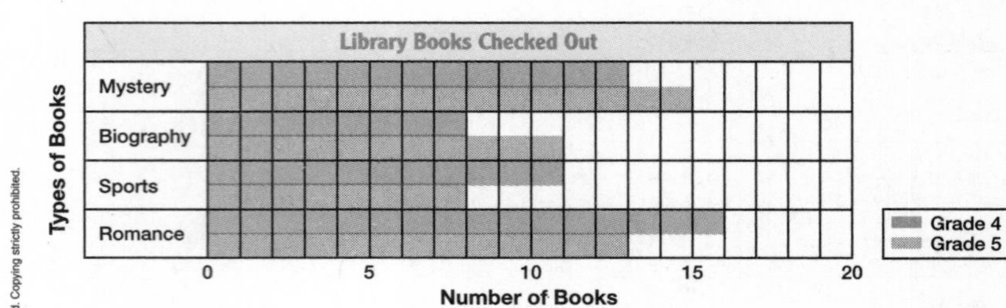

The most popular type of book checked out by fourth graders is romance.

The most popular type of book checked out by fifth graders is mystery.

Getting Started

Use the double bar graph above to solve each problem.

1. How many more sports books were checked out by fourth-grade students than by fifth-grade students? 3
2. How many books were checked out in all? 95

2 Teach

Introduce the Lesson Have students read the problem and tell what is to be done. (use the tally to complete the double bar graph and find the most popular type of book for each grade) Ask what information is given. (tallies of mystery, biography, sports, and romance books checked out by fourth graders and fifth graders)

- Have students complete the double bar graph. Then, have them fill in the solution sentences.

Develop Skills and Concepts Draw a double-column tally chart on the board and have students vote for their favorite breakfast food, such as eggs, cereal, or pancakes. List the girls' votes in one column and the boys' votes in the other column. Make a slash for each vote and cross each group of four with a fifth slash.

- Ask students to tell the boys' total votes for each food. Repeat this for the girls' votes. Explain to students that because they want to compare two groups of data, they need to put the information in a double bar graph.
- Display a sheet of large grid paper and make a horizontal double bar graph from the tally chart information.
- Remind students that a tally chart must be titled so that anyone seeing the work will know what information is being tallied or graphed. Using information from the double bar graph, ask questions that lead students to tell how many boys voted, how many boys prefer foods other than eggs, and so on. Have students make their own vertical double bar graph from the same information.

3 Practice

Have students complete page 322 independently.

Practice

Use the tally chart to complete the double bar graph and answer the questions.

Votes for Fourth-Grade President		
	4A	4B
Eduardo	卌 II	卌 III
Nick	卌 IIII	卌 卌
Janet	卌 卌 I	卌 卌 II
Kim	卌 卌	卌 卌 I

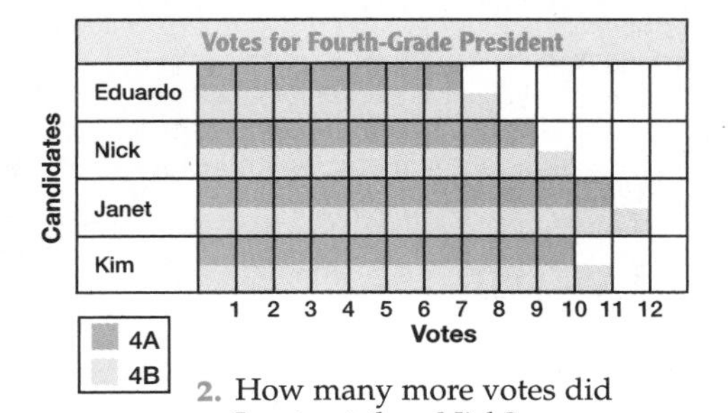

1. Who won the election?
   Janet
2. How many more votes did Janet get than Nick?
   4
3. How many votes were cast altogether?
   78
4. How many votes did the boys get altogether?
   34

Complete the double bar graphs using the information on the tally charts.

5.

Games Won		
	This Year	Last Year
Baily	卌 卌 III	卌 卌
Smith	卌 卌	卌 卌 II
Lincoln	卌 卌 卌	卌 卌 IIII

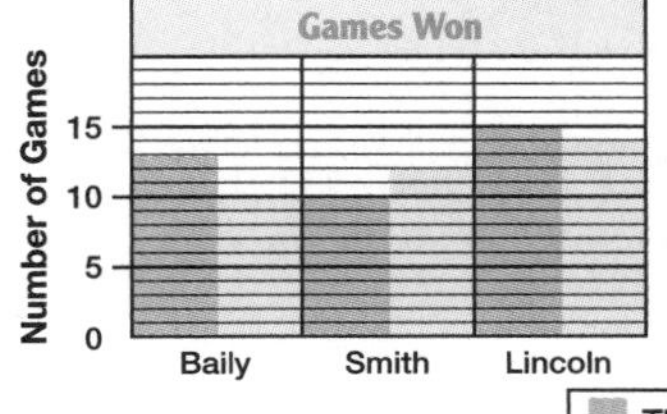

6.

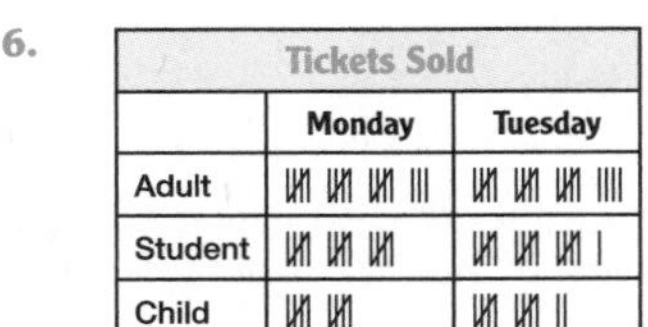

Tickets Sold		
	Monday	Tuesday
Adult	卌 卌 卌 III	卌 卌 卌 IIII
Student	卌 卌 卌	卌 卌 卌 I
Child	卌 卌	卌 卌 II

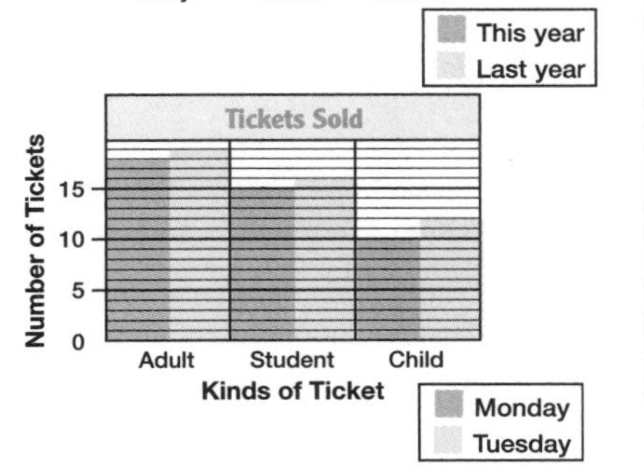

4 Assess

Ask students when they would use a double bar graph to display data. (when two groups of data are being compared)

5 Mixed Review

1. 953 ÷ 86 (11 R7)
2. 178 × 26 (4,628)
3. 35,789 + 2,527 (38,316)
4. 36.8 × 4 (147.2)
5. 4,061 − 2,948 (1,113)
6. 15.9 − 4.73 (11.17)
7. $2\frac{1}{3} + 3\frac{1}{4}$ ($5\frac{7}{12}$)
8. 7.95 + 12.835 + 6.6 (27.385)
9. 347 ÷ 8 (43 R3)
10. $\frac{3}{8} + \frac{3}{4}$ ($1\frac{1}{8}$)

For Mixed Abilities

Common Errors • Intervention

Some students may have difficulty giving the value of a bar that ends between numbers on the scale. Have them use a ruler or the edge of a sheet of paper to align the end of the bar with the scale and count over from the last given number to this edge.

Enrichment • Statistics

Have students tally the number of girls' families and the number of boys' families in the class who have vans, cars, and sport-utility vehicles. Tell them to make a double bar graph from their information. Then, have them write five questions for a classmate to answer from their graph.

More to Explore • Geometry

Give each student a circle cut from a sheet of plain paper. Ask students to fold the paper in half twice in one direction and twice in the other direction. Explain that they will have divided the circle into twelve equal arcs.

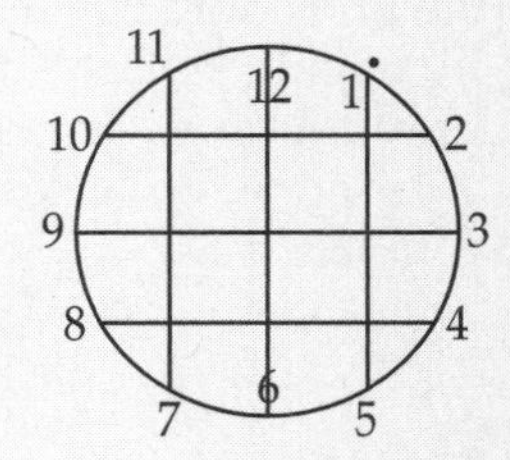

Have students label each fold point with a number (1 to 12) as though the circle were a clock. Ask them to use a straightedge to connect every other point. Ask students to name the figure formed. (hexagon)

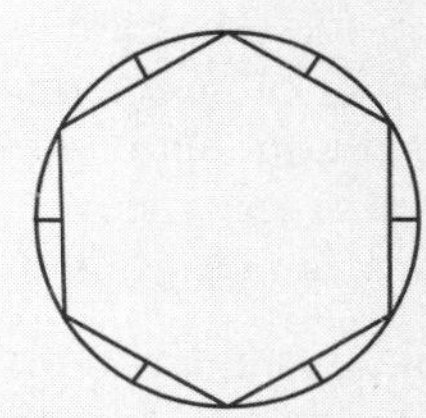

Have students connect every third point and name the figure. (square) Have them connect every fourth point (triangle) and every fifth point (dodecagon). Ask them to predict what will happen when they connect every sixth point. (They will get a straight line.)

16-2 Pictographs

pages 323–324

1 Getting Started

Objective

- To make and interpret pictographs

Vocabulary

pictograph

Warm Up • Mental Math

Have students compare the following numbers:

1. 2.5 ◯ 2.56 (<)
2. 68.02 ◯ 68 (>)
3. $\frac{4}{5}$ ◯ $\frac{16}{20}$ (=)
4. 0.75 ◯ 0.8 (<)
5. 2.068 ◯ 2.06 (>)
6. 42,676 ◯ 42,776 (<)
7. $4\frac{1}{2}$ ◯ $\frac{9}{2}$ (=)
8. 5.01 ◯ 5.10 (<)

Warm Up • Numeration

Review the procedure to find an average by having students find the average of numbers such as 10, 6, and 14. (10) Be sure each average is a whole number.

Name ______________________

Pictographs

A **pictograph** uses pictures to represent data on a graph. In this pictograph, we can see how the construction industry changes with the seasons. How many new houses were started in June?

New Housing Starts	
March	
April	
May	
June	
July	

Each ⌂ represents 10 houses.

We want to know how many new houses were started in June.

Each full house on the graph means 10 new houses started.

A half of a house means 5 new houses started.

There are 5 full houses and 1 half house pictured for June.

(5 × 10) + 5 = 55

There were 55 houses started in June.

Getting Started

Use the pictograph to answer each question.

1. Which month had the fewest new houses started? March
2. How many houses were started in April? 45
3. How many more houses were started in July than in March? 20
4. In which month were only 30 houses started? May
5. What was the monthly average of housing starts? 38
6. Which months were above average in number of housing starts? April, June, July

2 Teach

Introduce the Lesson Have students read the problem and tell what is to be found. (the number of new houses started in June) Ask what a **pictograph** does. (uses pictures to show data on a graph)

- Ask students what data is shown in the pictograph. (the number of new houses started in the months of March through July) Have students complete the information sentences. Then, have them solve the problem and complete the solution sentence.

Develop Skills and Concepts Tell students that the sales of baseballs by a small sporting goods store are as follows: April, $16\frac{1}{2}$ dozen; May, $15\frac{1}{2}$ dozen; June, 14 dozen; and July, 9 dozen. Tell students that these sales could be shown on a tally chart or a bar graph, but a pictograph is yet another way to show data on a graph.

- Draw a pictograph frame and write **Baseball Sales** above it. Have students name the months as you write them on the graph. Tell students that if we tried to show each individual ball sold, the graph would be huge, so we will group the balls in dozens. Tell students that we need to note that plan in a key at the bottom of the graph. Under the graph, write **Each ◯ is 1 dozen balls.**
- Ask students the number of sales in April as you draw 16 balls and $\frac{1}{2}$ of a ball. Tell students that each ball represents 1 dozen and the $\frac{1}{2}$ dozen is shown by $\frac{1}{2}$ of a ball. Have students draw the balls to represent sales in May, June, and July.
- Have students answer questions from the information on the graph.

3 Practice

Remind students that each picture on a pictograph may represent more than 1. Tell students that in Exercise 6, they are to find an average and in Exercise 9, they are to compare data to the average. Have students complete page 324 independently.

Practice

Use the pictograph to answer each question.

1. How many games did Baker win? 14
2. How many games did Jones win? 15
3. Which team won exactly 10 games? Williams

Baseball Games Won	
Bryant	
Baker	
Jones	
Williams	

Each represents 2 games.

Use the tally chart to make a pictograph. Then, answer each question.

4.

Birds Sighted	
Bluebird	卌 卌 卌 I
Robin	卌 卌 II
Cardinal	卌 卌 IIII
Blackbird	卌 卌

Birds Sighted	
Bluebird	
Robin	
Cardinal	
Blackbird	

Each represents 4 birds.

5. How many birds were sighted? 52
6. What is the average number of birds sighted? 13

7.

Bryant Band Members	
Grade 3	卌 III
Grade 4	卌 II
Grade 5	卌 卌 III
Grade 6	卌 卌 卌 I

Bryant Band Members	
Grade 3	
Grade 4	
Grade 5	
Grade 6	

Each represents 2 band members.

8. How many more Grade 6 band members are there than Grade 4 band members? 9
9. Which grades have more than the average number of members? 5, 6

4 Assess

Ask students to give an advantage of using a pictograph to display data. (Answers will vary. Some students might mention that it is easier to read the information in a pictograph.)

5 Mixed Review

1. 3.6×0.7 (2.52)
2. $\$1.95 \times 18$ (\$35.10)
3. $153 + 2,748 + 894$ (3,795)
4. $2.7 + 0.953 + 1.61$ (5.263)
5. $52,106 - 30,793$ (21,313)
6. $681 \div 27$ (25 R6)
7. 0.64×8 (5.12)
8. $\frac{4}{5} - \frac{3}{4}$ $(\frac{1}{20})$
9. $9.53 - 2.7$ (6.83)
10. $\frac{3}{5} + \frac{2}{3}$ $(1\frac{4}{15})$

For Mixed Abilities

Common Errors • Intervention

When interpreting pictographs, some students may ignore the key and read each picture as 1. To help them attend to the key, have them write the value of the picture as given in the key on at least one of the pictures in the graph.

Enrichment • Applications

Have students make a tally chart, a bar graph, and a pictograph to show the student absences for each grade in your school during 1 week.

More to Explore • Biography

"If I had a place to stand, I could move the earth," Archimedes once boasted. This ancient genius, who lived from 287 to 212 B.C. in Syracuse, Sicily, made this claim after discovering the principle of the lever and pulley. One popular story tells of Archimedes bathing and discovering that an object in water weighs exactly as much as the water whose place the object took. Archimedes was so excited about this discovery that he leaped from the tub, forgetting his clothes, and ran through the streets yelling, "Eureka!" meaning "I have found it!"

Archimedes invented the Archimedean screw, used to drain Egyptian fields flooded by the Nile River. Archimedes used his inventions to help defend Syracuse from the invading Romans. He devised catapults and even used a system of lenses and mirrors to set fire to a fleet of Roman ships. His tomb is engraved with the figure of a sphere inscribed in a cylinder, as a tribute to his work in geometry.

16-3 Line Graphs

pages 325–326

1 Getting Started

Objective
- To interpret line graphs

Vocabulary
line graph

Materials
line graph of spelling test scores from 80 to 100

Warm Up • Mental Math
Have students simplify the fractions.

1. $\frac{18}{2}$ (9)
2. $\frac{26}{21}$ $(1\frac{5}{21})$
3. $5\frac{4}{20}$ $(5\frac{1}{5})$
4. $\frac{19}{57}$ $(\frac{1}{3})$
5. $\frac{105}{5}$ (21)
6. $7\frac{9}{3}$ (10)
7. $82\frac{22}{20}$ $(83\frac{1}{10})$
8. $5\frac{7}{21}$ $(5\frac{1}{3})$

Warm Up • Numeration
Write the following on the board:

6 20 15 18 9

Ask students to put the numbers in order from least to greatest. (6, 9, 15, 18, 20) Repeat for more ordering of series of numbers. Include series of fractions or decimals.

Name ____________________

Line Graphs

A **line graph** is a good way to show changes in information. This line graph shows the high temperatures for each day for the first two weeks in May. What was the high temperature on May 6?

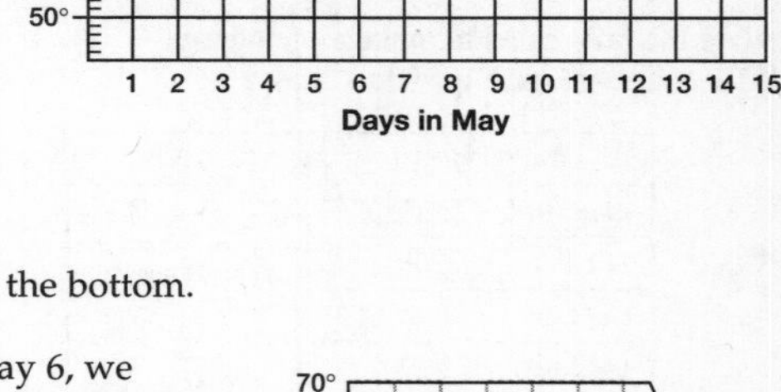

We want to find the high temperature for May 6 on the line graph.

The temperature is shown on the left side of the graph.

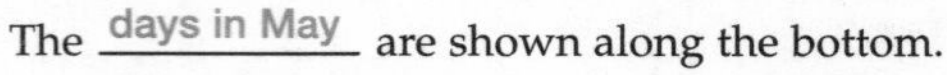

The days in May are shown along the bottom.

To find the high temperature for May 6, we move along the bottom of the graph until we reach the sixth day. Then, we go up the vertical line until we reach the dot. The temperature is the reading in degrees directly opposite that dot. On May 6, the high temperature was 68°.

Getting Started

Use the line graph above to answer each question.

1. What was the high temperature on May 10? 65°
2. On what day was the high temperature approximately 72°? May 13
3. On how many days was the high temperature exactly 70°? 2
4. What was the average high temperature for May 12, May 13, and May 14? 72°

---

2 Teach

Introduce the Lesson Have students read the problem and tell what is to be found. (the high temperature on May 6) Ask what information is given in the **line graph**. (high temperatures in May)

- Read the text and discuss the graph with students. Then, have them complete the sentences.

Develop Skills and Concepts Tell students that a line graph is another kind of graph used to show data.

- Show students a line graph of spelling test scores with scores from 80 to 100 listed on the left side and the following information represented: Test 1, 89; Test 2, 96; Test 3, 99; Test 4, 80; Test 5, 85; and Test 6, 92.
- Explain to students that the name of the line graph is Spelling Test Scores and that the numbers across the bottom are the different tests. Tell students that scores are shown on the left side of the graph. Tell students that we can find a score for a particular test by going across the bottom of the graph to the number of the test, follow that line upward to the dot and go across to the left of the graph to find the score.
- Have students find the score for Test 4, Test 6, and so on. Ask students other questions using the line graph.
- Help students understand why the range of scores is from 80 to 100 rather than from 1 to 100.

3 Practice

Remind students to go across the bottom of the graph, up the vertical line to the dot, and then to the left of the graph to find the number. Have them complete page 326 independently. Students may use a ruler to help them see across from a point to the scale on the left of the graph.

Practice

Use the line graph to answer each question.

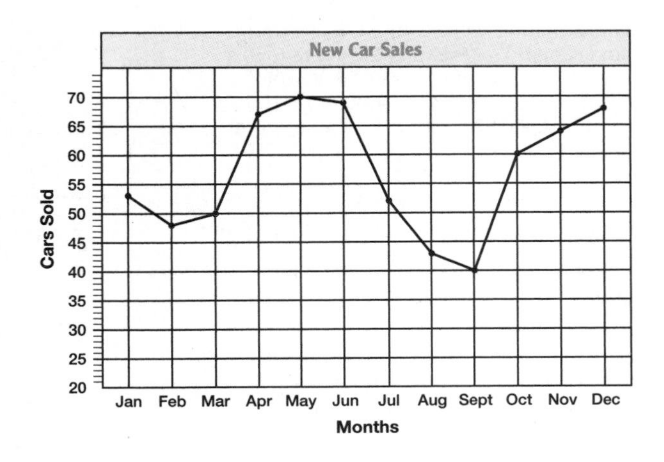

1. How many cars were sold in October? 60

2. How many cars were sold in February? 48

3. Did sales go up or down from June through September? down

4. In which month did the sales of cars increase the most over the month before? October

5. In which month were the most cars sold? May

6. In which month were the fewest cars sold? September

7. How many more cars were sold in April than in September? 27

8. What was the average number of cars sold during the last four months? 58 cars

4 Assess

Have students use the line graph on page 325 and tell the temperature on May 8 and 9. (56°)

5 Mixed Review

1. 5,671 × 23 (130,433)
2. $\frac{7}{10} + \frac{2}{5}$ $(1\frac{1}{10})$
3. 9.4 × 5.7 (53.58)
4. $\frac{15}{16} - \frac{7}{8}$ $(\frac{1}{16})$
5. \$25.00 − \$14.37 (\$10.63)
6. 0.62 × 8 (4.96)
7. 392 ÷ 6 (65 R2)
8. 15,395 + 26,008 (41,403)
9. $3\frac{5}{8} + 7\frac{2}{3}$ $(11\frac{7}{24})$
10. 895 ÷ 72 (12 R31)

For Mixed Abilities

Common Errors • Intervention

Some students may not properly align the points on the graph with the points on the scale when they are interpreting a line graph. Have them use the edge of a ruler or sheet of paper to align the point on the graph with its corresponding point on the scale.

Enrichment • Statistics

Have students make a line graph to show their last five to ten test scores in one subject.

More to Explore • Measurement

Have each student use a timepiece with a second hand for this activity.

Have students make two columns on a sheet of paper, one titled Guess, the other titled Check. Tell them that they are first going to guess how long it will take them to complete each of the activities you list on the board and write their estimate in the Guess column. Then, working with a partner, students will do each activity while being timed by their partners and write that time under the Check column. Next, have students find the difference between their estimation and the actual time for each activity.

The exercise can be extended by adding activities to the list.

Activities:

1. Hop 10 times.
2. Snap your fingers 20 times.
3. Count backward from 20.
4. Count by 5s to 100.
5. Write the numbers counting by 2s to 50.
6. Tie your shoe 10 times.
7. Write 40 Xs on the board.
8. Cut out 10 circles.

16-4 Mean, Median, Mode, and Range

pages 327–328

1 Getting Started

Objective

- To find the mean, median, mode, and range of a set of numbers

Vocabulary

mean, median, mode, range

Warm Up • Mental Math

Have students round the numbers to the nearest thousand.

1. 21,620 (22,000)
2. 4,378 (4,000)
3. 32,899 (33,000)
4. 65,111 (65,000)
5. 864,625 (865,000)
6. 726,451 (726,000)
7. 250,987 (251,000)

Warm Up • Calculators

Have students work the following on their calculators:

1. $16 \times 2 \times 0$ (0)
2. $16 \times 2 \times 1$ (32)
3. 2,000 + 200 - 800 (1,400)
4. 5×111 (555)
5. 486 − 6 tens + 2 hundreds (626)
6. 37,260 − 1,001 (36,259)
7. $250 \times 10 - 0$ (2,500)
8. $9 \times 9 \times 2$ (162)

Name ____________________

Mean, Median, Mode, and Range

We can examine information by looking at the mean, median, mode, and range. The **mean** is the average of a group of numbers, the **median** is the middle number, and the **mode** is the number or numbers that occur most frequently. The difference between the greatest number and the least number is called the **range**.

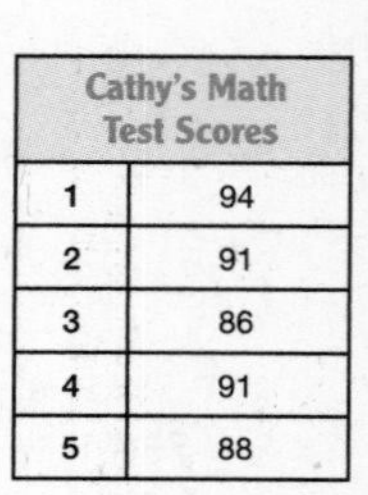

Cathy's Math Test Scores	
1	94
2	91
3	86
4	91
5	88

To find the mean, we find the sum of the test scores first.

94 + 91 + 86 + 91 + 88 = <u>450</u>

Then, we divide the sum of the test score by the number of scores.

450 ÷ <u>5</u> **=** <u>90</u>

To find the median, we list the scores in order from lowest to highest.

86, 88, 91, 91, 94

We select the middle score. If there are an even number of scores, the median is the mean of the two middle scores.

The median is <u>91</u>.

To find the mode, we look for the most frequent score.

<u>91</u> occurs more often than any other score.

The mode is <u>91</u>.

To find the range, we subtract the least score from the greatest score.

94 − 86 = <u>8</u> The range is <u>8</u>.

Getting Started

Find the mean, median, mode, and range of each set of numbers. Show your work.

1. 59, 25, 45, 61, 45
   mean <u>47</u>
   median <u>45</u>
   mode <u>45</u>
   range <u>36</u>

2. \$415, \$559, \$999, \$643, \$999
   mean <u>\$723</u>
   median <u>\$643</u>
   mode <u>\$999</u>
   range <u>\$584</u>

2 Teach

Introduce the Lesson Read the introductory paragraph aloud. Explain how to find the mean while reading the accompanying section in the text.

- Copy the test scores horizontally on the board. Ask a volunteer to go to the board and add the test scores. (450) Write **450 ÷ ___ = ___**. Have the class count the number of test scores as the volunteer writes the number 5 after the division sign. Then, have the class complete the division. (90) Explain that the average of all the scores is 90. Tell students that this is called the **mean** score.
- Next, explain that the **median** is the middle score. Read aloud the section about the median in the text. Ask a student to list the numbers on the board from lowest to highest. (86, 88, 91, 91, 94) Ask a volunteer to name the middle score. (91)
- Explain that the **mode** is the most frequent score. Have students look at the list of numbers in their text and find the most frequent score. (91)
- Finally, explain that **range** is the difference between the greatest score and the lowest score. Have a student write and solve the subtraction to find the range. (94 − 86 = 8)

Develop Skills and Concepts Write the following amounts of money on the board: **\$27, \$58, \$59, \$61, \$65, \$68,** and **\$68**. Tell students that these are amounts of money earned at a part-time job.

- Have students look at the amounts of money and estimate the average. (Answers will vary but should be around \$58.) Now, have them find the average by adding the amounts and dividing by the number of amounts. (\$58)

Practice

Find the mean, median, mode, and range of each set of numbers.

1. 40, 90, 61, 33, 61
   mean 57
   median 61
   mode 61
   range 57

2. $229, $304, $527, $636, $304
   mean $400
   median $304
   mode $304
   range $407

3. 888, 867, 902, 735, 888
   mean 856
   median 888
   mode 888
   range 167

4. 941, 946, 939, 815, 939
   mean 916
   median 939
   mode 939
   range 131

Problem Solving

Use the table to answer Exercises 5 to 9.

Distances Run in the Fifth Annual Race	
Runner	**Distance in Meters**
Hannah	44
Roger	40
Sally	45
Ted	47
Iris	39

5. Who ran the farthest?
   Ted

6. Who ran the shortest distance?
   Iris

7. How much further did Ted run than Iris?
   8 meters

8. What is the mean distance?
   43 meters

9. What is the median distance?
   44 meters

10. Complete the bar graph of distances run.

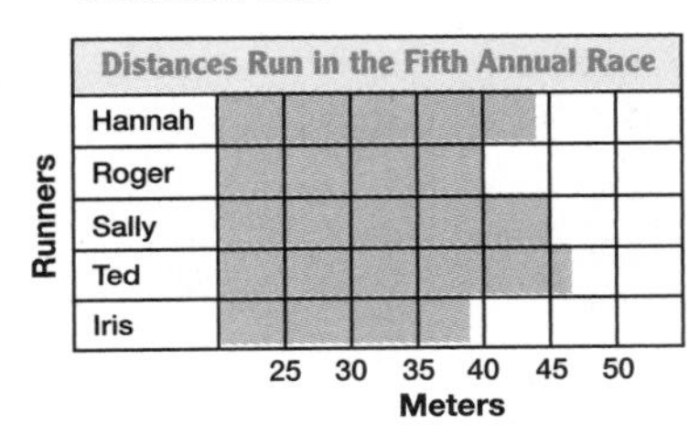

- Next, have students find the median, or middle amount. ($61) Explain that the mean and median are not the same because the mean is an average pulled down by the lowest amount, $27.
- Have students find the mode, or most frequent amount. ($68) Finally, have students calculate the range. ($41)

3 Practice

Have students complete all the exercises on the page. Remind them to align columns carefully when adding.

4 Assess

Have students write instructions telling someone in another class how to find the mean and the median of a set of numbers. (Answers will vary but should reflect the lesson concepts.)

For Mixed Abilities

Common Errors • Intervention

Some students may confuse *mean* and *median*. Put a number line on the board from 20 to 40. Have different volunteers go to the board and circle the following numbers on the line: 35, 42, 29, 41, and 38. Ask students which number is the middle number. (38) Mark it with an arrow. Discuss how "middle score" and "median" mean the same thing. Then, have students find the average. (37) Discuss how the average, or mean, is the number that if all the numbers were this number, the sum would be the same as the sum of the original numbers.

Enrichment • Applications

Tell students to follow these steps: record the odometer readings of 4 different cars, estimate how many thousand miles each has traveled, and estimate in thousands how much farther the most-traveled car has been driven than each of the others.

More to Explore • Numeration

Have students devise their own system of numeration, drawing original symbols to represent numerical values 0 to 50.

Ask students to write and answer questions concerning themselves or the class using their system of numeration; for example, How many students are in the class? How old are they?

Invite students to compose arithmetic problems using their systems and then solve them.

16-5 Ordered Pairs

pages 329–330

1 Getting Started

Objective
- To locate ordered pairs on a grid

Vocabulary
origin, ordered pairs

Materials
*line graph of high temperatures in May used in Lesson 16-3; grid paper

Warm Up • Mental Math
Have students tell the total price if they pay 6¢ tax on every dollar.

1. $7 ($7.42)
2. $80 ($84.80)
3. $20 ($21.20)
4. $5 ($5.30)
5. $10 ($10.60)
6. $100 ($106)
7. $2 ($2.12)
8. $9 ($9.54)

Warm Up • Graphing
Have students tell which days had a temperature of 60° and 64° as shown on the High Temperatures in May line graph used in Lesson 16-3. Have students locate more points on the line graph as you give other possible temperatures.

Name ____________________

Ordered Pairs

Ordered pairs of numbers can be used to give locations on a number grid. The ordered pair (4, 3) shows the location of point *A*. What ordered pair of numbers locates point *D*?

Second Number
First Number
Origin 0 1 2 3 4 5 6 7 8 9

We want to know what ordered pair identifies point *D*.

We know that point *A* is _4_ units over and _3_ units up.

REMEMBER To find an ordered pair, we start at the origin on the grid. The first number tells how far over to move. The second number tells how far up to move.

Point *D* is _2_ over and _5_ up.

The ordered pair (_2_, _5_) locates point *D*.

We write ***D*(2, 5)**.
We say **point *D* is the ordered pair two, five.**

Getting Started

Write the letter identified by each ordered pair. Use the grid above.

1. _G_ (6, 1) 2. _J_ (9, 2) 3. _H_ (3, 6) 4. _B_ (6, 6)

Write the ordered pair that identifies each letter. Use the grid above.

5. *I*(_1_, _1_) 6. *E*(_8_, _9_) 7. *C*(_8_, _3_) 8. *F*(_6_, _9_)

2 Teach

Introduce the Lesson Have students read the problem and tell what is to be found. (the ordered pair that locates point *D* on the grid) Ask what information is known about point *A*. (The ordered pair (4, 3) shows its location.)

- Discuss the grid and then have students complete the sentences to answer the question.

Develop Skills and Concepts Display a number grid similar to that on page 329 but with no points identified. Point to the zero and tell students that the zero point in the lower left corner of the grid is called the **origin**, meaning beginning. Locate other points as you tell students that every point on a grid also has a name and that an **ordered pair** names each point's location.

- Tell students that we always start at the origin to find the ordered pairs of any point on a grid. Locate (3, 6) as point *A* and tell students that to find point *A*'s ordered pair, we begin at the origin and go across the bottom of the grid to the line point *A* is on.
- Write **3 over** on the board. Tell students that we then go up line 3 to point *A* and look for the line that intersects line 3 at point *A*.
- Write **6 up** on the board. Write **A (3, 6)** on the board and tell students that we read this as **three, six**. Have students locate other points and write and read their ordered pairs.

3 Practice

Remind students that the first number tells how far over to go and the second number tells how far up to go on the grid. Have students complete page 330 independently.

Practice

Write the letter identified by each ordered pair. Use the grid.

1. L (5, 2)
2. B (6, 4)
3. K (2, 5)
4. Z (1, 6)
5. A (4, 6)
6. F (2, 8)
7. O (7, 2)
8. C (1, 1)
9. N (7, 4)
10. H (3, 4)

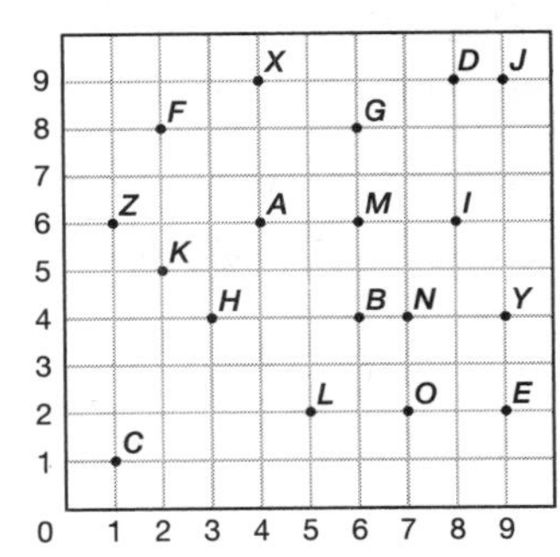

Write the ordered pair that identifies each letter. Use the grid above.

11. M(6, 6)
12. X(4, 9)
13. I(8, 6)
14. O(7, 2)
15. E(9, 2)
16. G(6, 8)
17. J(9, 9)
18. Y(9, 4)

Now Try This!

Take the numbers from the first square and rearrange them in the second square so that each column, row, and diagonal has the same sum.

15	1	11
5	9	13
7	17	3

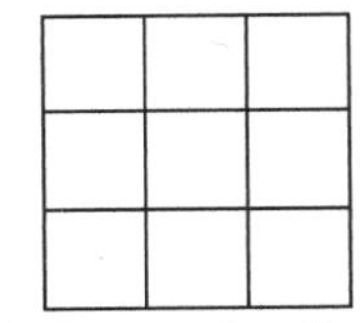

Arrangements will vary.

Now Try This! New squares can be created by rotating the numbers along the outside. Have students rotate the square one more time.

4 Assess

Ask students what the first number in an ordered pair tells. Then, ask what the second number tells. (The first number tells how far over to move. The second number tells how far up to move.)

For Mixed Abilities

Common Errors • Intervention

Some students may interpret ordered pairs incorrectly and think of (2, 3) as 2 up and 3 over. Correct students by having them read an ordered pair with the words "over" and "up." For example, they would read (2, 3) not as "two, three," but as "2 over, 3 up."

Enrichment • Ordered Pairs

Have students locate and label ten points on a grid. Then, have them exchange papers with a friend and write the ordered pair for each point.

More to Explore • Applications

Clear a bulletin board. Use cardboard as woodwork to outline windows and a door on it.

Invite a wallpaper hanger to your class to guide students in measuring, cutting, and hanging wallpaper to cover the bulletin board, working around the door and windows. Ask the professional to take students step-by-step through the measurement and math processes needed for the job.

Students can help write the figures and make computations on the board or on calculators. Estimation is also a part of the wallpaper hanger's job. Ask your guest to take the class step by step through the process of estimating how much paper it would take to cover one wall of your classroom.

ESL/ELL STRATEGIES

Explain the meaning of the terms *ordered pairs*, *number grid*, and *origin*. Then, have students practice describing several ordered pairs in sentence form. For example, write on the board **C (4, 5)** and guide students to respond, "Point C is the ordered pair four, five."

16-6 Graphing Ordered Pairs

pages 331–332

1 Getting Started

Objective
- To graph ordered pairs to draw figures

Materials
*large grid paper; grid paper; rulers

Warm Up • Mental Math
Ask students what is the number if A is 1 and Z is 26.

1. C + C (6)
2. Z ÷ A (26)
3. $\frac{2}{3}$ of X (16)
4. E × D (20)
5. Y ÷ E (5)
6. average of F and B (4)
7. 2.2 + Z (28.2)
8. A + B + C + D (10)

Warm Up • Geometry
Review the properties of a rectangle by having students tell about its sides (opposite sides are equal) and angles (four right angles). Have students draw and label different-sized rectangles on grid paper. Remind students to use a straightedge to draw line segments.

Name ______________________

Graphing Ordered Pairs

Graph the points $C(3, 6)$ and $D(7, 6)$. Use a ruler to draw $\overline{AB}$, $\overline{BD}$, $\overline{DC}$, and $\overline{CA}$. What kind of figure did you draw?

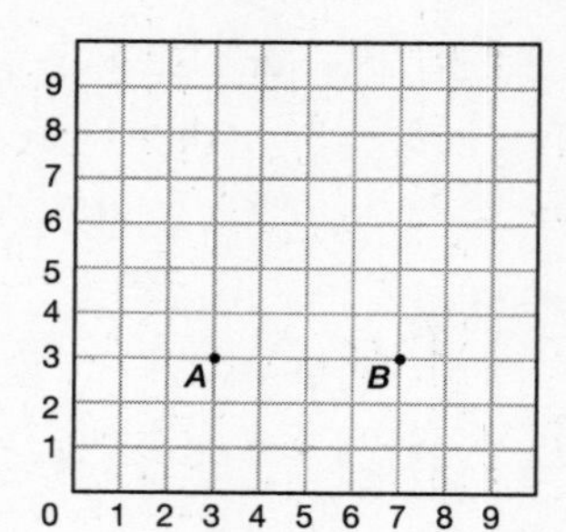

We graph and label points C and D.

Point $C(3, 6)$ is __3__ units over and __6__ units up.

Point $D(7, 6)$ is __7__ units over and __6__ units up.

We draw line segments __AB__, __BD__, __DC__, and __CA__.

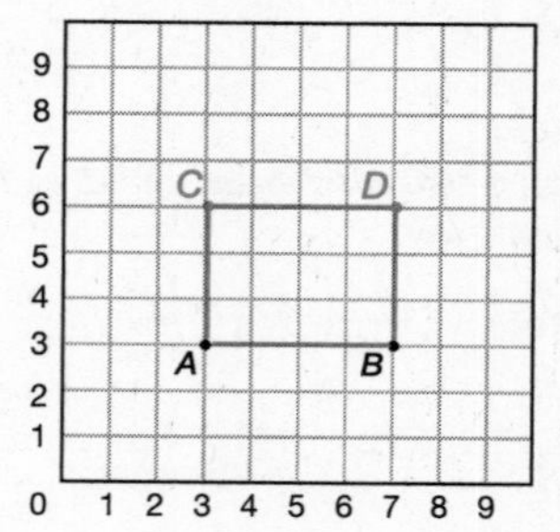

REMEMBER $\overline{AB}$ is a line segment from point A to point B.

The figure ABCD is a __rectangle__.

Getting Started

Graph each point. Draw line segments *AB*, *BC*, and *CA*.

1. $A(2, 9)$
2. $B(6, 9)$
3. $C(4, 5)$
4. The figure ABC is a __triangle__.

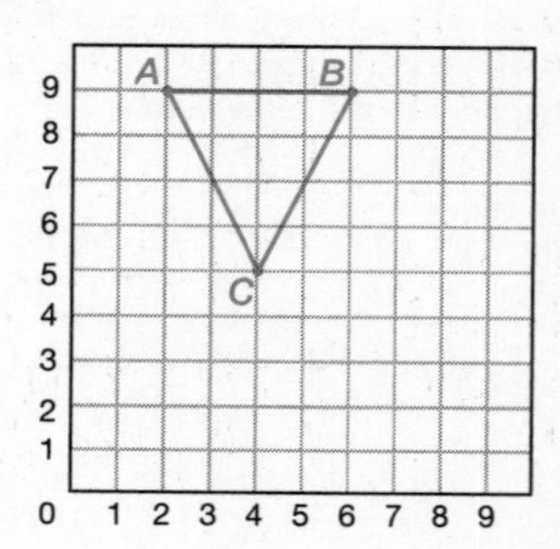

2 Teach

Introduce the Lesson Have students read the problem and tell what is to be solved. (the kind of figure formed by points *A*, *B*, *C*, and *D*) Ask students what is known. (Points *A* and *B* are located, point *C* is (3, 6), and point *D* is (7, 6).)

- Have students complete the sentences and the grid to solve the problem. Have students tell the number of units on each side of their figure to check their answer.

Develop Skills and Concepts Display a grid and have students write the numbers from 0 to 10 along the bottom and up the left side.

- Have a student locate *A* (1, 1) and *B* (2, 3), connect points *A* and *B*, and tell the figure formed. (line segment $\overline{AB}$)
- Tell students that ordered pairs can be used to describe other figures or pictures.
- Now, have a student locate *X* (3, 1) and *Y* (4, 3), draw line segment *XY*, and name the figure formed by $\overline{AB}$ and $\overline{XY}$. (parallel lines)
- Continue for students to locate *R* (2, 5), *S* (2, 8), *T* (4, 5), and *U* (4, 8) and tell the figure formed. (more parallel lines)

3 Practice

Have students complete page 332 independently.

Practice

Graph and label each point.

1. $A(3, 6)$ $B(4, 8)$ $C(5, 6)$
   $D(4, 6)$ $F(3, 4)$
   $E(4, 5)$ $G(5, 4)$

   Use a ruler to draw $\overline{AB}$, $\overline{BC}$, $\overline{AC}$, $\overline{DE}$, $\overline{FE}$, and $\overline{EG}$.

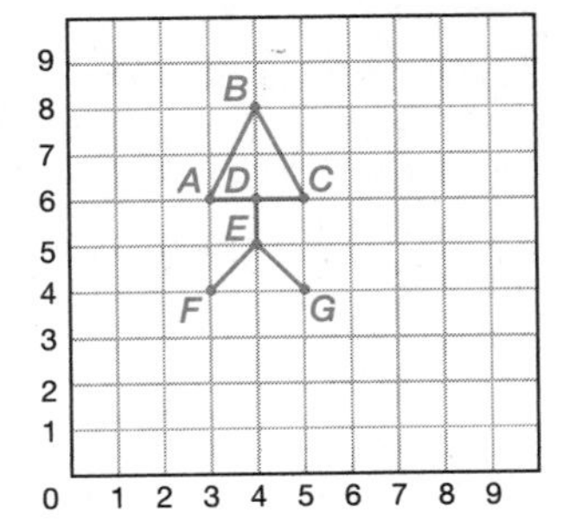

2. $A(2, 2)$ $B(2, 6)$
   $C(4, 2)$ $D(4, 6)$
   $E(2, 4)$ $F(4, 4)$
   $G(7, 2)$ $H(7, 6)$

   Use a ruler to draw $\overline{AB}$, $\overline{EF}$, $\overline{DC}$, and $\overline{HG}$.

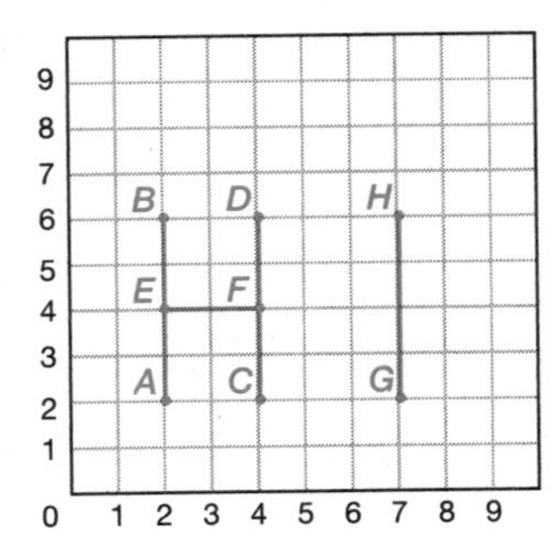

3. $A(1, 1)$ $J(4, 8)$ $K(5, 8)$
   $D(5, 1)$ $E(3, 4)$ $B(2, 4)$
   $C(5, 4)$ $H(4, 4)$ $G(4, 6)$
   $F(3, 6)$ $L(5, 7)$ $M(4, 7)$

   Use a ruler to draw $\overline{AB}$, $\overline{BE}$, $\overline{CD}$, $\overline{CH}$, $\overline{EF}$, $\overline{FG}$, $\overline{HJ}$, $\overline{JK}$, $\overline{KL}$, and $\overline{LM}$.

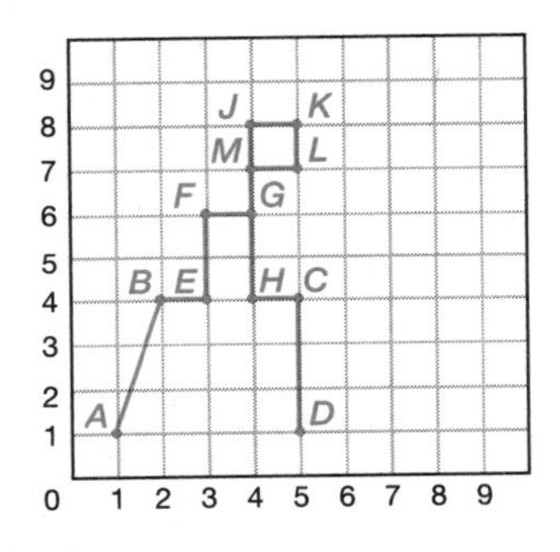

4 Assess

Have students graph the following points on graph paper: D (1, 5), E (2, 3), F (5, 3), and G (6, 5). Then, have them draw line segments DE, EG, GF, and FD. (Students should draw a trapezoid.)

5 Mixed Review

1. 16,008 − 12,973 (3,035)
2. 2.3 × 8.5 (19.55)
3. 1,278 × 9 (11,502)
4. $\frac{2}{5}$ of 15 (6)
5. 959 ÷ 72 (13 R23)
6. 32.76 + 21.204 (53.964)
7. Simplify $\frac{25}{35}$. ($\frac{5}{7}$)
8. 9.5 − 6.238 (3.262)
9. $3\frac{1}{4} + 7\frac{1}{3} + 8\frac{1}{2}$ ($19\frac{1}{12}$)
10. $\frac{8}{9} - \frac{5}{6}$ ($\frac{1}{18}$)

For Mixed Abilities

Common Errors • Intervention

Some students may reverse over and up when graphing an ordered pair. Have them work with partners, taking turns graphing ordered pairs that form simple patterns such as (2, 3), (2, 4), (2, 5), and (2, 6), which lie on a vertical line, and (3, 2), (4, 2), and (5, 2), which lie on a horizontal line.

Enrichment • Graphing

Have students graph ordered pairs to draw their initials or a creative picture. Have them list the coordinates for their picture for a partner to duplicate.

More to Explore • Geometry

Have students take a short piece of string and wrap it around a wooden block. Provide tempera paint in a shallow dish so that students can dip the string/block in paint and use it as a stamp. Let students make patterns on paper with the blocks. Explain that the block can be turned in many directions. The pattern will look different as it is rotated. For example,

 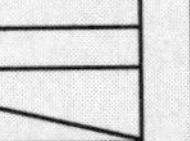  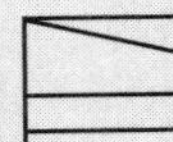

Point out that if they stamp the pattern several times in a row without turning the block, they have "translated" the pattern. However, if students rotate the block and move it, they are performing a "rotational translation."

16-7 Probability

pages 333–334

1 Getting Started

Objective
- To understand probability

Vocabulary
probability, likely, equally likely, impossible, certain, more likely, less likely, possible outcome

Materials
*paper bag; *red, yellow, green, and blue crayons; *math and reading books

Warm Up • Mental Math
Have students multiply by 6, divide by 3, and double the number.

1. 6 (24)
2. 5 (20)
3. 4 (16)
4. 10 (40)
5. 12 (48)
6. 50 (200)
7. 40 (160)
8. 15 (60)

Warm Up • Ratio
Review ratio by displaying 4 math books and 6 reading books. Remind students that the ratio of math books to reading books is 4 to 6 or $\frac{4}{6}$. Ask the ratio of reading books to math books ($\frac{6}{4}$), reading books to all the books ($\frac{6}{10}$), and so on.

Name ______________________

Probability

Jean is working a **probability** experiment by drawing numbers from a hat. What is the likelihood that she will pick an even number?

We want to know how **likely** it is that Jean will pick an even number from the hat.

There are __6__ numbers in the hat.

The even numbers are __2, 4, 6__.

There are __3__ even numbers.

There are __3__ odd numbers.

It is **equally likely** that Jean will pick an even number as an odd number.

It is **impossible** for Jean to pick a 7.

It is **certain** that Jean will pick a number less than 7.

It is **more likely** that Jean will pick a number less than 5 rather than greater than 5.

It is **less likely** that Jean will pick a number greater than 5 rather than less than 5.

Each number in the hat is called a **possible outcome**.

There are 6 possible outcomes. We say the probability of Jean picking an even number is **3 out of 6**. We can write this as a fraction $\frac{3}{6}$ or $\frac{1}{2}$.

The probability of Jean picking an even number is __3 out of 6__ or __$\frac{1}{2}$__.

Getting Started

Use the terms *more likely, equally likely, certain, less likely,* or *impossible* to describe the probability. Use the numbers in the hat.

1. 2 or 4 __equally likely__
2. number > 0 __certain__

2 Teach

Introduce the Lesson Have students read the problem aloud and tell what is to be found. (the probability that Jean will pick an even number) Ask students what information is given in the picture. (There are 6 pieces of paper and each has a different number from 1 to 6.) Have students complete the sentences to solve the problem.

- Read the remaining text aloud, pausing to answer questions as needed. Have students complete the solution sentence.

Develop Skills and Concepts Tell students that every time a baby is born, there is a 1 out of 2 chance that it will be a boy and a 1 out of 2 chance that it will be a girl. Tell students that a boy birth or a girl birth is called the **outcome** and because there are 2 possible outcomes, we say the **probability** of a boy is 1 of 2 and the probability of a girl is 1 of 2.

- Show students 1 red, 1 green, 1 blue, and 1 yellow crayon.
- Place the crayons in a paper bag and tell students that the chance or probability of picking a yellow crayon from the bag is 1 out of 4 because there is only 1 yellow crayon. Tell students that the probability of picking a crayon other than yellow is 3 out of 4 because 3 of the 4 crayons are not yellow.
- Ask students to tell the probability of picking a blue or green crayon. (2 out of 4)
- Ask additional probability questions.

3 Practice

Help students answer the first two or three questions about the cube and the spinner and then have them complete page 334 independently.

Practice

A cube has the letters *A, B, C, D, E,* and *F* printed on its faces. The outcome is the letter printed on the top of the cube after a toss.

1. What are the possible outcomes?
   A, B, C, D, E, and F

2. How many possible outcomes are there?
   6

3. Which term describes tossing a vowel compared to a consonant: *more likely, equally likely, certain, less likely,* or *impossible*?
   less likely

4. Which term describes the probability of tossing an *H*: *more likely, equally likely, certain, less likely,* or *impossible*?
   impossible

5. What is the probability of tossing an *A*?
   1 out of 6

6. What is the probability of tossing a vowel?
   2 out of 6

A spinner is divided into three equal parts numbered 1, 2, and 3.

7. What are the possible outcomes?
   1, 2, 3

8. How many different outcomes are possible?
   3

9. Which term describes the probability of spinning a number less than 4: *more likely, equally likely, certain, less likely,* or *impossible*?
   certain

10. Which term describes the probability of spinning each of the numbers: *more likely, equally likely, certain, less likely,* or *impossible*?
    equally likely

11. What is the probability of spinning a 2 or 3?
    2 out of 3

12. What is the probability of spinning an even number?
    1 out of 3

4 Assess

Have students use the information on page 333 and ask them how likely it would be for Jean to pick a number greater than 6. (impossible)

For Mixed Abilities

Common Errors • Intervention

Students may forget to count all the possibilities for a particular outcome and always write a probability as "1 out of . . . " When working a problem, have students first count to find all the possible outcomes and write the number. Next, they can count to find all the outcomes that satisfy the given condition and write the number. They then should use only the two numbers that have been written to state the probability.

Enrichment • Probability

Have students write as many examples of probability in daily living as they can. If possible, they should give the probability of the outcome. For example, the probability of wearing brown shoes is 2 out of 3 if you have 3 pairs of shoes and 2 pairs are brown.

More to Explore • Probability

Give each pair of students a page from a newspaper. Ask pairs to find a section with 100 words. Have them count the number of letters in each word and record it. Have one student count the word lengths while the other makes a tally.

Then, have pairs list the frequency with which words of differing lengths appeared in the 100-word section. Have them express the probability of finding a two-letter word. (Answers will vary.)

ESL/ELL STRATEGIES

When introducing probability, review these terms and ask students to explain what each one means in their own words: *certain, likely, equally likely, more likely, less likely, impossible,* and *outcome.* Elicit original sentences from students; for example, "I am certain that Eva will come to class today."

16-8 Listing Outcomes

pages 335–336

1 Getting Started

Objective

- To list all possible outcomes

Materials

*6 crayons of different colors;
*5 objects

Warm Up • Mental Math

Have students give a ratio.

1. months to weeks in a year ($\frac{12}{52}$)
2. weeks to months in a year ($\frac{52}{12}$)
3. 2s in 10 to 2s in 12 ($\frac{5}{6}$)
4. days in May to days in 1 year ($\frac{31}{365}$)
5. consonants to vowels ($\frac{21}{5}$)
6. vowels to consonants ($\frac{5}{21}$)

Warm Up • Probability

Discuss the probability of students' parents saying yes to a movie request. Help students see that yes is 1 out of 3 possible outcomes, with the other 2 being no and maybe. Have students discuss how the probability of a no answer may be 3 out of 3 if the movie is not for children, time does not allow, and so on. Discuss other situations of probability in daily living.

Name ____________________

Listing Outcomes

Tim is playing a game in which he spins the wheel twice each turn. His move on the board is determined by the two numbers the arrow points to. List all the possible outcomes that Tim could spin in one turn.

We want a list of all the possible number combinations Tim could spin.

We know the numbers on the wheel are 1, 2, 3, 4, 5, 6, 7, and 8.

Tim could spin the following combinations:

1, 1	2, 1	3, 1	4, 1	5, 1	6, 1	7, 1	8, 1
1, 2	2, 2	3, 2	4, 2	5, 2	6, 2	7, 2	8, 2
1, 3	2, 3	3, 3	4, 3	5, 3	6, 3	7, 3	8, 3
1, 4	2, 4	3, 4	4, 4	5, 4	6, 4	7, 4	8, 4
1, 5	2, 5	3, 5	4, 5	5, 5	6, 5	7, 5	8, 5
1, 6	2, 6	3, 6	4, 6	5, 6	6, 6	7, 6	8, 6
1, 7	2, 7	3, 7	4, 7	5, 7	6, 7	7, 7	8, 7
1, 8	2, 8	3, 8	4, 8	5, 8	6, 8	7, 8	8, 8

There are 64 different possible outcomes.

Getting Started

List all the possible outcomes.

Tracy tosses a nickel and a penny in the air to see if she will get heads or tails.

NH PH   NT PH   NH PT   NT PT

2 Teach

Introduce the Lesson Have students read the problem and tell what is to be done. (list all possible outcomes of Tim's spins in 1 turn) Ask what information is known from the problem and the picture. (Tim spins 2 times for 1 turn and the possible numbers are 1 through 8 for each spin.)

- Together with students, complete the sentence and the list to find and count the possible outcomes.
- Have students check to be sure their solution is complete. Note with students that the systematic recording of the outcomes allows for easier checking.

Develop Skills and Concepts Display 4 crayons. Tell students that we want to find all the possible outcomes, or ways, of combining 2 crayons at a time if we replace those 2 each time.

- Have a student pick 2 crayons and record the outcome on the board. Have the student replace the crayons and continue to combine and record until all outcomes have been found. (6)
- Add 2 more crayons and have students find all the outcomes. (15) Remind students that a systematic way of finding and recording the outcomes prevents error and saves time.

3 Practice

Remind students to make a systematic recording of their outcomes. Have students complete page 336 independently.

Practice

1. Each of the letters in the word COURAGE is written on a piece of paper and put into a box. List all the possible outcomes if two papers are pulled out at a time. (All papers are returned to the box after each turn.)
   C,O C,G O,A U,A R,G G,E
   C,U C,E O,G U,G R,E
   C,R O,U O,E U,E A,G
   C,A O,R U,R R,A A,E

2. List all the possible partner combinations for a chess game if Team A plays Team B.
   A,D A,G B,F C,E
   A,E B,D B,G C,F
   A,F B,E C,D C,G

Chess Game	
Team A	**Team B**
Alexi	Danielle
Bruce	Erica
Carl	Flo
	Gerry

3. Complete the diagrams to show all possible combinations of candidates for president and vice-president of the safety squad. The people running for the jobs are Greg, Lee, Kristine, Lin, and Heather.

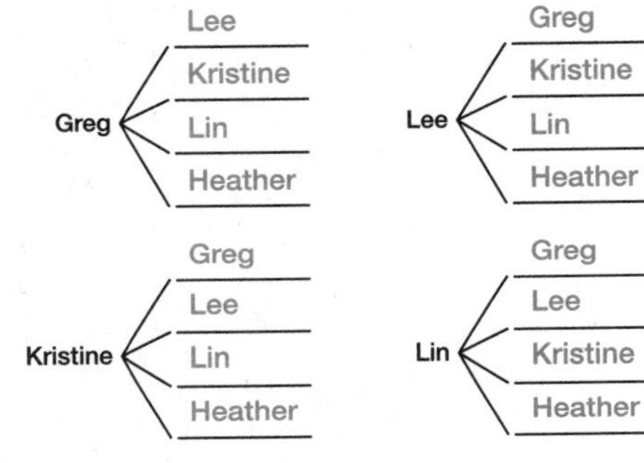

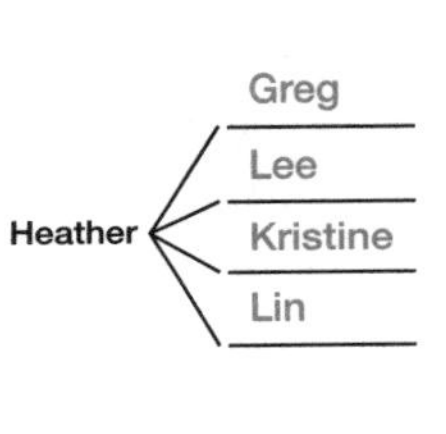

For Mixed Abilities

Common Errors • Intervention

Some students may miss some of the possible outcomes because they are not listing them in an organized way. Have students work with partners, using 5 pieces of paper numbered 1 to 5. They should pretend that they are selecting two pieces of paper from a box. Have them write all the possible pairs of numbers, starting with all the numbers that could be paired with 1, then with 2, then 3, then 4, and finally 5.

Enrichment • Probability

Tell students that different kinds of ice-cream sundaes can be made with 3 flavors of ice cream, 2 kinds of nuts, 4 syrup toppings, and 2 kinds of sprinkles. Using one of each ingredient for every sundae, have students list all possible outcomes.

More to Explore • Numeration

Have students write three word problems that include four different mathematical operations each. Have them exchange with a partner to work each other's problems. Extend the activity by having students write a word problem that involves the use of only one operation used four times.

4 Assess

Put the following chart on the board:

Team 1	Team 2
Monica	Sonnie
Jeryl	Carol
Mark	Amelia

Have students list all possible partner combinations for a game of checkers if Team 1 plays Team 2.

(Monica, Sonnie; Monica, Carol; Monica, Amelia; Jeryl, Sonnie; Jeryl, Carol, Jeryl, Amelia; Mark, Sonnie; Mark, Carol; Mark, Amelia)

5 Mixed Review

1. 576×25 (14,400)
2. $\frac{7}{8} + \frac{1}{2}$ ($1\frac{3}{8}$)
3. \$150.00 − \$127.31 (\$22.69)
4. $62{,}975 + 398$ (63,373)
5. $7 - 3.56$ (3.44)
6. $698 \div 27$ (25 R23)
7. $0.75 + 2.3 + 4.871$ (7.921)
8. $\frac{5}{3} - \frac{5}{6}$ ($\frac{5}{6}$)
9. 0.8×0.9 (0.72)
10. \$7.52 ÷ 8 (\$0.94)

16-9 Problem Solving: Work Backward

pages 337–338

1 Getting Started

Objective
- To solve problems by working backward

Warm Up • Mental Math
Have students simplify the fractions.

1. $\frac{2}{6}$ ($\frac{1}{3}$)
2. $\frac{4}{8}$ ($\frac{1}{2}$)
3. $\frac{4}{10}$ ($\frac{2}{5}$)
4. $\frac{3}{9}$ ($\frac{1}{3}$)
5. $\frac{6}{9}$ ($\frac{2}{3}$)
6. $\frac{5}{10}$ ($\frac{1}{2}$)
7. $\frac{4}{24}$ ($\frac{1}{6}$)
8. $\frac{8}{20}$ ($\frac{2}{5}$)

Name ____________________

Problem Solving
Lesson 16-9

Problem Solving: Work Backward

Mrs. Tomich withdraws $375 each week from the bank for her household expenses. She records her weekly expenses on a graph. She forgot to make an entry for Wednesday. How much money did she spend on Wednesday if she has $15 left at the end of the week?

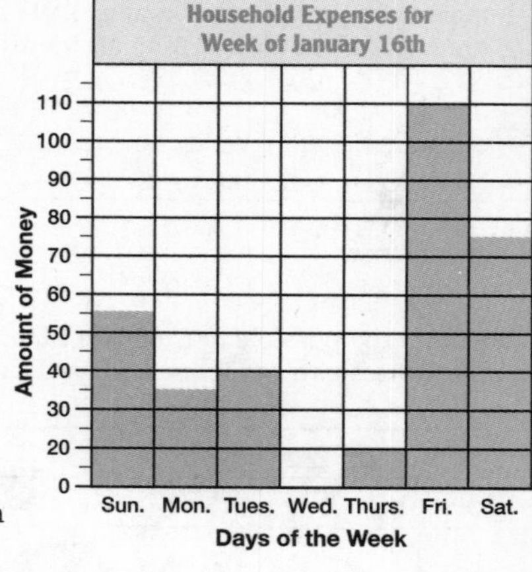

SEE

We want to know how much money Mrs. Tomich spent on Wednesday.

PLAN

Because we know the amount Mrs. Tomich started with, and how much money she has left, we can work backward to find out how much she spent on Wednesday. We can subtract the amount she has left from the amount she withdrew. Then, we can find the total amount Mrs. Tomich spent for the six days she recorded. Finally, we subtract that amount from the total amount of money she spent for the entire week.

DO

$375 − $15 = $360

$55 + $35 + $40 + $20 + $110 + $75 = $335

$360 − $335 = $25

Mrs. Tomich spent $25 on Wednesday.

CHECK

We can check our answer by working forward.

$375 − $55 − $35 − $40 − $25 − $20 − $110 − $75 = $15

The amount of money Mrs. Tomich has left is $15.

2 Teach

Introduce the Lesson Remind students that problems can be solved by the four-step problem-solving strategy.

Develop Skills and Concepts Have a student read the problem aloud and tell what is to be done. (find out how much money Mrs. Tomich spent on Wednesday if she has $15 left at the end of the week)

- Read the SEE and PLAN sections aloud. Explain that to find what Mrs. Tomich spent on Wednesday, students will have to work backward from what she had in the end. Explain that some problems can be solved working them in reverse.
- Direct students' attention to the DO section, where the problem has been laid out. Have students read this section aloud as volunteers give the results of the calculations. ($360; $335; $335; $25; $25) Ask a student to read the solution sentence. (Mrs. Tomich spent $25 on Wednesday.)
- Have students work the CHECK section together, going forward through Mrs. Tomich's week to see that she has $15 left if she starts with $375.

3 Apply

Solution Notes

1. Have students start with $7.00, the amount of change Kyle received. Then, have them add the $8.00 for the cost of the ball and $5.00 for the cost of the bat.
2. Have students find the amount of the purchase, 14 × $0.37. ($5.18) Add the price and the change to find the amount given. ($5.18 + $4.82 = $10.00)
3. Start with 54. Subtract 50. (4) Multiply 4 by 5 to get 20.
4. Start with 100. Subtract 36. (64) Divide 64 by 8 to get 8.
5. Start with 90. Subtract 69. (21) Multiply 21 by 2 to get 42.

Apply

Solve each problem.

1. Kyle bought a bat for $5.00. Then, he bought a ball costing $3.00 more than the bat. If he received $7.00 in change, how much money did he give the cashier?
   $20.00

2. Georgiana bought fourteen 37¢ stamps. If she received $4.82 in change, how much money did she give the clerk?
   $10.00

3. If I divide my age by 5 and add 50, I get 54. How old am I?
   20

4. If my sister multiplies her age by 8 and adds 36, she gets 100. How old is my sister?
   8

5. If my dad divides his age by 2 and adds 69, he gets 90. How old is my dad?
   42

Problem-Solving Strategy: Using the Four-Step Plan

- ★ **SEE** What do you need to find?
- ★ **PLAN** What do you need to do?
- ★ **DO** Follow the plan.
- ★ **CHECK** Does your answer make sense?

6. The perimeter of a rectangle is 78 centimeters. One-half of the width is 12 centimeters. What is the length of the rectangle?
   15 centimeters

7. Andy worked a subtraction problem with decimals and found the correct answer to be 57.6. If he increased each of the two numbers in the subtraction problem by 6.4, what answer would he get?
   57.6

8. Sandy drinks 8.5 glasses of water a day. Does Sandy drink more or less than 260 glasses of water a month? Explain how you got your answer.
   See Solution Notes.

For Mixed Abilities

More to Explore • Creative Drill

Have each student bring an old calendar to class. Have students complete any or all of the following activities using their calendars:

1. Draw a circle around the multiples of 7. What pattern do you see? (The multiples of 7 are in a row.) Will this pattern always be the same? (yes)
2. Complete the following column addition: the sum of all Tuesdays, the sum of all Fridays, and the sum of the last full week of the month.
3. Box in a 3 × 3 square on the calendar and find the average of the numbers in the box. Repeat for another 3 × 3 square. What pattern do you see? (The average is the number in the center square.) Test the pattern in another 3 × 3 square.

Challenge students to come up with their own calendar activities.

Higher-Order Thinking Skills

6. **Analysis:** The perimeter is twice the length plus twice the width. The perimeter is 78 inches. One-half of the width is 12 inches. Then, the width is 24 inches. Twice the width is 48 inches and the difference between the perimeter and twice the width will be twice the length. ($78 - 48 = 30$) So, the length is 15 centimeters.
7. **Synthesis:** Increasing both subtrahend and minuend by the same amount has no effect on the difference.
6. **Evaluation:** Because $8.5 \times 31 = 263.5$ and $8.5 \times 30 = 255$, it is less than 260 glasses of water per month for every month except January, March, May, July, August, October, and December.

4 Assess

Have students create a problem that involves finding their age by working backward. (Answers will vary.)

page 339

Items	Objectives
1–3	To find the mean (see pages 327–328)
4	To interpret a pictograph (see pages 323–324)
5	To interpret a line graph (see pages 325–326)
6–9	To locate and name ordered pairs on a grid (see pages 329–330)
10–11	To find the probability of drawing objects from a set (see pages 333–334)

Name ____________________

Find the mean.

1. 18, 26 22
2. 16, 47, 51 38
3. 23, 35, 45, 15, 32 30

Use the graphs to answer Exercises 4 and 5.

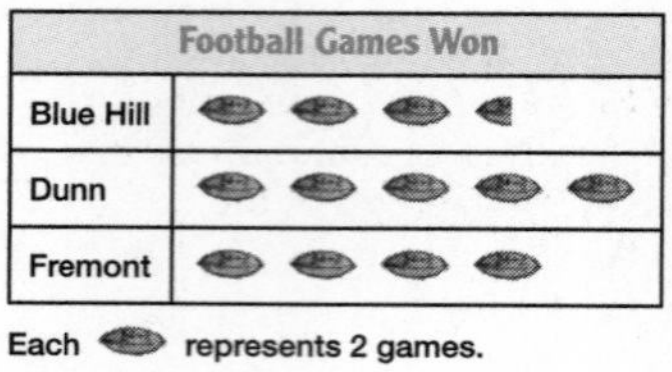
Football Games Won
Blue Hill
Dunn
Fremont
Each represents 2 games.

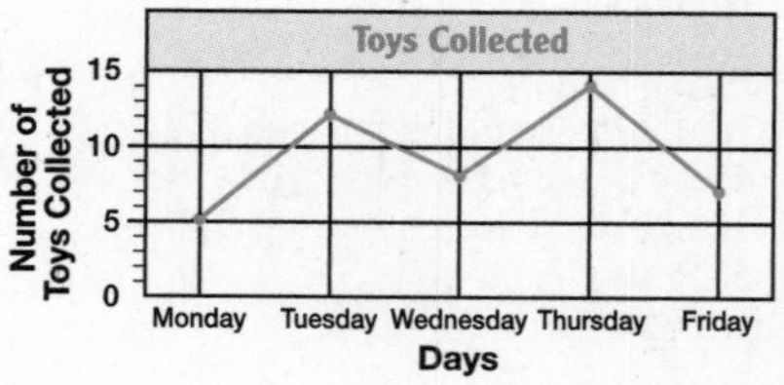
Toys Collected
Number of Toys Collected
15
10
5
0
Monday
Tuesday
Wednesday
Thursday
Friday
Days

4. How many games did Fremont win? 8
5. How many toys were collected on Thursday? 14

Use the graph to complete Exercise 6 to 9.

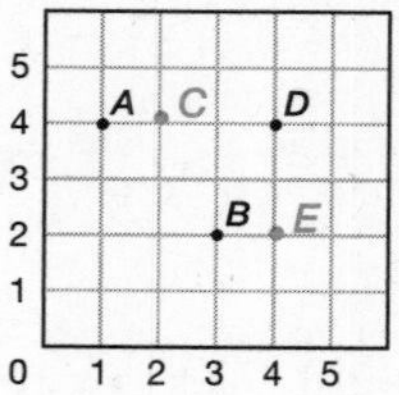
A
C
D
B
E
0
1
2
3
4
5

6. Give the letter for (3, 2). B
7. Give the ordered pair for point *A*. (1, 4)
8. Graph and label point *C* (2, 4).
9. Graph and label point *E* (4, 2).

Ten lettered pieces of paper are placed in a box. One letter is drawn.

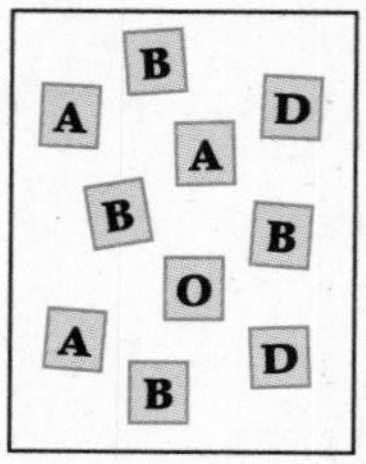
B
A
D
A
B
B
O
A
D
B

10. What is the probability the letter will be a *B*? 4 out of 10
11. What is the probability the letter will be a *C*? 0 out of 10

Alternate Chapter Test

You may wish to use the Alternate Chapter Test on page 376 of this book for further review and assessment.

Circle the letter of the correct answer.

1. $\begin{array}{r} 16{,}245 \\ +\ 45{,}682 \\ \hline \end{array}$
   a. 51,827
   (b.) 61,927
   c. 62,927
   d. NG

2. $\begin{array}{r} 7{,}046 \\ -\ 4{,}895 \\ \hline \end{array}$
   (a.) 2,151
   b. 3,851
   c. 11,941
   d. NG

3. 36 × 28
   a. 360
   b. 908
   c. 1,308
   (d.) NG

4. $8\overline{)109}$
   a. 1 R35
   (b.) 13 R5
   c. 135
   d. NG

5. Find the perimeter.

   a. 215 units
   (b.) 217 units
   c. 225 units
   d. NG

6. Find the area.

   a. 48 in.
   b. 28 sq in.
   (c.) 48 sq in.
   d. NG

7. $\frac{2}{3}$ of 24 = n
   n = ?
   a. 8
   b. 12
   (c.) 16
   d. NG

8. $\begin{array}{r} 2\frac{2}{3} \\ +\ 3\frac{1}{2} \\ \hline \end{array}$
   a. $5\frac{1}{6}$
   b. $5\frac{3}{5}$
   (c.) $6\frac{1}{6}$
   d. NG

9. $\begin{array}{r} \frac{15}{8} \\ -\ \frac{3}{4} \\ \hline \end{array}$
   a. $\frac{14}{4}$
   b. $\frac{1}{8}$
   (c.) $1\frac{1}{8}$
   d. NG

10. 7.9 + 9.8
   a. 1.77
   (b.) 17.7
   c. 177
   d. NG

11. $\begin{array}{r} 15.9 \\ -\ 8.73 \\ \hline \end{array}$
   (a.) 7.17
   b. 7.23
   c. 7.27
   d. NG

12. $\begin{array}{r} 5.9 \\ \times\ 0.3 \\ \hline \end{array}$
   (a.) 1.77
   b. 17.7
   c. 177
   d. NG

13. Find the average.
   16, 24, 23
   a. 6
   b. 20
   c. 23
   (d.) NG

STOP

score

Cumulative Assessment

page 340

Items	Objectives
1	To add two 5-digit numbers (see pages 47–48)
2	To subtract 4-digit numbers (see pages 61–62)
3	To multiply two 2-digit numbers with regrouping (see pages 103–104)
4	To divide 3-digit numbers by 1-digit numbers (see pages 193–194)
5	To find perimeter (see pages 239–240)
6	To find area (see pages 241–242)
7	To find the fraction of a number (see pages 253–254)
8	To add two mixed numbers (see pages 289–290)
9	To subtract fractions with unlike denominators (see pages 285–286)
10–12	To add, subtract, and multiply decimals (see pages 307–316)
13	To find average (see pages 207–208)

Alternate Cumulative Assessment

Circle the letter of the correct answer.

1. $\begin{array}{r} 27{,}124 \\ +\ 65{,}983 \\ \hline \end{array}$
   a 92,107
   b 93,007
   (c) 93,107
   d NG

2. $\begin{array}{r} 4{,}027 \\ -\ 2{,}783 \\ \hline \end{array}$
   a 1,234
   b 1,344
   c 2,744
   (d) NG

3. $\begin{array}{r} 26 \\ \times\ 45 \\ \hline \end{array}$
   a 1,070
   (b) 1,170
   c 1,180
   d NG

4. $6\overline{)215}$
   a 30 R5
   (b) 35 R5
   c 36 R1
   d NG

5. Find the perimeter of a rectangle that is 69 in. by 17 in.
   (a) 172 in.
   b 178 in.
   c 182 in.
   d NG

6. Find the area of a rectangle that is 7 m by 5 m.
   (a) 35 sq m
   b 40 sq m
   c 42 sq m
   d NG

7. $\frac{2}{5}$ of 75 =
   a 15
   b 20
   (c) 30
   d NG

8. $\frac{14}{9} - \frac{2}{3}$ =
   (a) $\frac{8}{9}$
   b 1
   c $1\frac{1}{9}$
   d NG

9. $3\frac{2}{5} + 5\frac{2}{3}$ =
   a $8\frac{1}{15}$
   b $8\frac{4}{5}$
   (c) $9\frac{1}{15}$
   d NG

10. 6.3 + 4.7 =
   a 10
   (b) 11
   c 110
   d NG

11. 17.6 − 9.56 =
   (a) 8.04
   b 8.14
   c 8.16
   d NG

12. $\begin{array}{r} 6.4 \\ \times\ 0.5 \\ \hline \end{array}$
   (a) 3.20
   b 32.0
   c 320
   d NG

13. Find the average of 19, 25, 32, and 16.
   a 20
   b 22
   c 24
   (d) NG

Glossary

A

acute angle an angle that has less of an opening than a right angle (p. 217)

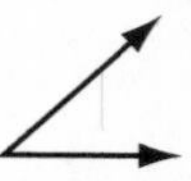

addend a number that is added to another number (p. 1)

In 3 + 4 = 7, 3 and 4 are both addends.

angle a figure formed by two rays that have the same endpoint; the endpoint is called the vertex. (p. 217)

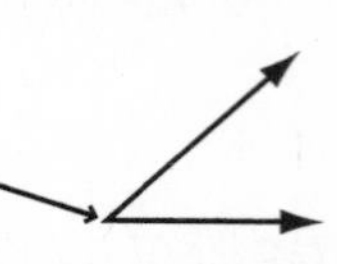

area the number of square units needed to cover a region (p. 241)

Associative (Grouping) Property When the grouping of three or more addends or factors is changed, the sum or product remains the same. (pp. 5, 97)

$(2 + 5) + 1 = 2 + (5 + 1)$

$(6 \times 3) \times 2 = 6 \times (3 \times 2)$

B

bar graph a graph that uses columns or bars to show data; a double-bar graph compares two groups of data. (p. 321)

C

Celsius scale a metric temperature scale naming 0° as the freezing point of water and 100° as its boiling point (p. 157)

centimeter (cm) a metric unit of length (p. 149)

100 centimeters = 1 meter

chord a line segment that connects two points on a circle (p. 225)

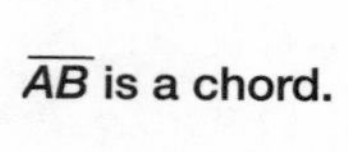

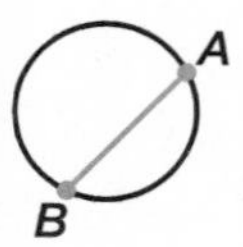

circle a closed figure that is not a polygon (p. 225)

common factor a number that is a factor of two or more given numbers (p. 262)

Commutative (Order) Property The order of addends or factors does not change the sum or product. (pp. 5, 79)

$5 + 7 = 7 + 5$

cone a solid figure with a circle for its base and a curved surface forming a point (p. 227)

congruent a word describing figures, sides, or angles having the same shape and size (p. 233)

cube a solid figure with six equal, square sides (p. 227)

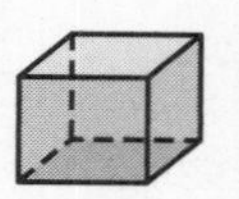

customary system a system of measurement that measures length, volume, capacity, weight, and temperature (pp. 143, 145, 147, 157)

Inches, miles, pounds, cubic feet, and ounces are examples of customary units.

cylinder a solid figure with two bases that are congruent circles (p. 227)

D

decagon a plane figure with ten sides and ten angles (p. 219)

decimal a fractional part that uses place value and a decimal point to show tenths, hundredths, and so on (p. 297)

0.6 is the decimal equivalent for the fraction $\frac{3}{5}$.

decimeter (dm) a metric unit of length (p. 151)

1 decimeter = 10 centimeters

denominator the number below the line in a fraction (p. 249)

In $\frac{3}{5}$, 5 is the denominator.

diameter a chord that passes through the center of a circle (p. 225)

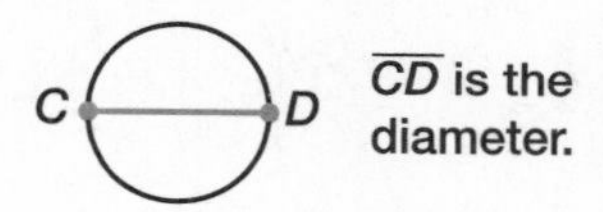

difference the answer in a subtraction problem (p. 1)

In 14 − 2 = 12, 12 is the difference.

digit any of the symbols used to write numbers; 0, 1, 2, 3, 4, 5, 6, 7, 8, and 9 are digits. (p. 15)

Distributive Property Multiplication can be distributed over addition. (p. 97)

$4 \times (3 + 7) = (4 \times 3) + (4 \times 7)$

dividend the number that is being divided in a division problem (p. 111)

In 42 ÷ 7 = 6, 42 is the dividend.

divisor the number that is being divided into the dividend (p. 111)

In 42 ÷ 7 = 6, 7 is the divisor.

E

edge a line segment where two faces of a solid figure meet (p. 227)

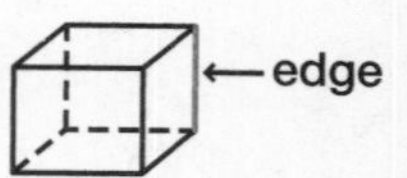

elapsed time the total amount of time that passes from the starting time to the ending time (p. 139)

equilateral triangle a triangle with three equal sides (p. 221)

equivalent fractions fractions that name the same number (p. 255)

expanded form a number written as the sum of its place values (p. 25)

426 is 400 + 20 + 6 or (4 × 100) + (2 × 10) + (6 × 1).

F

face a plane figure making up part of a solid figure (p. 227)

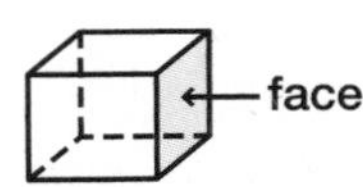

fact family related facts using the same numbers (p. 7)

7 − 4 = 3   7 − 3 = 4

3 + 4 = 7   4 + 3 = 7

factor a number to be multiplied (p. 73)

Fahrenheit scale a temperature scale naming 32° as the freezing point of water and 212° as its boiling point (p. 157)

flip (reflection) the change in the position of a figure that is the result of picking it up and turning it over (p. 237)

formula a rule expressed using symbols (p. 241)

fraction a number that names a part of a whole (p. 249)

$\frac{1}{2}$ is a fraction.

G

gram (g) a metric unit of mass (p. 155)

1,000 grams = 1 kilogram

greater than a comparison of two numbers with the number of greater value written first (p. 29)

10 > 5

H

hexagon a plane figure with six sides and six angles (p. 219)

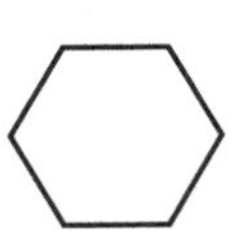

I

Identity Property Any number added to zero is that number; any number from which zero is subtracted is that number; any number multiplied by one is that number. (pp. 5, 7, 79)

7 + 0 = 7   7 − 0 = 7   7 × 1 = 7

intersecting lines lines that meet at one point (p. 215)

Line *AB* intersects line *CD* at point *P*.

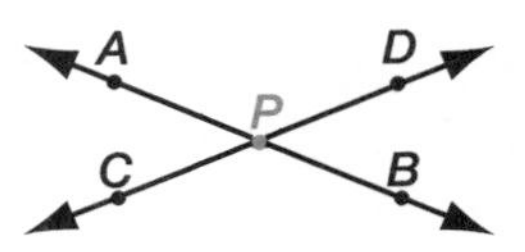

inverse operations operations that undo each other (p. 7)

isosceles triangle a triangle with two equal sides (p. 221)

K

kilogram (kg) a metric unit of weight (p. 155)

1 kilogram = 1,000 grams

kilometer (km) a metric unit of length (p. 151)

1 kilometer = 1,000 meters

L

least common multiple (LCM) the smallest number that is a multiple of two or more numbers (p. 287)

less than a comparison of two numbers with the number of lesser value written first (p. 15)

3 < 10

line a set of points that go on indefinitely in both directions (p. 215)

line graph a representation of numerical facts using points and lines on a grid to show information (p. 325)

line of symmetry a line that equally divides a figure to produce a mirror image (p. 235)

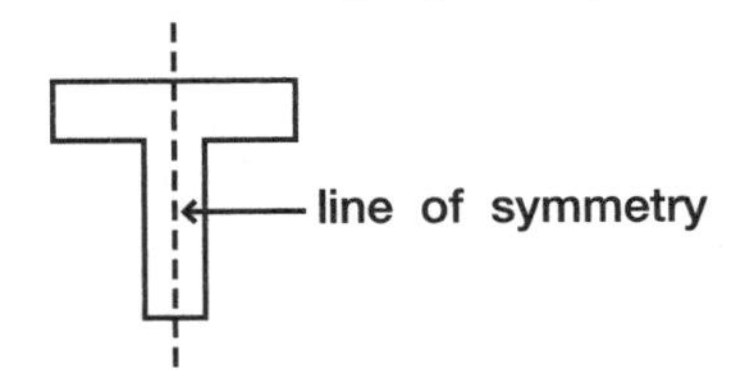

line segment a part of a line having two endpoints (p. 215)

liter (L) a metric unit of capacity (p. 153)

1 liter = 1,000 milliliters

M

mean a number representing the average of a group of numbers (p. 327)

median the middle number in a series of numbers (p. 327)

meter (m) a metric unit of length (p. 151)

1 meter = 100 centimeters

metric system a system of measurement that measures length, weight, volume, mass, and temperature based on the decimal system (pp. 149, 151, 153, 155, 157)

Meters, grams, and liters are basic metric units.

mile (mi) a customary unit of length (p. 143)

1 mile = 5,280 feet

milliliter (mL) a metric unit of liquid measure (p. 153)

minuend a number or quantity from which another is subtracted (p. 1)

mixed number a fractional number greater than 1 that is written as a whole number and a fraction (p. 263)

$5\frac{2}{3}$ is a mixed number.

mode the most frequent number in a set of data (p. 327)

multiple the product of any given number and a whole number (p. 95)

N

numerator the number above the line in a fraction (p. 249)

In $\frac{3}{5}$, 3 is the numerator.

O

obtuse angle an angle that has a greater opening than a right angle but less than a straight line (p. 217)

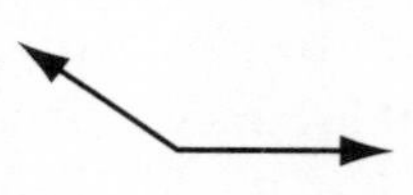

octagon a plane figure with eight sides and eight angles (p. 219)

ordered pair two numbers that define one point on a grid; the first number names the distance to the right or left, and the second names the distance up or down. (p. 329)

Order of Operations the order in which we do operations (p. 81)

Do the operations inside parentheses.
Multiply and divide in order from left to right.
Add and subtract in order from left to right.

ordinal number a number that shows the position of things in order, such as first, second, third (p. 33)

ounce (oz) a customary unit of weight (p. 147)

outcome a possible result of an experiment or a game (p. 333)

P

parallel lines lines in the same plane that do not intersect (p. 215)

parallelogram a quadrilateral whose opposite sides are parallel and the same length (p. 223)

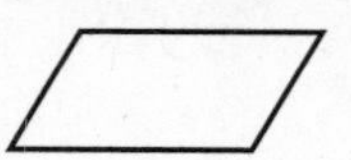

pentagon a plane figure with five straight sides and five angles (p. 219)

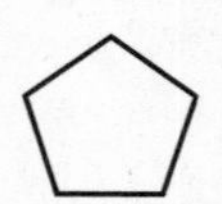

perimeter the distance around a shape that is the sum of the lengths of all of its sides (p. 239)

perpendicular lines lines that form right angles where they intersect (p. 215)

pictograph a way to show numbers or amounts using symbols or pictures (p. 323)

place value the value of the place where a digit appears in a number (pp. 15, 17, 29, 31)

In 137,510, the 7 is in the thousands place and stands for 7,000.

plane figure a shape that appears on a flat surface (p. 219)

point an exact position in space, usually shown by a dot (p. 215)

polygon a closed plane figure having three or more angles or sides (p. 219)

probability the chance an event will happen (p. 333)

product the answer to a multiplication problem (p. 73)

pyramid a solid figure whose base is a polygon and whose faces are triangles with a common vertex (p. 227)

Q

quadrilateral a polygon with four sides and four angles (p. 219)

quotient the answer to a division problem (p. 111)

R

radius a line segment that connects the center of the circle to any point on the circle (p. 225)

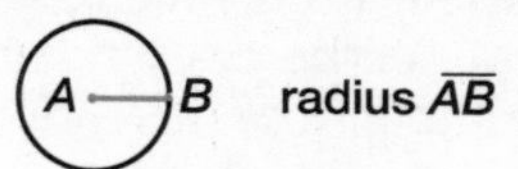

range the difference between the greatest number and the least number (p. 327)

ray a part of a line that has one endpoint (p. 215)

rectangle a quadrilateral with four right angles in which pairs of opposite sides are the same length (p. 223)

rectangular prism a solid figure with six rectangular sides (p. 227)

remainder the number left over in a division problem (p. 123)

```
       16 R4
In 6)100
     - 6
       40
     - 36
        4
```

In this problem, 4 is the remainder.

rhombus a quadrilateral whose opposite sides are parallel and all sides are the same length (p. 193)

right angle an angle that makes a square corner; the symbol used to show a right angle is ⌉. (p. 217)

rounding estimating a number's value by raising or lowering any of its place values (pp. 31, 43, 305)

S

scalene triangle a triangle with no sides the same length (p. 221)

similar figures plane figures that have the same shape but not necessarily the same size or position (p. 233)

Figures *A* and *B* are similar.

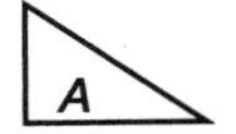

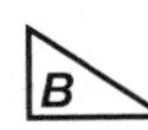

simplest form a fraction or mixed number whose numerator and denominator cannot be divided by any common factor other than 1 (p. 261)

simplest form ↓ $\frac{12}{36} = \frac{1}{3}$ simplest form ↓ $\frac{34}{6} = 5\frac{4}{6} = 5\frac{2}{3}$

slide (translation) a move that slides a figure a given distance in a given direction (p. 237)

solid figure a figure with three dimensions: length, width, and height (p. 227)

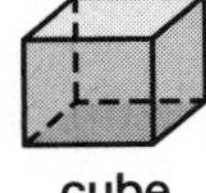
cube

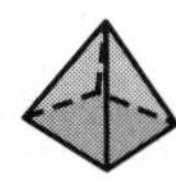
pyramid

cylinder

sphere a solid figure in the shape of a ball (p. 227)

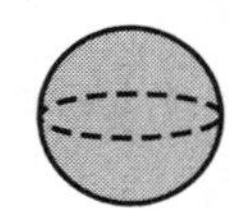

square a parallelogram with four right angles and opposite sides equal (p. 223)

standard form a number written using the symbols 0 through 9 in place-value form (pp. 15, 17, 29, 31)

4,036 is in standard form.

straight angle an angle that measures 180° (p. 217)

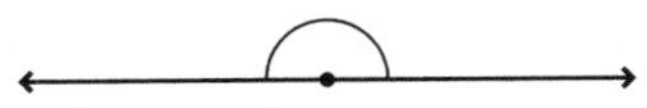

subtrahend the number that is subtracted from the minuend (p. 1)

sum the answer to an addition problem (pp. 1, 3)

symmetry A figure has symmetry if it can be folded along a line so that both parts match exactly. (p. 235)

T

tally a mark used to count (p. 321)

transformation sliding, flipping, or turning a plane figure (p. 237)

trapezoid a quadrilateral with only one pair of parallel sides (p. 223)

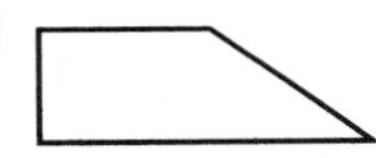

triangle a polygon with three sides (p. 219)

turn (rotation) a move that rotates a figure clockwise or counterclockwise and at a certain angle around a point (p. 237)

V

vertex (pl. vertices) the point where two sides of an angle, two sides of a plane figure, or three or more sides of a solid figure meet (pp. 217, 227)

← vertex

volume the number of cubic units needed to fill a solid figure (p. 243)

Z

Zero Property of Multiplication If a factor is zero, the product will be zero. (p. 79)

Chapter 1 Alternate Test

page 346

Items	Objectives
1–11	To compute basic addition facts (see pages 1–2)
12–16	To add three, four, or five 1-digit numbers (see pages 3–4)
17–22	To understand the Grouping Property and the Zero Property of Addition (see pages 5–6)
23–34	To compute basic subtraction facts (see pages 1–2)
35–38	To find missing addends (see pages 9–10)

Name ______________________

Add.

1. 3 + 9 = 12
2. 5 + 4 = 9
3. 9 + 0 = 9
4. 1 + 8 = 9
5. 6 + 4 = 10
6. 8 + 8 = 16
7. 7 + 6 = 13
8. 6 + 9 = 15
9. 2 + 8 = 10
10. 7 + 5 = 12
11. 9 + 8 = 17
12. 4 + 5 + 8 = 17
13. 3 + 7 + 6 = 16
14. 6 + 2 + 8 + 7 = 23
15. 7 + 1 + 6 + 1 = 15
16. 2 + 6 + 3 + 4 + 1 = 16
17. (6 + 3) + 2 = 11
18. 3 + (7 + 1) = 11
19. (8 + 1) + 4 = 13
20. 4 + (0 + 7) = 11
21. (2 + 5) + 8 = 15
22. 9 + (4 + 7) = 20

Subtract.

23. 8 − 5 = 3
24. 16 − 7 = 9
25. 12 − 7 = 5
26. 13 − 5 = 8
27. 1 − 1 = 0
28. 15 − 6 = 9
29. 9 − 8 = 1
30. 10 − 9 = 1
31. 11 − 6 = 5
32. 14 − 7 = 7
33. 9 − 5 = 4
34. 6 − 6 = 0

Write each missing addend.

35. $4 + n = 7$ $n =$ 3
36. $n + 6 = 14$ $n =$ 8
37. $9 + n = 12$ $n =$ 3
38. $6 + n = 12$ $n =$ 6

Name ____________________

Add.

1. $3 + 9 = $ ____ 2. $5 + 4 = $ ____ 3. $9 + 0 = $ ____ 4. $1 + 8 = $ ____

5. $6 + 4 = $ ____ 6. $8 + 8 = $ ____ 7. $7 + 6 = $ ____ 8. $6 + 9 = $ ____

9. $\begin{array}{r} 2 \\ +\ 8 \\ \hline \end{array}$ 10. $\begin{array}{r} 7 \\ +\ 5 \\ \hline \end{array}$ 11. $\begin{array}{r} 9 \\ +\ 8 \\ \hline \end{array}$ 12. $\begin{array}{r} 4 \\ 5 \\ +\ 8 \\ \hline \end{array}$

13. $\begin{array}{r} 3 \\ 7 \\ +\ 6 \\ \hline \end{array}$ 14. $\begin{array}{r} 6 \\ 2 \\ 8 \\ +\ 7 \\ \hline \end{array}$ 15. $\begin{array}{r} 7 \\ 1 \\ 6 \\ +\ 1 \\ \hline \end{array}$ 16. $\begin{array}{r} 2 \\ 6 \\ 3 \\ 4 \\ +\ 1 \\ \hline \end{array}$

17. $(6 + 3) + 2 = $ ____ 18. $3 + (7 + 1) = $ ____ 19. $(8 + 1) + 4 = $ ____

20. $4 + (0 + 7) = $ ____ 21. $(2 + 5) + 8 = $ ____ 22. $9 + (4 + 7) = $ ____

Subtract.

23. $8 - 5 = $ ____ 24. $16 - 7 = $ ____ 25. $12 - 7 = $ ____ 26. $13 - 5 = $ ____

27. $1 - 1 = $ ____ 28. $15 - 6 = $ ____ 29. $9 - 8 = $ ____ 30. $10 - 9 = $ ____

31. $\begin{array}{r} 11 \\ -\ 6 \\ \hline \end{array}$ 32. $\begin{array}{r} 14 \\ -\ 7 \\ \hline \end{array}$ 33. $\begin{array}{r} 9 \\ -\ 5 \\ \hline \end{array}$ 34. $\begin{array}{r} 6 \\ -\ 6 \\ \hline \end{array}$

Write each missing addend.

35. $4 + n = 7$
$n = $ ____

36. $n + 6 = 14$
$n = $ ____

37. $\begin{array}{r} 9 \\ +\ n \\ \hline 12 \end{array}$
$n = $ ____

38. $\begin{array}{r} 6 \\ +\ n \\ \hline 12 \end{array}$
$n = $ ____

Chapter 2 Alternate Test

page 348

Items	Objectives
1–8	To identify the place value of digits in numbers up to 9 digits (see pages 17–18, 25–28)
9–13	To read and write numbers less than 1 billion (see pages 17–18, 25–28)
14–16	To compare and order numbers less than 10,000 (see pages 29–30)
17–19	To round numbers to the nearest ten (see pages 31–32)
20–25	To round numbers to the nearest hundred or dollar (see pages 31–32)

Name ______________________

Write the place value of each green digit.

1. 4,621 tens
2. 11,624 hundreds
3. 7,243 thousands
4. 127,640 hundred thousands
5. 64,327 ones
6. 4,681,000 millions
7. 161,121 ten thousands
8. 34,691,200 ten millions

Write each number.

9. two hundred ninety-five 295
10. seven thousand, four hundred twenty-nine 7,429
11. six million, four hundred thousand, fifty-one 6,400,051

Write each missing word.

12. 27,653 twenty-seven thousand, six hundred fifty-three
13. 34,006,126 thirty-four million, six thousand, one hundred twenty-six

Compare. Write < or > in the circle.

14. 7,826 (>) 3,271
15. 4,362 (<) 4,623
16. 6,327 (>) 6,273

Round each number to the nearest ten.

17. 4,565 4,570
18. 6,212 6,210
19. 9,287 9,290

Round each number to the nearest hundred or dollar.

20. 657 700
21. 387 400
22. $5.27 $5
23. 4,582 4,600
24. $6.72 $7
25. 349 300

Name ______________________________

Write the place value of each green digit.

1. 4,621 ______________
2. 11,624 ______________
3. 7,243 ______________
4. 127,640 ______________
5. 64,327 ______________
6. 4,681,000 ______________
7. 161,121 ______________
8. 34,691,200 ______________

Write each number.

9. two hundred ninety-five __________
10. seven thousand, four hundred twenty-nine __________
11. six million, four hundred thousand, fifty-one __________

Write each missing word.

12. 27,653 twenty-seven __________, six __________ fifty-three
13. 34,006,126 thirty-four __________, six __________, one __________ twenty-six

Compare. Write $<$ or $>$ in the circle.

14. 7,826 ◯ 3,271
15. 4,362 ◯ 4,623
16. 6,327 ◯ 6,273

Round each number to the nearest ten.

17. 4,565 __________
18. 6,212 __________
19. 9,287 __________

Round each number to the nearest hundred or dollar.

20. 657 __________
21. 387 __________
22. $5.27 __________
23. 4,582 __________
24. $6.72 __________
25. 349 __________

Chapter 3 Alternate Test

page 350

Items	Objectives
1–4	To add two 2-digit numbers (see pages 37–38)
5–8	To add two 3-digit numbers (see pages 39–40)
9–16	To add two 4-digit numbers (see pages 41–42)
17–24	To estimate sums up to 4 digits (see pages 43–44)
25–28	To add more than two numbers (see pages 49–50)

Name ______________________

Add.

1. 47 + 32 = 79	2. 61 + 17 = 78	3. 93 + 28 = 121	4. 27 + 84 = 111
5. 246 + 221 = 467	6. 302 + 459 = 761	7. 364 + 252 = 616	8. 974 + 668 = 1,642
9. 1,324 + 3,361 = 4,685	10. 5,536 + 3,528 = 9,064	11. 8,317 + 3,835 = 12,152	12. 4,635 + 7,977 = 12,612
13. 6,452 + 8,693 = 15,145	14. 7,958 + 4,655 = 12,613	15. 2,976 + 5,493 = 8,469	16. 9,498 + 3,784 = 13,282

Round each addend to the nearest hundred or dollar. Estimate the sum.

17. 683 + 326 → 700 + 300 = 1,000	18. 179 + 420 → 200 + 400 = 600	19. 6,877 + 2,416 → 6,900 + 2,400 = 9,300	20. \$23.50 + 19.64 → \$24 + 20 = \$44

Round each addend to the nearest thousand. Estimate the sum.

21. 6,426 + 2,397 → 6,000 + 2,000 = 8,000	22. 4,956 + 2,631 → 5,000 + 3,000 = 8,000	23. 2,751 + 3,284 → 3,000 + 3,000 = 6,000	24. 6,421 + 3,791 → 6,000 + 4,000 = 10,000

Add.

25. 736 + 254 + 323 = 1,313	26. 6,436 + 1,527 + 3,375 = 11,338	27. 13,276 + 3,184 + 27,722 = 44,182	28. \$45.67 + 8.23 + 36.51 = \$90.41

Add.

1. 47 + 32	**2.** 61 + 17	**3.** 93 + 28	**4.** 27 + 84
5. 246 + 221	**6.** 302 + 459	**7.** 364 + 252	**8.** 974 + 668
9. 1,324 + 3,361	**10.** 5,536 + 3,528	**11.** 8,317 + 3,835	**12.** 4,635 + 7,977
13. 6,452 + 8,693	**14.** 7,958 + 4,655	**15.** 2,976 + 5,493	**16.** 9,498 + 3,784

Round each addend to the nearest hundred or dollar. Estimate the sum.

17. 683 + 326	**18.** 179 + 420	**19.** 6,877 + 2,416	**20.** $23.50 + 19.64

Round each addend to the nearest thousand. Estimate the sum.

21. 6,426 + 2,397	**22.** 4,956 + 2,631	**23.** 2,751 + 3,284	**24.** 6,421 + 3,791

Add.

25. 736 254 + 323	**26.** 6,436 1,527 + 3,375	**27.** 13,276 3,184 + 27,722	**28.** $45.67 8.23 + 36.51

Chapter 4 Alternate Test

page 352

Items	Objectives
1–4	To subtract two 2-digit numbers and check subtraction with addition (see pages 55–56)
5–8	To subtract two 3-digit numbers and check subtraction with addition (see pages 57–58)
9–12	To subtract two 3-, 4-, or 5-digit numbers with regrouping (see pages 59–60)
13–20	To subtract two 4- or 5-digit numbers and check subtraction with addition (see pages 61–64)
21–28	To estimate differences of 3- and 4-digit numbers (see pages 65–66)

Name ______________________

Subtract and check.

1. 75 − 62 = 13
2. 47 − 32 = 15
3. 94 − 37 = 57
4. 67 − 48 = 19
5. 659 − 324 = 335
6. 763 − 256 = 507
7. 927 − 565 = 362
8. 715 − 287 = 428
9. 702 − 431 = 271
10. 503 − 218 = 285
11. $6.04 − 3.48 = $2.56
12. $7.00 − 4.95 = $2.05
13. 8,483 − 3,674 = 4,809
14. $96.43 − 48.27 = $48.16
15. 27,466 − 19,598 = 7,868
16. $654.18 − 247.73 = $406.45
17. 8,005 − 4,679 = 3,326
18. $60.00 − 15.49 = $44.51
19. 50,003 − 36,357 = 13,646
20. $700.36 − 257.38 = $442.98

Round each number to the nearest thousand. Estimate the difference.

21. 6,421 − 2,716 → 6,000 − 3,000 = 3,000
22. 7,056 − 2,194 → 7,000 − 2,000 = 5,000
23. 4,809 − 3,674 → 5,000 − 4,000 = 1,000
24. 9,362 − 3,212 → 9,000 − 3,000 = 6,000

Round each number to the nearest hundred. Estimate the difference.

25. 697 − 356 → 700 − 400 = 300
26. 387 − 126 → 400 − 100 = 300
27. 428 − 287 → 400 − 300 = 100
28. 949 − 655 → 900 − 700 = 200

Name ________________________________

Subtract and check.

1. 75 − 62	2. 47 − 32	3. 94 − 37	4. 67 − 48
5. 659 − 324	6. 763 − 256	7. 927 − 565	8. 715 − 287
9. 702 − 431	10. 503 − 218	11. $6.04 − 3.48	12. $7.00 − 4.95
13. 8,483 − 3,674	14. $96.43 − 48.27	15. 27,466 − 19,598	16. $654.18 − 247.73
17. 8,005 − 4,679	18. $60.00 − 15.49	19. 50,003 − 36,357	20. $700.36 − 257.38

Round each number to the nearest thousand. Estimate the difference.

21. 6,421 − 2,716	22. 7,056 − 2,194	23. 4,809 − 3,674	24. 9,362 − 3,212

Round each number to the nearest hundred. Estimate the difference.

25. 697 − 356	26. 387 − 126	27. 428 − 287	28. 949 − 655

Chapter 5 Alternate Test

page 354

Items	Objectives
1–35	To recall multiplication facts through 9 (see pages 73–86)
36–41	To find the value of expressions with parentheses (see pages 81–82)
42–47	To find missing factors (see pages 89–90)

Name ____________________

Multiply.

1. $8 \times 6 = 48$
2. $9 \times 0 = 0$
3. $8 \times 3 = 24$
4. $7 \times 7 = 49$
5. $0 \times 8 = 0$
6. $8 \times 9 = 72$
7. $9 \times 3 = 27$
8. $7 \times 8 = 56$
9. $6 \times 2 = 12$
10. $6 \times 6 = 36$
11. $2 \times 6 = 12$
12. $7 \times 9 = 63$
13. $6 \times 7 = 42$
14. $2 \times 7 = 14$
15. $1 \times 6 = 6$
16. $6 \times 5 = 30$
17. $5 \times 7 = 35$
18. $7 \times 6 = 42$
19. $6 \times 9 = 54$
20. $5 \times 9 = 45$
21. $2 \times 8 = 16$
22. $6 \times 3 = 18$
23. $5 \times 5 = 25$
24. $9 \times 6 = 54$
25. $4 \times 7 = 28$
26. $8 \times 4 = 32$
27. $4 \times 5 = 20$
28. $0 \times 0 = 0$
29. $3 \times 5 = 15$
30. $4 \times 9 = 36$
31. $7 \times 4 = 28$
32. $9 \times 5 = 45$
33. $3 \times 6 = 18$
34. $8 \times 8 = 64$
35. $4 \times 8 = 32$

Solve for *n*.

36. $(6 \times 4) + 3 = n$ $n = $ 27
37. $(8 - 5) \times 3 = n$ $n = $ 9
38. $5 \times (3 + 4) = n$ $n = $ 35
39. $30 + (6 \times 5) = n$ $n = $ 60
40. $45 - (9 \times 5) = n$ $n = $ 0
41. $(6 \times 6) - 26 = n$ $n = $ 10
42. $6 \times n = 42$ $n = $ 7
43. $n \times 9 = 45$ $n = $ 5
44. $n \times 4 = 36$ $n = $ 9
45. $n \times 7 = 49$ $n = $ 7
46. $n \times 3 = 24$ $n = $ 8
47. $8 \times n = 0$ $n = $ 0

Name ______________________________

Multiply.

1. 8×6
2. 9×0
3. 8×3
4. 7×7
5. 0×8
6. 8×9
7. 9×3
8. 7×8
9. 6×2
10. 6×6
11. 2×6
12. 7×9
13. 6×7
14. 2×7
15. 1×6
16. 6×5
17. 5×7
18. 7×6
19. 6×9
20. 5×9
21. 2×8
22. 6×3
23. 5×5
24. 9×6
25. 4×7
26. 8×4
27. 4×5
28. 0×0
29. 3×5
30. 4×9
31. 7×4
32. 9×5
33. 3×6
34. 8×8
35. 4×8

Solve for *n*.

36. $(6 \times 4) + 3 = n$ $n =$ _____
37. $(8 - 5) \times 3 = n$ $n =$ _____
38. $5 \times (3 + 4) = n$ $n =$ _____
39. $30 + (6 \times 5) = n$ $n =$ _____
40. $45 - (9 \times 5) = n$ $n =$ _____
41. $(6 \times 6) - 26 = n$ $n =$ _____
42. $6 \times n = 42$ $n =$ _____
43. $n \times 9 = 45$ $n =$ _____
44. $n \times 4 = 36$ $n =$ _____
45. $n \times 7 = 49$ $n =$ _____
46. $n \times 3 = 24$ $n =$ _____
47. $8 \times n = 0$ $n =$ _____

Chapter 6 Alternate Test

page 356

Items	Objectives
1–5	To recall basic multiplication facts (see page 95–96)
6–11	To solve for *n* while following the rule of order (see pages 97–98)
12–15	To multiply 2-digit numbers by 1-digit numbers without regrouping (see pages 99–100)
16–19	To multiply 2-digit numbers by 1-digit numbers with one regrouping (see pages 101–102)
20–23	To multiply 2-digit numbers by 1-digit numbers including money with two regroupings (see pages 103–106)

Name ______________________

Write the first nine multiples of each number.

1. 2 0, 2, 4, 6, 8, 10, 12, 14, 16
2. 6 0, 6, 12, 18, 24, 30, 36, 42, 48
3. 8 0, 8, 16, 24, 32, 40, 48, 56, 64

Skip-count by 9.

4. 27, 36, 45, 54, 63, 72, 81, 90, 99

Skip-count by 7.

5. 21, 28, 35, 42, 49, 56, 63, 70, 77

Solve for *n*.

6. $9 \times (0 \times 2) = n$ $n = 0$
7. $8 \times 6 = n$ $n = 48$
8. $(6 \times 3) + (9 \times 2) = n$ $n = 36$
9. $5 \times (7 \times 2) = n$ $n = 70$
10. $(7 \times 0) + (7 \times 0) = n$ $n = 0$
11. $11 \times 7 = n$ $n = 77$

Multiply.

12. $43 \times 2 = 86$
13. $13 \times 3 = 39$
14. $22 \times 4 = 88$
15. $11 \times 6 = 66$
16. $27 \times 3 = 81$
17. $24 \times 4 = 96$
18. $13 \times 5 = 65$
19. $37 \times 2 = 74$
20. $0.46 \times 5 = $2.30
21. $0.63 \times 7 = $4.41
22. $0.84 \times 8 = $6.72
23. $0.49 \times 6 = $2.94

Name ______________________________

Write the first nine multiples of each number.

1. 2 ____, ____, ____, ____, ____, ____, ____, ____, ____

2. 6 ____, ____, ____, ____, ____, ____, ____, ____, ____

3. 8 ____, ____, ____, ____, ____, ____, ____, ____, ____

Skip-count by 9.

4. 27, ____, ____, ____, ____, ____, ____, ____, 99

Skip-count by 7.

5. 21, ____, ____, ____, ____, ____, ____, ____, 77

Solve for *n*.

6. $9 \times (0 \times 2) = n$ $n =$ ____

7. $8 \times 6 = n$ $n =$ ____

8. $(6 \times 3) + (9 \times 2) = n$ $n =$ ____

9. $5 \times (7 \times 2) = n$ $n =$ ____

10. $(7 \times 0) + (7 \times 0) = n$ $n =$ ____

11. $11 \times 7 = n$ $n =$ ____

Multiply.

12. 43×2

13. 13×3

14. 22×4

15. 11×6

16. 27×3

17. 24×4

18. 13×5

19. 37×2

20. 0.46×5 ($\$0.46 \times 5$)

21. $\$0.63 \times 7$

22. $\$0.84 \times 8$

23. $\$0.49 \times 6$

Chapter 7 Alternate Test

page 358

Items	Objectives
1–8, 10–12, 14, 16–18	To master division facts using 2 through 9 as divisors (see pages 111–118)
9, 13, 15	To master special division properties of 1 and 0 (see pages 119–120)
19–22	To solve basic division problems with remainders (see pages 123–124)
23–26	To divide 2–digit numbers by 1-digit numbers to get 2-digit quotients without a remainder (see pages 125–126)
27–34	To divide 2-digit numbers by 1-digit numbers to get 2-digit quotients with a remainder (see pages 127–128)

Name ____________________

Divide.

1. $6\overline{)48}$ 8
2. $7\overline{)28}$ 4
3. $6\overline{)30}$ 5
4. $3\overline{)12}$ 4
5. $2\overline{)18}$ 9
6. $5\overline{)20}$ 4
7. $2\overline{)18}$ 9
8. $9\overline{)45}$ 5
9. $6\overline{)0}$ 0
10. $7\overline{)49}$ 7
11. $72 \div 9 =$ 8
12. $35 \div 5 =$ 7
13. $8 \div 8 =$ 1
14. $40 \div 8 =$ 5
15. $6 \div 6 =$ 1
16. $63 \div 7 =$ 9
17. $72 \div 8 =$ 9
18. $64 \div 8 =$ 8
19. $6\overline{)50}$ 8 R2
20. $9\overline{)46}$ 5 R1
21. $8\overline{)34}$ 4 R2
22. $7\overline{)37}$ 5 R2
23. $6\overline{)66}$ 11
24. $3\overline{)96}$ 32
25. $4\overline{)96}$ 24
26. $2\overline{)36}$ 18
27. $5\overline{)77}$ 15 R2
28. $4\overline{)98}$ 24 R2
29. $3\overline{)97}$ 32 R1
30. $2\overline{)73}$ 36 R1
31. $6\overline{)62}$ 10 R2
32. $7\overline{)86}$ 12 R2
33. $4\overline{)67}$ 16 R3
34. $8\overline{)86}$ 10 R6

Name ____________________

Divide.

1. $6\overline{)48}$ 2. $7\overline{)28}$ 3. $6\overline{)30}$ 4. $3\overline{)12}$ 5. $2\overline{)18}$

6. $5\overline{)20}$ 7. $2\overline{)18}$ 8. $9\overline{)45}$ 9. $6\overline{)0}$ 10. $7\overline{)49}$

11. $72 \div 9 =$ ____ 12. $35 \div 5 =$ ____ 13. $8 \div 8 =$ ____ 14. $40 \div 8 =$ ____

15. $6 \div 6 =$ ____ 16. $63 \div 7 =$ ____ 17. $72 \div 8 =$ ____ 18. $64 \div 8 =$ ____

19. $6\overline{)50}$ 20. $9\overline{)46}$ 21. $8\overline{)34}$ 22. $7\overline{)37}$

23. $6\overline{)66}$ 24. $3\overline{)96}$ 25. $4\overline{)96}$ 26. $2\overline{)36}$

27. $5\overline{)77}$ 28. $4\overline{)98}$ 29. $3\overline{)97}$ 30. $2\overline{)73}$

31. $6\overline{)62}$ 32. $7\overline{)86}$ 33. $4\overline{)67}$ 34. $8\overline{)86}$

Chapter 8 Alternate Test

page 360

Items	Objectives
1	To use A.M. and P.M. notation To find time before (see pages 137–138)
2	To use A.M. and P.M. notation To find time after (see pages 137–138)
3–4, 6–7	To determine the appropriate customary unit of length (see pages 143–144)
5	To determine the appropriate customary unit of capacity (see pages 145–146)
8	To determine the appropriate metric unit of capacity (see pages 153–154)
9, 11	To determine the appropriate metric unit of mass (see pages 155–156)
10	To determine the appropriate metric unit of length (see pages 149–152)
12	To determine the appropriate customary or metric unit of mass (see pages 147–148; 155–156)

Name ____________________

Write the times. Include A.M. or P.M.

1. 25 minutes before nine in the morning.

2. 2 hours and 25 minutes after 7:25 P.M.

Circle the best estimate.

3. 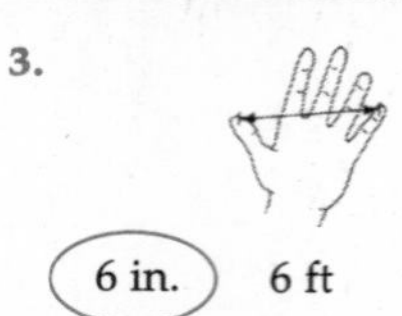

6 in. 6 ft 6 mi

4. 

7 in. 7 ft **7 yd**

5. 

1 pt 1 qt 1 gal

6. 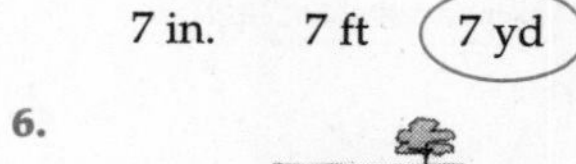

10 in. **10 yd** 10 mi

7. 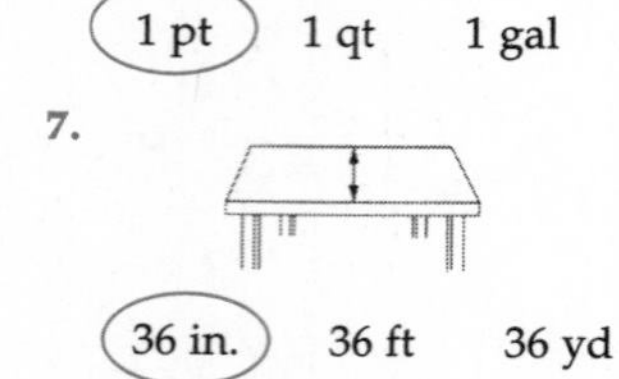

36 in. 36 ft 36 yd

8. 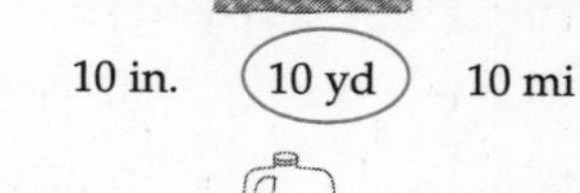

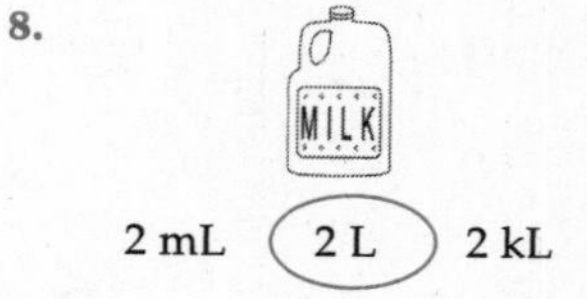

2 mL **2 L** 2 kL

9. 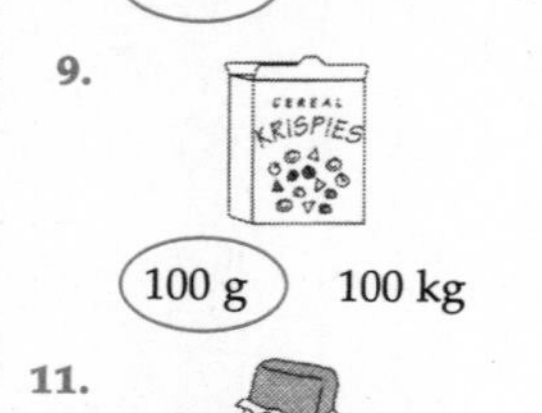

100 g 100 kg

10. 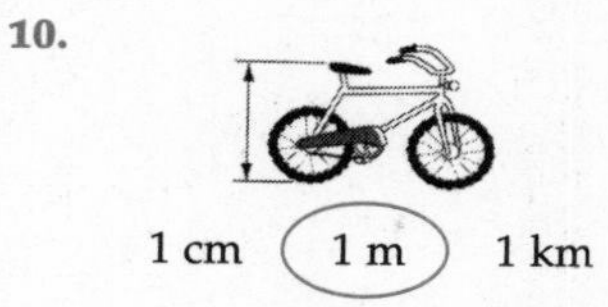

1 cm **1 m** 1 km

11. GRANOLA

50 g 50 kg

12. 

2 kg **2 T**

Name ____________________

CHAPTER 8 ALTERNATE TEST

Write the times. Include A.M. or P.M.

1. 25 minutes before nine in the morning.

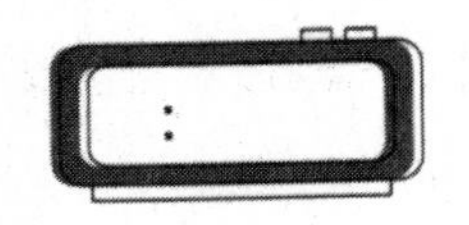

2. 2 hours and 25 minutes after 7:25 P.M.

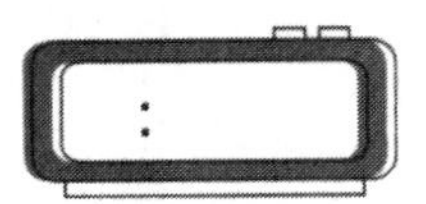

Circle the best estimate.

3.

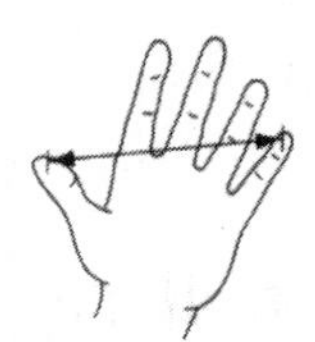

6 in. 6 ft 6 mi

4.

7 in. 7 ft 7 yd

5.

1 pt 1 qt 1 gal

6.

10 in. 10 yd 10 mi

7.

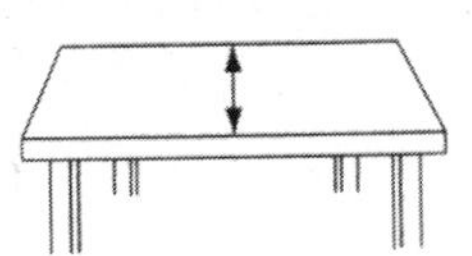

36 in. 36 ft 36 yd

8.

2 mL 2 L 2 kL

9.

100 g 100 kg

10.

1 cm 1 m 1 km

11.

50 g 50 kg

12.

2 kg 2 T

Chapter 9 Alternate Test

page 362

Items	Objectives
1, 3, 4	To multiply 3-digit numbers by 1-digit numbers with regrouping (see pages 165–166)
2	To multiply 3-digit amounts of money by 1-digit numbers (see pages 167–168)
5–8	To multiply 4-digit numbers or money by 1-digit numbers with or without regrouping (see pages 171–172)
9–11	To multiply two 2-digit numbers without regrouping (see pages 175–176)
12–16	To multiply two 2-digit numbers with regrouping (see pages 177–178)
17–28	To multiply 3-digit numbers or money by 2-digit numbers with regrouping (see pages 179–180)

Name ______________________

CHAPTER 9 ALTERNATE TEST

Multiply.

1. 214 × 3 = 642
2. $5.16 × 5 = $25.80
3. 747 × 6 = 4,482
4. 637 × 5 = 3,185
5. 6,307 × 4 = 25,228
6. 3,652 × 8 = 29,216
7. $27.32 × 3 = $81.96
8. 8,456 × 5 = 42,280
9. 12 × 24 = 288
10. 62 × 13 = 806
11. 42 × 22 = 924
12. 64 × 43 = 2,752
13. 83 × 45 = 3,735
14. 67 × 76 = 5,092
15. 93 × 68 = 6,324
16. 37 × 75 = 2,775
17. 527 × 48 = 25,296
18. $6.27 × 56 = $351.12
19. 548 × 36 = 19,728
20. 458 × 77 = 35,266
21. 752 × 63 = 47,376
22. $6.84 × 39 = $266.76
23. 498 × 75 = 37,350
24. 596 × 87 = 51,852
25. 397 × 54 = 21,438
26. 429 × 36 = 15,444
27. $9.89 × 47 = $464.83
28. 434 × 38 = 16,492

Name ______________________________

CHAPTER 9 ALTERNATE TEST

Multiply.

1. $\begin{array}{r} 214 \\ \times \quad 3 \\ \hline \end{array}$

2. $\begin{array}{r} \$5.16 \\ \times \quad 5 \\ \hline \end{array}$

3. $\begin{array}{r} 747 \\ \times \quad 6 \\ \hline \end{array}$

4. $\begin{array}{r} 637 \\ \times \quad 5 \\ \hline \end{array}$

5. $\begin{array}{r} 6{,}307 \\ \times \quad 4 \\ \hline \end{array}$

6. $\begin{array}{r} 3{,}652 \\ \times \quad 8 \\ \hline \end{array}$

7. $\begin{array}{r} \$27.32 \\ \times \quad 3 \\ \hline \end{array}$

8. $\begin{array}{r} 8{,}456 \\ \times \quad 5 \\ \hline \end{array}$

9. $\begin{array}{r} 12 \\ \times 24 \\ \hline \end{array}$

10. $\begin{array}{r} 62 \\ \times 13 \\ \hline \end{array}$

11. $\begin{array}{r} 42 \\ \times 22 \\ \hline \end{array}$

12. $\begin{array}{r} 64 \\ \times 43 \\ \hline \end{array}$

13. $\begin{array}{r} 83 \\ \times 45 \\ \hline \end{array}$

14. $\begin{array}{r} 67 \\ \times 76 \\ \hline \end{array}$

15. $\begin{array}{r} 93 \\ \times 68 \\ \hline \end{array}$

16. $\begin{array}{r} 37 \\ \times 75 \\ \hline \end{array}$

17. $\begin{array}{r} 527 \\ \times \quad 48 \\ \hline \end{array}$

18. $\begin{array}{r} \$6.27 \\ \times \quad 56 \\ \hline \end{array}$

19. $\begin{array}{r} 548 \\ \times \quad 36 \\ \hline \end{array}$

20. $\begin{array}{r} 458 \\ \times \quad 77 \\ \hline \end{array}$

21. $\begin{array}{r} 752 \\ \times \quad 63 \\ \hline \end{array}$

22. $\begin{array}{r} \$6.84 \\ \times \quad 39 \\ \hline \end{array}$

23. $\begin{array}{r} 498 \\ \times \quad 75 \\ \hline \end{array}$

24. $\begin{array}{r} 596 \\ \times \quad 87 \\ \hline \end{array}$

25. $\begin{array}{r} 397 \\ \times \quad 54 \\ \hline \end{array}$

26. $\begin{array}{r} 429 \\ \times \quad 36 \\ \hline \end{array}$

27. $\begin{array}{r} \$9.89 \\ \times \quad 47 \\ \hline \end{array}$

28. $\begin{array}{r} 434 \\ \times \quad 38 \\ \hline \end{array}$

Chapter 10 Alternate Test

page 364

Items	Objectives
1–2	To divide 3-digit numbers by 1-digit numbers to get a 3-digit quotient without a remainder (see pages 187–188)
3–5	To divide 3-digit numbers by 1-digit numbers to get a 3-digit quotient with a remainder (see pages 189–190)
6–8	To divide 3-digit numbers by 1-digit numbers to get a 3-digit quotient with zero (see pages 191–192)
9–12	To divide 3-digit numbers by 1-digit numbers to get a 2-digit quotient with or without a remainder (see pages 193–194)
13–16	To divide money by a 1-digit number (see pages 195–196)
17–20	To divide 3-digit numbers by 2-digit numbers to get a 1-digit quotient with or without a remainder (see pages 201–202)
21–24	To divide 3-digit numbers and amounts of money by 2-digit numbers to get a 2-digit quotient with or without a remainder (see pages 201–202)
25–28	To find the average of a set of numbers (see pages 207–208)

Name ____________________

Divide.

1. 4)488 — 122
2. 3)666 — 222
3. 5)557 — 111 R2
4. 7)918 — 131 R1
5. 3)703 — 234R1
6. 2)800 — 400
7. 3)622 — 207 R1
8. 4)835 — 208 R3
9. 5)365 — 73
10. 8)344 — 43
11. 7)505 — 72 R1
12. 7)651 — 93
13. 3)$6.18 — $2.06
14. 9)$7.38 — $0.82
15. 3)$8.22 — $2.74
16. 6)$3.66 — $0.61
17. 22)176 — 8
18. 42)260 — 6 R8
19. 32)236 — 7 R12
20. 64)259 — 4 R3
21. 24)288 — 12
22. 35)$7.35 — $0.21
23. 47)953 — 20 R13
24. 37)555 — 15

Find the average.

25. 6, 4, 8, 3, 4 — 5
26. 39, 42, 65, 26 — 43
27. 74, 68, 116, 92, 45 — 79
28. 262, 434, 204 — 300

Name ______________________________

CHAPTER 10 ALTERNATE TEST

Divide.

1. $4\overline{)488}$
2. $3\overline{)666}$
3. $5\overline{)557}$
4. $7\overline{)918}$
5. $3\overline{)703}$
6. $2\overline{)800}$
7. $3\overline{)622}$
8. $4\overline{)835}$
9. $5\overline{)365}$
10. $8\overline{)344}$
11. $7\overline{)505}$
12. $7\overline{)651}$
13. $3\overline{)\$6.18}$
14. $9\overline{)\$7.38}$
15. $3\overline{)\$8.22}$
16. $6\overline{)\$3.66}$
17. $22\overline{)176}$
18. $42\overline{)260}$
19. $32\overline{)236}$
20. $64\overline{)259}$
21. $24\overline{)288}$
22. $35\overline{)\$7.35}$
23. $47\overline{)953}$
24. $37\overline{)555}$

Find the average.

25. 6, 4, 8, 3, 4 ______
26. 39, 42, 65, 26 ______
27. 74, 68, 116, 92, 45 ______
28. 262, 434, 204 ______

Chapter 11 Alternate Test

page 366

Items	Objectives
1–2, 4–6	To identify and name points, lines, perpendicular lines, intersecting lines, and parallel lines (see pages 215–216)
3, 7–9	To identify and name angles, right angles, acute angles, obtuse angles, and straight angles (see pages 217–218)
10–12	To identify polygons by the number of sides (see pages 219–220)
13–15	To identify quadrilaterals (see pages 223–224)

Name ______________________

Write the name of the figure.

1. X Y

ray *XY* or $\overrightarrow{XY}$

2. X Y

line *XZ* or $\overleftrightarrow{XZ}$

3. O M N

angle *MNO* or ∠*MNO*

Write *intersecting*, *parallel*, or *perpendicular* to describe each pair of lines.

4. Q S R T

parallel

5. U W X V

perpendicular

6. 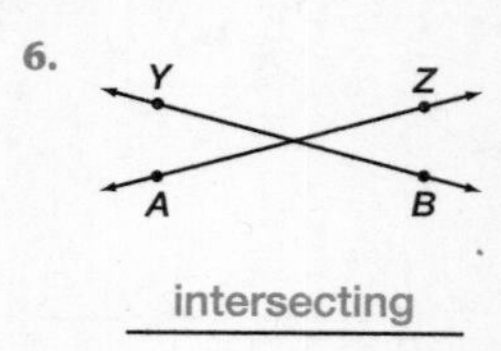

intersecting

Name and classify the angle as *acute*, *right*, *obtuse*, or *straight*.

7. F E D

∠*DEF*; obtuse

8. G H I

∠*GHI*; straight

9. K L J

∠*JKL*; acute

Name the kind of polygon.

10. 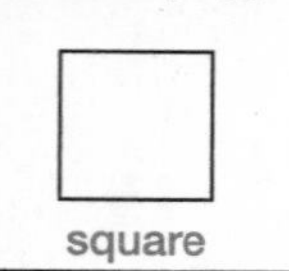

square

11. 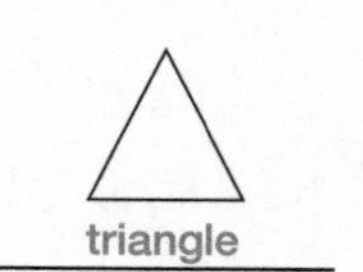

triangle

12. 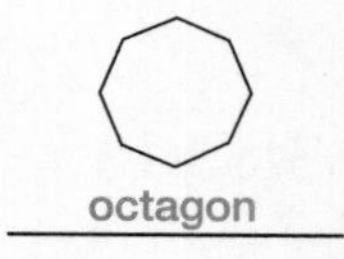

octagon

Name the kind of quadrilateral.

13. 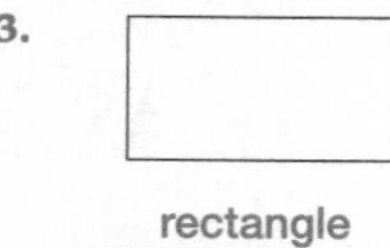

rectangle

14. 

trapezoid

15. 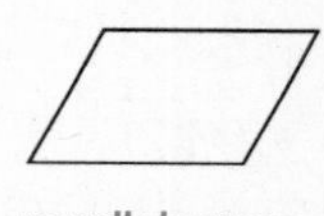

parallelogram

Name ______________________________

CHAPTER 11
ALTERNATE TEST

Write the name of the figure.

1.
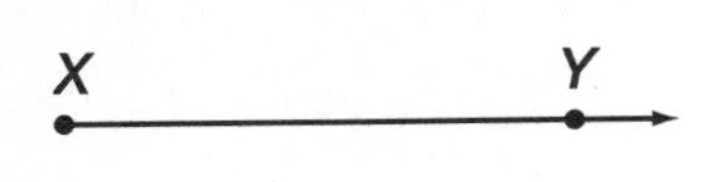

____

2.
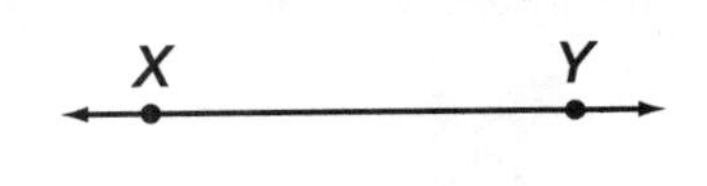

____

3.
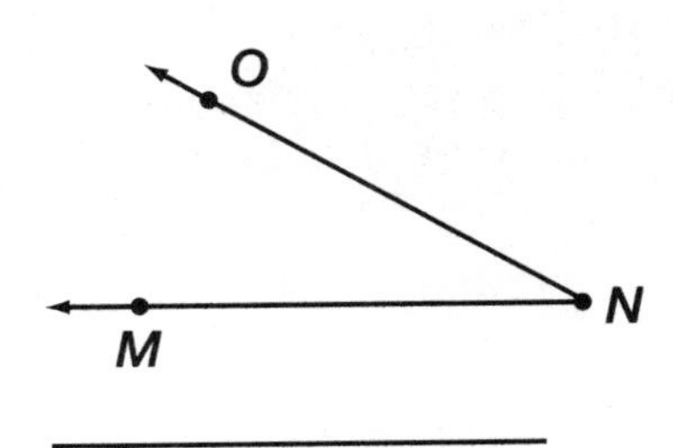

____

Write *intersecting*, *parallel*, or *perpendicular* to describe each pair of lines.

4.
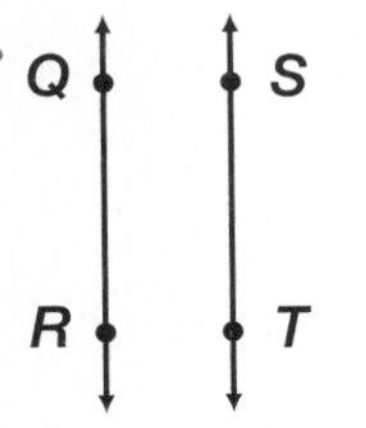

____

5.
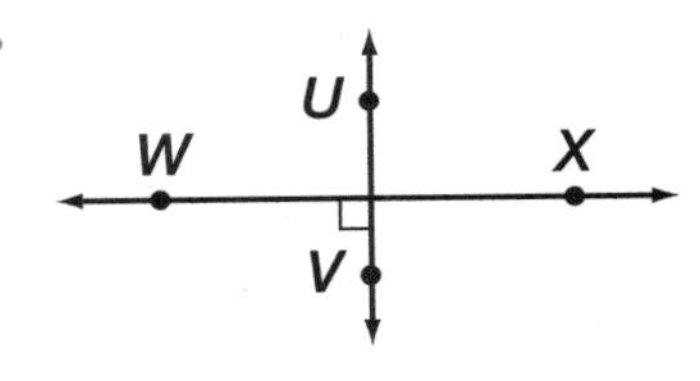

____

6.
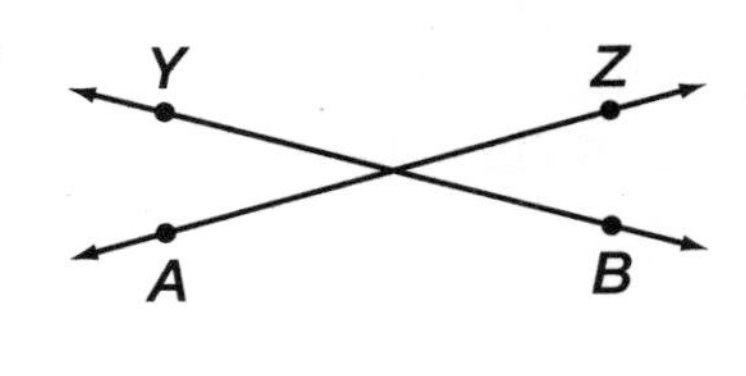

____

Name and classify the angle as *acute*, *right*, *obtuse*, or *straight*.

7.
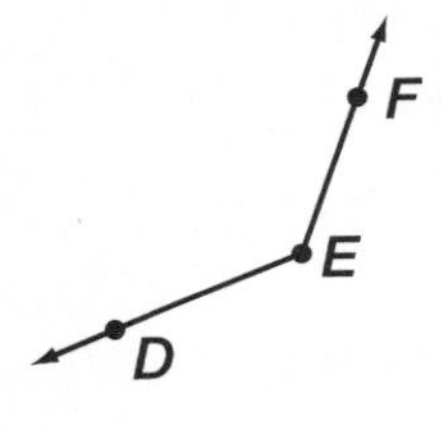

____

8.
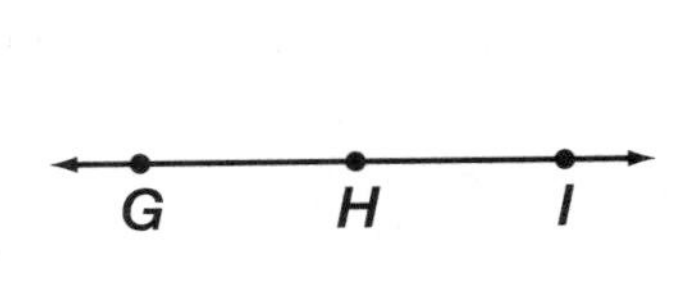

____

9.
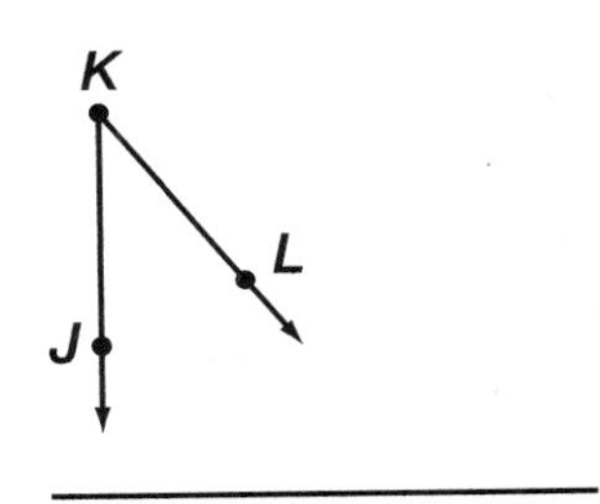

____

Name the kind of polygon.

10.

____

11.
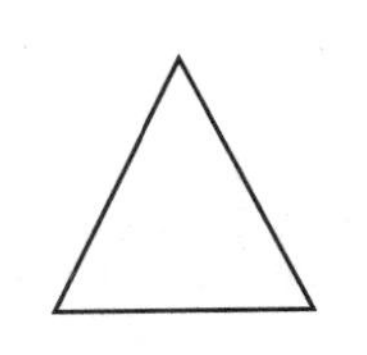

____

12.
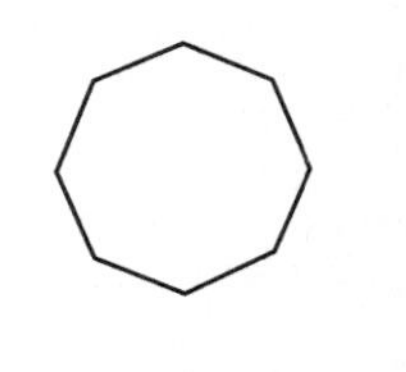

____

Name the kind of quadrilateral.

13.
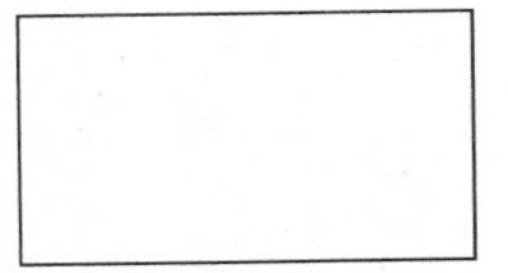

____

14.
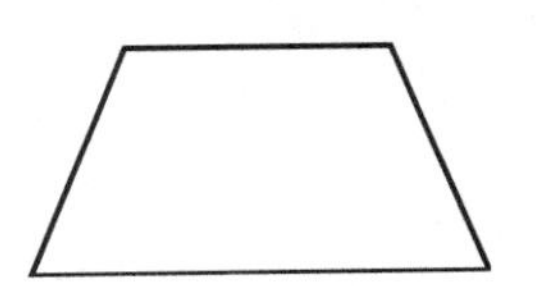

____

15.
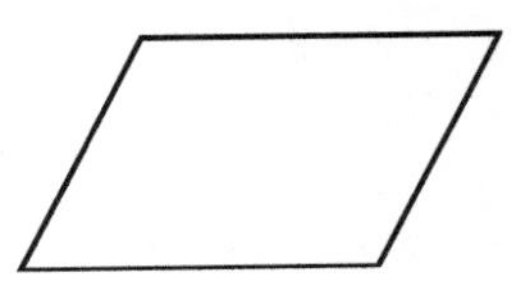

____

Chapter 12 Alternate Test

page 368

Items	Objectives
1–3	To identify figures that are congruent or similar (see pages 233–234)
4–6	To draw lines of symmetry (see pages 235–236)
7–8	To find the perimeter of a polygon (see pages 239–240)
9–10	To find the area of rectangles and squares (see pages 241–242)
11–12	To find the volume of a box (see pages 243–244)

Name ______________________

CHAPTER 12 ALTERNATE TEST

Write *congruent* or *similar* to describe the pair of figures.

1. similar
2. congruent
3. similar

Draw all lines of symmetry.

4. 5. 6.

Find the perimeter.

7. 11 cm — 44 cm

8. 9 m, 20 m — 58 m

Find the area.

9. 12 cm, 10 cm — 120 sq cm

10. 9 m — 81 sq m

Find the volume.

11. 27 cubic units

12. 3 yd, 5 yd, 4 yd — 60 cubic yards

Name ______________________________

Write *congruent* or *similar* to describe the pair of figures.

1. 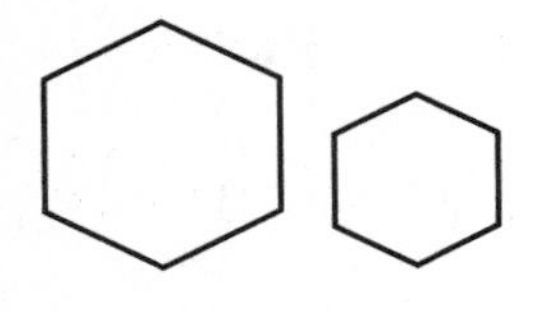

____

2. 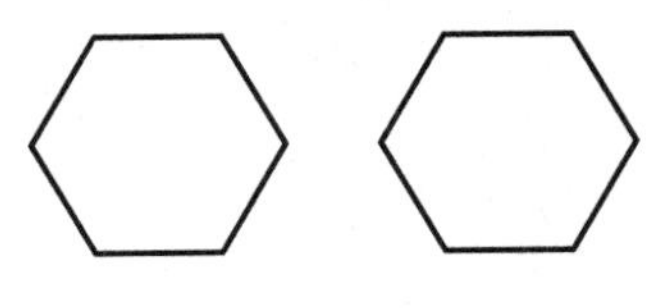

____

3. 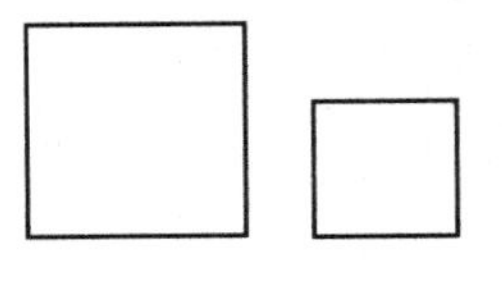

____

Draw all lines of symmetry.

4. 

5. 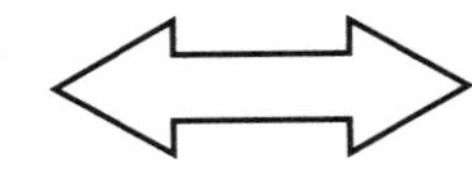

6. 

Find the perimeter.

7. 11 cm

____

8. 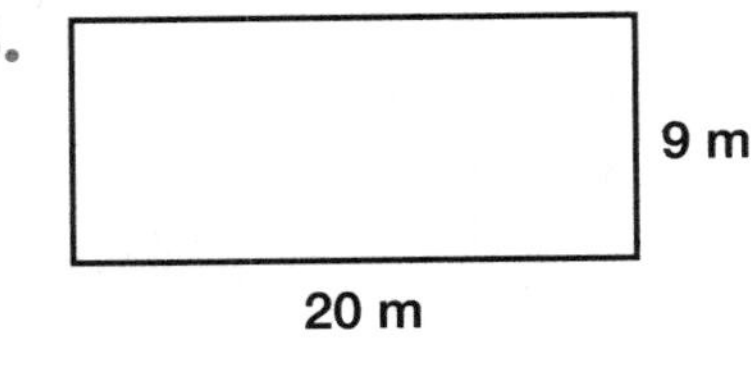

____

Find the area.

9. 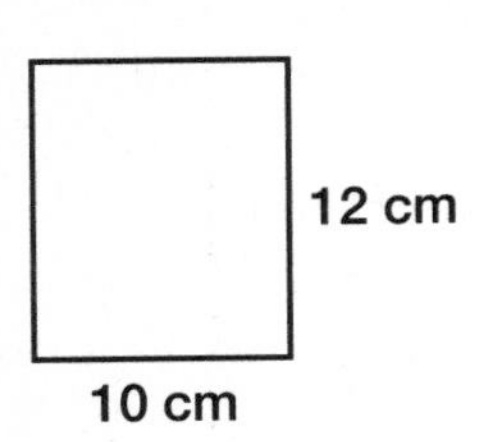

____

10. 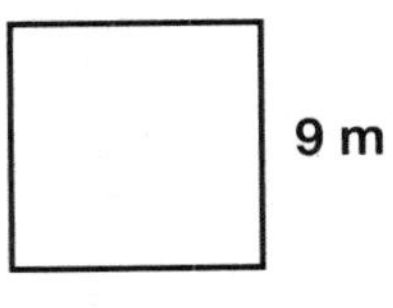

____

Find the volume.

11. 

____

12. 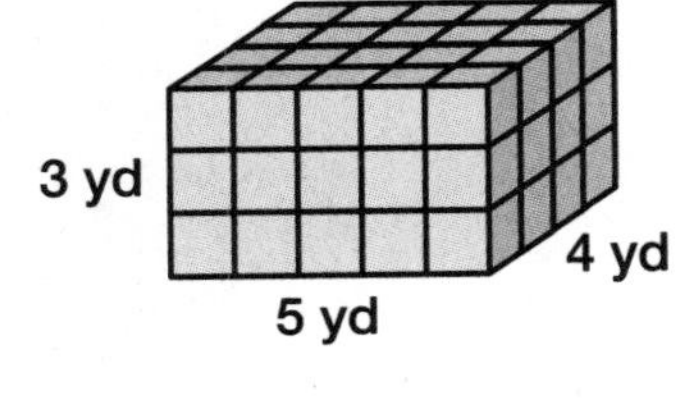

____

Chapter 13 Alternate Test

page 370

Items	Objectives
1–2	To write a fraction for part of a whole or part of a set (see pages 249–250)
3–4	To find the unit fraction of a number (see pages 253–254)
5	To find a fraction of a number by multiplying and dividing (see pages 253–254)
6–9	To find equivalent fractions by multiplying (see pages 255–258)
10–13	To compare and order fractions (see pages 259–260)
14–17	To reduce a fraction to simplest terms (see pages 261–262)
18–21	To write an improper fraction as a mixed number (see pages 263–264)
22–23	To write mixed numbers for lengths measured to the nearest quarter inch (see pages 267–268)
24–25	To use ratios to compare two quantities (see pages 269–270)

Name ______________________

Write the fraction.

1. What part of the square is green?

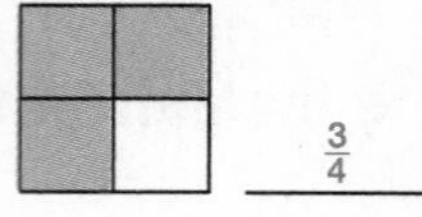

$\frac{3}{4}$

2. What part of the coins are pennies?

$\frac{3}{5}$

Find the part.

3. $\frac{1}{5}$ of 30 = 6
4. $\frac{1}{3}$ of 45 = 15
5. $\frac{3}{4}$ of a price of \$56 — \$42

Write the missing numerator.

6. $\frac{2}{5} = \frac{4}{10}$
7. $\frac{5}{8} = \frac{10}{16}$
8. $\frac{1}{3} = \frac{3}{9}$
9. $\frac{1}{6} = \frac{4}{24}$

Write < or > in the circle.

10. $\frac{3}{4} > \frac{3}{8}$
11. $\frac{5}{8} > \frac{1}{2}$
12. $\frac{4}{12} < \frac{1}{2}$
13. $\frac{4}{5} > \frac{3}{4}$

Simplify.

14. $\frac{4}{10} = \frac{2}{5}$
15. $\frac{2}{8} = \frac{1}{4}$
16. $\frac{8}{12} = \frac{2}{3}$
17. $\frac{12}{18} = \frac{2}{3}$

Write each fraction as a mixed number.

18. $\frac{9}{2} = 4\frac{1}{2}$
19. $\frac{7}{3} = 2\frac{1}{3}$
20. $\frac{17}{5} = 3\frac{2}{5}$
21. $\frac{18}{12} = 1\frac{1}{2}$

Measure each figure to the nearest quarter inch.

22. 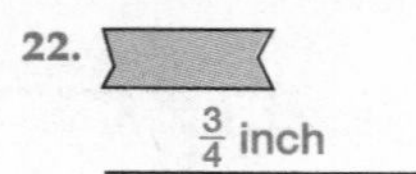

$\frac{3}{4}$ inch

23. 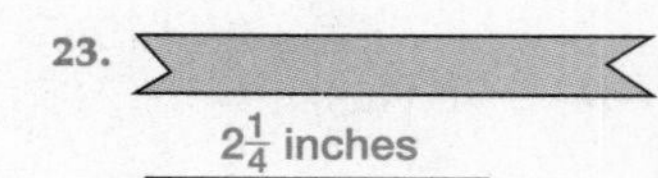

$2\frac{1}{4}$ inches

Write the comparison as a fraction.

24. black counters to gray counters $\frac{4}{5}$

25. green counters to all counters $\frac{5}{14}$

Name ____

CHAPTER 13 ALTERNATE TEST

Write the fraction.

1. What part of the square is green?

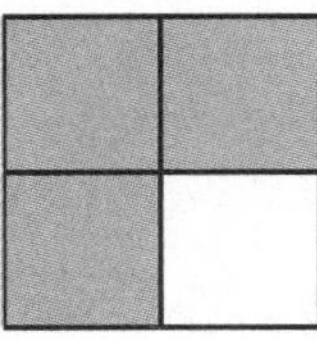

____

2. What part of the coins are pennies?

____

Find the part.

3. $\frac{1}{5}$ of 30 = ________

4. $\frac{1}{3}$ of 45 = ________

5. $\frac{3}{4}$ of a price of $56

____

Write the missing numerator.

6. $\frac{2}{5} = \frac{\quad}{10}$

7. $\frac{5}{8} = \frac{\quad}{16}$

8. $\frac{1}{3} = \frac{\quad}{9}$

9. $\frac{1}{6} = \frac{\quad}{24}$

Write < or > in the circle.

10. $\frac{3}{4} \bigcirc \frac{3}{8}$

11. $\frac{5}{8} \bigcirc \frac{1}{2}$

12. $\frac{4}{12} \bigcirc \frac{1}{2}$

13. $\frac{4}{5} \bigcirc \frac{3}{4}$

Simplify.

14. $\frac{4}{10} =$

15. $\frac{2}{8} =$

16. $\frac{8}{12} =$

17. $\frac{12}{18} =$

Write each fraction as a mixed number.

18. $\frac{9}{2} =$

19. $\frac{7}{3} =$

20. $\frac{17}{5} =$

21. $\frac{18}{12} =$

Measure each figure to the nearest quarter inch.

22. ________

23. ________

Write the comparison as a fraction.

24. black counters to gray counters

25. green counters to all counters

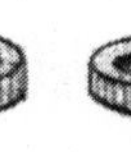
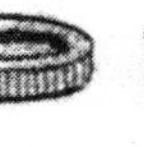

Chapter 14 Alternate Test

page 372

Items	Objectives
1, 6	To add fractions with like denominators (see pages 275–276; 279–280)
2–5, 7–8	To subtract fractions with like denominators (see pages 277–280)
9, 11–13, 15–16, 21–22	To add or subtract two fractions with unlike denominators (see pages 281–282; 285–286)
10, 14	To add or subtract two fractions when one denominator is a multiple of the other denominator (see pages 285–286; 287–288))
17–20, 23–24	To add two mixed numbers (see pages 289–290)

Name ____________________

Add or subtract. Simplify if necessary.

1. $\frac{3}{7} + \frac{2}{7}$ = $\frac{5}{7}$
2. $\frac{6}{8} - \frac{3}{8}$ = $\frac{3}{8}$
3. $\frac{7}{9} - \frac{3}{9}$ = $\frac{4}{9}$
4. $\frac{7}{10} - \frac{3}{10}$ = $\frac{2}{5}$
5. $\frac{10}{11} - \frac{7}{11}$ = $\frac{3}{11}$
6. $\frac{3}{8} + \frac{3}{8}$ = $\frac{3}{4}$
7. $\frac{5}{6} - \frac{2}{6}$ = $\frac{1}{2}$
8. $\frac{3}{6} - \frac{1}{6}$ = $\frac{1}{3}$
9. $\frac{1}{6} + \frac{1}{4}$ = $\frac{5}{12}$
10. $\frac{5}{12} - \frac{1}{4}$ = $\frac{1}{6}$
11. $\frac{10}{14} - \frac{2}{7}$ = $\frac{3}{7}$
12. $\frac{6}{8} + \frac{8}{12}$ = $1\frac{5}{12}$
13. $\frac{9}{12} - \frac{3}{8}$ = $\frac{3}{8}$
14. $\frac{6}{8} - \frac{3}{16}$ = $\frac{9}{16}$
15. $\frac{3}{4} + \frac{5}{6}$ = $1\frac{7}{12}$
16. $\frac{11}{9} - \frac{2}{6}$ = $\frac{8}{9}$
17. $4\frac{4}{9} + 1\frac{6}{18}$ = $5\frac{7}{9}$
18. $6\frac{4}{6} + 3\frac{1}{4}$ = $9\frac{11}{12}$
19. $24\frac{2}{8} + 16\frac{3}{4}$ = 41
20. $3\frac{3}{5} + 8\frac{1}{3}$ = $11\frac{14}{15}$
21. $\frac{2}{3} + \frac{1}{5} =$ $\frac{13}{15}$
22. $\frac{3}{4} - \frac{1}{7} =$ $\frac{17}{28}$
23. $2\frac{1}{6} + 3\frac{3}{4} =$ $5\frac{11}{12}$
24. $8\frac{1}{2} + 3\frac{1}{6} =$ $11\frac{2}{3}$

Name ______________________

Add or subtract. Simplify if necessary.

1. $\frac{3}{7} + \frac{2}{7}$

2. $\frac{6}{8} - \frac{3}{8}$

3. $\frac{7}{9} - \frac{3}{9}$

4. $\frac{7}{10} - \frac{3}{10}$

5. $\frac{10}{11} - \frac{7}{11}$

6. $\frac{3}{8} + \frac{3}{8}$

7. $\frac{5}{6} - \frac{2}{6}$

8. $\frac{3}{6} - \frac{1}{6}$

9. $\frac{1}{6} + \frac{1}{4}$

10. $\frac{5}{12} - \frac{1}{4}$

11. $\frac{10}{14} - \frac{2}{7}$

12. $\frac{6}{8} + \frac{8}{12}$

13. $\frac{9}{12} - \frac{3}{8}$

14. $\frac{6}{8} - \frac{3}{16}$

15. $\frac{3}{4} + \frac{5}{6}$

16. $\frac{11}{9} - \frac{2}{6}$

17. $4\frac{4}{9} + 1\frac{6}{18}$

18. $6\frac{4}{6} + 3\frac{1}{4}$

19. $24\frac{2}{8} + 16\frac{3}{4}$

20. $3\frac{3}{5} + 8\frac{1}{3}$

21. $\frac{2}{3} + \frac{1}{5} =$ ____

22. $\frac{3}{4} - \frac{1}{7} =$ ____

23. $2\frac{1}{6} + 3\frac{3}{4} =$ ____

24. $8\frac{1}{2} + 3\frac{1}{6} =$ ____

Chapter 15 Alternate Test

page 374

Items	Objectives
1–8	To identify the place value of digits in a decimal through thousandths (see pages 297–302)
9–16	To compare and order decimals through thousandths (see pages 303–304)
17–20	To add decimals through thousandths (see pages 307–308; 311–312)
21–24	To subtract decimals through thousandths (see pages 309–312)
25, 30, 32	To multiply a whole number and a decimal (see pages 313–314)
26–29, 31	To multiply two decimals (see pages 315–316)

Name ____________________

Write the place value of the 4 in each number.

1. 27.43 tenths
2. 621.84 hundredths
3. 164.32 ones
4. 297.364 thousandths

Write the place value of the 7 in each number.

5. 937.63 ones
6. 432.71 tenths
7. 562.437 thousandths
8. 263.07 hundredths

Write <, =, or > in each circle.

9. 2.97 (>) 2.79
10. 6.19 (>) 6.1
11. 0.246 (<) 2.46
12. 9.04 (<) 9.40
13. 6.7 (>) 6.07
14. 0.15 (>) 0.051
15. 8.20 (=) 8.2
16. 72.08 (<) 72.1

Add.

17. 5.4 + 2.7 = 8.1
18. 17.36 + 36.255 = 53.615
19. 47.3 + 84.91 = 132.21
20. 86.24 + 19.8 = 106.04

Subtract.

21. 6.4 − 3.1 = 3.3
22. 26.147 − 15.369 = 10.778
23. 47.6 − 28.29 = 19.31
24. 67.13 − 15.4 = 51.73

Multiply.

25. 4.5 × 4 = 18
26. 23.4 × 0.3 = 7.02
27. 15.3 × 0.7 = 10.71
28. 7.6 × 3.4 = 25.84
29. 32.6 × 4.3 = 140.18
30. 8.9 × 4 = 35.6
31. 6.1 × 0.05 = 0.305
32. 0.07 × 8 = 0.56

Name ______________________________

Write the place value of the 4 in each number.

1. 27.43 ____________
2. 621.84 ____________
3. 164.32 ____________
4. 297.364 ____________

Write the place value of the 7 in each number.

5. 937.63 ____________
6. 432.71 ____________
7. 562.437 ____________
8. 263.07 ____________

Write $<$, $=$, or $>$ in each circle.

9. 2.97 ◯ 2.79
10. 6.19 ◯ 6.1
11. 0.246 ◯ 2.46
12. 9.04 ◯ 9.40
13. 6.7 ◯ 6.07
14. 0.15 ◯ 0.051
15. 8.20 ◯ 8.2
16. 72.08 ◯ 72.1

Add.

17. $\begin{array}{r} 5.4 \\ +\ 2.7 \\ \hline \end{array}$

18. $\begin{array}{r} 17.36 \\ +\ 36.255 \\ \hline \end{array}$

19. $\begin{array}{r} 47.3 \\ +\ 84.91 \\ \hline \end{array}$

20. $\begin{array}{r} 86.24 \\ +\ 19.8 \\ \hline \end{array}$

Subtract.

21. $\begin{array}{r} 6.4 \\ -\ 3.1 \\ \hline \end{array}$

22. $\begin{array}{r} 26.147 \\ -\ 15.369 \\ \hline \end{array}$

23. $\begin{array}{r} 47.6 \\ -\ 28.29 \\ \hline \end{array}$

24. $\begin{array}{r} 67.13 \\ -\ 15.4 \\ \hline \end{array}$

Multiply.

25. $\begin{array}{r} 4.5 \\ \times\ 4 \\ \hline \end{array}$

26. $\begin{array}{r} 23.4 \\ \times\ 0.3 \\ \hline \end{array}$

27. $\begin{array}{r} 15.3 \\ \times\ 0.7 \\ \hline \end{array}$

28. $\begin{array}{r} 7.6 \\ \times\ 3.4 \\ \hline \end{array}$

29. $\begin{array}{r} 32.6 \\ \times\ 4.3 \\ \hline \end{array}$

30. $\begin{array}{r} 8.9 \\ \times\ 4 \\ \hline \end{array}$

31. $\begin{array}{r} 6.1 \\ \times\ 0.05 \\ \hline \end{array}$

32. $\begin{array}{r} 0.07 \\ \times\ 8 \\ \hline \end{array}$

Chapter 16 Alternate Test

page 376

Items	Objectives
1–3	To find the mean (see pages 327–328)
4	To interpret a pictograph (see pages 323–324)
5	To interpret a line graph (see pages 325–326)
6–9	To locate, name, and graph ordered pairs on a grid (see pages 329–332)
10–11	To understand probability (see pages 333–334)

Name ______________________

Find the mean.

1. 32, 46 39
2. 24, 27, 72 41
3. 15, 21, 42, 62 35

Use the graphs to answer Exercises 4 and 5.

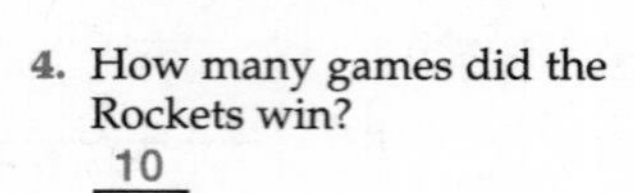

Games Won	
Jets	
Bulls	
Rockets	

Each represents 2 games.

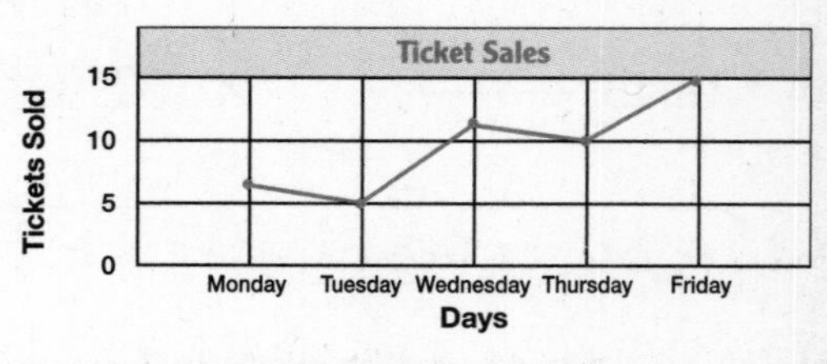

4. How many games did the Rockets win? 10
5. How many tickets were sold on Tuesday? 5

Use the graph to complete Exercises 6 to 9.

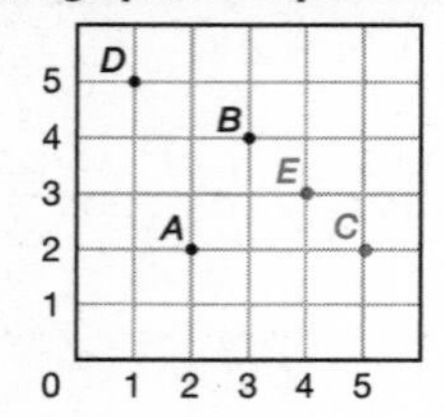

6. Give the letter for (2, 2). *A*
7. Give the ordered pair for point *B*. (3, 4)
8. Graph and label point *C* (5, 2).
9. Graph and label point *E* (4, 3).

Ten lettered pieces of paper are placed in a box. One letter is drawn.

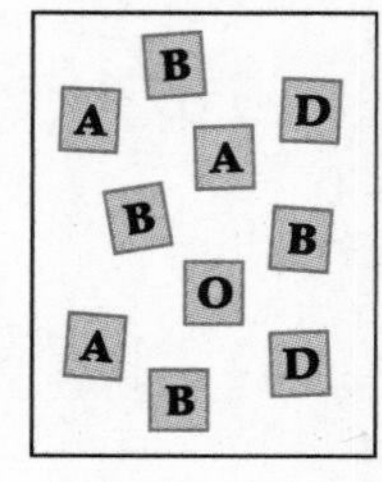

10. What is the probability the letter will be an *A*? 3 out of 10
11. What is the probability the letter will be an *O*? 1 out of 10

Name ______________________________

Find the mean.

1. 32, 46 ____ 2. 24, 27, 72 ____ 3. 15, 21, 42, 62 ____

Use the graphs to answer Exercises 4 and 5.

Games Won	
Jets	● ● ● ● ◐
Bulls	● ● ●
Rockets	● ● ● ● ●

Each ● represents 2 games.

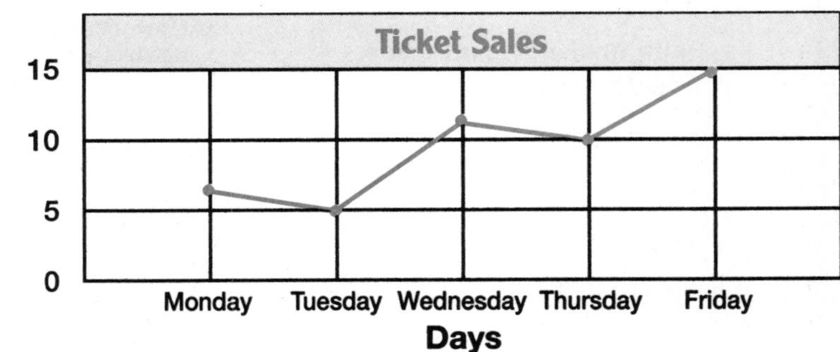

4. How many games did the Rockets win?

____

5. How many tickets were sold on Tuesday?

____

Use the graph to complete Exercises 6 to 9.

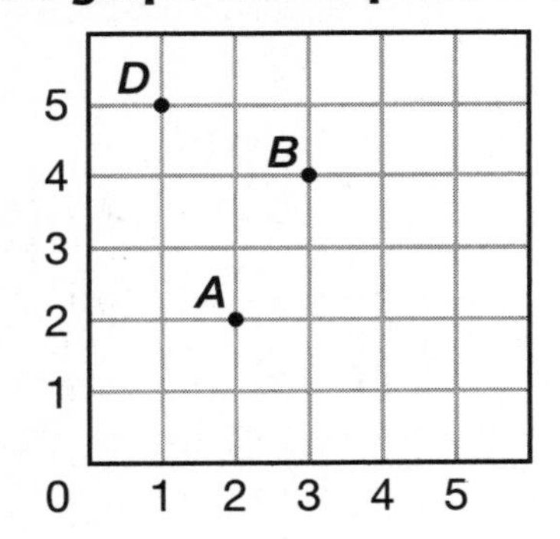

6. Give the letter for (2, 2). ____

7. Give the ordered pair for point *B*. ____

8. Graph and label point *C* (5, 2).

9. Graph and label point *E* (4, 3).

Ten lettered pieces of paper are placed in a box. One letter is drawn.

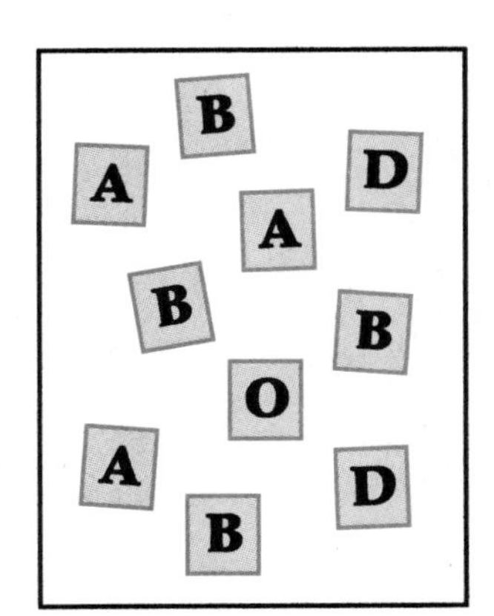

10. What is the probability the letter will be an *A*?

____

11. What is the probability the letter will be an *O*?

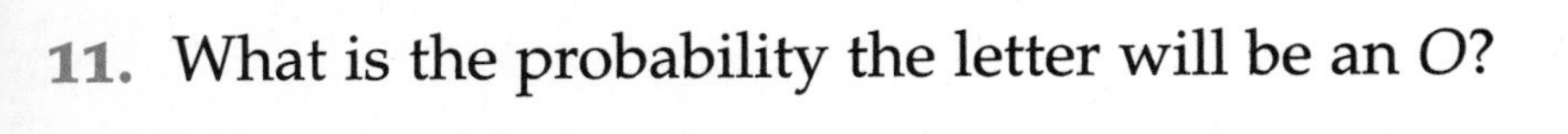

____

Index